# Acoustic and Ultrasonic Sensing Technology in Non-Destructive Testing

# Acoustic and Ultrasonic Sensing Technology in Non-Destructive Testing

Guest Editors

**Sergio Castiñeira-Ibáñez**
**Daniel Tarrazó-Serrano**
**Constanza Rubio Michavila**

MDPI

Basel • Beijing • Wuhan • Barcelona • Belgrade • Novi Sad • Cluj • Manchester

*Guest Editors*
Sergio Castiñeira-Ibáñez
Centro de Tecnologías Físicas: Acústica, Metamateriales y Astrofísica
Universitat Politècnica de València
València
Spain

Daniel Tarrazó-Serrano
Centro de Tecnologías Físicas: Acústica, Metamateriales y Astrofísica
Universitat Politècnica de València
València
Spain

Constanza Rubio Michavila
Centro de Tecnologías Físicas: Acústica, Metamateriales y Astrofísica
Universitat Politècnica de València
València
Spain

*Editorial Office*
MDPI AG
Grosspeteranlage 5
4052 Basel, Switzerland

This is a reprint of the Special Issue, published open access by the journal *Sensors* (ISSN 1424-8220), freely accessible at: https://www.mdpi.com/journal/sensors/special_issues/I057K828MK.

For citation purposes, cite each article independently as indicated on the article page online and as indicated below:

Lastname, A.A.; Lastname, B.B. Article Title. *Journal Name* **Year**, *Volume Number*, Page Range.

**ISBN 978-3-7258-4487-6 (Hbk)**
**ISBN 978-3-7258-4488-3 (PDF)**
**https://doi.org/10.3390/books978-3-7258-4488-3**

Cover image courtesy of Sergio Castiñeira-Ibáñez, Daniel Tarrazó-Serrano and Constanza Rubio Michavila

# Contents

**About the Editors** . . . . . . . . . . vii

**Preface** . . . . . . . . . . ix

**Sergio Castiñeira-Ibáñez, Daniel Tarrazó-Serrano and Constanza Rubio**
Acoustic and Ultrasonic Sensing Technology in Non-Destructive Testing
Reprinted from: *Sensors* **2025**, *25*, 3271, https://doi.org/10.3390/s25113271 . . . . . . 1

**Yunxiu Ma, Linzhi Hu, Yuxuan Dong, Lei Chen and Gang Liu**
Bottom Plate Damage Localization Method for Storage Tanks Based on Bottom Plate-Wall Plate Synergy
Reprinted from: *Sensors* **2025**, *25*, 2515, https://doi.org/10.3390/s25082515 . . . . . . 7

**Fatahlla Moreh, Yusuf Hasan, Bilal Zahid Hussain, Mohammad Ammar, Frank Wuttke and Sven Tomforde**
MicrocrackAttentionNext: Advancing Microcrack Detection in Wave Field Analysis Using Deep Neural Networks Through Feature Visualization
Reprinted from: *Sensors* **2025**, *25*, 2107, https://doi.org/10.3390/s25072107 . . . . . . 26

**Artem T. Turov, Yuri A. Konstantinov, Ekaterina E. Totmina, Anna G. Votinova, Grigoriy F. Masich, Dmitry A. Korobko and Andrei A. Fotiadi**
Registration of Sounds Emitted by the Madagascar Hissing Cockroach Using a Distributed Acoustic Sensor
Reprinted from: *Sensors* **2025**, *25*, 2101, https://doi.org/10.3390/s25072101 . . . . . . 49

**Olgirdas Tumšys, Lina Draudvilienė and Egidijus Žukauskas**
Detailed Determination of Delamination Parameters in a Multilayer Structure Using Asymmetric Lamb Wave Mode
Reprinted from: *Sensors* **2025**, *25*, 539, https://doi.org/10.3390/s25020539 . . . . . . 67

**Xing Zhu, Hui Chen, Zhanglei Wu, Shumei Yang, Xiaopeng Li and Tiantao Li**
An Experimental Study of the Acoustic Signal Characteristics of Locked-Segment Damage Evolution in a Landslide Model
Reprinted from: *Sensors* **2024**, *24*, 4947, https://doi.org/10.3390/s24154947 . . . . . . 81

**Jianbin Luo, Shaofei Jiang, Yamian Zeng and Changqin Lai**
Three-Dimensional Reconstruction and Visualization of Underwater Bridge Piers Using Sonar Imaging
Reprinted from: *Sensors* **2024**, *24*, 4732, https://doi.org/10.3390/s24144732 . . . . . . 98

**Ivan Malashin, Vadim Tynchenko, Dmitry Martysyuk, Nikita Shchipakov, Nikolay Krysko, Maxim Degtyarev, et al.**
Assessment of Anisotropic Acoustic Properties in Additively Manufactured Materials: Experimental, Computational, and Deep Learning Approaches
Reprinted from: *Sensors* **2024**, *24*, 4488, https://doi.org/10.3390/s24144488 . . . . . . 125

**Maarten Adams, Arnaud Huijer, Christos Kassapoglou, Johannes A. A. Vaders and Lotfollah Pahlavan**
In Situ Non-Destructive Stiffness Assessment of Fiber Reinforced Composite Plates Using Ultrasonic Guided Waves
Reprinted from: *Sensors* **2024**, *24*, 2747, https://doi.org/10.3390/s24092747 . . . . . . 148

**Ziyan Zeng, Jing Wu, Mingfang Zheng and Hongwei Ma**
Rail Flaw Detection via Kolmogorov Entropy of Chaotic Oscillator Based on Ultrasonic Guided Waves
Reprinted from: *Sensors* **2024**, *24*, 2730, https://doi.org/10.3390/s24092730 . . . . . . . . . . . . . **168**

**Philipp Schulmeyer, Manfred Weihnacht and Hagen Schmidt**
A Dual-Mode Surface Acoustic Wave Delay Line for the Detection of Ice on 64°-Rotated Y-Cut Lithium Niobate
Reprinted from: *Sensors* **2024**, *24*, 2292, https://doi.org/10.3390/s24072292 . . . . . . . . . . . . . **187**

**Dmitry Olegovich Dolmatov and Vadim Yurevich Zhvyrblya**
Optimal Design of Sparse Matrix Phased Array Using Simulated Annealing for Volumetric Ultrasonic Imaging with Total Focusing Method
Reprinted from: *Sensors* **2024**, *24*, 1856, https://doi.org/10.3390/s24061856 . . . . . . . . . . . . . **199**

**Felix Brand and Klaus Stefan Drese**
Frequency-Resolved High-Frequency Broadband Measurement of Acoustic Longitudinal Waves by Laser-Based Excitation and Detection
Reprinted from: *Sensors* **2024**, *24*, 1630, https://doi.org/10.3390/s24051630 . . . . . . . . . . . . . **214**

**Yuchi Shen, Jing Wu, Junfeng Chen, Weiwei Zhang, Xiaolin Yang and Hongwei Ma**
Quantitative Detection of Pipeline Cracks Based on Ultrasonic Guided Waves and Convolutional Neural Network
Reprinted from: *Sensors* **2024**, *24*, 1204, https://doi.org/10.3390/s24041204 . . . . . . . . . . . . . **237**

**Linsen Xiong and Zhi-mei Qi**
A Grating Interferometric Acoustic Sensor Based on a Flexible Polymer Diaphragm
Reprinted from: *Sensors* **2023**, *23*, 9912, https://doi.org/10.3390/s23249912 . . . . . . . . . . . . . **254**

**M. S. Vechera, S. I. Konovalov, R. S. Konovalov, B. I, V. M. Tsaplev, A. D. Soloveva and J. Lee**
Assessment of the Influence of the Geometrical Shape of the Damper on the Efficiency of an Ultrasonic Operation Piezoelectric Transducer
Reprinted from: *Sensors* **2023**, *23*, 9662, https://doi.org/10.3390/s23249662 . . . . . . . . . . . . . **271**

**Baochun Xu, Jiacai Huang and Yachun Jie**
Application of the Lamb Wave Mode of Acoustic Emission for Monitoring Impact Damage in Plate Structures
Reprinted from: *Sensors* **2023**, *23*, 8611, https://doi.org/10.3390/s23208611 . . . . . . . . . . . . . **281**

**Jie Huang, Pengchao Chen, Rui Li, Kuan Fu, Yanan Wang, Jinyao Duan and Zhenlin Li**
Systematic Evaluation of Ultrasonic In-Line Inspection Techniques for Oil and Gas Pipeline Defects Based on Bibliometric Analysis
Reprinted from: *Sensors* **2024**, *24*, 2699, https://doi.org/10.3390/s24092699 . . . . . . . . . . . . . **305**

# About the Editors

**Sergio Castiñeira-Ibáñez**

Sergio Castiñeira-Ibáñez (Mislata, Spain) holds a degree (Universitat de València, UV) and PhD in Physical Sciences (Universitat Politècnica de València, UPV ) and is an Associate Professor in the Department of Applied Physics at the UPV. He also worked for an extended period of his academic and teaching career at the UV. His research focuses on studying the physical properties of periodic systems applied to acoustic waves and lenses for ultrasound transmission and is carried out at the Centro de Tecnologías Físicas: Acústica, Metamateriales y Astrofísica, part of the UPV. He has participated in more than ten competitive research projects in his field, contributing to the advancement of knowledge and innovation in applied physics. Throughout his research career, he has authored over thirty articles published in JCR-indexed journals, reflecting a broad dissemination of his scientific results. He is also the co-author of more than thirty papers presented at international conferences, several of which were invited contributions. He has co-authored two patents and contributed two chapters in technology books.

**Daniel Tarrazó-Serrano**

Daniel Tarrazó-Serrano (València, Spain) graduated in 2012 with a B.Sc. in Telecommunications Engineering, with a specialization in sound and images, and in 2013 with an M.Sc. in Acoustics from the Universitat Politècnica de València (UPV). He obtained two awards while he was studying: an Academic Award in his second course and an Excellence Student Award for obtaining the best results. In 2014 he completed his second B.Sc. in Telecommunications Systems, and he obtained an M.Sc. in Biomedical Engineering from the Universitat de València and UPV in 2016. He started his Ph.D. in 2017 as a senior researcher at the Physics Technologies Center. His line of research focuses on applied acoustics, particularly ultrasonics. He was a collaborating professor in the Applied Physics Department at the UPV and a visiting researcher at the School of Non-Destructive Testing and Security (National Research Tomsk Polytechnic University), located in Siberia, Russia. In 2020, he obtained the title of Doctor Cum Laude with a special mention awarded for his extraordinary thesis developed as part of the "Technologies for Health and Welfare" Ph.D. program, "Design, modelling, characterization and implementation of acoustic lenses for modulation of ultrasound beams." Currently, he is a Lecturer and Professor in the Department of Applied Physics at the UPV (Spain). He actively participates in multidisciplinary R&D (in agronomy, environmental acoustics, ultrasound, and numerical modeling), teaching research, public R&D, and transfer projects associated with companies. He has published scientific articles, book chapters, and congress papers in R&D and educational publications and participates in several university educational innovation projects.

**Constanza Rubio Michavila**

Constanza Rubio Michavila (Castellón de la Plana, Spain) received an MS Physics degree (1990) from the Universitat de València (UV) and a PhD in Applied Physics (1995) from the Universitat Politècnica de València (UPV). She is a Full Professor in the Applied Physics Department at the UPV and the author of more than 60 research articles published in journals. She has participated in more than 20 public and transfer research projects and has also made numerous contributions to conferences and book chapters. She is the author of three patents developed at the UPV and actively participates in developing educational improvements and innovative material for the subjects she teaches. She also studies the behavior of waves in the optical range and its influence on monitoring the environment and crops through Remote Sensing, allowing for more sustainable treatments and policies throughout the entire value chain.

# Preface

We are pleased to present this Special Issue reprint on acoustic and ultrasonic sensing technologies applied to non-destructive testing (NDT), which aims to capture recent advancements in ultrasonic methodologies and their expanding role in safety-critical monitoring, infrastructure evaluation, and intelligent sensing. The seventeen articles collected in this Special Issue cover a broad spectrum of topics, including guided wave-based damage detection, novel acoustic transducers, AI-enhanced signal processing, and cross-disciplinary case studies spanning civil, environmental, and biological domains. These contributions reflect not only technical excellence but also the innovative spirit driving this evolving field. The intended audience includes researchers, engineers, and professionals working in sensing technologies, structural health monitoring, and smart materials. We would like to thank all the contributing authors for their high-quality work and dedication, as well as the peer reviewers and editorial staff for their invaluable support in bringing this Special Issue to publication. We hope this collection serves as both a reference and a source of inspiration for future research.

**Sergio Castiñeira-Ibáñez, Daniel Tarrazó-Serrano, and Constanza Rubio Michavila**
*Guest Editors*

sensors

MDPI

*Editorial*

# Acoustic and Ultrasonic Sensing Technology in Non-Destructive Testing

**Sergio Castiñeira-Ibáñez, Daniel Tarrazó-Serrano * and Constanza Rubio**

Centro de Tecnologías Físicas, Universitat Politècnica de València, Camí de Vera s/n, 46022 València, Spain; sercasib@fis.upv.es (S.C.-I.); crubiom@fis.upv.es (C.R.)
* Correspondence: dtarrazo@fis.upv.es

## 1. Introduction

The physics of acoustics, particularly the high-frequency domain of ultrasonics, remains a dynamic area of innovation in sensing technologies. Ultrasonic waves are mechanical oscillations propagating through matter at frequencies beyond the range of human hearing. They offer a unique ability to interrogate the interior of materials and structures non-invasively. Because they are sensitive to mechanical, structural, and elastic properties, ultrasonic waves are an ideal tool for applications that require internal inspection without direct access or physical intrusion. This makes them especially valuable in the context of non-destructive testing (NDT), structural health monitoring, and, increasingly, biological and environmental sensing. Ultrasonic wave propagation is governed by classical elastic wave theory, where energy can travel through solids in various modes such as longitudinal, shear, surface, or guided wave modes. Guided wave modes, such as Lamb waves, are particularly well suited for plate-like or multilayered structures, as they can propagate over long distances while maintaining high sensitivity to geometric or material discontinuities. These characteristics have made guided waves a cornerstone of modern NDT approaches, especially in large infrastructure systems such as storage tanks, pipelines, and railway tracks [1,2]. Among the most established ultrasonic techniques, acoustic emission (AE) is lauded for its ability to detect transient elastic waves generated by the rapid release of energy in materials under stress. Unlike conventional ultrasonic testing methods that rely on externally introduced signals, AE leverages the material's own response to structural changes such as crack initiation, fiber breakage, or corrosion. This passive approach enables the continuous, real-time monitoring of critical components and has been widely adopted in applications ranging from pressure vessels and pipelines to aerospace structures [3]. AE complements guided wave and imaging techniques, particularly in damage localization and early failure detection scenarios [4].

Numerous contributions to this Special Issue reflect the central role of ultrasonic-guided waves in contemporary sensing applications. Advances include their use in detecting damage in the bottom plates of storage tanks, assessing internal stiffness in fiber-reinforced composites, identifying flaws in multilayer structures, and monitoring defects in railway systems. The versatility of guided waves, particularly when combined with advanced signal processing and sensor arrays, enables the efficient inspection of difficult-to-reach areas with minimal intervention [5]. Parallel to improvements in wave propagation techniques, significant sensor and transducer design developments continue to shape the field. Piezoelectric ceramics such as PZT remain widely used, but new materials, such as flexible polymers and polymer composites, are gaining ground due to their adaptability and mechanical robustness. Some articles in this Special Issue explore novel damping geometries, surface acoustic wave devices, and optimized array configurations. This shows how hardware design remains crucial to improving signal-to-noise ratios and imaging resolution [6].

Beyond classical inspection, integrating computational intelligence into ultrasonic systems is redefining the research landscape. Several studies in this Special Issue apply deep learning architectures, such as convolutional neural networks, to ultrasonic signal interpretation. These data-driven methods offer superior performance in identifying and quantifying defects, particularly in noisy environments or when handling complex, high-dimensional data [7]. For example, recent research applies neural networks to extract microcrack information from wavefield interactions [8] or to perform quantitative crack sizing in pipelines based on one-dimensional time-series data [9].

Advances in materials science, particularly in additive manufacturing, have introduced complex and often anisotropic structures that demand tailored acoustic evaluation strategies. At the same time, ultrasonic methods are expanding beyond traditional solid media to include aquatic and volumetric environments, requiring new imaging and reconstruction techniques suited for submerged or irregular geometries [10]. NDT is also relevant in environmental monitoring, where acoustic measurements can capture structural changes in soils or geological formations and provide early indicators of instability or failure. Moreover, the high sensitivity of new acoustic systems to minute vibrations and emissions has enabled novel applications in biological and ecological contexts, such as detecting activity in insects, plants, or small animals. Recent advances include configurable ultrasonic lenses designed for subwavelength resolution and beam shaping, which leverage geometric design and material properties to control wave propagation with high precision [11]. Techniques for tunable beam bending and directional focusing, such as those using Janus-type acoustic structures, have been explored to enable flexible energy steering in complex scenarios [12]. These innovations are beneficial for scanning hidden regions or navigating obstacles and are applicable in both industrial and medical contexts. Among the emerging concepts in subwavelength focusing, photonic nanojets represent a promising direction for future ultrasonic imaging enhancement. Generated by dielectric microspheres under optical illumination, these highly localized and high-intensity beams can be coupled with acoustic systems to improve spatial resolution and sensing accuracy. While not yet featured in the contributions of this Special Issue, their potential integration into optoacoustic or hybrid systems is likely to influence next-generation sensor designs [13].

Altogether, the articles presented in this Special Issue reflect a growing trend toward integrating ultrasonic technologies with innovative sensing strategies, data-driven processing, and interdisciplinary applications. Whether used for inspecting the integrity of critical infrastructure, analyzing manufactured components, or monitoring complex biological and geological systems, ultrasonic sensors are becoming more adaptive, accurate, and multifunctional. The convergence of classical acoustic physics with artificial intelligence, flexible materials, and advanced imaging techniques ensures that acoustic and ultrasonic sensing technologies will remain at the forefront of scientific and technological development.

## 2. An Overview of Published Articles

This Special Issue brings together seventeen original contributions that reflect the state-of-the-art in acoustic and ultrasonic sensing for NDT. Throughout this section, the numbers in parentheses (e.g., (1), (2), etc.) refer to the individual articles included in this issue, as listed in Section "List of Contributions". The research spans sensor design, advanced signal processing, machine learning integration, and novel applications across civil, mechanical, and biological domains. For clarity, we have organized the papers into thematic categories and summarize key results.

### 2.1. Structural Health Monitoring and Damage Detection

Several papers focus on damage identification using guided waves and acoustic emission (AE). Xu et al. (1) applied the Lamb Wave Mode of AE to monitor impact damage in epoxy glass fiber plates, showing that damage-induced AE can be directly extracted from impact signals in real time. Shen et al. (4) integrated guided wave detection with a one-dimensional convolutional neural network (CNN) to detect pipeline cracks, achieving errors below 2% in both the simulation and experiment.

Adams et al. (10) used guided waves to assess the stiffness of fiber-reinforced composites non-destructively. Their inverse numerical model estimated matrix components with less than 10.4% error and tensile/flexural stiffness with average errors of 3.6% and 9.0%, respectively. Ma et al. (17) introduced a collaborative method using sensor arrays on a tank bottom and wall plates to locate corrosion-induced defects in storage tanks. A correlation-based signal processing approach improved image clarity and achieved localization errors as low as 5.4%.

In rail systems, Zeng et al. (9) developed a chaotic oscillator model based on the Duffing system to identify ultrasonic guided wave signals. Notches of 0.46 mm and 1.78 mm were detected at the rail base and head, respectively, with Kolmogorov entropy used as a quantitative index. Tumšys et al. (14) investigated delamination in multilayer composites using the asymmetric Lamb wave $A_0$ mode, demonstrating its sensitivity to layer separation.

### 2.2. Sensor and Transducer Development

Sensor design is a recurring theme. Vechera et al. (2) studied the effect of snubber geometry in piezoelectric ultrasonic transducers (PETs). A truncated cone-shaped damper increased the transmitter's bandwidth and transient performance by modifying the inclination angle $\alpha$ of the generatrix. Xiong and Qi (3) proposed a grating interferometric acoustic sensor based on a flexible polymer diaphragm. It achieved a minimum detectable pressure of 164.8 $\mu Pa/\sqrt{Hz}$ and a flat frequency response with less than 3.2% jitter in the speech range, suggesting applications in acoustic monitoring and voice acquisition.

Brand and Drese (5) introduced a laser-induced optoacoustic system using a photorefractive interferometer. Phase velocities and attenuations of longitudinal waves were measured across a 3–55 MHz range with relative errors under 0.2% for materials, such as silicon and aluminum. Schulmeyer et al. (7) developed a dual-mode surface acoustic wave (SAW) delay line on a 64° Y-cut lithium niobate substrate, capable of differentiating between liquid water and ice while simultaneously measuring temperature, proving effective in harsh environments.

### 2.3. Signal Processing and Machine Learning

Several articles showcase the power of data-driven analysis. Shen et al. (4) leveraged CNNs for crack detection, while Dolmatov and Zhvyrblya (6) optimized the design of sparse matrix phased arrays for the total focusing method (TFM). Their approach reduced the data volume by up to 84% without sacrificing image quality.

Malashin et al. (11) proposed a hybrid experimental–computational method for evaluating acoustic anisotropy in additively manufactured materials. A new anisotropy coefficient, derived from transverse wave velocities, showed a strong correlation (0.97) with echo amplitude variations. A neural network optimized via genetic algorithms efficiently predicted this coefficient. Moreh et al. (16) designed a deep learning model for detecting microcracks from wavefield data. Their asymmetric encoder–decoder network, enhanced with feature reuse and manifold visualization, reached a detection accuracy of 87.74%.

### 2.4. Cross-Disciplinary and Emerging Applications

This issue also highlights innovative uses of ultrasonic sensing. Huang et al. (8) performed a bibliometric review of ultrasonic In-Line Inspection (ILI) for oil and gas pipelines. The paper visualizes research trends and classifies defect detection techniques, offering a comparative framework from lab testing to industrial deployment.

Luo et al. (12) presented a sonar-based 3D reconstruction system for submerged bridge abutments. Using MS1000 sonar and automated processing, they extracted coordinates of contours and reconstructed abutments with volume estimation errors of 13.56% (holes) and 10.65% (spalling). Zhu et al. (13) simulated landslides in the lab to analyze AE and micro-seismic signal patterns. They proposed a classification method to separate burst and continuous AE signals, revealing their evolution as a precursor to failure.

Finally, Turov et al. (15) used distributed acoustic sensors (DASs) to monitor the sounds of the Madagascar hissing cockroach. Both the insect hissing and mechanical interactions with optical fibers were recorded, which could lead to potential new innovations in agricultural or ecological monitoring using fiber-optic acoustic sensing.

## 3. Conclusions

This Special Issue, devoted to acoustic and ultrasonic sensing technology in non-destructive testing, presents a comprehensive and up-to-date overview of state-of-the-art research in the relevant field. The seventeen contributions not only demonstrate significant progress in conventional techniques, but also incorporate emerging approaches such as signal processing through artificial intelligence, innovative sensor and transducer design, and interdisciplinary applications. The topics covered in this collection clearly align with the thematic areas outlined in the section, "An Overview of Published Articles", including structural health monitoring and damage detection, sensor and transducer development, advanced signal processing, and machine learning, as well as novel applications across civil, mechanical, biological, and environmental domains. This thematic diversity underscores the ongoing evolution of ultrasonic sensing toward more adaptive, accurate, and multifunctional solutions, reaffirming its central role in current scientific and technological innovation. We believe that these seventeen contributions add great value to the scientific advancement of non-destructive testing. We thank all authors for the meticulous reviewing of their research.

**Conflicts of Interest:** The authors declare no conflicts of interest.

## List of Contributions

1. Xu, B.; Huang, J.; Jie, Y. Application of the Lamb Wave Mode of Acoustic Emission for Monitoring Impact Damage in Plate Structures. *Sensors* **2023**, *23*, 8611. https://doi.org/10.3390/s23208611.
2. Vechera, M.S.; Konovalov, S.I.; Konovalov, R.S.; Tsaplev, V.M.; Soloveva, A.D.; Lee, J. Assessment of the Influence of the Geometrical Shape of the Damper on the Efficiency of an Ultrasonic Operation Piezoelectric Transducer. *Sensors* **2023**, *23*, 9662. https://doi.org/10.3390/s23249662.
3. Xiong, L.; Qi, Z.-m. A Grating Interferometric Acoustic Sensor Based on a Flexible Polymer Diaphragm. *Sensors* **2023**, *23*, 9912. https://doi.org/10.3390/s23249912.
4. Shen, Y.; Wu, J.; Chen, J.; Zhang, W.; Yang, X.; Ma, H. Quantitative Detection of Pipeline Cracks Based on Ultrasonic Guided Waves and Convolutional Neural Network. *Sensors* **2024**, *24*, 1204. https://doi.org/10.3390/s24041204.

5. Brand, F.; Drese, K.S. Frequency-Resolved High-Frequency Broadband Measurement of Acoustic Longitudinal Waves by Laser-Based Excitation and Detection. *Sensors* **2024**, *24*, 1630. https://doi.org/10.3390/s24051630.
6. Dolmatov, D.O.; Zhvyrblya, V.Y. Optimal Design of Sparse Matrix Phased Array Using Simulated Annealing for Volumetric Ultrasonic Imaging with Total Focusing Method. *Sensors* **2024**, *24*, 1856. https://doi.org/10.3390/s24061856.
7. Schulmeyer, P.; Weihnacht, M.; Schmidt, H. A Dual-Mode Surface Acoustic Wave Delay Line for the Detection of Ice on 64°-Rotated Y-Cut Lithium Niobate. *Sensors* **2024**, *24*, 2292. https://doi.org/10.3390/s24072292.
8. Huang, J.; Chen, P.; Li, R.; Fu, K.; Wang, Y.; Duan, J.; Li, Z. Systematic Evaluation of Ultrasonic In-Line Inspection Techniques for Oil and Gas Pipeline Defects Based on Bibliometric Analysis. *Sensors* **2024**, *24*, 2699. https://doi.org/10.3390/s24092699.
9. Zeng, Z.; Wu, J.; Zheng, M.; Ma, H. Rail Flaw Detection via Kolmogorov Entropy of Chaotic Oscillator Based on Ultrasonic Guided Waves. *Sensors* **2024**, *24*, 2730. https://doi.org/10.3390/s24092730.
10. Adams, M.; Huijer, A.; Kassapoglou, C.; Vaders, J.A.A.; Pahlavan, L. In Situ Non-Destructive Stiffness Assessment of Fiber Reinforced Composite Plates Using Ultrasonic Guided Waves. *Sensors* **2024**, *24*, 2747. https://doi.org/10.3390/s24092747.
11. Malashin, I.; Tynchenko, V.; Martysyuk, D.; Shchipakov, N.; Krysko, N.; Degtyarev, M.; Nelyub, V.; Gantimurov, A.; Borodulin, A.; Galinovsky, A. Assessment of Anisotropic Acoustic Properties in Additively Manufactured Materials: Experimental, Computational, and Deep Learning Approaches. *Sensors* **2024**, *24*, 4488. https://doi.org/10.3390/s24144488.
12. Luo, J.; Jiang, S.; Zeng, Y.; Lai, C. Three-Dimensional Reconstruction and Visualization of Underwater Bridge Piers Using Sonar Imaging. *Sensors* **2024**, *24*, 4732. https://doi.org/10.3390/s24144732.
13. Zhu, X.; Chen, H.; Wu, Z.; Yang, S.; Li, X.; Li, T. An Experimental Study of the Acoustic Signal Characteristics of Locked-Segment Damage Evolution in a Landslide Model. *Sensors* **2024**, *24*, 4947. https://doi.org/10.3390/s24154947.
14. Tumšys, O.; Draudvilienė, L.; Žukauskas, E. Detailed Determination of Delamination Parameters in a Multilayer Structure Using Asymmetric Lamb Wave Mode. *Sensors* **2025**, *25*, 539. https://doi.org/10.3390/s25020539.
15. Turov, A.T.; Konstantinov, Y.A.; Totmina, E.E.; Votinova, A.G.; Masich, G.F.; Korobko, D.A.; Fotiadi, A.A. Registration of Sounds Emitted by the Madagascar Hissing Cockroach Using a Distributed Acoustic Sensor. *Sensors* **2025**, *25*, 2101. https://doi.org/10.3390/s25072101.
16. Moreh, F.; Hasan, Y.; Hussain, B.Z.; Ammar, M.; Wuttke, F.; Tomforde, S. MicrocrackAttentionNext: Advancing Microcrack Detection in Wave Field Analysis Using Deep Neural Networks Through Feature Visualization. *Sensors* **2025**, *25*, 2107. https://doi.org/10.3390/s25072107.
17. Ma, Y.; Hu, L.; Dong, Y.; Chen, L.; Liu, G. Bottom Plate Damage Localization Method for Storage Tanks Based on Bottom Plate-Wall Plate Synergy. *Sensors* **2025**, *25*, 2515. https://doi.org/10.3390/s25082515.

## References


1. Rose, J.L. *Ultrasonic Guided Waves in Solid Media*; Cambridge University Press: Cambridge, UK, 2004.
2. Barzdenas, V.; Vasjanov, A. A method of optimizing characteristic impedance compensation using cut-outs in high-density pcb designs. *Sensors* **2022**, *22*, 964. [CrossRef] [PubMed]
3. Krautkrämer, J.; Krautkrämer, H. *Ultrasonic Testing of Materials*, 4th ed.; Springer: Berlin/Heidelberg, Germany, 1990.

4. Gu, A.; Liu, Z.; Pan, R.; Xu, B. Damage Identification in Concrete Using Instantaneous Dominant Frequency of Acoustic Emission Signals. *Buildings* **2024**, *14*, 3784. [CrossRef]
5. Shah, J.K.; Wang, H.; El-Hawwat, S. Detection of internal crack growth in polyethylene pipe using guided wave ultrasonic testing. *Earthq. Eng. Eng. Vib.* **2024**, *23*, 319–329. [CrossRef]
6. Drinkwater, B.W.; Wilcox, P.D. Ultrasonic arrays for non-destructive evaluation: A review. *NDT & E Int.* **2006**, *39*, 525–541.
7. Cantero-Chinchilla, S.; Wilcox, P.D.; Croxford, A.J. Deep learning in automated ultrasonic NDE—Developments, axioms and opportunities. *NDT & E Int.* **2022**, *131*, 102703.
8. Moreh, F.; Lyu, H.; Rizvi, Z.H.; Wuttke, F. Deep neural networks for crack detection inside structures. *Sci. Rep.* **2024**, *14*, 4439. [CrossRef] [PubMed]
9. Shen, Y.; Wu, J.; Chen, J.; Zhang, W.; Yang, X.; Ma, H. Quantitative detection of pipeline cracks based on ultrasonic guided waves and convolutional neural network. *Sensors* **2024**, *24*, 1204. [CrossRef] [PubMed]
10. Amin, A.; Jack, D.A.; Ravindranath, P.K.; Fleck, T.J. Nondestructive quantification of internal raster path for additively manufactured components via ultrasonic testing. *Sci. Rep.* **2024**, *14*, 11416. [CrossRef] [PubMed]
11. Castiñeira-Ibáñez, S.; Tarrazó-Serrano, D.; Uris, A.; Rubio, C.; Minin, O.V.; Minin, I.V. Cylindrical 3D printed configurable ultrasonic lens for subwavelength focusing enhancement. *Sci. Rep.* **2020**, *10*, 20279. [CrossRef] [PubMed]
12. Castiñeira-Ibáñez, S.; Tarrazó-Serrano, D.; Uris, A.; Rubio, C. Tunable acoustic hooks from Janus cylinder. *Results Phys.* **2021**, *24*, 104134. [CrossRef]
13. Heifetz, A.; Kong, J.A.; Sahakian, A.V.; Taflove, A.; Backman, V. Photonic nanojets. *J. Comput. Theor. Nanosci.* **2009**, *6*, 1979–1992. [CrossRef] [PubMed]

*sensors*

MDPI

*Article*

# Bottom Plate Damage Localization Method for Storage Tanks Based on Bottom Plate-Wall Plate Synergy

**Yunxiu Ma [1,2], Linzhi Hu [1], Yuxuan Dong [1], Lei Chen [1,*] and Gang Liu [1]**

1 College of Pipeline and Civil Engineering, China University of Petroleum (East China), Qingdao 266580, China; yunxiu611@126.com (Y.M.); hulz22@163.com (L.H.); b23060018@s.upc.edu.cn (Y.D.); liugang@upc.edu.cn (G.L.)

2 Pipe-Network Group (Xuzhou) Pipeline Inspection and Testing Co., Xuzhou 221000, China

* Correspondence: leo@upc.edu.cn; Tel.: +86-175-6005-3657

**Abstract:** Ultrasonic guided waves can be employed for in-service defect detection in storage tank bottom plates; however, conventional single-array approaches face challenges from boundary scattering noise at side connection welds. This study proposes a collaborative bottom plate-wall plate detection methodology to address these limitations. Sensor arrays were strategically deployed on both the bottom plate and wall plate, achieving multidimensional signal acquisition through bottom plate array excitation and dual-array reception from both the bottom plate and tank wall. A correlation coefficient-based matching algorithm was developed to distinguish damage echoes from weld-induced scattering noise by exploiting path-dependent signal variations between the two arrays. The investigation revealed that guided wave signals processed through data matching effectively preserved damage echo signals while substantially attenuating boundary scattering signals. Building upon these findings, correlation matching was implemented on guided wave signals received by corresponding array elements from both the bottom plate and wall plate, followed by total focusing imaging (TFM) using the processed signals. Results demonstrate that the collaborative bottom plate-wall plate detection imaging cloud maps, after implementing signal correlation matching, effectively suppress artifacts compared with imaging results obtained solely from bottom plate arrays. The maximum relative localization error was measured as 5.4%, indicating superior detection accuracy.

**Keywords:** t-shaped structure; ultrasonic guided waves; damage detection; total focusing imaging (TFM); collaborative bottom plate-wall plate detection

## 1. Introduction

Storage tanks, as critical storage equipment in the petrochemical industry, are prone to localized corrosion during their long-term service due to factors such as construction quality, erosion from stored media, and surrounding environmental conditions [1–3]. Consequently, there is an urgent need for an online inspection technology capable of rapidly locating defects in the tank floor.

Traditional non-destructive testing (NDT) techniques are mostly invasive, characterized by low efficiency and high consumption of human and material resources [4–7]. Ultrasonic guided waves, as a type of mechanical elastic wave constrained and guided by structural boundaries [8–11], exhibit excellent detection capabilities for both surface and internal defects. Moreover, guided waves in different modes demonstrate unique propagation characteristics due to their specific wave structures [12,13], leading to their widespread application in NDT [14–16]. Fakih et al. [17] conducted experimental and finite element studies on defects in friction stir welds and proposed an adaptive noise-assisted empirical mode decomposition (EMD) data processing method. This method evaluates the extent of damage to welds based on variations in captured signal replicas. Zhang et al. [18] achieved detection of a 5 mm $\times$ 1 mm crack defect behind a "T"-shaped pipeline support structure by exciting the $CSH_0$ mode guided waves using electromagnetic transducers. Wu et al. [19] investigated a Rayleigh-type wave (RTW) characteristic guided wave in "T"-shaped welded joints and achieved long-distance detection of small defects in the "T"-shaped weld joint area by exciting this modal guided wave at the weld end face. Raišutis et al. [20] combined transmission tomography with Lamb waves excited on the outer side of the tank bottom plate, enabling damage detection within the tank floor. Lowe et al. [21] investigated the propagation characteristics of Lamb and $SH_0$ mode guided waves in the bottom plate and wall plate of storage tanks through finite element modeling and experimental validation. Their findings revealed that $SH_0$ mode guided waves exhibit a higher signal-to-noise ratio (SNR) during tank inspections. Wang et al. [22] developed a magnetostrictive $SH_0$ mode electromagnetic transducer for tank bottom plates and implemented a rotational focusing imaging method, achieving multi-defect detection within the inspection area.

In damage imaging processes, high-quality defect visualization and accurate localization can typically be achieved when the input signals exhibit a high signal-to-noise ratio (SNR). However, in large-scale containment structures such as petroleum storage tanks, the bottom plate assemblies fabricated through welding of multiple steel plates inevitably contain heterogeneous welded joints (e.g., butt welds and lap joints) with varying geometric configurations. During non-destructive testing (NDT) of the outermost edge plates in tank bottom structures, the discontinuous boundary scattering echoes caused by geometric and material discontinuities at sidewall weld connections are practically unavoidable, as illustrated in Figure 1. These scattering signals, when focused through transducer arrays, generate ghost echoes resembling structural artifacts within the inspection zone.

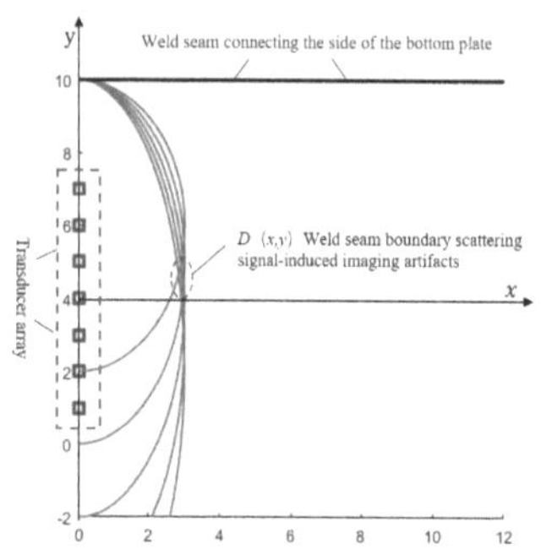


**Figure 1.** Schematic illustration of interference effects caused by side connection welds on the storage tank bottom plate.

Figure 2 delineates the peripheral edge plate arrangement. Conventional guided wave testing methodologies position transducers on these exterior edge plates, leveraging wave transmission through wall structures for internal defect detection. The inevitable presence of ambient noise during field inspections introduces positioning inaccuracies in damage imaging, rendering single-modal detection techniques insufficient for reliable assessment. Consequently, recent advancements focus on boundary scattering mitigation strategies [23–25]. Yu et al. [26] developed a wavelet reconstruction time-reversal (WR-TR) method integrating multimodal superposition finite-difference time-domain (MS-FDTD) simulations with maximum energy frame (MEF) analysis, achieving $\leq$2.3 mm localization accuracy for plate defects under strong boundary scattering interference. Huang et al. [27]

innovatively treated boundary-reflected signals as virtual excitation sources, enabling composite plate defect detection through probabilistic damage imaging. Hall et al. [28] proposed sparse-array multipath guided wave imaging utilizing structural reverberations to construct signal dictionaries, validated on both aluminum and composite plates. Zhang et al. [29] proposed a multipath edge-reflected defect imaging methodology. This approach employs mirror points from four edges of a plate structure as additional virtual transmitters and receivers. By systematically identifying all potential propagation paths of edge-reflected wave packets, the method detects defect-induced damage paths through comprehensive path tracing. The final imaging synthesis is achieved through a multiplicative fusion strategy, which effectively integrates the multi-path detection results. This innovative technique enables reliable damage detection in conventional blind zones that are inaccessible to traditional inspection methods.

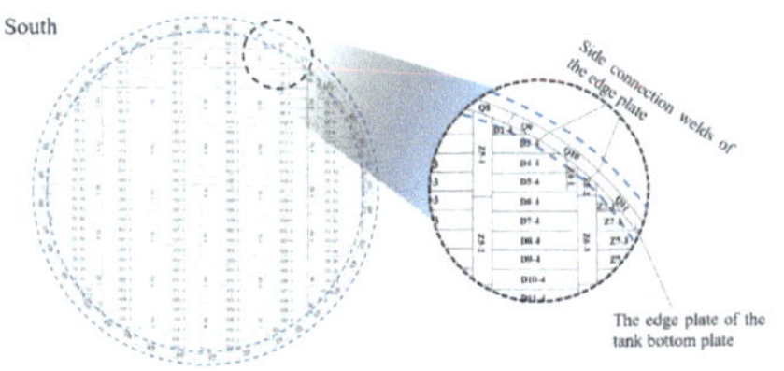


**Figure 2.** Schematic of the storage tank bottom plate structure.

This paper addresses the phenomenon of artifacts caused by the reflection signals of guided waves from the side connection welds during array-based detection of the tank bottom. A collaborative detection method for the tank bottom and wall plate, tailored to the special "T"-shaped structure of the tank bottom, is proposed. By arranging transducer arrays on both the tank bottom and the wall plate, multidimensional data matrices from both components are acquired. Based on the differences in the propagation paths of boundary scattering echoes to the tank bottom and wall plate, the information received from both components is matched and fused. This approach effectively reduces the impact of boundary scattering signals and ultimately achieves defect detection behind the "T"-shaped structure of the tank bottom using the total focusing imaging method.

## 2. Dispersion Curve Construction for Storage Tank Baseplate

The Wave Finite Element Method (WFE), a numerical approach specifically designed for wave propagation analysis, combines analytical wave propagation principles with conventional finite element advantages. This method effectively addresses dynamic problems in periodic structures and gradually varying non-uniform configurations. By utilizing established commercial software to model only periodic substructures, WFE achieves full waveguide analysis while substantially reducing model dimensions and computational expenses compared to finite element eigenvalue frequency approaches. The Floquet Boundary Condition (Floquet BC) method [30], originally developed for deriving dispersion curves of periodic waveguides, embeds periodicity characteristics into computations, thereby limiting solutions to a single unit cell and simplifying problem complexity. See Figure 3.

Numerical simulations were conducted using the COMSOL Multiphysics 6.1 Solid Mechanics module, where a 0.2 mm $\times$ 0.2 mm $\times$ 20 mm unit cell of the storage tank baseplate was modeled. A free tetrahedral meshing strategy was implemented with strictly identical mesh configurations on periodic boundary interfaces, employing the predefined "Extremely fine" mesh size setting to ensure precision.

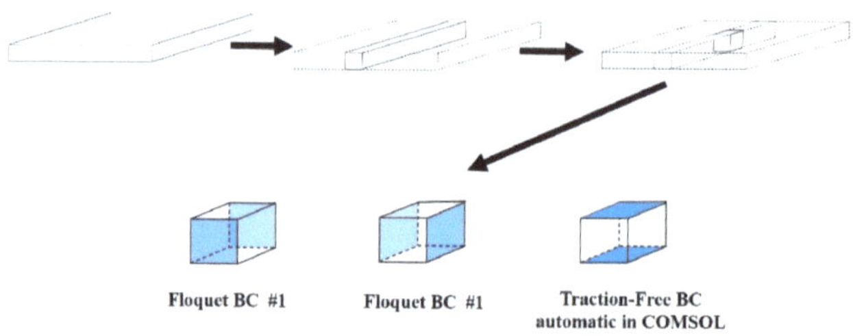


**Figure 3.** Schematic of the unit cell model implemented in the Floquet boundary condition (Floquet BC) method.

The materials used for both the tank bottom plate and wall plate are Q235 carbon steel, with an elastic modulus of 208 GPa, a density of 7850 $kg/m^3$, and a Poisson's ratio of 0.3. By employing the wave finite element method, the dispersion curves of guided waves in the tank bottom plate and wall plate structure were calculated. The results are shown in Figure 4 below:

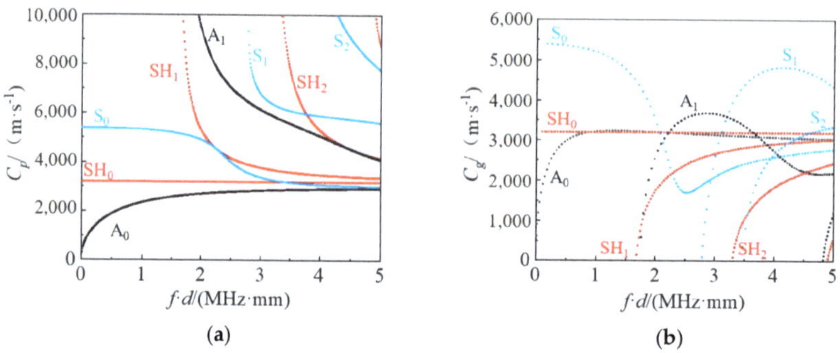


**Figure 4.** Dispersion characteristics of storage tank bottom plate structure (**a**) Phase velocity spectrum (**b**) Group velocity profile.

From the dispersion curves of the tank bottom plate shown in Figure 4, it can be observed that, compared to the $A_0$ and $S_0$ mode Lamb waves, the $SH_0$ mode guided waves maintain a stable propagation velocity across different frequency-thickness products due to their unique vibration propagation characteristics. This stability makes $SH_0$ mode guided waves particularly ideal for damage detection applications. Therefore, this study will focus on the stable propagation characteristics of $SH_0$ mode guided waves to conduct further investigations for damage detection.

## 3. Experimental Setup

The "T"-shaped structure in the experimental setup represents a 100,000-cubic-meter tank bottom model. Both the bottom plate and wall plate are made of Q235 structural carbon steel. The thickness of the bottom plate is 20 mm, while the wall plate has a thickness of 30 mm. The length of the edge plate on the outer side of the tank is 120 mm, and the overall width of the bottom plate is 3 m. The specific experimental platform is shown in Figure 5. In this study, the signal generator used is a Tektronix AFG1062 function generator (Tektronix, Beaverton, OR, USA), which is capable of effectively generating sine waves, square waves, and arbitrary waveforms. During the experiment, the guided wave signal excited is a five-cycle Hanning window-modulated sine wave with a center frequency of 65 kHz. The specific modulation method is shown in Equation (1). The signal amplifier is an ATA-2021H (Aigtek, Xi'an, China), with a controllable voltage gain of up to 60 times and a maximum output voltage of 200 V. The oscilloscope is a RIGOL DS1104 (RIGOL, Suzhou, China), which supports multi-channel data acquisition with a maximum real-time sampling rate of 1 GSa/s in full-channel mode. In this experiment, the sampling rate of the oscilloscope is set to 12.5 MSa/s. The transducers used are d15-type $10 \times 10 \times 3$ mm piezoelectric shear plates, which can effectively excite $SH_0$ mode guided waves when excited perpendicular to the polarization direction of the transducer. The coupling agent used is a metal adhesive, which exhibits excellent compatibility between ceramics and metals. Additionally, it possesses fast-drying properties, allowing for rapid and high-strength coupling of the transducer with the wall plate.

**Figure 5.** Experimental setup diagram of the "T"-shaped structure of the tank bottom plate.

Hanning window function expression:

$$A_t = \begin{cases} u(t) = 0.5\left[1 - cos(2\pi f \times \frac{t}{n})\right] \cdot sin(2\pi f t), 0 \le t \le n/f \\ 0, t \ge n/f \end{cases} \quad (1)$$

where $f$ represents the central frequency of the signal in kHz; $t$ is the time interval in seconds (s); $n$ denotes the number of periods of the signal.

## 4. Damage Localization Methodology

### *4.1. Total Focusing Imaging-Based Localization Using Bottom Plate Array*

Total focusing method (TFM), as an advanced ultrasonic array inspection technique, involves sequential excitation and reception of array elements. During the data acquisition phase, it enables the collection of data from all excitation and reception channels within the array, providing a rich dataset for subsequent processing. The TFM algorithm divides the imaging area into a grid and calculates the acoustic propagation path between each array element and each grid point. Based on the group velocity of the guided waves, the corresponding time is computed, and the signal amplitude at that time in the acquired data is assigned to the corresponding grid point. This process is repeated iteratively, and the accumulated amplitude information is used as the grayscale value for the pixel, thereby achieving full-focus imaging of the region. The schematic diagram of the principle of Total Focus Imaging is shown in Figure 6 below.

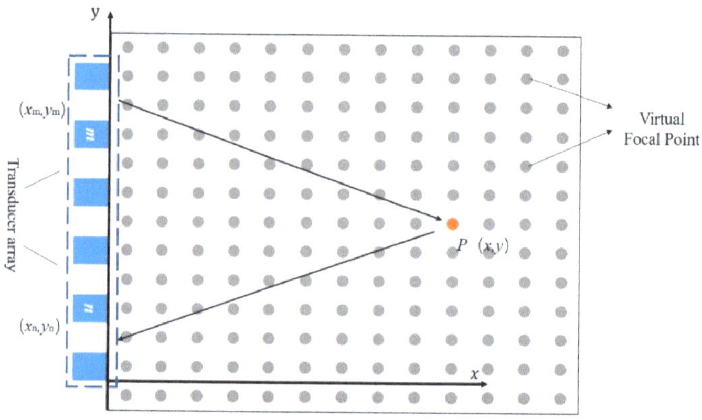


**Figure 6.** Schematic of the Total Focus Imaging Principle.

During the online inspection of storage tank bottoms, non-invasive methods are typically employed, and the transducer array can only be positioned on the exterior of the tank. Therefore, the most straightforward approach in practical detection is to arrange the transducer array on the outer edge plate of the tank bottom. By exciting and receiving array elements, damage information from different regions of the bottom plate is collected, enabling the imaging and localization of damage defects on the bottom plate.

Taking the array inspection of the tank bottom as an example, consider a single propagation path involving an excitation element ($x_m$, $y_m$), a reception element ($x_n$, $y_n$), and a damage point $P$ ($x$, $y$). The time-of-flight $T$ ($x$, $y$) along the propagation path is calculated based on their geometric relationship.

$$T(x,y) = \frac{\sqrt{(x_m - x)^2 + (y_m - y)^2} + \sqrt{(x_n - x)^2 + (y_n - y)^2}}{C_g} \tag{2}$$

where $x_m$ represents the horizontal coordinate of the excitation transducer, in millimeters (mm); $y_m$ denotes the vertical coordinate of the excitation transducer, in millimeters (mm); $x_n$ is the horizontal coordinate of the receiving transducer, in millimeters (mm); $y_n$ is the vertical coordinate of the receiving transducer, in millimeters (mm); $C_g$ is the group velocity of the excited guided wave, in meters per second (m/s); $T$ ($x$, $y$) is the time-of-flight of the guided wave signal from the excitation element to the receiving transducer after being reflected by the damage, in seconds (s).

After calculating the time-of-flight of the damage signal along the propagation path, the amplitude information of the corresponding echo is extracted. This amplitude information is assigned to the grid point at the corresponding distance. By sequentially superimposing these values, the final amplitude at the focal point is obtained, enabling the focused imaging of the damage signal.

$$I(x,y) = \sum_{i=1}^{N} \sum_{j=1}^{N} S_{ij}(T_{ij}(x,y)) \tag{3}$$

where $T_{ij}$ represents the time taken for the guided wave excited by transducer $i$ to be received by transducer $j$, in seconds (s); $S_{ij}$ denotes the characteristic amplitude information extracted from the signal received by transducer $j$ when excited by transducer $i$, in volts (V); and $I$ ($x$, $y$) is the final amplitude at the focal point after superimposing the amplitude information from multiple damage paths, in volts (V).

### 4.2. Collaborative Bottom Plate-Wall Plate Damage Detection Methodology

When inspecting the bottom plate of a storage tank, leveraging its unique "T"-shaped structure, in addition to acquiring guided wave signals from the bottom plate for detection, a transducer array can also be deployed on the tank wall to obtain more comprehensive data. During the experiment, the transducers on the tank bottom are sequentially excited, and the signals are received by the transducers on both the bottom plate and the wall plate. This dual-array data acquisition from the bottom plate and wall plate enables the collection of multidimensional information.

In practical non-destructive testing processes, since the tank bottom plate is constructed by welding multiple steel plates, it is inevitable to receive non-damage echo noise signals reflected from the side connection welds of the bottom plate. These signals can undoubtedly introduce certain negative effects on the detection process. In this experiment, a d15-type piezoelectric shear transducer is used. When excited perpendicular to the polarization direction of the transducer, the $SH_0$ mode is dominant, while excitation along the polarization direction generates the $S_0$ mode. Due to the non-dispersive and mode-conversion-free propagation characteristics of the $SH_0$ mode guided waves, they are utilized for damage detection in the tank bottom plate. During the detection process, the damage reflection signals within the detection area in front of the array are $SH_0$ mode guided waves, whereas the boundary reflections are $S_0$ mode guided waves.

When the guided wave signals propagate from the bottom plate to the tank wall, the plate thickness changes, causing the frequency-thickness product to shift from 1.3 MHz · mm to 1.95 MHz·mm. Due to the non-dispersive nature of the $SH_0$ mode, its velocity remains unchanged. Additionally, since the damage echoes propagate the same distance to the bottom plate and the tank wall, the time-domain signals received by the bottom plate and the tank wall maintain high correlation. In contrast, the $S_0$ mode guided waves propagating toward the side of the array undergo boundary reflection, and their wave packets propagate from both the bottom plate and the wall plate to the receiving array. When the frequency-thickness product of the $S_0$ mode changes from 1.3 MHz·mm to 1.95 MHz·mm, its propagation speed decreases from 4970 m/s on the bottom plate to 3657 m/s on the wall plate. Due to this velocity difference, the time-domain discrepancy between the two signals increases with distance, leading to a corresponding reduction in their correlation.

Therefore, based on the differences in the propagation paths of damage echoes and boundary-scattered noise signals to the bottom plate and wall plate arrays, a data processing method involving mutual matching and calibration of signals from the bottom plate and wall plate is proposed. By extracting the guided wave signals from the corresponding array elements on the bottom plate and wall plate, and aligning the initial reference points of the signals, the amplitude signals of the bottom plate ($A_0$) and the wall plate ($A_1$) are obtained. The amplitude signals from the bottom plate and wall plate are divided into multiple windows, and the correlation coefficient $\rho_{XY}$ between the signals from the bottom plate and wall plate is calculated sequentially using a sliding window approach combined with Equation (4). The correlation between the two waveform signals is determined based on the magnitude of the correlation coefficient in the corresponding window. During this process, a predefined correlation coefficient threshold $\eta_0$, combined with Equation (5), is used to enhance the parts of the bottom plate signal with high correlation and attenuate those with low correlation. The flowchart of the correlation matching data processing is shown in Figure 7. After processing, the data are subjected to total focusing imaging, thereby achieving collaborative detection of the bottom plate and wall plate.

$$\rho_{XY} = \frac{Cov(A_0, A_1)}{\sigma_0 \sigma_1} \tag{4}$$

where $\rho_{XY}$ represents the correlation coefficient between the bottom plate and wall plate signals within the current window; $Cov(A_0, A_1)$ denotes the covariance between the amplitude signals of the bottom plate ($A_0$) and wall plate ($A_1$) within the current window; $\sigma_0$ and $\sigma_1$ are the standard deviations of the amplitude signals ($A_0$ and $A_1$, respectively) within the current window.

$$A_0' = \left[ \begin{array}{l} A_0(1 + \frac{|\rho_{XY} - \eta_0|}{1 - \eta_0}), \rho_{XY} \geq \eta_0 \\ A_0(1 - \frac{|\rho_{XY} - \eta_0|}{\eta_0}), \rho_{XY} < \eta_0 \end{array} \right. \quad (5)$$

where $A_0'$ represents the processed amplitude signal of the tank bottom plate in the current window, in volts (V); $A_0$ denotes the raw amplitude signal of the tank bottom plate in the current window, in volts (V); $\rho_{XY}$ is the correlation coefficient in the current window; $\eta_0$ is the correlation processing threshold.

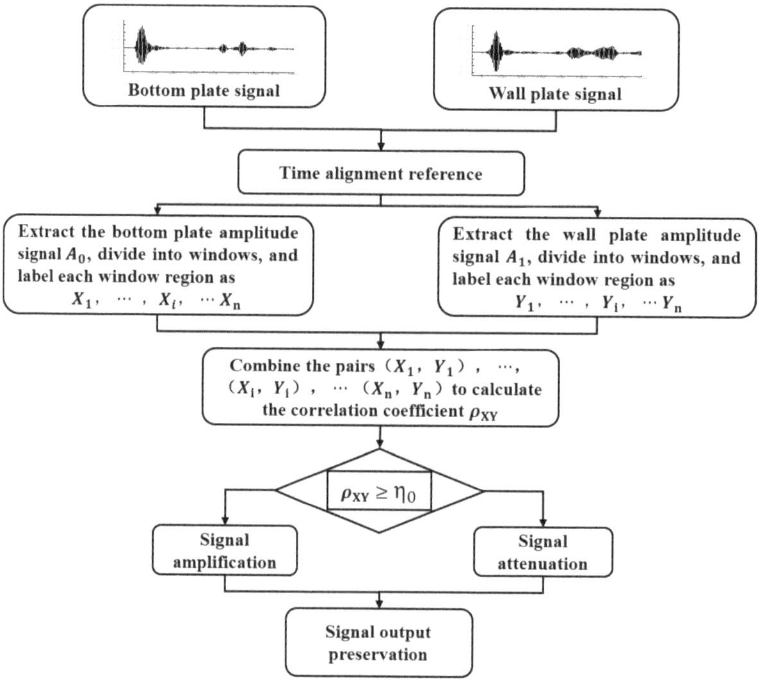


**Figure 7.** Flowchart of the correlation-based processing workflow for guided wave signals from the bottom plate and wall plate.

Based on the theoretical analysis above, a correlation analysis between the bottom plate and the wall plate was conducted. The guided wave signals received by the transducers located at symmetrical positions relative to the large fillet weld on the outer bottom plate and wall plate of the storage tank in array $D_1(1330, 120)$ were acquired. The guided wave signals received by the bottom plate and wall plate of the storage tank were processed using the aforementioned correlation matching method. After optimization, the data processing window size was set to 100 data points per window, and the correlation processing threshold was set to 0.7. The results of the data processing are shown in Figure 8 below:

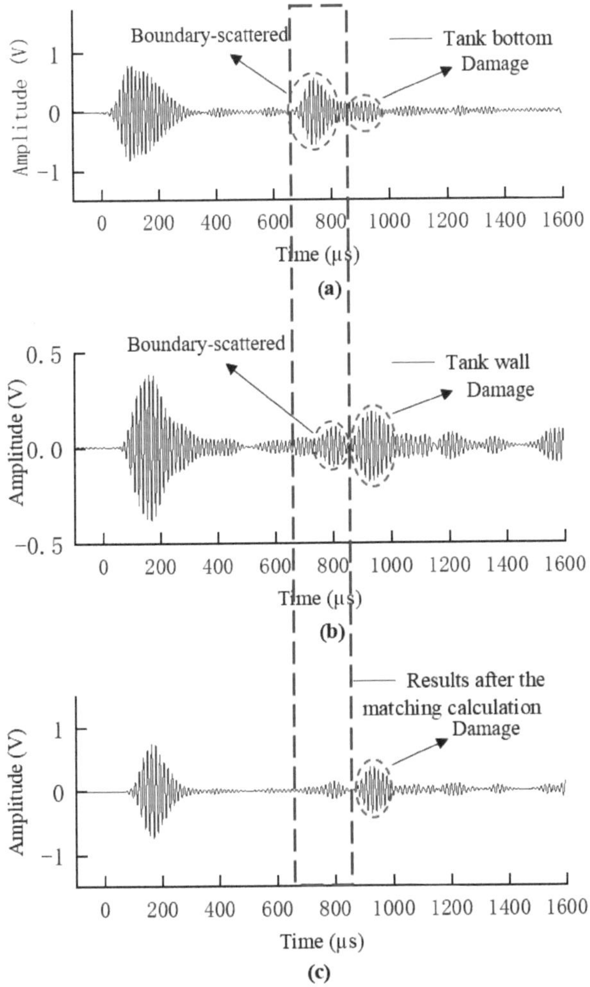


**Figure 8.** Signal correlation processing results for guided waves from the tank bottom plate and wall plate. (**a**) Detection waveform received by the tank bottom array element; (**b**) Detection waveform received by the tank wall array element; (**c**) Waveform obtained after the matching and fusion of the waveform data from the tank bottom and wall array elements.

As shown in Figure 8 above, the time-domain detection signals of the storage tank bottom plate, wall plate, and the bottom plate after correlation matching processing are displayed. From Figure 8a,b, it can be observed that the damage echo signals on the bottom plate and wall plate, due to the non-dispersive nature of the $SH_0$ mode guided waves, exhibit consistent time-of-flight when propagating the same distance to the bottom plate and wall plate. In contrast, the boundary-scattered $S_0$ mode guided waves propagate partially along the bottom plate and are received by the bottom plate array, while the other portion propagates along the wall plate and is received by the wall plate array. Due to the difference in thickness between the bottom plate and the wall plate, the frequency-thickness product of the guided waves changes along different propagation paths, leading to variations in the group velocity of the boundary-reflected guided waves on the bottom plate and wall plate. Because of the differences in propagation paths and velocities of the boundary-scattered signals, the echo signals of the boundary scattering gradually diverge in the time domain between the bottom plate and the wall plate. The correlation between the boundary-scattered echo signals received by the bottom plate array and those received by the wall plate array progressively decreases. The results after correlation matching calculations are shown in Figure 8c, where the damage echo signals are well preserved, while the boundary-scattered echo signals are effectively attenuated.

## 5. Results and Analysis

### *5.1. Damage Detection via Bottom Plate Array Total Focusing Imaging*

#### 5.1.1. Analysis of the Interference Effect of Side Boundary Scattering Signals on the Tank Bottom Damage Detection

To further validate the impact of scattering signals from the side welds of the storage tank bottom plate on the detection results during actual inspection processes, this study employs COMSOL 6.1 to establish a finite element model. In this model, the end-face scattering signals are used to simulate the influence of scattering signals from the side connection welds of the tank bottom plate during detection. A "T"-shaped structure model of the tank bottom plate, as shown in Figure 9, is constructed. The bottom plate thickness is set to 20 mm, and the tank wall thickness is set to 30 mm, consistent with the experimental setup. The material properties are set as shown in Table 1 below. A damage defect is placed 450 mm away from the array position. The $SH_0$ mode guided wave is excited using a line source, and the guided wave signals are received using point probes. In this section, by modifying the boundary absorption layer of the finite element model, the influence of scattering signals from the side connection welds on the damage detection of the tank bottom plate is investigated.

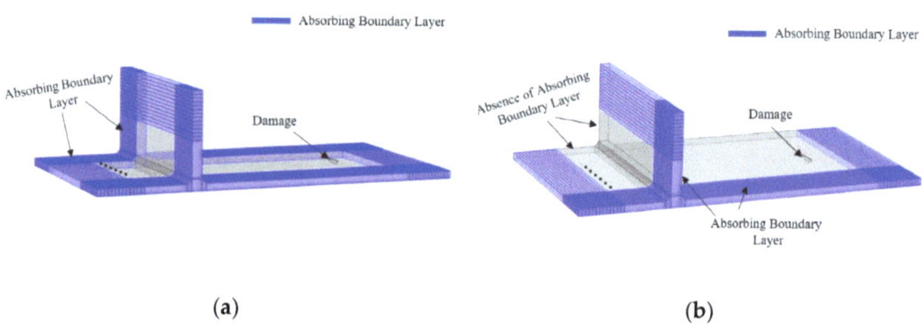


**Figure 9.** Finite element models for studying the influence of side connection welds on tank bottom plate detection. (**a**) Model without side connection weld reflections; (**b**) Model with side connection weld reflections.

**Table 1.** Material Properties of the T-Joint Finite Element Model for Storage Tank Bottom Plate.

| Component | Young's Modulus $E$ (GPa) | Density $\rho$ (kg/m$^3$) | Poisson's Ratio $\nu$ |
|---|---|---|---|
| Bottom Plate | 208 | 7850 | 0.3 |
| Wall Plate | 208 | 7850 | 0.3 |
| Weld Seam | 208 | 7900 | 0.3 |

To ensure the accuracy and stability of the solution process while reducing the computational data volume, the grid size should satisfy the spatial resolution criterion, i.e., the maximum grid size should be less than 1/6 of the minimum wavelength, and the time step should be less than 1/20 of the system's highest response frequency. The specific calculation formulas are provided in (6) and (7).

$$\Delta_{x,y,z} \leq \frac{\lambda}{6} \tag{6}$$

$$\Delta t \leq \frac{1}{20f} \tag{7}$$

where, $\Delta_{x,y,z}$ denotes the tetrahedral element size (mm) in the finite element model, $\lambda$ represents the minimum wavelength (mm) of the excited guided wave, $\Delta t$ indicates the computational time step (s), and $f$ specifies the central frequency (kHz) of the excitation signal. This study employed a 65 kHz $SH_0$ mode guided wave excitation. The maximum element size derived from Equation (6) was $8.1 \times 10^{-3}$ m, while the maximum time step calculated via Equation (7) was $7.6 \times 10^{-7}$ s. The global mesh size was set to $8.1 \times 10^{-3}$ m, with local refinement zones at $4.05 \times 10^{-3}$ m, using free tetrahedral mesh elements with a time step of $7.6 \times 10^{-7}$ s.

This subsection investigates the influence of side-connection weld scattering signals on damage detection in storage tank bottom plates by modifying the absorbing boundary layer configuration in the finite element model. The ABLs adopt a multi-tiered configuration based on mesh gradation, with a total length $L \geq 2\lambda$ and maximum absorption parameter $\alpha_{\max} = 10f$. To mitigate acoustic impedance discontinuities, the thickness of the damping layers progressively increases from the inner to outer regions. The governing equations are provided in Equation (8).

$$\alpha = \alpha_{\max}\left(\frac{l}{L}\right)^n \tag{8}$$

where, $l$ denotes the length of each absorbing boundary layer (ABL) tier; L represents the total ABL length; a indicates the current-tier damping coefficient, while A corresponds to the maximum damping coefficient; $n$ signifies the power exponent (assigned as 2).

The TFM results of the signals received by the two model arrays are shown in Figure 10 below:

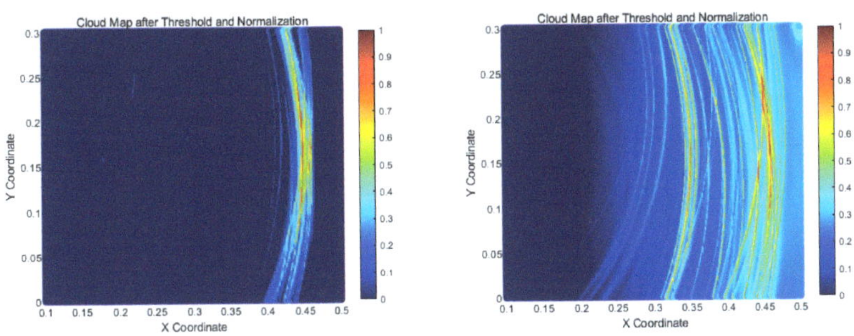


(**a**) (**b**)

**Figure 10.** Total focusing imaging results from the bottom plate array in finite element models. (**a**) TFM results without interference from side connection weld reflections; (**b**) TFM results with interference from side connection weld reflections.

From the above finite element study, it is evident that under the interference of scattering signals from the side connection welds, the damage signal received by the bottom plate array can be focused and imaged at the 447 mm position. However, with the removal of one side's absorbing boundary layer, Figure 10b shows that, in addition to the damage echo being focused and imaged at the 458 mm position, a non-damage artifact also appears at 351 mm. Furthermore, compared to Figure 10a, Figure 10b exhibits significantly more noise interference, which undoubtedly introduces a certain level of disturbance to the detection results.

#### 5.1.2. Damage Detection Experiment for Storage Tank Bottom Plate

In the detection of damage behind the "T"-shaped structure of the storage tank bottom plate, to simulate the actual online inspection process as closely as possible, neither the excitation transducer nor the receiving sensors can be placed inside the tank. The damage detection experiment was conducted by arranging a transducer array on the outer edge plate of the tank. Each array consists of ten transducers, positioned 100 mm from the large fillet weld, with a center-to-center spacing of 40 mm between each transducer element.

In this study, three sets of experiments were conducted to detect damage on the storage tank bottom plate at different locations and distances. The arrangement of the transducer arrays was consistent with the description above. Taking the first transducer position from right to left as the origin, a planar coordinate system was established with the $y$-axis parallel to the large fillet weld and the $x$-axis perpendicular to the large fillet weld, extending radially inward along the tank bottom plate. The imaging area was divided into an $8 \times 8$ mm grid. Among the three sets of experiments, the first set involved detecting pitting corrosion, while the other two sets focused on crack defects. The position coordinates and dimensions of the defects are listed in Table 2 below:

**Table 2.** Location coordinates and dimensional parameters of defects on the storage tank bottom plate.

| Group | Location Coordinates (mm) | Dimensions (mm) | Depth (mm) |
|---|---|---|---|
| 1# | $D_1$ (1330, 120) | $\varphi = 10$ | 9.3 |
| 2# | $D'_1$ (1130, 250) | $63 \times 2$ | 5.3 |
|  | $D'_2$ (1530, 150) | $73 \times 2$ | 7.7 |
| 3# | $D''_1$ (460, 200) | $39 \times 2$ | 3.1 |
|  | $D''_2$ (1070, 150) | $58 \times 2$ | 4.2 |

The schematic diagrams of the three detection arrays and the locations of the damage defects are shown in Figure 11 below:

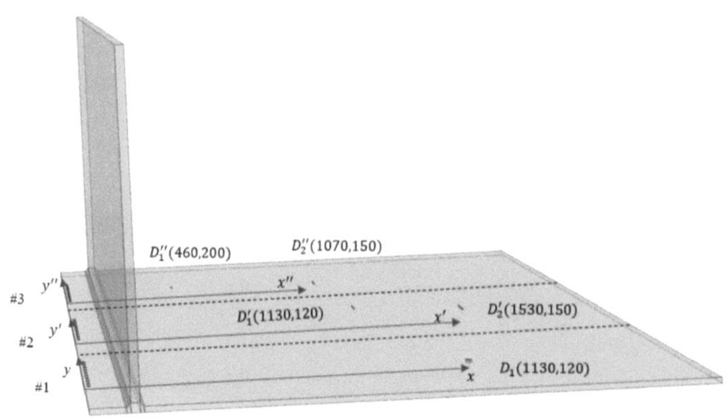


**Figure 11.** Schematic diagram of detection array and relative positions of damage defects on the tank bottom plate.

As shown in Figure 11, transducer arrays were arranged on the outer bottom plate of the storage tank for all three sets of parallel experiments. The data collected from the tank bottom plate were processed using the TFM, and the imaging results are presented in Figure 12 below:

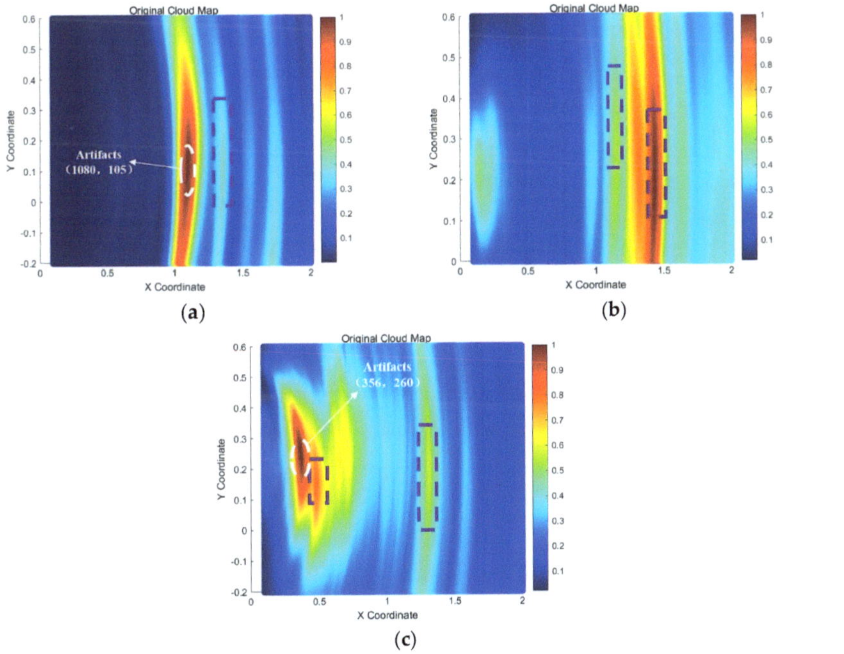


**Figure 12.** TFM cloud maps from bottom plate array detection. (**a**) Bottom plate array results for $D_1$ (1330, 120); (**b**) Bottom plate array results for $D'_1$ (1130, 250) and $D'_2$ (1530, 150); (**c**) Bottom plate array results for $D''_1$ (460, 200) and $D''_2$ (1070, 150).

In Figure 12, the purple squares represent the images of the damage defects. From the experimental results above, it is evident that by arranging a transducer array on the tank bottom plate and employing the total focusing method, defect signals can be successfully focused and imaged. However, due to the inevitable presence of boundary scattering signals from the bottom plate in the experimental environment, some unnecessary noise is introduced, leading to the appearance of artifacts in the images that do not correspond to actual defects. The most notable examples are the artifacts circled in white in Figure 12a,c, located at (1080, 105) and (356, 260), respectively. These artifacts exhibit significantly higher grayscale values than the actual damage defects, introducing strong interference and reducing the reliability of the detection results.

### 5.2. Collaborative Detection with Bottom Plate-Wall Plate Signal Fusion

Based on the theoretical findings above, it is demonstrated that using the correlation matching processing method for signals from the tank bottom plate and wall plate can effectively preserve damage echo signals while attenuating scattering signals caused by discontinuous interfaces. Therefore, the following section will incorporate this data processing approach to conduct collaborative bottom plate-wall plate detection experiments. Building on the bottom plate array described earlier, a wall plate transducer array is added, aligned parallel to the large fillet weld and the edge plate end face. The array elements on both the bottom plate and the wall plate are symmetrically positioned relative to the large fillet weld, with a distance of 10 cm from the weld and an inter-element spacing of 4 cm, as illustrated in Figures 13 and 14. This setup is used to carry out collaborative detection research on the tank bottom plate and wall plate.

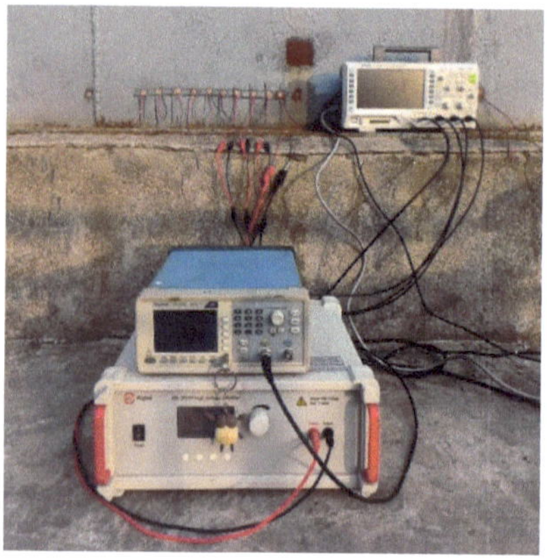

**Figure 13.** Experimental setup for collaborative bottom plate-wall plate damage detection.

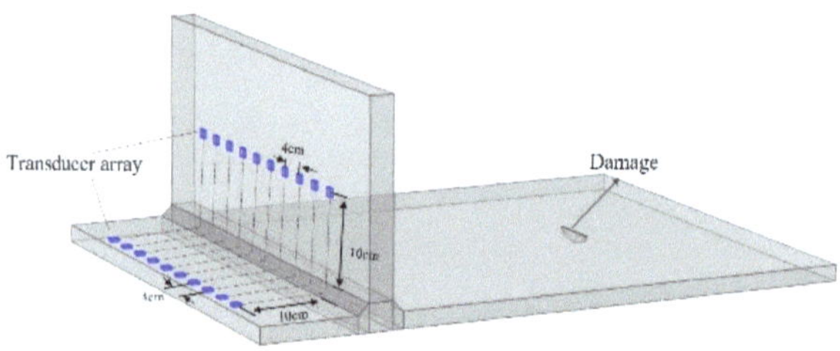


**Figure 14.** Schematic of collaborative bottom plate-wall plate damage detection.

The signals from the bottom plate and wall plate transducers at symmetric positions relative to the large fillet weld were acquired. The signals received by the corresponding transducers were processed according to the method described in Figure 7, completing the matching of the bottom plate and wall plate signals. The time window and correlation threshold remained consistent with the previous settings. The processed data were then subjected to the total focusing imaging algorithm to achieve collaborative bottom plate-wall plate detection.

The imaging results of the collaborative bottom plate-wall plate detection are shown in Figure 15. The left-side images (a), (c), and (e) in Figure 15 represent the original imaging cloud maps of the collaborative detection results. To ensure the clarity and accuracy of damage identification, the pixel grayscale values in the left-side cloud maps that were less than 0.8 were set to zero, and the resulting damage localization cloud maps are shown in Figure 15b,d,f.

Comparing the imaging results in Figures 12 and 15, the collaborative bottom plate-wall plate detection method effectively suppresses artifacts in the detection results compared to the independent imaging results obtained using only the bottom plate array. In all three parallel experiments, no artifacts with grayscale values higher than those of the actual damage defects were observed. In contrast to the single-array imaging results of the tank bottom plate, the collaborative bottom plate-wall plate detection method achieves clearer and more reliable localization and imaging of damage defects.

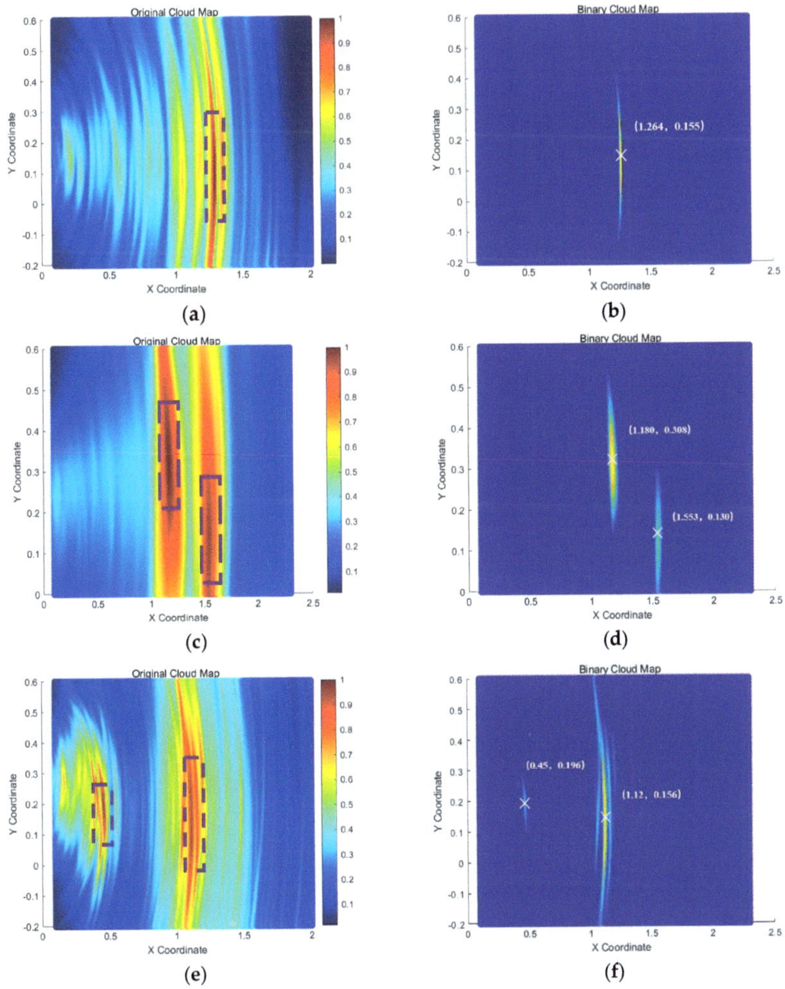


**Figure 15.** TFM cloud maps from collaborative bottom plate-wall plate array detection. (**a**) Collaborative results for $D_1$ (1330, 120); (**b**) Enhanced results for $D_1$ (1330, 120); (**c**) Collaborative results for $D'_1$ (1130, 250) and $D'_2$ (1530, 150); (**d**) Enhanced collaborative results for $D'_1$ (1130, 250) and $D'_2$ (1530, 150); (**e**) Collaborative results for $D''_1$ (460, 200) and $D''_2$ (1070, 150); (**f**) Enhanced collaborative localization results for $D''_1$ (460, 200) and $D''_2$ (1070, 150).

### 5.3. Error Analysis

In this experiment, 10 array transducers were arranged on the bottom plate and wall plate of the storage tank with a center-to-center spacing of 40 mm. Therefore, the central coordinate position of the detection array is (0, 220). The relative error was calculated by determining the distance ($r_1$) between the actual damage center coordinates and the array center coordinates, and the distance ($r_2$) between the detected damage coordinates and the array center coordinates. The specific calculation formula is as follows (9), (10) and (11).

$$r_1 = \sqrt{(x_1 - x_0)^2 + (y_1 - y_0)^2} \tag{9}$$

$$r_2 = \sqrt{(x_2 - x_0)^2 + (y_2 - y_0)^2} \tag{10}$$

$$\delta = \frac{|r_1 - r_2|}{r_1} \times 100\% \quad (11)$$

where $(x_0, y_0)$ represents the central coordinate position of the transducer array in millimeters (mm); $(x_1, y_1)$ denotes the central coordinate position of the actual damage location, and $(x_2, y_2)$ represents the central coordinate position of the detected damage, both in millimeters (mm). $r_1$ and $r_2$ are the relative distances from the actual damage and the detected damage to the central coordinate of the transducer array, respectively, in millimeters (mm). $\delta$ is the relative error between the relative distance from the actual damage defect to the central coordinate of the transducer array and the relative distance from the detected defect coordinates to the central coordinate of the transducer array. The specific calculation results are shown in Table 3 below:

**Table 3.** Damage detection results for the storage tank bottom plate.

| Actual Damage Location (mm) | Imaging Method | Localization Result (mm) | Relative Error $\delta$ | Number of Artifacts with Grayscale Exceeding Damage |
|---|---|---|---|---|
| $D_1$ (1330,120) | Bottom Plate Array Detection | (1380, 90) | 3.6% | 2 |
| | Collaborative Bottom Plate-Wall Plate Detection | (1264, 115) | 5.0% | 0 |
| $D'_1$ (1130,250) | Bottom Plate Array Detection | (1174, 235) | 3.5% | 1 |
| | Collaborative Bottom Plate-Wall Plate Detection | (1180, 308) | 5.4% | 0 |
| $D'_2$ (1530,150) | Bottom Plate Array Detection | (1431, 227) | 5.8% | 0 |
| | Collaborative Bottom Plate-Wall Plate Detection | (1553, 130) | 1.4% | 0 |
| $D''_1$ (460,200) | Bottom Plate Array Detection | (508, 147) | 5.4% | 1 |
| | Collaborative Bottom Plate-Wall Plate Detection | (450, 196) | 2.1% | 0 |
| $D''_2$ (1070,150) | Bottom Plate Array Detection | (1233, 180) | 15.3% | 2 |
| | Collaborative Bottom Plate-Wall Plate Detection | (1120, 155) | 4.6% | 0 |

The damage detection experimental results for the tank bottom plate indicate that the damage localization method based on the proposed bottom plate-wall plate collaborative detection can successfully identify damage defects at different locations on the bottom plate. Compared to full-focus imaging with the bottom plate array, the bottom plate-wall plate collaborative detection method not only successfully detects damage at three different positions and distances on the bottom plate but also demonstrates significant effectiveness in suppressing artifacts caused by boundary reflections. In the three experimental cases, the artifact suppression rate was 100%. As shown in Table 3, the maximum relative error of damage localization using the total focusing imaging method with the bottom plate array is 15.3%, whereas the maximum relative error using the collaborative bottom plate-wall plate detection method is only 5.4%, indicating excellent detection accuracy.

Compared to the conventional bottom plate array detection method, the proposed collaborative bottom plate-wall plate detection method enhances the detection process by deploying transducer arrays on both the bottom plate and the wall plate, thereby acquiring multidimensional information about the tank bottom plate. By incorporating a correlation matching processing approach, the signals from corresponding array elements on the bottom plate and wall plate are matched, effectively reducing boundary scattering noise. This approach significantly suppresses the generation of artifacts during detection. In all three parallel experiments of tank bottom plate damage detection, no artifacts with grayscale values higher than those of the actual damage defects were observed, providing a more reliable detection outcome for tank bottom plate damage inspection.

## 6. Conclusions

The collaborative bottom plate-wall plate ultrasonic guided wave damage localization method proposed in this study is designed for T-shaped storage tank structures where conventional bottom plate array detection faces challenges such as: significant detection errors caused by overlapping wave packets from boundary scattering signals at side connection welds ($D''_2$), and artifacts generated by the focusing of boundary scattering signals within the detection zone. To address these issues, transducer arrays are deployed on both the bottom plate and wall plate. By exciting the bottom plate and receiving signals simultaneously from both arrays, multidimensional information about the tank bottom plate is acquired. Guided wave signals from corresponding array elements on the bottom plate and wall plate are processed using a correlation coefficient matching algorithm to attenuate boundary scattering interference. Subsequent total focusing imaging (TFM) of the processed data demonstrates that the collaborative detection framework achieves higher signal-to-noise ratios (SNR) compared to single-array imaging. In three parallel experiments, no artifacts with grayscale values exceeding the actual damage were observed, effectively reducing false detection.

The core principle of boundary scattering suppression lies in leveraging the distinct velocity responses of $SH_0$ mode guided waves and Lamb waves under frequency-thickness product variations. Specifically, $SH_0$ waves maintain constant velocity despite thickness changes (e.g., from the bottom plate to the wall plate), while Lamb waves exhibit velocity dispersion. This disparity introduces asymmetric time-domain characteristics between damage echoes ($SH_0$-dominated) and boundary scattering signals (Lamb wave-dominated). The correlation-based processing exploits this asymmetry to selectively attenuate scattering-related wave packets. Consequently, the method achieves optimal performance in T-shaped structures with mismatched bottom plate-wall plate thicknesses, where frequency-thickness product transitions amplify velocity discrepancies. For practical applications, prior knowledge of plate thicknesses is recommended to calibrate the algorithm for specific geometric configurations.

**Author Contributions:** Conceptualization, Y.M. and L.C.; methodology, L.H.; software, L.H.; validation, L.H.; data curation, L.H. and Y.D.; writing—original draft preparation, L.H.; writing—review and editing, Y.M., Y.D. and L.C.; supervision, L.C. and Y.D.; project administration, L.C.; funding acquisition, G.L. All authors have read and agreed to the published version of the manuscript.

**Funding:** This research received no external funding.

**Institutional Review Board Statement:** Not applicable.

**Informed Consent Statement:** Not applicable.

**Data Availability Statement:** The authors confirm that the data supporting the findings of this study are available from the corresponding author upon reasonable request.

**Conflicts of Interest:** Author Yunxiu Ma was employed by the Pipe-Network Group (Xuzhou) Pipeline Inspection and Testing Co company. The remaining authors declare that the research was conducted in the absence of any commercial or financial relationships that could be construed as a potential conflict of interest.

## References

1. Yang, Z.J.; Chen, D.; Chen, L.; Liu, Y.Z.; An, R.; Wang, G. Finite Element Analysis and Research on Corrosion Defect of Tank Floor. *Appl. Mech. Mater.* **2017**, *853*, 514–518. [CrossRef]
2. Luo, F.W.; Ran, R.; Wang, L. Study on Corrosion Law of Large Crude Oil Storage Tank Floor and Risk-Based Inspection and Maintenance Technology. *Corros. Sci. Technol.* **2020**, *19*, 66–74.

3. Cirimello, P.G.; Otegui, J.L.; Ramajo, D.; Carfi, G. A Major Leak in a Crude Oil Tank: Predictable and Unexpected Root Causes. *Eng. Fail. Anal.* **2019**, *100*, 456–469. [CrossRef]
4. Shi, Y.; Zhang, C.; Li, R.; Cai, M.; Jia, G. Theory and Application of Magnetic Flux Leakage Pipeline Detection. *Sensors* **2015**, *15*, 31036–31055. [CrossRef]
5. Spurek, M.A.; Spierings, A.B.; Lany, M.; Revaz, B.; Santi, G.; Wicht, J.; Wegener, K. In-Situ Monitoring of Powder Bed Fusion of Metals Using Eddy Current Testing. *Addit. Manuf.* **2022**, *60*, 103259. [CrossRef]
6. Chen, Y.; Feng, B.; Kang, Y.; Liu, B.; Wang, S.; Duan, Z. A Novel Thermography-Based Dry Magnetic Particle Testing Method. *IEEE Trans. Instrum. Meas.* **2022**, *71*, 9505309. [CrossRef]
7. Raišutis, R.; Tumšys, O.; Žukauskas, E.; Samaitis, V.; Draudvilienė, L.; Jankauskas, A. An Inspection Technique for Steel Pipes Wall Condition Using Ultrasonic Guided Helical Waves and a Limited Number of Transducers. *Materials* **2023**, *16*, 5410. [CrossRef]
8. Shah, J.K.; Wang, H.; El-Hawwat, S. Detection of Internal Crack Growth in Polyethylene Pipe Using Guided Wave Ultrasonic Testing. *Earthq. Eng. Eng. Vib.* **2024**, *23*, 319–329. [CrossRef]
9. Diogo, A.R.; Moreira, B.; Gouveia, C.A.; Tavares, J.M.R. A Review of Signal Processing Techniques for Ultrasonic Guided Wave Testing. *Metals* **2022**, *12*, 936. [CrossRef]
10. Lyu, F.; Zhou, X.; Ding, Z.; Qiao, X.; Song, D. Application Research of Ultrasonic-Guided Wave Technology in Pipeline Corrosion Defect Detection: A Review. *Coatings* **2024**, *14*, 358. [CrossRef]
11. Zang, X.; Xu, Z.D.; Lu, H.; Zhu, C.; Zhang, Z. Ultrasonic Guided Wave Techniques and Applications in Pipeline Defect Detection: A Review. *Int. J. Press. Vessel. Pip.* **2023**, *206*, 105033. [CrossRef]
12. Yu, X.; Qin, R.; Deng, M. New Insights into Topographically Feature Guided Waves (FGW) Propagation in Non-Uniform Elastic Waveguides. *Wave Motion* **2022**, *109*, 102866. [CrossRef]
13. Zhu, H.; Kotousov, A.; Ng, C.T. Defects Evaluation Near Edges of Structural Elements Using the Fundamental Mode of Edge Waves. *J. Sound Vib.* **2023**, *557*, 117753. [CrossRef]
14. Li, F.; Liu, Z.; Sun, X.; Li, H.; Meng, G. Propagation of Guided Waves in Pressure Vessel. *Wave Motion* **2015**, *52*, 216–228. [CrossRef]
15. Huang, S.Y.; Guo, Y.Q.; Zang, X.L.; Xu, Z.D. Design of Ultrasonic Guided Wave Pipeline Non-Destructive Testing System Based on Adaptive Wavelet Threshold Denoising. *Electronics* **2024**, *13*, 2536. [CrossRef]
16. Cheng, Y.; Deng, Y.; Cao, J.; Xiong, X.; Bai, L.; Li, Z. Multi-Wave and Hybrid Imaging Techniques: A New Direction for Nondestructive Testing and Structural Health Monitoring. *Sensors* **2013**, *13*, 16146–16190. [CrossRef]
17. Fakih, M.A.; Mustapha, S.; Tarraf, J.; Ayoub, G.; Hamade, R. Detection and Assessment of Flaws in Friction Stir Welded Joints Using Ultrasonic Guided Waves: Experimental and Finite Element Analysis. *Mech. Syst. Signal Process.* **2018**, *101*, 516–534. [CrossRef]
18. Zhang, X.; Zhou, J.; Hu, Y.; Liu, Y.; Shen, X. Electromagnetic Acoustic Detection of Pipe Defects Hidden Above T-Type Support Structures with Circumferential Shear Horizontal Guided Wave. *Micromachines* **2024**, *15*, 550. [CrossRef]
19. Wu, J.; Jiang, C.; Fang, H.; Ng, C.T. Damage Detection in the T-Welded Joint Using Rayleigh-Like Feature Guided Wave. *NDT E Int.* **2023**, *135*, 102806. [CrossRef]
20. Raišutis, R.; Kažys, R.; Mažeika, L.; Žukauskas, E.; Šliteris, R.; Vladišauskas, A. Application of Ultrasonic Guided Waves for Non-Destructive Testing of Large and Complex Geometry Engineering Structures. *Vibroeng. Procedia* **2017**, *14*, 87–90. [CrossRef]
21. Lowe, P.S.; Duan, W.; Kanfoud, J.; Gan, T.H. Structural Health Monitoring of Above-Ground Storage Tank Floors by Ultrasonic Guided Wave Excitation on the Tank Wall. *Sensors* **2017**, *17*, 2542. [CrossRef] [PubMed]
22. Wang, H.; Wu, B.; Gao, X.; Liu, Y.; Li, X.; Liu, X. Ultrasonic Guided Wave Defect Detection Method for Tank Bottom Plate Based on SH0 Mode Multichannel Magnetostrictive Sensor. *Measurement* **2023**, *223*, 113790. [CrossRef]
23. Du, F.; Zeng, L.; Sha, B.; Huang, L.; Xiao, J. Damage Imaging in Composite Laminates Using Broadband Multipath Lamb Waves. *IEEE Trans. Instrum. Meas.* **2022**, *71*, 6008014. [CrossRef]
24. Han, C.; Yang, G.; Wang, J.; Guo, X. The Research on Propagation Characteristics of Acoustic Emission Signals in Stiffened Plates Based on the Multipath Propagation Model. *Ultrasonics* **2020**, *108*, 106177. [CrossRef]
25. Zeng, L.; Huang, L.; Luo, Z.; Lin, J. Damage Imaging that Exploits Multipath Scattered Lamb Waves. *Struct. Health Monit.* **2020**, *19*, 1629–1644. [CrossRef]
26. Yu, S.; Fan, C.; Zhang, M.; Zhao, Y. A Lamb wave time-reversal field reconstruction method for damage detection with automatic focusing determination. *Ultrasonics* **2023**, *133*, 107030. [CrossRef]
27. Huang, L.; Zeng, L.; Lin, J.; Zhang, N. Baseline-free damage detection in composite plates using edge-reflected Lamb waves. *Compos. Struct.* **2020**, *247*, 112423. [CrossRef]
28. Hall, J.S.; Michaels, J.E. Multipath ultrasonic guided wave imaging in complex structures. *Struct. Health Monit.* **2015**, *14*, 345–358. [CrossRef]

29. Zhang, N.; Xu, C.; Zeng, L.; Deng, M. Blind zone defect imaging using multipath edge-reflected Lamb waves. *Ultrasonics* **2025**, *145*, 107473. [CrossRef]
30. Hakoda, C.; Rose, J.; Shokouhi, P.; Lissenden, C. Using Floquet Periodicity to Easily Calculate Dispersion Curves and Wave Structures of Homogeneous Waveguides. *AIP Conf. Proc.* **2018**, *1949*, 020016.

sensors

MDPI

Article

# MicrocrackAttentionNext: Advancing Microcrack Detection in Wave Field Analysis Using Deep Neural Networks Through Feature Visualization

**Fatahlla Moreh [1,*], Yusuf Hasan [2], Bilal Zahid Hussain [3], Mohammad Ammar [2], Frank Wuttke [1] and Sven Tomforde [4]**

[1] Department of Geo-Science, Christian Albrechts University, 24118 Kiel, Germany; frank.wuttke@ifg.uni-kiel.de
[2] Department of Computer Engineering, Aligarh Muslim University, Aligarh 202001, India; yusufhasan@zhcet.ac.in (Y.H.); mohammadammar@zhcet.ac.in (M.A.)
[3] Department of Electrical and Computer Engineering, Texas A&M University, College Station, TX 77840, USA; zahidhussain909@tamu.edu
[4] Institute of Computer Science, Christian Albrechts University, 24118 Kiel, Germany; st@informatik.uni-kiel.de
* Correspondence: fatahlla.moreh@ifg.uni-kiel.de; Tel.: +49-178-5801290

**Abstract:** Microcrack detection using deep neural networks (DNNs) through an automated pipeline using wave fields interacting with the damaged areas is highly sought after. However, these high-dimensional spatio–temporal crack data are limited. Moreover, these datasets have large dimensions in the temporal domain. The dataset presents a substantial class imbalance, with crack pixels constituting an average of only 5% of the total pixels per sample. This extreme class imbalance poses a challenge for deep learning models with different microscale cracks, as the network can be biased toward predicting the majority class, generally leading to poor detection accuracy for the under-represented class. This study proposes an asymmetric encoder–decoder network with an adaptive feature reuse block for microcrack detection. The impact of various activation and loss functions are examined through feature space visualisation using the manifold discovery and analysis (MDA) algorithm. The optimized architecture and training methodology achieved an accuracy of 87.74%.

**Keywords:** microcrack detection; CNN; spatio–temporal data; feature space visualization; acoustic emission; segmentation; attention

## 1. Introduction

Microcrack detection in materials is of significant importance due to the potential for catastrophic failures, which can lead to substantial financial losses and safety hazards in industries [1,2]. In a number of areas, including materials science, aerospace, and infrastructure, where the existence of small fissures might jeopardize the structural integrity of materials, non-destructive techniques (NDTs) for microcrack detection is imperative [3–5]. Conventional techniques for detecting microcracks are frequently labor-intensive and not very scalable [6,7]. Some NDTs, such as visual inspection and dye-penetrant testing, are fairly simple but are limited to surface-level cracks and require intensive labor. Others, such as magnetic particle testing and eddy current testing, are exclusive for some materials. More advanced techniques such as X-ray computed tomography, although quite effective, are relatively expensive and limited to smaller components [8]. Moreover, detecting cracks in complex structures, such as aircraft bodies or intricate machinery components, poses a substantial challenge using conventional methods. The use of acoustic wave-based approaches for crack detection offers a powerful solution, as these methods allow for the analysis of structures that are not easily accessible or too complex to inspect manually. However, acoustic emission-based approaches are prone to background noises, which makes their analysis untractable through data analysis techniques. Deep learning techniques provide a solution to this problem by analyzing huge amounts of data to discern meaningful patterns. Convolutional neural networks (CNNs) are especially good at processing spatial data due to their ability to capture local spatial correlations within an image [9,10]. Segmentation through CNNs is highly sought after as these techniques help localize the objects of interest in the image. These techniques have been extensively used in medical image segmentation and showed a high degree of reliability and accuracy [11].

Nevertheless, standard segmentation methods demonstrate limited performance on this particular dataset, due to the complex spatio–temporal nature of the crack patterns. This becomes even more significant when the cracks represent a minority class in the dataset, leading to poor detection accuracy. The dataset not only presents a severe class imbalance but also introduces additional complexities, including multiple crack sizes, varying crack shapes, and the presence of cracks in different locations. These variations significantly impact the wave propagation behavior, making it challenging for traditional segmentation models to generalize effectively. The model must learn to differentiate between subtle variations in wave interactions caused by different types of cracks while overcoming the bias introduced by dominant non-crack regions. This issue is enhanced when dealing with very small cracks, as they not only lead to data imbalance but may also cause minimal disruption in underlying wave behavior. In such cases, the waves may exhibit minimal changes, making it difficult for the model to detect the cracks accurately. Moreover well-known segmentation models such as U-Net are not compatible with these datasets, as these have both temporal and spatial dimensions. The model not only has to detect patterns across spatial dimensions but also across temporal domains, and the U-Net model is used for the image to image translation problem where the input and output dimensions are equal.

This challenge necessitates the development of a more tailored custom model. The proposed MicrocrackAttentionNext is designed to overcome the limitations of vanilla models such as U-Net by incorporating enhanced spatial and temporal feature extraction. Unlike U-Net [12] where the input and target share the same modality (image-to-image translation). The proposed model processes spatio–temporal input data and outputs spatial crack predictions, enabling it to handle more complex data while improving micro-scale detection accuracy. The asymmetric encoder–decoder structure with attention layers is particularly effective, as it focuses on capturing critical crack patterns rather than relying heavily on skip connections. The attention mechanism ensures that the model prioritizes the time steps when the waves interact with the cracks, improving detection precision. To further evaluate the effect of activation functions and different losses, the proposed work employs feature map analysis. The extent of the influence of different activations is difficult to determine against conventional metrics such as accuracy and F1 score. Hence, it is imperative to analyze the internal dynamics of the model.

Methods such as principal component analysis, t-SNE [13], and UMAP [14] are used to analyze the higher dimensional feature maps of these black box models against the target. However, these methods provide little to no insight when used on segmentation problems. This study uses the recently proposed manifold discovery analysis (MDA) [15] to qualitatively assess the impacts of various activation functions. Moreover, through this, the proposed study is able to analyze the effect activations had on the feature maps of the model, allowing us to choose the best activation function for the given problem. These activation functions aim to strike a delicate balance between adaptability and computational efficiency, essential considerations in the micro-material domain, where capturing fine details is crucial for accurate crack detection. Empirical exploration and meticulous fine-tuning of these activation functions is imperative to identify the optimal choice that aligns with the distinctive characteristics of micro-material images. Ultimately, an effective approach to crack detection in micro-materials relies on the thoughtful selection and optimization of activation functions within the CNN architecture.

The primary contributions of this paper are the following:

- Introducing MicrocrackAttentionNext—an improvement over [16]—and introduction of Adaptive Feature Utilization Block for efficient feature utilization.
- Analysis of the impact of activation functions on the performance of MicrocrackAttentionNext through Manifold Discovery and Analysis in the context of microcrack detection.

The paper's structure is outlined in the following manner: Section 2 provides a concise overview of relevant studies. Section 3 deals with the dataset used and the proposed methodology. Section 4 explains the experiments conducted. The assessment of the performance of the proposed system and the results obtained are included in Section 5. The concluding remarks and future works are presented in Section 6.

## 2. Related Works

This section primarily examines non-invasive methods for the detection of microcracks in numerical data. The first two paragraphs present methodologies rooted in classical foundations, followed by an exploration of more recent deep-learning approaches.

In their study, Punnose et al. [17] conducted an experimental evaluation of the time-of-flight diffraction (TOFD) technique for accurately sizing surface-breaking cracks. Using steel test blocks with vertical and inclined slits of varying heights, the researchers employed TOFD equipment with 45° longitudinal angle beam probes to scan and analyze these simulated defects. The findings demonstrated that TOFD could measure crack depths with an average error of ±0.13 mm for vertical slits and ±0.05 mm for inclined slits, and crack lengths with an average error of ±0.36 mm and ±0.29 mm, respectively. However, the study also identified challenges in detecting defects less than 2 mm deep due to lateral wave interference and limited time resolution near the surface. The findings demonstrate TOFD's capability for accurate crack measurement and point to the necessity of enhancing its sensitivity to shallow flaws.

Kou et al. [18] proposed a fully noncontact nonlinear ultrasonic testing (NUT) method using laser ultrasonics for detecting closed surface cracks. Traditional ultrasonic techniques struggle with closed cracks due to minimal wave scattering. This study employs a pulsed laser grating source to generate narrowband surface acoustic waves (SAWs) and uses a nonlinear numerical model based on the finite element method (FEM) to simulate higher harmonic generation.

Conventional techniques for detecting microcracks are frequently labour-intensive and not very scalable. Deep learning and convolutional neural networks (CNNs), in particular, have become a potent and effective technique for microcrack detection automation in recent years [19–21].

Tran et al. [22] applied 1D convolutional neural networks (1D CNNs) for structural damage detection, utilizing acceleration signals to detect cracks in numerical steel beam models. Their approach showed high detection accuracy, comparable to more complex methods, by processing time-series data and extracting key features related to structural changes. While they focused on single-dimensional data, the proposed research extends this by using multi-dimensional spatio-temporal data, which includes wave propagation across the material. This allows for more detailed analysis, capturing both spatial and temporal interactions crucial for detecting microcracks.

Jiang et al. [23] combined 1D CNNs with support vector machines (SVMs) to enhance structural damage detection. The 1D CNNs were used to localize damage, while SVMs focused on classifying the severity, benefiting from the strengths of both models in feature extraction and small-sample learning.

Barbosh et al. [24] used acoustic emission (AE) waveforms and DenseNet to detect and localize the crack. The localization was conducted to determine whether the crack was close to the sensor or far away. The cracks were also classified into severe and less severe cracks.

Moreover, Li et al. [25] proposed a GM-ResNet-based approach to enhance crack detection, utilizing ResNet-34 as the foundational network. To address challenges in global and local information assimilation, a global attention mechanism was incorporated for optimized feature extraction. Recognizing limitations in ResNet-34, the fully connected layer was replaced with a multilayer fully connected neural network (MFCNN), featuring multiple layers, including batch normalization and Leaky ReLU nonlinearity. This innovative substitution significantly improved the model's ability to capture complex data distributions and patterns, enhancing feature extraction and representation capabilities while preventing overfitting during training.

Moreh et al. [16] explore the use of DNN for automated crack detection in structures using seismic wave signals. The authors improve on previous asymmetric encoder–decoder models by experimenting with different encoder backbones and decoder layers. The best combination was found to be the 1D-DenseNet encoder and the Transpose Convolutions as decoders. The proposed model achieved an accuracy of 83.68% with a total parameter count of 1.393 million.

This study builds upon the foundational contributions of Moreh et al. [16,26,27], extending their methodologies to a broader scope. The existing body of work in this field remains relatively sparse, with few studies addressing crack segmentation through the specific approach employed in this research.

## 3. Method

The proposed work targets the detection of various microcrack sizes and locations within seismic wave field numerical data. For this purpose, MicrocrackAttentionNext extracts crucial signals from the data to identify and detect those cracks. This is done by learning the temporal representations, followed by spatial representations. These encoded data are then passed through the decoder to achieve semantic spatial segmentation.

The following section describes the seismic wave data followed by the architecture of the proposed MicrocrackAttentionNext model and, subsequently, the training procedure used.

### 3.1. Wave Field Data

The wave field dataset utilized in this work, while effective for crack detection, presents some limitations in terms of data dimensionality. These datasets are characterized by large temporal dimensions, which increases the complexity of data processing and model training. The dataset consists of wave field numerical data for homogeneous 2D plates, where each plate is modeled with lattice particles that share consistent properties, such as density and Young's Modulus. The modeling of structural systems is achieved using Voronoi–Delaunay meshing algorithms within the lattice element method (LEM). Lattice nodes, representing unit cell centers, are connected by beams capable of handling normal forces ($N$), shear forces ($V$), and bending moments ($M$). If the strain energy $U_e$ in an element exceeds a predefined threshold $U_{th}$, the element undergoes stiffness reduction or removal, simulating failure states. To simulate wave propagation through the material, an external force of 1000 N is applied at the midpoint of the left boundary (Figure 1), ensuring that the waves propagate across the entire plate, interacting with both non-crack (Figure 2b) and crack regions (Figure 2a), the wave deflects when it interacts with the crack, and the displacement is recorded by the sensors. The resulting displacements in both the x- and y-directions are recorded over 2000 time steps, capturing detailed temporal changes in the wave field. Figure 3 shows 2D visualization of the time signal received by sensor 9 in a sample. These displacements are measured by a 9 × 9 (81) sensor grid (Figure 1) uniformly distributed across the material, resulting in a wave field dataset with dimensions of 2 × 81 × 2000. This approach provides spatio-temporal data that captures the interaction between the propagating waves and the cracks, allowing for in-depth analysis of crack detection model performance. This approach provides spatio-temporal data that capture the interaction between the propagating waves and the cracks, allowing for in-depth analysis of crack detection model performance. The dataset used for training and evaluation consists of a total of 7480 samples, with 4810 allocated for training, 910 for validation, and 1760 for testing.

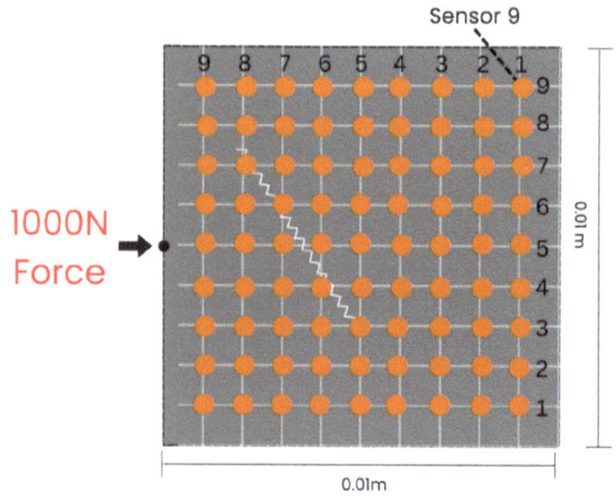


**Figure 1.** Arrangement of the 9 × 9 sensor array in the sample domain.

A major challenge arises from the severe class imbalance present in the dataset. On average, only 5% of the total pixels represent cracks, with the remaining majority belonging to intact, non-crack regions. This imbalance poses a substantial obstacle for deep learning models, which are prone to bias toward the majority class. As a result, the models tend to predict non-crack regions more frequently, leading to suboptimal detection accuracy for the minority class (crack regions). Addressing this issue requires careful design of the model and training process to ensure that the network can effectively learn from the minority class, and accurately identify crack regions without being overpowered by the majority class imbalance. Moreover, the effect of different activations and loss functions are assessed both quantitatively and qualitatively.

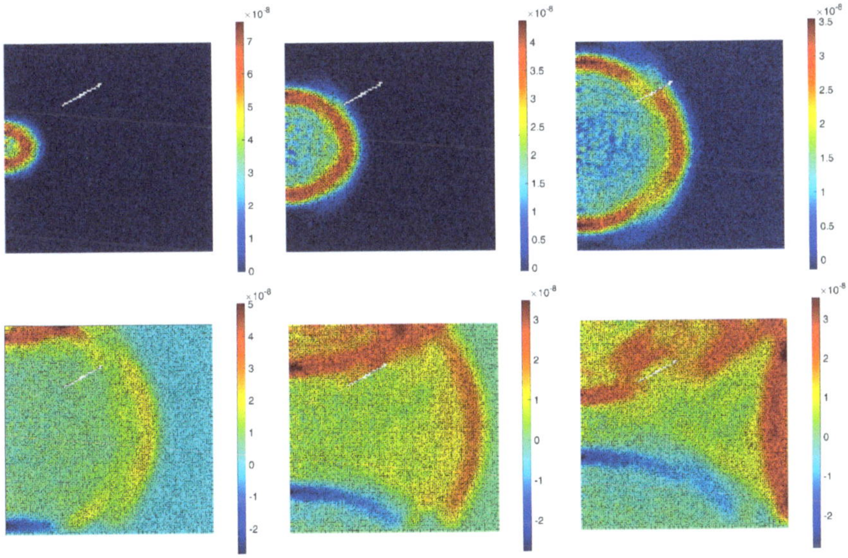

(a)

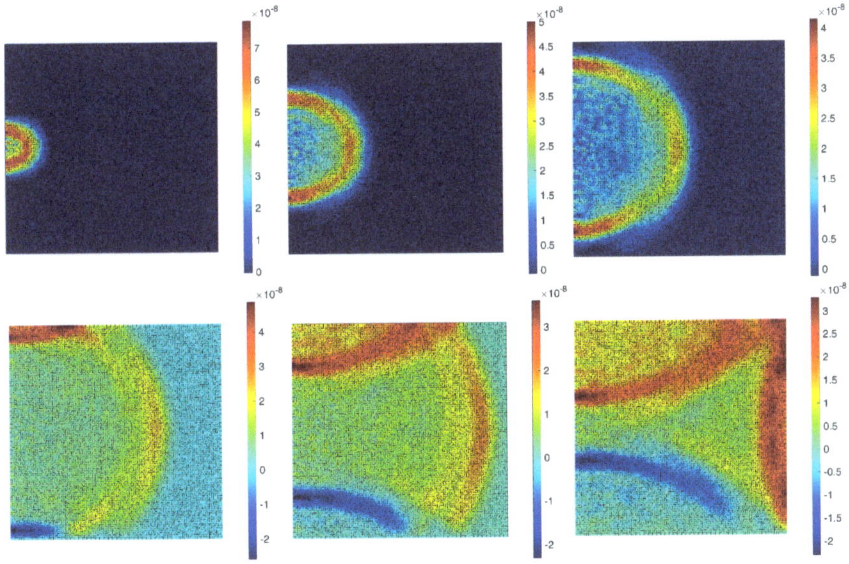

(b)

**Figure 2.** The 6 frames (100 time-steps interval, from left to right) of a displacement wave propagation inside the defined plate with [28] (a) crack and (b) no crack.

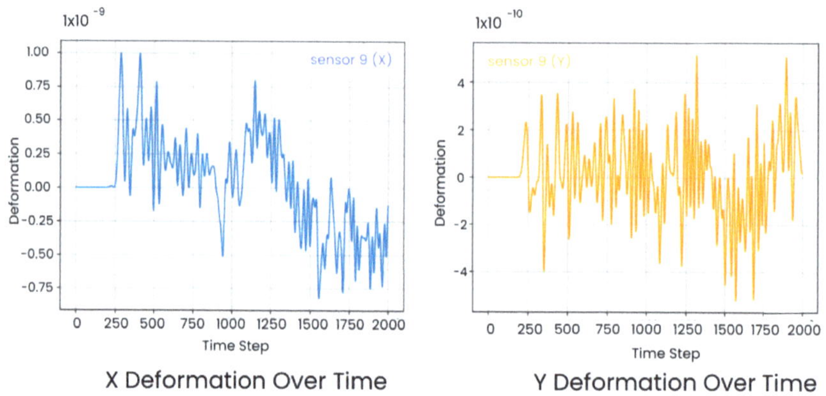


**Figure 3.** The 2D—visualization of time signal received by sensor 9 in a sample.

### 3.2. MicrocrackAttentionNext Model Architecture

MicrocrackAttentionNext, shown in Figure 4, is an asymmetric encoder–decoder network. The input to the model is a tensor with shape $\mathbf{X} \in \mathbb{R}^{C_{\text{in}} \times T \times S}$, where $C_{\text{in}} = 2$ represents the input channels corresponding to the $x$ and $y$ components of wave data, $T = 2000$ is the temporal dimension, and $S = 81$ corresponds to the spatial dimension, which is a flattened $9 \times 9$ sensor grid. To reduce computational complexity and focus on salient temporal features, the network uses an initial max pooling layer with a kernel size of $(4, 1)$. This layer transforms the input tensor $\mathbf{X}$ to $\mathbf{X}_1 \in \mathbb{R}^{2 \times 500 \times 81}$ by downsampling the temporal dimension from 2000 to $T_1 = 500$. This reduction is crucial as it reduces the amount of data the subsequent layers need to process.

The encoder is composed of four convolutional blocks [26,27], each designed to progressively extract higher-level features from the input data. The first convolutional block applies two 1D convolutional layers with kernel sizes $(3, 1)$ and padding $(1, 0)$, which maintain the spatial dimensions while expanding the channel dimension from 2 to 16. These layers are followed by batch normalization and activation functions, introducing non-linearity. A squeeze-and-excitation (SE) module is then applied, which recalibrates channel-wise feature responses by modeling interdependencies between channels. This module enhances the representational power of the network by allowing it to focus on the most informative features. Following the first 1D convolutional block, a max pooling layer with a kernel size of $(2, 1)$ further reduces the temporal dimension from 500 to $T_2 = 250$. Group normalization is applied to the data, normalizing across channels and improving convergence during training. An AttentionLayer computes self-attention over the temporal and spatial dimensions, enabling the network to weigh different parts of the input differently. This attention mechanism is essential for focusing on relevant features and capturing dependencies across the data. A residual connection adds the attention output back to the original input, facilitating better gradient flow and mitigating issues such as vanishing gradients [29,30].

This pattern repeats in the subsequent convolutional blocks, with each block increasing the number of channels (from 16 to 32, 32 to 64, and 64 to 128) and further reducing the temporal dimension (from 250 to 125, 125 to 62, and 62 to 31) through additional pooling layers. The consistent use of $(3, 1)$ kernels ensures effective temporal feature extraction while preserving spatial dimensions. SE modules and attention mechanisms are integrated throughout. Feature maps are upsampled and reintroduced to the Conv1 block through a self-attention module (SAM)-inspired mechanism [26], enabling the decoder backbone to reuse features and increase model performance. The feedback mechanism employs bilinear interpolation for resizing and utilizes Conv2D layers to selectively regulate the features propagated back into the network giving it the alias of Adaptive Feature Reutilization block.

At the bottleneck of the network, a convolutional layer with a large kernel size of $(31, 1)$ is employed, covering the entire temporal dimension $T_5 = 31$. This layer transforms the tensor to $\mathbf{X}_{\text{bottleneck}} \in \mathbb{R}^{B \times 128 \times 1 \times 81}$, capturing long-range temporal dependencies and encapsulating high-level temporal information into a compact form. Batch normalization and activation are applied to maintain training stability and introduce non-linearity.

The decoder begins by reshaping this bottleneck tensor into a spatial grid $\mathbf{X}_{\text{reshaped}} \in \mathbb{R}^{128 \times 9 \times 9}$, reorganizing the data for spatial processing. A point-wise convolution reduces the channel dimension from 128 to 16, preparing the data for upsampling. The network then uses transposed convolutional layers to reconstruct the spatial dimensions progressively. The first transposed convolution upsamples the spatial dimensions from $9 \times 9$ to $18 \times 18$ and reduces the channel dimension from 16 to 8. The second transposed convolution further upsamples the dimensions to $16 \times 16$, maintaining the channel count at 8. Each transposed convolution is followed by batch normalization to ensure stable learning and effective non-linear transformations.

Finally, a point-wise convolution reduces the channel dimension from 8 to 1, and a sigmoid activation function scales the output to the range $[0, 1]$. The output tensor $\mathbf{Y} \in \mathbb{R}^{1 \times 16 \times 16}$ represents the reconstructed spatial data, which are then flattened into a vector $\mathbf{Y}_{\text{flat}} \in \mathbb{R}^{256}$ (since $16 \times 16 = 256$), making it suitable for downstream tasks. Figure 4 shows the proposed model architecture.

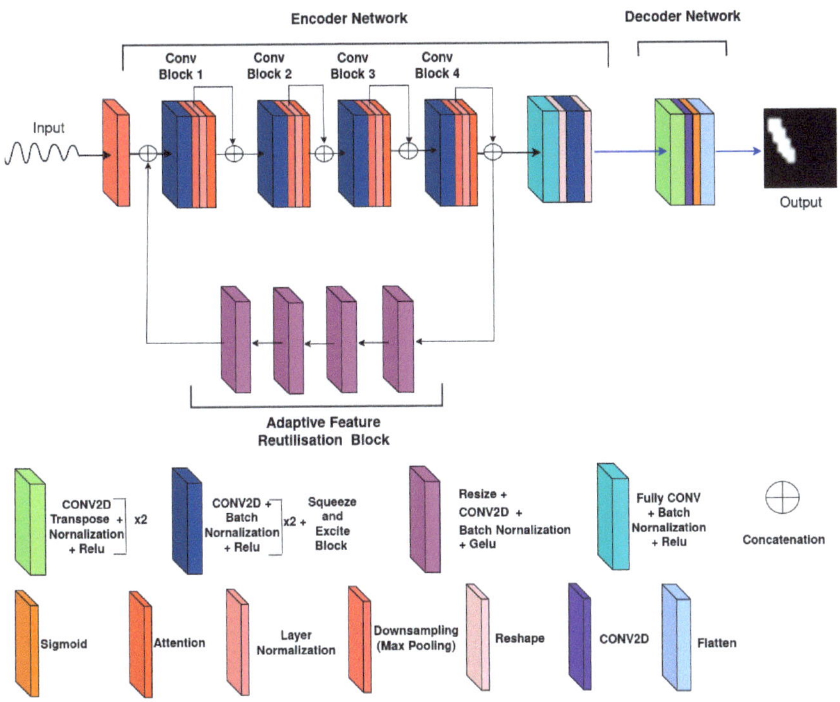


**Figure 4.** MicrocrackAttentionNext model architecture.

The architectural choices in MicrocrackAttentionNext are designed to balance feature extraction capability and computational efficiency. The initial temporal downsampling reduces the data size, allowing the network to process longer sequences without excessive computational overhead. The 1D convolutional blocks with increasing channel dimensions enable the extraction of hierarchical features in the temporal domain without mixing the spatial component. It is found that learning temporal and spatial components separately enables the model to learn better representations while being computationally efficient. The squeeze-and-excitation layers optimize the network's focus on informative channels, improving feature quality [31–33]. Using a large kernel size in the bottleneck layer is an intentional choice to capture long-range temporal dependencies, which are important in sequences where connected events are separated by large time steps [34,35]. The reshaping and upsampling in the decoder reconstruct the spatial dimensions effectively, ensuring that the high-level features extracted by the encoder are used to generate outputs. These architectural choices are arrived at through rigorous experimentation. The following section dives into these experiments in detail.

## 4. Experiments

### 4.1. Architectural Choices

The baseline model architecture, MicrocrackAttentionNext, as illustrated in Figure 5, serves as the foundation for evaluating the impact of architectural adjustments. Each modification's effect on accuracy across crack sizes is analyzed through corresponding accuracy vs. crack size graphs. $M_x$ refers to the specific model variant, where $x$ represents the variant index.

**M1: Vanilla (Baseline)** The performance of the vanilla model [27] is depicted in Figure 5, which plots accuracy against architectural modifications for different crack sizes. This has a simple convolution-based encoder. It serves as the baseline for further investigation of the impact of various architectural changes. The vanilla encoder shows satisfactory performance across various crack sizes.

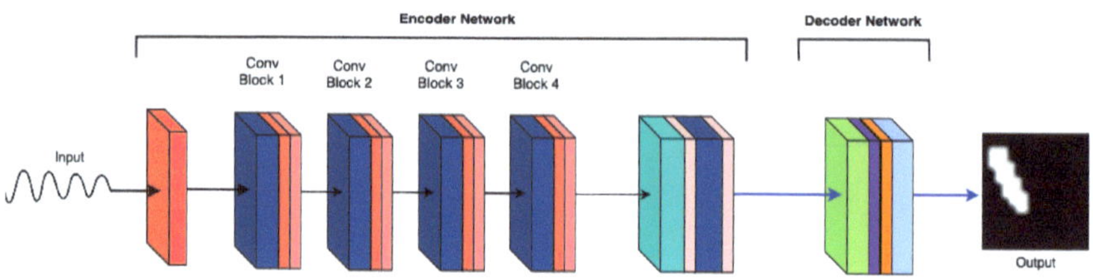


**Figure 5.** M1: Vanilla (Baseline) model architecture.

**M2: Adding Adaptive Feature Reuse Block:** The adaptive feature reuse block shows marginal improvement in accuracies across crack sizes. This is due to the increased discriminative nature of the block because of the sigmoid layer in the block's last layer.

#### 4.1.1. Attention Mechanism Placement

**M3: Self Attention Layers:** It is observed that self-attention after convolution blocks (Figure 6) does not affect the accuracies of the model. This is after the skip connections.

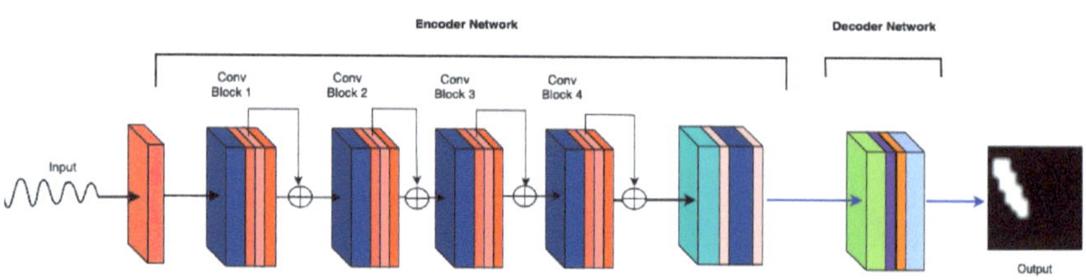


**Figure 6.** M3: Self-attention layer after convolution block.

**M4: Attention before Max Pooling in the First Layer:** Figure 7 illustrates the accuracy trends for attention applied before pooling. At Epoch 50, the performance remains suboptimal across different crack sizes. This is probably due to the information bottleneck caused by the attention layers.

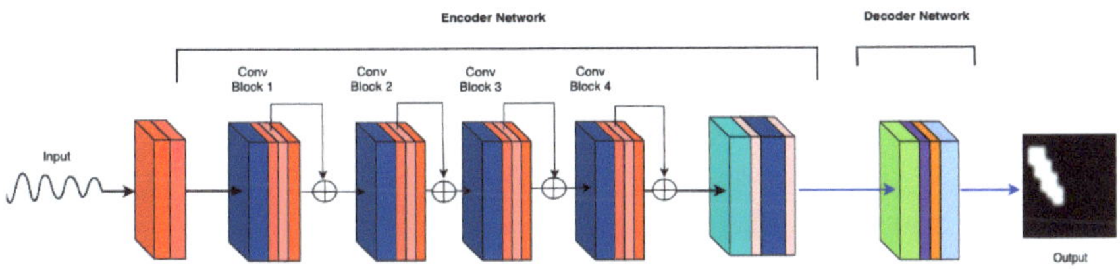


**Figure 7.** M4: Attention before max pooling in the first layer.

**M5: Attention before Max Pooling (Prolonged training):** Prolonged training to Epoch 100 results in even poorer performance across all crack sizes, suggesting that early attention causes severe information bottlenecks for the subsequent layers.

**M6: Attention after Max Pooling:** Performance remains suboptimal for attention after max pooling configuration. Figure 8 shows the performance of the attention after max pooling. This is also caused by the bottleneck issue.

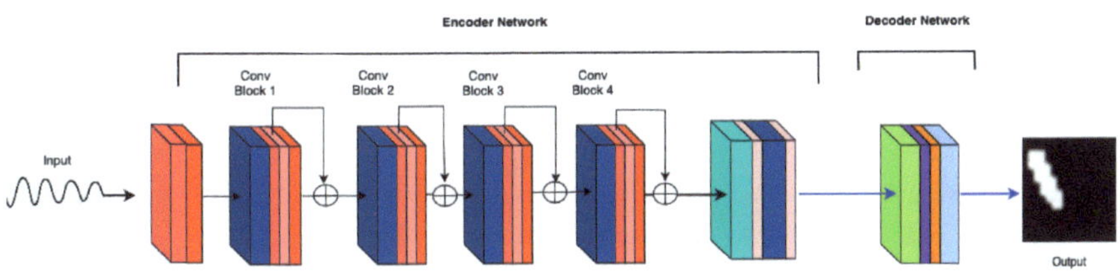


**Figure 8.** M6: Attention after max pooling in the first layer.

4.1.2. Pooling Variants

**M7: 4x1 Max Pooling and Average Pooling Hybrid:** Using max pooling for the first layer and average pooling in the subsequent layers, the graph indicates virtually no change in the accuracy. This implies that the effects of average pooling is not substantially different from max pooling in this case.

**M8: All Average Pooling:** The results for this configuration reveal slightly higher accuracy for crack sizes > 1 µm and >2 µm. Otherwise, the graph indicates virtually no change in the accuracy.

**M9: Convolutional Pooling Layers:** Figure 9 compares the performance of convolutional pooling layers at Epoch 50. The results demonstrate decreased accuracy across all crack sizes, with more dips for smaller cracks. This highlights the advantage of max pooling over convolutional downsampling spatially, as it ensures the preservation of dominant features within each pooling region without introducing additional learnable parameters. By focusing on the maximum value, max pooling effectively captures critical localized details, such as the subtle variations indicative of smaller cracks, which are often diluted by the averaging effect of convolutional downsampling.

**M10: Convolutional Pooling Layers (Prolonged Training):** Prolonged training to Epoch 100 enhances performance slightly, emphasizing the trainable pooling's ability to adaptively refine features. Overall, the deviation remains within a maximum of 2% for all crack sizes, indicating no major improvements to the model.

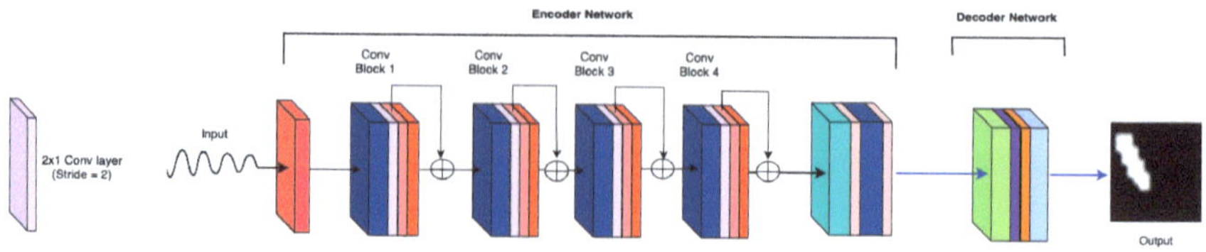


**Figure 9.** M9: Convolutional pooling layers.

4.1.3. Consecutive Attention Layers in the Encoder

**M11: Two Attention Layers (Epoch 50):** The graph in Figure 10 reveals improved accuracy for smaller crack sizes due to the enhanced focus provided by stacked attention layers. The benefits are more pronounced at finer resolutions.

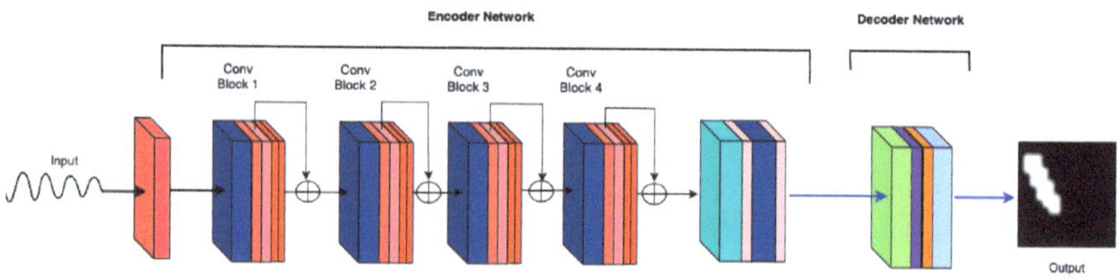


**Figure 10.** M11: Two attention layers.

**M12: Two Attention Layers (Prolonged Training):** Prolonged training of this configuration results in improved accuracies; however, it still remains the same across the board.

**M13: MicrocrackAttentionNext with single FeatureReuse:** The model shown in Figure 11 demonstrates strong performance across all thresholds. It achieves an overall accuracy of 85.53% for all cracks ($\geq 0$) and progressively higher accuracies for larger cracks, reaching 98.29% for cracks $> 4$ µm. The single feature reuse mechanism effectively balances the retention of critical features, aiding in accurate detection of both small and large cracks.

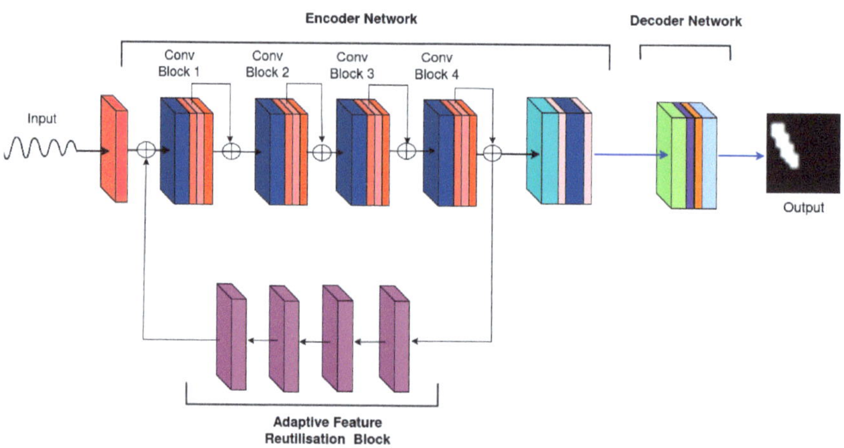


**Figure 11.** M13: MicrocrackAttentionNext with single FeatureReuse model architecture.

**M14: MicrocrackAttentionNext with dual FeatureReuse:** Incorporating feature reuse twice results in slightly lower accuracy compared to the single feature reuse model. The accuracy for all cracks ($\geq 0$) drops to 83.24%, with corresponding reductions observed across all thresholds. For cracks $> 4$ µm, the model achieves 97.2%, showing its capability for detecting larger cracks but indicating potential redundancy or noise in the reused features, which may hinder performance on smaller cracks.

**M15: MicrocrackAttentionNext with dual FeatureReuse and Extended Training:** Extending the training to 100 epochs with two feature reuse steps does not significantly improve performance. The overall accuracy further drops to 82.27%, and the accuracy for larger cracks is >4 μm is 96.3%. The results suggest that prolonged training does not compensate for the challenges introduced by additional feature reuse, particularly in handling smaller cracks effectively.

Figure 12 shows detection outcomes from the various network variants developed. The performance measures are provided at different threshold levels, spanning from relaxed to stringent detection criteria. At the strictest threshold (>4 μm), the performance values cover a wide range, from a low of 11.09 in configuration M6 to a top value of 98.2 in configuration M1. We select the configuration with the highest metric, M1, as it delivers the best detection performance.

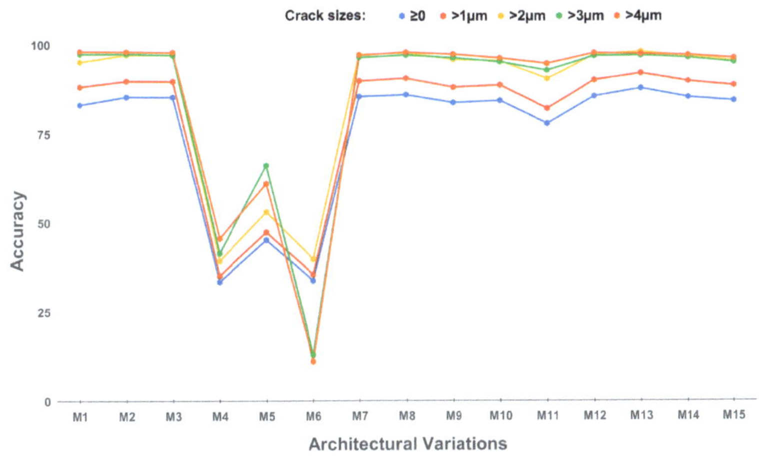


**Figure 12.** Performance comparison of architectural configurations for crack detection: **M1**: Baseline (MicrocrackAttentionNext 50E), **M2**: Adding Adaptive Feature Reuse Block, **M3**: Self-Attention Layers, **M4**: Attention before Max Pooling in the First Layer, **M5**: Attention before Max Pooling (Prolonged training), **M6**: Attention after Max Pooling, **M7**: 4x1 Max Pooling and Average Pooling Hybrid, **M8**: All Average Pooling, **M9**: Convolutional Pooling Layers, **M10**: Convolutional Pooling Layers (Prolonged training), **M11**: Two Attention Layers (Epoch 50), **M12**: Two Attention Layers (Prolonged Training), **M13**: MicrocrackAttentionNext with single FeatureReuse, **M14**: MicrocrackAttentionNext with dual FeatureReuse, **M15**: MicrocrackAttentionNext with dual FeatureReuse and Extended Training.

### *4.2. Training Procedure*

The model was trained using the Adam optimizer [36] with a learning rate of 0.001 for a total of 50 epochs. Multiple experiments were run on different activation functions and loss functions. The experiments involved evaluating four different activation functions against four loss metrics, resulting in a total of 16 experiments. The activation functions and loss metrics are outlined below. All experiments were conducted on NVIDIA RTX 4090 GPU accelerated systems having Intel Xeon CPU with Tensorflow 2.10 and CUDA 12.0.

#### 4.2.1. Activation Functions

Activation functions are used to introduce non-linearity within neural networks, each offering different advantages for a DL model. The rectified linear unit (ReLU) [37] is defined as $\text{ReLU}(x) = \max(0, x)$, outputting the input if positive and zero otherwise, thus avoiding vanishing gradient issues. The scaled exponential linear unit (SELU) [38] normalizes outputs automatically, scaling negative inputs with an exponential function and multiplying positive inputs by a fixed constant, where $\lambda = 1.0507$ and $\alpha = 1.67326$. The Gaussian error linear unit (GELU) [39] employs the Gaussian cumulative distribution function, $\Phi(x) = \frac{1}{2}\left[1 + \text{erf}\left(\frac{x}{\sqrt{2}}\right)\right]$, to probabilistically weigh input significance. GELU smoothly blends linear and non-linear behavior, making it more flexible in capturing complex patterns. The exponential linear unit (ELU) [40] applies $\text{ELU}(x) = x$ for positive inputs and $\alpha(e^x - 1)$ for negatives, mitigating vanishing gradients more effectively than ReLU, and accelerating convergence, with $\alpha$ typically set to 1. Each function enhances network performance through tailored non-linear transformations.

#### 4.2.2. Loss Functions

1. **Dice Loss [41]**:
   Dice loss is based on the Dice coefficient and is commonly used for segmentation tasks. It measures the overlap between the predicted and true labels, focusing on improving performance for imbalanced datasets.
   $$\text{Dice Loss} = 1 - \frac{2|X \cap Y|}{|X| + |Y|} \tag{1}$$
   where $X$ and $Y$ are the predicted and true sets, respectively.
2. **Focal Loss [41]**:
   Focal loss is designed to address class imbalance by down-weighting the loss assigned to well-classified examples, making the model focus more on hard-to-classify instances.
   $$\text{Focal Loss}(p_t) = -\alpha(1 - p_t)^{\gamma} \log(p_t) \tag{2}$$
   where $p_t$ is the predicted probability, $\alpha$ is a weighting factor, and $\gamma$ is a focusing parameter.
3. **Weighted Dice Loss [42]**:
   Weighted Dice loss is a variation of Dice loss that assigns different weights to different classes, enhancing performance on datasets with imbalanced class distributions by penalizing certain classes more.
   $$\text{Weighted Dice Loss} = 1 - \frac{2\sum w_i x_i y_i}{\sum w_i x_i^2 + \sum w_i y_i^2} \tag{3}$$
   where $w_i$ is the weight assigned to class $i$, and $x_i$, $y_i$ are the predicted and true values for class $i$.
4. **Combined Weighted Dice Loss [43]**:
   This is a hybrid loss that combines weighted Dice loss and CrossEntropy loss, allowing the model to balance overall performance while addressing class imbalances by tuning the contribution of each component.
   $$\text{CWDL} = \alpha \cdot \text{WDL} + (1 - \alpha) \cdot \text{CrossEntropy Loss} \tag{4}$$
   where CWDL is combined weighted Dice loss, WDL is weighted Dice loss, and $\alpha$ is a weighting factor to balance the two loss components.

It is found that the combination of combined weighted Dice loss and GeLU are the best performing. The combined weighted Dice loss performed the best across all activations (Table 1). However, it was found that the proposed study is able to squeeze more accuracy through the GeLU function.

**Table 1.** Comparison against previous works on the same dataset.

| Literature | Accuracy | IoU | DSC | Precision | Recall |
|---|---|---|---|---|---|
| 1D-DenseNet-TConv (200 epochs) | 0.8368 | 0.7601 | 0.8637 | 0.8753 | 0.8523 |
| MicroCracksAttNet (50 epochs) | 0.8601 | 0.7666 | 0.8684 | 0.8811 | 0.8888 |
| MicroCracksMetaNet (50 epochs) | 0.867 | 0.8082 | 0.8943 | 0.9066 | 0.8911 |
| **Proposed MicrocrackAttentionNect (50 epochs)** | **0.8777** | **0.8521** | **0.9145** | **0.8601** | **0.8518** |

### 4.3. Evaluation Metrics

For the evaluation part, the proposed study utilized the same metrics as in [16], namely Dice similarity coefficient (DSC) and accuracy, which frequently employed the performance of models. The DSC measures the overlap between predicted and actual results, particularly in segmentation tasks. Its mathematical formulation is given by:

$$\mathrm{DSC} = \frac{2 \cdot \mathrm{TP}}{2 \cdot \mathrm{TP} + \mathrm{FP} + \mathrm{FN}} \tag{5}$$

The accuracy threshold determines whether the IoU score is adequate to consider a prediction correct.

$$\mathrm{Accuracy} = \begin{cases} 1 & \text{if } \mathrm{IoU}(y_{\mathrm{true}}, \hat{y}_{\mathrm{pred}}) > t_{\mathrm{IOU}} \\ 0 & \text{otherwise} \end{cases} \tag{6}$$

The IoU metric is calculated using the following equation (Equation (7)).

$$\mathrm{IoU} = \frac{\mathrm{TP}}{\mathrm{TP} + \mathrm{FP} + \mathrm{FN}}. \tag{7}$$

## 5. Results and Discussion

A key challenge in wave-based crack detection is differentiating between actual cracks and reflections caused by the sample's border. The proposed model successfully learned to distinguish between these two cases, despite their similar wave behavior. When a wave encounters a crack, its interaction leads to a change in propagation, similar to when it reaches a boundary and reflects back. This distinction is particularly challenging because the returning wave from the border behaves similarly to one encountering a crack. Our model's ability to differentiate these cases highlights its capability to extract nuanced temporal and spatial features. However, the model struggles when all crack sizes are included in the evaluation. This may be attributed to the scarcity of the smallest cracks in the dataset, leading to poor representation in the feature space. Additionally, smaller cracks may cause only subtle disruptions in wave behavior, making them harder to detect.

### 5.1. MDA Analysis

Manifold discovery and analysis (MDA) helps visualize the higher dimensional manifolds formed by the intermediate layers of the model in lower dimension [15]. These plots help visualise the learned features in the $\ell$th layer with respect to the output manifold. Unlike methods such as t-SNE and UMAP, which only work on classification tasks, MDA works on regression tasks, where the output manifold can have a complex shape. MDA also preserves the geodesic distances between higher dimensional feature points, preserving both local and global structure.

In a nutshell, MDA works as follows: First, distance is computed between the estimated outputs of the DNN, from this distance, the farthest point is chosen to construct the boundary of the output manifold. All the points are sorted w.r.t the farthest point in k bins using the optimal histogram bin count. These bins become the labels that will be used in the second step. Second, the high-dimensional features from an intermediate layer are projected to the manifold using the Bayesian manifold projection (BMP) approach. BMP computes a posterior distribution over the low-dimensional space by combining the prior (based on pseudo-labels and manifold structure) with a likelihood (based on the observed data). Finally, a DNN trained on predicting the location of uncertain Bayesian points on a 2D embedding space is used to visualise the results. The plots are assessed qualitatively on the following points:

- Feature Separation and Continuity: The MDA visualization shows a curved shape, indicating that the features extracted from the neural network follow a smooth continuum along the manifold. This suggests that the neural network is capturing meaningful information.
- Color Gradient: A spectrum of gradients is shown, implying that the model has learned to separate different features.

The MDA plot Figure 13a for the untrained model shows a relatively disorganized and diffuse clustering of points. This suggests that the feature representations at Layer 64 are not yet structured in a meaningful way to distinguish between patterns within the dataset. The absence of clear separation or distinct clustering patterns indicates that the untrained model has not yet learned to capture the underlying structure of the data, which is expected at the initial stages of training. At this stage, the network's representations are largely random, as it has not yet learned the task-specific features. The spread-out nature of the points highlights that the model is treating all inputs similarly, without any differentiation based on the features it should detect. In contrast, the MDA plot Figure 13b for the trained model reveals a much more structured and organized distribution. The $\ell$th layer of a well-trained model shows the cluster with a smooth arch-like structure and a gradient of colors differentiating the two output extremums; red representing 0 and blue representing 1. The analysis of various layer depths and activation functions in the MicrocrackAttentionNext reveals a clear pattern in the network's ability to form distinguishable manifold curves. Figure 14(1) visualizes the manifold at different layers of the network. All the layers show consistent and smooth arch-like shapes. This implies that all the layers have learned good representations. In Figure 14(2), effects of various activations are plotted, and it is found that ELU shows a more spread out cluster especially toward the light colors (towards crack class) and ReLU shows a good arch-like structure with tightly packed dots of red color (no crack class). Still, the light color dots are much more incoherent and less compact. SeLU shows a poorly defined structure compared to all other activations. This is also reflected in the results table, where SeLU performs worse in all cases. This behavior of SeLU can be attributed to its self-normalizing property, as it forces all outputs to behave similarly, "dampening" the importance of smaller, rare patterns. This makes the model less discriminative, making the plot less defined and incoherent. In terms of cluster shape and compactness, GeLU is similar to ReLU, which should be the case, as GeLU is the smoothed version of ReLU. Very similar performance in the results table is also observed. Among all the activations, GeLU performed the best; this fact also reflects on the MDA plot where the cluster is very smooth and the lighter points are more compactly packed relative to other activation functions. Its smooth probabilistic gating mechanism helps in finely controlling how information is passed through the network, allowing the model to focus more on the minority class. Figure 15 shows MDA visualization of the 1D Densenet proposed in [16] compared with the proposed MicrocrackAttentionNext model. One thing to note in all of the above MDA plots is the absence of a full spectrum of colors in the plot. This is mainly attributed to the severe class imbalance in the data. This class imbalance leads to very few values in the feature map strongly correlating to the strong value of predicting the crack class. This is further aggravated by the dimensionality reduction, which renders even fewer points corresponding to higher confidence values. Hence, only one blue point is seen. The proposed MicrocrackAttentionNext achieved a DSC of 0.91. Table 2 shows the comparison of different loss functions used to train MicrocrackAttentionNext.

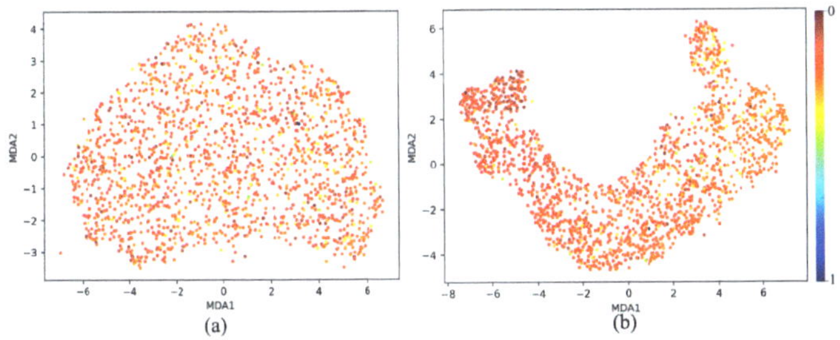

**Figure 13.** MDA visualization of Layer 64 comparing (**a**) untrained model and (**b**) trained model.

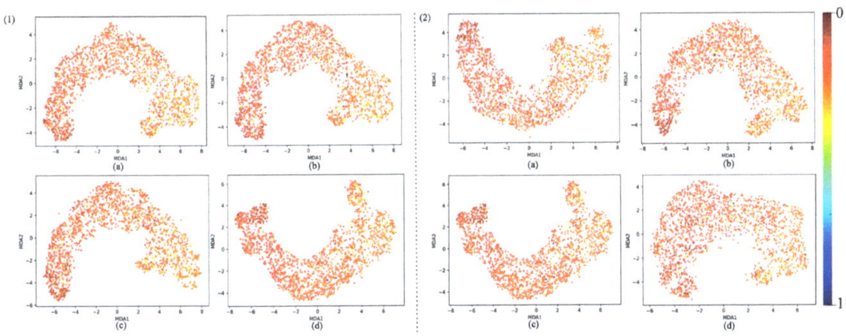

**Figure 14.** (**1**) MDA visualizations of layers using Gelu activation and Dice loss, shown for (**a**) Layer 22, (**b**) Layer 25, (**c**) Layer 34, and (**d**) Layer 64 and, (**2**) MDA visualization of Layer 64 utilizing different activation functions: (**a**) ELU, (**b**) ReLU, (**c**) GELU, and (**d**) SELU.

**Table 2.** Comparison of accuracies using different loss functions for multiple crack sizes. FL: focal loss, DL: Dice loss, WDL: weighted Dice loss, CWDL: combined weighted Dice loss.

| Activation Function | Loss Function | > 0 µm | > 1 µm | > 2 µm | > 3 µm | > 4 µm |
|---|---|---|---|---|---|---|
| GeLU | FL | 0.8275 | 0.8612 | 0.9354 | 0.9501 | 0.9541 |
| | DL | 0.8633 | 0.9012 | 0.9585 | 0.9701 | 0.9802 |
| | WDL | 0.8381 | 0.8798 | 0.9415 | 0.9670 | 0.9793 |
| | CWDL | **0.8774** | **0.9211** | **0.9814** | **0.9808** | **0.9848** |
| ReLU | FL | 0.8252 | 0.8632 | 0.9456 | 0.9701 | 0.9802 |
| | DL | 0.8553 | 0.8902 | 0.9646 | 0.9770 | 0.9829 |
| | WDL | 0.8213 | 0.8687 | 0.9293 | 0.9524 | 0.9703 |
| | CWDL | 0.8678 | 0.9134 | 0.9673 | 0.9808 | 0.9866 |
| ELU | FL | 0.8313 | 0.8797 | 0.9558 | 0.9839 | 0.9911 |
| | DL | 0.8502 | 0.9011 | 0.9673 | 0.9831 | 0.9884 |
| | WDL | 0.8563 | 0.9034 | 0.9605 | 0.9739 | 0.9829 |
| | CWDL | 0.8515 | 0.9041 | 0.9673 | 0.9847 | **0.9920** |
| SeLU | FL | 0.8206 | 0.8671 | 0.9503 | 0.9793 | 0.9902 |
| | DL | 0.8412 | 0.8993 | 0.9707 | **0.9870** | 0.9893 |
| | WDL | 0.8201 | 0.8664 | 0.9307 | 0.9555 | 0.9712 |
| | CWDL | 0.8443 | 0.8910 | 0.9625 | 0.9854 | 0.9929 |

Best result for a crack size is highlighted with bold.

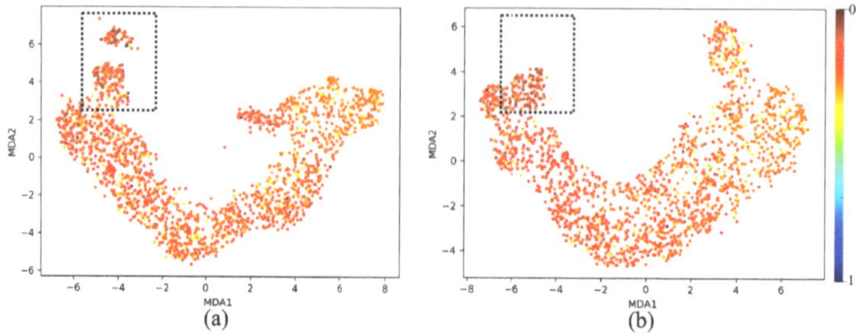


**Figure 15.** MDA visualization comparing (**a**) 1D Densenet [16] and (**b**) proposed model—MicrocrackAttentionNext. The highlighted region in black indicates the region where the cluster is broken in 1D Densenet. In contrast, the same region in MicrocrackAttentionNext shows coherency, implying that the MicrocrackAttentionNext learned good feature representations for microcracks.

### 5.2. Thresholding Analysis in Prediction Accuracy

Pixel-level predictions were binarized using a threshold of 0.5, where values exceeding this threshold indicated the presence of a crack. Prediction accuracy was assessed based on the intersection over union (IoU) metric, with a threshold of 0.5 selected to determine accurate crack identification. This threshold ensures a balance between minimizing false positives and maintaining practical accuracy.

Figure 16 demonstrates the relationship between varying IoU thresholds and overall accuracy. An IoU value approaching 1 requires near-perfect pixel classification but results in a significant accuracy drop due to its stringent nature. Conversely, an IoU value near 0 allows for minimal correct pixel identification but fails to filter out false positives. The chosen IoU threshold of 0.5 provides a practical compromise, achieving reasonable accuracy while maintaining meaningful crack identification.

Figure 17 explores the influence of both IoU and binarizing thresholds on accuracy, holding one threshold constant at 0.5. Variations in the binarizing threshold showed minimal impact on accuracy, except at extreme values (near 0 or 1), indicating strong confidence in the model's pixel-level predictions.

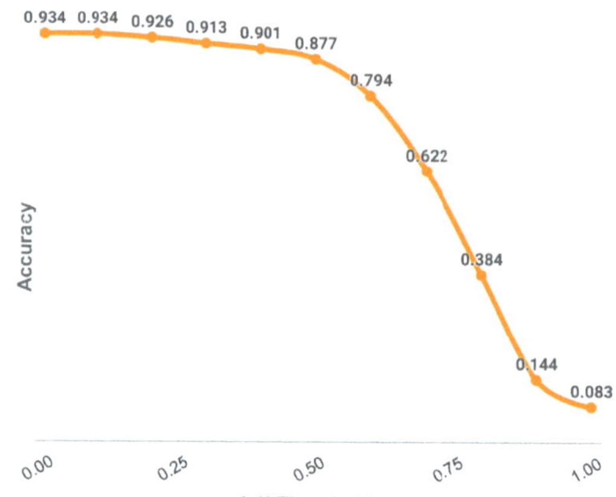


**Figure 16.** Effect of IOU threshold on accuracy for bin threshold at 0.5 and inclusive of all crack size.

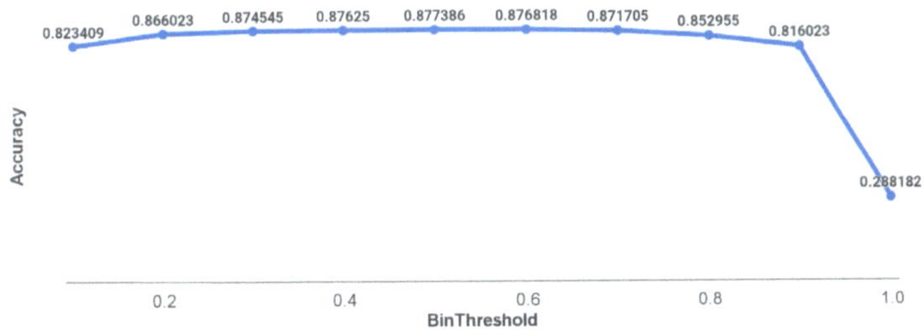


**Figure 17.** Binarising threshold when IoU threshold is 0.5.

### 5.3. Comparison with Similar Works

The uniqueness of this work lies in the dataset used and the applied deep learning algorithms. The data, which were exclusively collected at the researcher's university, consist of purely numerical values, distinguishing them from the image data commonly used in the literature. To the best of our knowledge, no deep learning models have been trained on this type of numerical data. The goal is to develop models that not only detect cracks but also precisely segment them. Additionally, the presence of multiple crack sizes increases complexity, as the behavior of the wave differs depending on the crack dimensions.

The earlier implemented models can serve as benchmarks for comparison in this work. These previous models were applied to the same dataset and provide a foundation for evaluating the performance of the proposed deep learning approaches. The results from these models allow us to objectively assess the progress and improvements made with the new algorithms.

Table 1 presents a comparison of the performance of the earlier models [16,26,27] with the current results.

The proposed method, **MicrocrackAttentionNext**, achieves an accuracy of **0.8777**, an IoU of **0.8521**, and a DSC of **0.9145**. These results outperform previous models on key metrics, suggesting that our approach is not only highly accurate but also effective at localizing cracks. A higher IoU indicates that the predicted crack regions overlap closely with the ground truth, while the DSC improvement reflects enhanced segmentation performance. However, the model's precision (**0.8601**) and recall (**0.8518**) indicate room for improvement. In particular, there remains scope to reduce false positives and false negatives. Future work will involve further tuning of the network architecture, incorporating additional data, and exploring advanced loss functions to address these shortcomings.

The proposed model provides better performance on DSC and IoU metrics, suggesting that the proposed model is not only able to detect the crack with high accuracy but also localize it better. A higher IoU score also suggests that the predicted crack region and the ground truth region have a high overlap, meaning the model is drawing the crack boundary closer to where the crack actually exists. With better DSC and IoU the model is also less likely to make mistakes and detect cracks where there are no cracks present (false positives) or miss cracks that should be detected (false negatives). This is important for real-world applications, such as structural health monitoring where accuracy is critical. Figure 18a shows successful segmentation from MicrocrackAttentionNext while Figure 18b shows some failure segmentation samples.

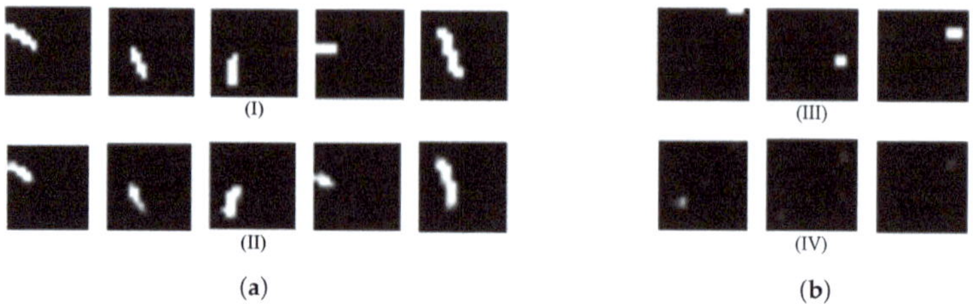

Figure 18. (a) Positive examples from MicrocrackAttentionNext (I) represents the ground truth and (II) the corresponding model prediction. (b) Failure sample (III) represents the ground truth, while (IV) shows the corresponding model prediction.

Table 3 compares the models on the basis of training time, number of layers, and parameters. This table illustrates the computational complexity and efficiency of each architecture. For instance, while some models (such as 1D-DenseNet-TConv originally trained for 200 epochs) have a significantly larger number of layers and require more training time, our proposed model maintains competitive performance with considerably fewer layers and reduced training time. Such efficiency is critical for deployment in environments with limited computational resources.

Table 3. Training performance and architectural complexity comparison among four deep learning models: MicroCracksAttNet, MicroCracksMetaNet, 1D-DenseNet-TConv, and MicrocrackAttentionNext.

| Attributes | MicroCracksAttNet | MicroCracksMetaNet | 1D-DenseNet-TConv | MicrocrackAttentionNext |
|---|---|---|---|---|
| Layers | 58 | 72 | 444 | 94 |
| Epochs | 50 | 50 | 200 | 50 |
| Total params | 1,129,209 | 1,131,690 | 1,393,429 | 1,280,299 |
| Trainable params | 1,127,321 | 1,129,722 | 1,376,137 | 1,277,355 |
| Non-trainable params | 1888 | 1968 | 17,292 | 2944 |
| Time taken by first Epoch | 26.45 s | 42.35 s | 89.14 s | 36.98 s |
| Total training time | 906.33 s | 2117.50 s | 15,560.56 s | 1849.54 s |

Furthermore, Table 4 provides a breakdown of the segmentation accuracy with respect to various crack sizes (measured in micro meters). This detailed analysis shows that the proposed model maintains high accuracy across all crack size ranges. For instance, while the accuracy in the 0–3 µm range is moderate (0.5214), it improves significantly for larger cracks, reaching up to 0.9917 in the 9–14 µm range. This indicates that the model is particularly effective at localizing larger cracks, although there remains potential to improve performance on very fine cracks. Addressing this could be a focus for future model refinement, possibly through specialized training strategies or additional data augmentation techniques.

**Table 4.** Crack detection accuracy measured across crack size intervals (in micrometers).

| Model | Crack Sizes μm | | | |
|---|---|---|---|---|
| | **0–3 μm** | **3–6 μm** | **6–9 μm** | **9–14 μm** |
| MicrocrackAttentionNext (50 epochs) | 0.5214 | 0.9787 | 0.9879 | 0.9917 |
| MicroCracksMetaNet (50 epochs) | 0.4513 | 0.9549 | 0.9753 | 0.9862 |
| MicroCracksAttNet (50 epochs) | 0.4420 | 0.9587 | 0.9778 | 0.9862 |
| 1D-DenseNet-TConv (50 epochs) | 0.3719 | 0.9268 | 0.9778 | 0.9835 |
| 1D-DenseNet-TConv (200 epochs) | 0.4682 | 0.9793 | 0.9852 | 0.9972 |

## 6. Conclusions and Future Work

### 6.1. Conclusions

In this study, we introduced MicrocrackAttentionNext, an advanced deep learning model designed for microcrack detection in wave field analysis. The key novelty of this model lies in its Adaptive Feature Reuse Block, which significantly improves feature utilization across different crack sizes. This block helps the model focus on the most informative features, enhancing detection accuracy while reducing the risk of overfitting, particularly when dealing with small and subtle cracks. Through a series of experiments, we compared various model configurations, including baseline models and models with different architectural adjustments. The experimental results indicated that MicrocrackAttentionNext outperforms previous models in multiple metrics, with notable improvements in the Dice similarity coefficient (DSC) and intersection over union (IoU), which directly correlate with better crack localization and segmentation. Specifically, the model achieved an accuracy of 87.77% overall, surpassing the previous benchmarks, and demonstrated a DSC of 0.9145 and IoU of 0.8521, suggesting a superior ability to precisely detect and delineate cracks.

However, despite these significant advancements, the precision and recall values of the proposed model (0.8601 and 0.8518, respectively) indicate room for further refinement, particularly in reducing false positives and false negatives. To this end, future work will focus on fine-tuning the network architecture and exploring the integration of additional data sources to address these gaps. One possible avenue for improvement is the introduction of more advanced loss functions tailored to handle class imbalance more effectively, especially for detecting small cracks that introduce minor perturbations in the wave field. Moreover, extending the model's training with more epochs and experimenting with alternative attention mechanisms could potentially yield further improvements in both the overall performance and handling of smaller crack sizes.

The model is capable of segmenting the microcracks, allowing us to determine their spatial locations in the material. The qualitative examination of the activation functions using the manifold discovery and analysis (MDA) algorithm allowed the evaluation of the impact of different activation and loss functions on the model's performance. The proposed model and 1D-Densenet were analyzed using the MDA plots. The manifold of the proposed model was more compact with a much smoother arc than the 1D-Densenet.

### 6.2. Future Work

In future efforts to improve microcrack detection models, two primary strategies can be pursued: expanding datasets and refining model architectures. The dataset used presents a challenge due to severe class imbalance, which requires more advanced techniques for data generation and augmentation to mitigate the bias introduced. Moreover, the segmentation output is limited by low resolution and without appropriate upscaling techniques critical details may be lost. Although the encoder architecture performs well enough, more changes are necessary in the decoder section of the segmentation model to achieve improved results and maintain consistency with high-quality input features. Expanding the dataset to include a wider range of crack patterns and sizes will further enhance the robustness of the model. Larger and more diverse datasets with multiple cracks in a single sample will help address class imbalance, improve generalization, and refine the model's ability to distinguish subtle crack patterns. Additionally, a more balanced dataset containing a greater number of small cracks would help the model better capture small wave disturbances. Since smaller cracks may introduce only minor perturbations in wave behavior, their detection is more challenging. Data from multiple excitation points for a single sample will enhance the quality of the dataset. To enhance the encoder's ability to capture long-range dependencies, a state space model can be used, integrating the recently proposed Mamba architecture in particular. This adjustment could improve the model's ability to handle complex spatial relationships, thereby strengthening feature extraction and contributing to overall performance gains in the segmentation task.

**Author Contributions:** F.M. led the research, taking primary responsibility for the development of the methodology, conducting experiments, and driving the overall direction of the project. F.M. and Y.H. collaboratively built and optimized multiple models, refining interpretations and improving the results. M.A. and B.Z.H. contributed significantly to feature space visualization and M.A. analysis, enhancing the understanding of the data. F.M., Y.H., M.A., and B.Z.H. jointly wrote the manuscript, while F.W. and S.T. provided crucial proofreading and feedback. All authors have read and agreed to the published version of the manuscript.

**Funding:** Funding for this research was provided by Land Schleswig-Holstein within the funding programme Open Access Publikationsfonds.

**Data Availability Statement:** The data supporting the findings of this study are not publicly available due to privacy restrictions. Access to the data can be considered upon reasonable request to the corresponding author, subject to institutional and ethical guidelines.

**Acknowledgments:** We acknowledge financial support by Land Schleswig-Holstein within the funding programme Open Access Publikationsfonds.

**Conflicts of Interest:** The authors declare no conflicts of interest. The funders had no role in the design of the study; in the collection, analyses, or interpretation of data; in the writing of the manuscript; or in the decision to publish the results.

## Abbreviations

The following abbreviations are used in this manuscript:

| | |
|---|---|
| AE | Acoustic Emission |
| CNN | Convolutional Neural Network |
| CWDL | Combined Weighted Dice Loss |
| DL | Dice Loss |
| DSC | Dice Similarity Coefficient |
| DNN | Deep Neural Network |
| FL | Focal Loss |
| GELU | Gaussian Error Linear Unit |
| IoU | Intersection over Union |
| LEM | Lattice Element Method |
| MDA | Manifold Discovery and Analysis |
| MDPI | Multidisciplinary Digital Publishing Institute |
| NDT | Non-Destructive Testing |
| ReLU | Rectified Linear Unit |
| SAW | Surface Acoustic Wave |

| | |
|---|---|
| SE | Squeeze-and-Excitation |
| SELU | Scaled Exponential Linear Unit |
| TConv | Transposed Convolution |
| TOFD | Time-of-Flight Diffraction |
| WDL | Weighted Dice Loss |

## References

1. Malekloo, A.; Ozer, E.; AlHamaydeh, M.; Girolami, M. Machine learning and structural health monitoring overview with emerging technology and high-dimensional data source highlights. *Struct. Health Monit.* **2022**, *21*, 1906–1955. [CrossRef]
2. Golewski, G.L. The Phenomenon of Cracking in Cement Concretes and Reinforced Concrete Structures: The Mechanism of Cracks Formation, Causes of Their Initiation, Types and Places of Occurrence, and Methods of Detection—A Review. *Buildings* **2023**, *13*, 765. [CrossRef]
3. Chen, B.; Yang, J.; Zhang, D.; Liu, W.; Li, J.; Zhang, M. The Method and Experiment of Micro-Crack Identification Using OFDR Strain Measurement Technology. *Photonics* **2024**, *11*, 755. [CrossRef]
4. Yuan, B.; Ying, Y.; Morgese, M.; Ansari, F. Theoretical and Experimental Studies of Micro-Surface Crack Detections Based on BOTDA. *Sensors* **2022**, *22*, 3529. [CrossRef]
5. Retheesh, R.; Samuel, B.; Radhakrishnan, P.; Mujeeb, A. Detection and analysis of micro cracks in low modulus materials with thermal loading using laser speckle interferometry. *Russ. J. Nondestruct. Test.* **2017**, *53*, 236–242. [CrossRef]
6. Choi, Y.; Park, H.W.; Mi, Y.; Song, S. Crack Detection and Analysis of Concrete Structures Based on Neural Network and Clustering. *Sensors* **2024**, *24*, 1725. [CrossRef]
7. Animashaun, D.; Hussain, M. Automated Micro-Crack Detection within Photovoltaic Manufacturing Facility via Ground Modelling for a Regularized Convolutional Network. *Sensors* **2023**, *23*, 6235. [CrossRef]
8. Andleeb, I.; Hussain, B.Z.; Joncas, J.; Barchi, S.; Roy-Beaudry, M.; Parent, S.; Grimard, G.; Labelle, H.; Duong, L. Automatic Evaluation of Bone Age Using Hand Radiographs and Pancorporal Radiographs in Adolescent Idiopathic Scoliosis. *Diagnostics* **2025**, *15*, 452. [CrossRef]
9. LeCun, Y.; Bengio, Y.; Hinton, G. Deep learning. *Nature* **2015**, *521*, 436–444. [CrossRef]
10. Simonyan, K.; Zisserman, A. Very Deep Convolutional Networks for Large-Scale Image Recognition. *arXiv* **2015**, arXiv:1409.1556.
11. Moorthy, J.; Gandhi, U.D. A Survey on Medical Image Segmentation Based on Deep Learning Techniques. *Big Data Cogn. Comput.* **2022**, *6*, 117. [CrossRef]
12. Ronneberger, O.; Fischer, P.; Brox, T. U-Net: Convolutional Networks for Biomedical Image Segmentation. *arXiv* **2015**, arXiv:1505.04597.
13. van der Maaten, L.; Hinton, G. Visualizing Data using t-SNE. *J. Mach. Learn. Res.* **2008**, *9*, 2579–2605.
14. McInnes, L.; Healy, J.; Melville, J. UMAP: Uniform Manifold Approximation and Projection for Dimension Reduction. *arXiv* **2020**, arXiv:1802.03426.
15. Islam, M.T.; Zhou, Z.; Ren, H.; Khuzani, M.B.; Kapp, D.; Zou, J.; Tian, L.; Liao, J.C.; Xing, L. Revealing hidden patterns in deep neural network feature space continuum via manifold learning. *Nat. Commun.* **2023**, *14*, 8506. [CrossRef] [PubMed]
16. Moreh, F.; Lyu, H.; Rizvi, Z.H.; Wuttke, F. Deep neural networks for crack detection inside structures. *Sci. Rep.* **2024**, *14*, 4439. [CrossRef] [PubMed]
17. Punnose, S.; Balasubramanian, T.; Pardikar, R.; Palaniappan, M.; Subbaratnam, R. Time-of-flight diffraction (TOFD) technique for accurate sizing of surface-breaking cracks. *Insight* **2003**, *45*, 426–430. [CrossRef]
18. Kou, X.; Pei, C.; Chen, Z. Fully noncontact inspection of closed surface crack with nonlinear laser ultrasonic testing method. *Ultrasonics* **2021**, *114*, 106426. [CrossRef]
19. Su, C.; Wang, W. Concrete Cracks Detection Using Convolutional NeuralNetwork Based on Transfer Learning. *Math. Probl. Eng.* **2020**, *2020*, 7240129. [CrossRef]
20. Chen, F.C.; Jahanshahi, M. NB-CNN: Deep Learning-based Crack Detection Using Convolutional Neural Network and Naïve Bayes Data Fusion. *IEEE Trans. Ind. Electron.* **2018**, *65*, 4392–4400. [CrossRef]
21. Hamishebahar, Y.; Guan, H.; So, S.; Jo, J. A Comprehensive Review of Deep Learning-Based Crack Detection Approaches. *Appl. Sci.* **2022**, *12*, 1374. [CrossRef]
22. Tran, V.L.; Vo, T.C.; Nguyen, T.Q. One-dimensional convolutional neural network for damage detection of structures using time series data. *Asian J. Civ. Eng.* **2024**, *25*, 827–860. [CrossRef]
23. Jiang, C.; Zhou, Q.; Lei, J.; Wang, X. A Two-Stage Structural Damage Detection Method Based on 1D-CNN and SVM. *Appl. Sci.* **2022**, *12*, 10394. [CrossRef]
24. Barbosh, M.; Ge, L.; Sadhu, A. Automated crack identification in structures using acoustic waveforms and deep learning. *J. Infrastruct. Preserv. Resil.* **2024**, *5*, 10.

25. Li, X.; Xu, X.; He, X.; Wei, X.; Yang, H. Intelligent Crack Detection Method Based on GM-ResNet. *Sensors* **2023**, *23*, 8369. [CrossRef] [PubMed]
26. Moreh, F.; Hasan, Y.; Rizvi, Z.H.; Wuttke, F.; Tomforde, S. MCMN Deep Learning Model for Precise Microcrack Detection in Various Materials. In Proceedings of the 2024 International Conference on Machine Learning and Applications (ICMLA), Miami, FL, USA, 18–20 December 2024; pp. 1928–1933. [CrossRef]
27. Moreh, F.; Hasan, Y.; Rizvi, Z.H.; Wuttke, F.; Tomforde, S. Wave-Based Neural Network with Attention Mechanism for Damage Localization in Materials. In Proceedings of the 2024 International Conference on Machine Learning and Applications (ICMLA), Miami, FL, USA, 18–20 December 2024; pp. 122–129. [CrossRef]
28. Wuttke, F.; Lyu, H.; Sattari, A.S.; Rizvi, Z.H. Wave based damage detection in solid structures using spatially asymmetric encoder–decoder network. *Sci. Rep.* **2021**, *11*, 20968. [CrossRef]
29. Raghu, M.; Unterthiner, T.; Kornblith, S.; Zhang, C.; Dosovitskiy, A. Do Vision Transformers See Like Convolutional Neural Networks? *arXiv* **2022**, arXiv:2108.08810.
30. He, K.; Zhang, X.; Ren, S.; Sun, J. Deep Residual Learning for Image Recognition. *arXiv* **2015**, arXiv:1512.03385.
31. Yuan, S. SECNN: Squeeze-and-Excitation Convolutional Neural Network for Sentence Classification. *arXiv* **2023**, arXiv:2312.06088.
32. Hong, S.B.; Kim, Y.H.; Nam, S.H.; Park, K.R. S3D: Squeeze and Excitation 3D Convolutional Neural Networks for a Fall Detection System. *Mathematics* **2022**, *10*, 328. [CrossRef]
33. Hu, J.; Shen, L.; Albanie, S.; Sun, G.; Wu, E. Squeeze-and-Excitation Networks. *arXiv* **2019**, arXiv:1709.01507.
34. Lea, C.; Vidal, R.; Reiter, A.; Hager, G.D. Temporal Convolutional Networks: A Unified Approach to Action Segmentation. *arXiv* **2016**, arXiv:1608.08242.
35. Zhang, Y.; Ding, X.; Yue, X. Scaling Up Your Kernels: Large Kernel Design in ConvNets towards Universal Representations. *arXiv* **2024**, arXiv:2410.08049.
36. Kingma, D.P.; Ba, J. Adam: A Method for Stochastic Optimization. *arXiv* **2017**, arXiv:1412.6980.
37. Nair, V.; Hinton, G.E. Rectified linear units improve restricted boltzmann machines. In Proceedings of the 27th International Conference on International Conference on Machine Learning, Haifa, Israel, 21–24 June 2010; pp. 807–814.
38. Klambauer, G.; Unterthiner, T.; Mayr, A.; Hochreiter, S. Self-Normalizing Neural Networks. *arXiv* **2017**, arXiv:1706.02515.
39. Hendrycks, D.; Gimpel, K. Gaussian Error Linear Units (GELUs). *arXiv* **2023**, arXiv:1606.08415.
40. Clevert, D.A.; Unterthiner, T.; Hochreiter, S. Fast and Accurate Deep Network Learning by Exponential Linear Units (ELUs). *arXiv* **2016**, arXiv:1511.07289.
41. Lin, T.Y.; Goyal, P.; Girshick, R.; He, K.; Dollár, P. Focal Loss for Dense Object Detection. *arXiv* **2018**, arXiv:1708.02002.
42. Yeung, M.; Sala, E.; Schönlieb, C.B.; Rundo, L. Unified Focal loss: Generalising Dice and cross entropy-based losses to handle class imbalanced medical image segmentation. *arXiv* **2021**, arXiv:2102.04525.
43. Jadon, S. A survey of loss functions for semantic segmentation. In Proceedings of the 2020 IEEE Conference on Computational Intelligence in Bioinformatics and Computational Biology (CIBCB), Via del Mar, Chile, 27–29 October 2020. [CrossRef]

sensors

MDPI

*Article*

# Registration of Sounds Emitted by the Madagascar Hissing Cockroach Using a Distributed Acoustic Sensor

**Artem T. Turov [1,2], Yuri A. Konstantinov [1], Ekaterina E. Totmina [2], Anna G. Votinova [3], Grigoriy F. Masich [1], Dmitry A. Korobko [4] and Andrei A. Fotiadi [4,5,*]**

1 Perm Federal Research Center of the Ural Branch of the Russian Academy of Sciences (PFRC UB RAS), 13a Lenin St., 614000 Perm, Russia; artemtur442@gmail.com (A.T.T.); yuri.al.konstantinov@ro.ru (Y.A.K.); masich@icmm.ru (G.F.M.)

2 General Physics Department, Applied Mathematics and Mechanics Faculty, Perm National Research Polytechnic University, Prospekt Komsomolsky 29, 614990 Perm, Russia; totmina_katya@mail.ru

3 Faculty of Biology, Perm State National Research University, Bukirev Street 15, 614990 Perm, Russia

4 S.P. Kapitsa Research Institute of Technology, Ulyanovsk State University, 42 Leo Tolstoy Street, 432970 Ulyanovsk, Russia; korobkotam@rambler.ru

5 Electromagnetism and Telecommunication Department, University of Mons, B-7000 Mons, Belgium

* Correspondence: andrei.fotiadi@umons.ac.be

**Abstract:** Recent advancements have expanded the applications of fiber-optic distributed acoustic sensors (DAS), including their use in monitoring the acoustic activity of insects, which can be either harmful or beneficial to agriculture. Previous studies have demonstrated the capability of DAS to record and analyze insect-generated acoustic signals in real-world conditions; however, these studies primarily involved large insect colonies. In this work, a fiber-optic DAS is used for the first time to record the sounds produced by a single insect under controlled laboratory conditions. This was achieved using an optimized and cost-effective experimental setup designed and assembled, including a specially developed and manufactured sensing element. The results demonstrate that the fiber-optic DAS effectively captures the acoustic signals of the Madagascar hissing cockroach (*Gromphadorhina portentosa*), including both the mechanical interactions of the insect with the optical fiber and the characteristic hissing sound produced in response to external stimulation.

**Keywords:** fiber optic sensor; distributed acoustic sensor; DAS; insect; hissing cockroach; *Gromphadorhina portentosa*; agriculture

## 1. Introduction

In the late 20th and early 21st centuries, distributed fiber-optic sensors became increasingly important in modern science and technology [1–4]. Distributed Acoustic Sensors (DAS), first introduced in 1977, have proven to be highly effective in applications such as oil and gas exploration, transportation, and processing [5–7], as well as in structural health monitoring [8–12] and perimeter security [13,14]. While DAS technology has primarily been developed for these industries, ongoing advancements have led to the emergence of new potential applications.

For instance, DAS has been reported as an indispensable tool for the early detection of red palm weevil infestations [15], outperforming traditional methods such as trained dogs, X-ray imaging, and visual inspection in terms of efficiency and scalability. Additionally, DAS technology holds great promise for monitoring individual plants on large-scale plantations [16], tracking animals and their movement patterns [17], and detecting environmental events [18,19]. A common characteristic of these emerging fields—spanning biology, ecology, agriculture, flaw detection, voice recognition, and sound engineering—is that their budgets are typically smaller than those in industries such as oil and gas, defense, and engineering. Furthermore, the sensor characteristics required for these new applications, such as acoustic frequency range, sensitivity, and sensing element length, may differ significantly from those used in traditional DAS applications. As a result, despite its unique capabilities, DAS has not yet achieved widespread adoption in biology and agriculture, even though no comparable alternative exists for certain applications.

One of the most promising real-world implementations of DAS technology is its use for the early detection of red palm weevil larvae, as demonstrated by researchers in Saudi Arabia [15]. Their setup (Figure 1) consists of a narrowband laser source, an acousto-optic modulator, erbium-doped fiber optical amplifiers, optical circulators and filters, a photodetector, a data acquisition unit, and 1 km of standard single-mode telecommunication optical fiber.

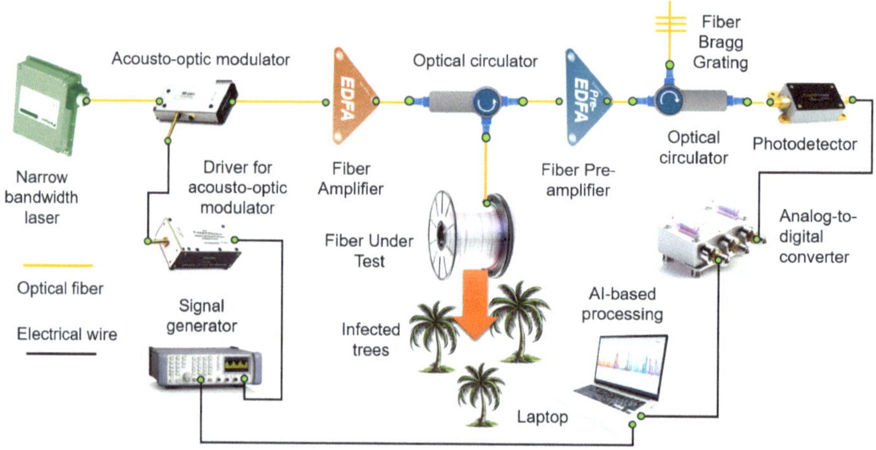


**Figure 1.** Distributed acoustic sensor (DAS) setup for the early detection of red palm weevil infestations. The system's key components include a laser source, acousto-optic modulator, erbium-doped fiber amplifiers, optical circulators and filters, a photodetector, and a data acquisition unit. Adapted from [15].

Each tested tree was wound with 5 m of optical fiber (Figure 2). Prior to field deployment, the acoustic emissions of palm tree trunks infested with weevil larvae were studied under laboratory conditions using the same DAS setup. A pattern of sounds emitted by the insect was recorded at each stage of its development, allowing for the characterization of infestation-related acoustic signatures.

The signals recorded in plantation conditions were processed using a neural network, achieving an infestation detection accuracy of 97%. This demonstrates that a fiber-optic distributed acoustic sensor (DAS) is a feasible and highly effective solution for detecting tree pests at an early stage, helping to mitigate tree destruction caused by insect larvae infestations.

Traditional detection methods, such as visual inspections, pheromone traps, and spot checks, often fail to identify infestations in their early stages. Other potential early detection methods, such as X-ray imaging or trained detection dogs, are impractical for large-scale plantations, such as those of oil palms, due to scalability limitations. Acoustic monitoring has gained popularity as a technique for detecting insect activity within wood, as it enables the identification of movement and feeding sounds. Research efforts have explored the use of contact microphones and vibration sensors for this purpose. However, deploying traditional microphones to monitor each tree individually would be economically unfeasible compared to a fiber-optic sensor.

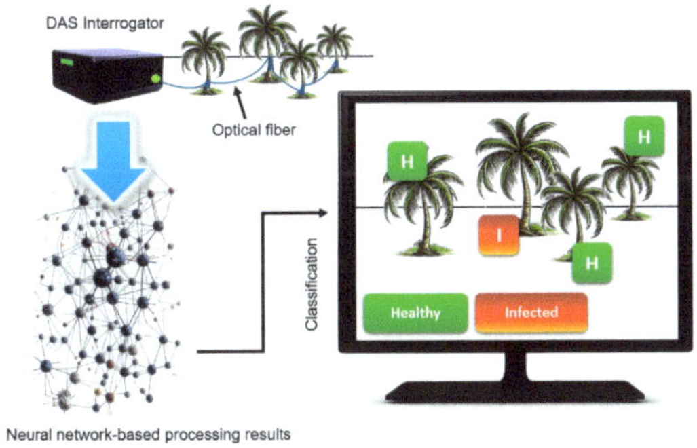


**Figure 2.** Experimental setup for the early detection of red palm weevil larvae in plantation conditions. The scheme illustrates the arrangement of optical fiber on trees and the deployment of the distributed acoustic sensor (DAS) system for monitoring acoustic emissions associated with insect activity. Adapted from [15].

Fiber-optic sensors and distributed acoustic sensing systems based on optical fibers are becoming increasingly popular in various fields, including seismic monitoring, infrastructure diagnostics, and security. One key advantage of DAS technology is its ability to monitor entire plantations or farms using a single sensor or a small number of sensing elements. However, the application of DAS for detecting insect activity in forest ecosystems remains a relatively new and promising research area. This is due to the fact that the sensor requirements in biology and agriculture differ significantly from those used in fields such as mineral extraction and perimeter security.

Currently, no reports have been found on the detection of individual insects using fiber-optic DAS through airborne acoustic signals or direct contact between the insect and the sensing element. Previous studies, such as the one described above, have focused on detecting infestations by groups of larvae through their acoustic emissions within the wood. Therefore, the objective of this study is to investigate the feasibility of registering the acoustic signature of a single insect using a fiber-optic DAS of a "classical" design outside of a woody environment.

This research will help evaluate the accuracy of the experimental setup and its parameters, as well as assess the potential of fiber-optic DAS for detecting insects through their acoustic emissions, not only in large colonies within sound-conducting environments but also in more diverse and less conventional settings.

The Madagascar hissing cockroach (*Gromphadorhina portentosa*) was selected as the subject of this study for several reasons:

- Large size. The Madagascar hissing cockroach is one of the largest cockroach species in the world, with some individuals reaching up to 10 cm in length;
- Ability to produce hissing sounds. This species uses hissing as its primary defense mechanism against insectivorous predators;
- Lack of wings. Unlike many other cockroach species, Madagascar hissing cockroaches have lost the ability to fly, as wings are unnecessary for burrowing in forest litter. Instead, they have developed a thick and durable chitinous exoskeleton;
- Suitability for experiments. Due to their large size and wingless nature, Madagascar hissing cockroaches have been widely used in scientific research. For in-

stance, Japanese scientists selected this species for experiments in developing remote-controlled cyborg cockroaches, as they are large enough to carry equipment without the interference of wings [20].

## 2. Materials and Methods

### 2.1. The Madagascar Hissing Cockroach

The Madagascar hissing cockroach (*Gromphadorhina portentosa*) (Figure 3) is a large tropical insect that lives on the trunks and branches of trees and bushes and feeds on green shoots and fruits. Individuals of this species move slowly and have no wings. When sensing danger, the cockroach freezes and makes a loud hiss. The sound is produced by a sharp contraction of the abdomen, forcing air through the spiracles. Sound signals are also used for intraspecific communication (for example, the fight between males for females) [21]. This cockroach is one of the largest representatives of the cockroach family in the world. Its body length is, on average, about 4–5 cm, but can reach 10 cm. The insect weighs up to 40 g.

**Figure 3.** A male *Gromphadorhina portentosa* (Madagascar hissing cockroach) was used in the study. The insect's acoustic emissions and movements were recorded using a fiber-optic distributed acoustic sensor (DAS).

The acoustic signature of the cockroach's hiss spans a broad frequency range (0.8–4.6 kHz), characterized by blurred edges, narrow peaks concentrated in the center, and compactly arranged triangular expansions. In some cases, two or three additional peaks appear, particularly along the high-frequency edge (Figure 4) [22].

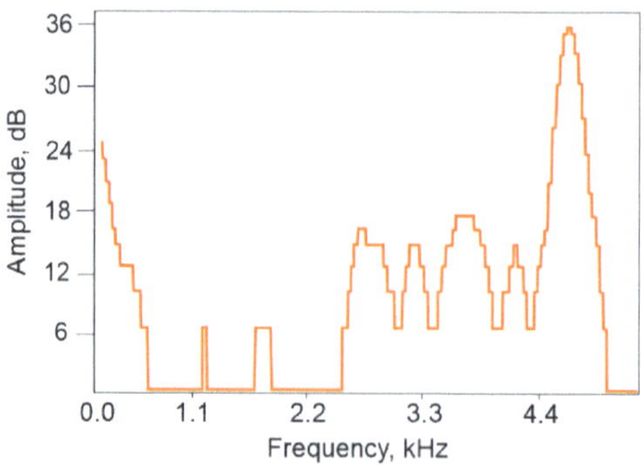


**Figure 4.** Amplitude-frequency dependence of *Gromphadorhina portentosa* hiss at a specific point in time. The spectrum illustrates the frequency components of the insect's acoustic emission. Adapted from [22].

The spectrum of the cockroach hiss slightly changes within its duration. Moreover, it depends on the size, age, and other features of a cockroach individual. Nevertheless, the spectrum of *Gromphadorhina portentosa* hiss is quite repeatable, especially in terms of main harmonics such as 4.9 kHz (Figure 4).

### *2.2. Distributed Acoustic Sensing*

To better understand the principles underlying the distributed acoustic sensor used in this study, this section presents the theoretical foundation of phase-sensitive optical time-domain reflectometry (Φ-OTDR). Distributed phase-sensitive optical time-domain reflectometry utilizes coherent light backscattered from an optical fiber to detect and analyze perturbations induced by external physical fields. This technique enables real-time, distributed acoustic sensing by continuously monitoring changes in the backscattered signal caused by external influences along the fiber length.

The Φ-OTDR technique enables distributed acoustic sensing by utilizing coherent Rayleigh backscattering from an optical fiber. This method is employed for detecting and analyzing perturbations induced by external physical fields along the fiber length. It involves injecting probe pulses from a highly coherent laser source into the sensing fiber, where each pulse interacts with inhomogeneities in the fiber's refractive index. These interactions result in a scattered signal that is recorded at the fiber input and analyzed to localize and quantify external forces acting on the fiber.

In a Φ-OTDR system similar to that used in the experiments, a probe pulse is injected into an optical fiber that exhibits low linear losses $\alpha(x)$ distributed along its length. At the initial moment $t = 0$, the leading edge of the pulse is positioned at the fiber input $x = 0$. The probe pulse is assumed to have a rectangular shape, with $I_1$ the peak pulse intensity and $T$ the pulse duration. As the pulse propagates through the fiber, it undergoes Rayleigh backscattering due to multiple reflections from microscopic refractive index variations that were frozen into the fiber during manufacturing. These variations can be modeled as Rayleigh reflection centers with a relative reflectivity $\sim \rho(x)$, randomly distributed along the fiber length.

At a given time $t_0 + T$, the leading edge of the pulse reaches the fiber position

$$x_1 = \frac{c}{n}(t_0 + T), \tag{1}$$

while the trailing edge is at

$$x_0 = \frac{c}{n} t_0. \tag{2}$$

The light backscattered by the distributed reflection centers creates an interference pattern within the pulse. The backscattered intensity measured at the pulse's trailing edge is denoted as $I_2(t_0, x_0)$. The complex amplitude $E_2(t_0 + T, x_0)$ of the electric field associated with $I_2(t_0 + T, x_0)$ is given by the superposition of all backscattered fields generated by the pulse at earlier times and arriving at $x_0$ precisely at $t_0 + T$. The backscattering events that contribute to this superposition occur over a time interval between $t_0 + \frac{T}{2}$ and $t_0 + T$. Thus, the total backscattered electric field is

$$E_2(t_0 + T, x_0) = \int_0^{\frac{T}{2}} E_1\left(t_0 + T - \tau, x_0 + \frac{c}{n}\tau\right)\rho\left(x_0 + \frac{c}{n}\tau\right)d\tau. \tag{3}$$

Here $E_1(t, x)$ represents the complex amplitude of the probe pulse, modeled as a segment of a monochromatic optical wave at frequency

$$\omega_0 = 2\pi f_0 = \frac{2\pi}{\lambda_0}\frac{c}{n}, \tag{4}$$

where $\lambda_0$ is the laser wavelength. The pulse propagates through the single-mode optical fiber with a group velocity of $\frac{c}{n}$, following

$$E_1(t,x) = \sqrt{\gamma(x)I_1}\exp\left\{i\omega_0\left(t - \frac{n}{c}x\right)\right\}\text{for } t \in \left[\frac{n}{c}x;\ \frac{n}{c}x + T\right],\ x \in [0,\ L], \tag{5}$$

where the integrated loss factor is expressed as

$$\gamma(x) = \exp\left(-\int_0^x \alpha(x')dx'\right). \tag{6}$$

Substituting this into the equation for $E_2(t_0 + T, x_0)$ results in the expressions for the backscattered signal amplitude and intensity at the probe pulse trailing edge:

$$E_2(t_0 + T, x_0) = \sqrt{\gamma(x_0)I_1}\exp\left\{i\left(\omega_0\left(t_0 + T - \frac{n}{c}x_0\right)\right)\right\}\int_0^{\frac{T}{2}} \exp\{-i2\omega_0\tau\}\rho\left(x_0 + \frac{c}{n}\tau\right)d\tau, \tag{7}$$

$$I_2(t_0 + T, x_0) = \gamma(x_0)I_1\int_0^{\frac{T}{2}}\int_0^{\frac{T}{2}} \exp\{-i2\omega_0(\tau - \tau')\}\rho\left(x_0 + \frac{c}{n}\tau\right)\rho^*\left(x_0 + \frac{c}{n}\tau'\right)d\tau\, d\tau'. \tag{8}$$

Once formed, these amplitude and intensity quantities propagate to the fiber input without modification of the interference structure just exhibiting linear losses.

The detected intensity $I_D(t_D)$, recorded at the fiber input at $t_D = (2t_0 + T)$, depends on the Rayleigh backscattering from a fiber segment $\left[x_D - \frac{\Delta}{2}, x_D + \frac{\Delta}{2}\right]$, where

$$x_D = \frac{c}{2n}\left(t_D - \frac{T}{2}\right) \tag{9}$$

is the mean interval point, and the probe pulse duration $T$ sets the spatial resolution

$$\Delta = \frac{c}{n}\frac{T}{2} \tag{10}$$

of the described sensing technique.

The intensity detected at the fiber input at the moment $t_D$ can be expressed as

$$I_D(t_D) = 2I_1\gamma^2(x_D)\int_\Lambda^{\frac{T}{2}} \text{Re}\{\exp\{-i2\omega_0\Delta\tau\}F(\Delta\tau,\ x_D)\}d\Delta\tau + I_1\Lambda F(0,\ x_D), \tag{11}$$

where the first and second terms describe the coherent and incoherent parts of the Rayleigh scattering process.

The function

$$F(\Delta\tau,\ x_D) = \int_{-\frac{T}{4}+\frac{\Delta\tau}{2}}^{\frac{T}{4}-\frac{\Delta\tau}{2}} \rho\left(x_D + \frac{c}{n}\left(\tau + \frac{\Delta\tau}{2}\right)\right)\rho^*\left(x_D + \frac{c}{n}\left(\tau - \frac{\Delta\tau}{2}\right)\right)d\tau \tag{12}$$

is the autocorrelation function that characterizes the distribution of Rayleigh backscattering centers in an undisturbed fiber. Importantly, for any given fiber, this function is specific and is commonly considered the "fingerprint" of the Rayleigh sensor.

One can see that the coherent part of the detected intensity signal $I_D(t_D)$ is very sensitive to fluctuations of the probe laser frequency $\omega_0$ and to changes in mutual positions of the reflection centers $\sim \rho(x)$. This inherent sensitivity underlies the use of the $\Phi$-OTDR for distributed sensing. The external perturbations affect the mutual positions of the reflection centers $\sim \rho(x)$ that, in turn, affect the signal $I_D(t_D)$. To measure these perturbations, the fiber is interrogated by periodic coherent probe pulses with the period $W$. When sensing fiber is perturbated, the distribution of the reflection centers $\sim \rho(x)$ slightly varies. Such perturbation does not affect the incoherent part of the detected signal but does affect its coherent part. Mathematically, it can be expressed as a substitution of

$$\Delta\tau \to (1 + s(x_D))\Delta\tau \tag{13}$$

into the function $F(\Delta\tau,\ x_D)$, where $s(x_D) << 1$ ($\sim 10^{-6}$) is a small variable parameter evaluating the effect of the external force on the fiber at point $x_D$. As the external force varies in time, it modulates $s(x_D, t)$. Therefore, the signals $I_D(t_{D,m})$ detected from different consequent probe pulses express the modulation

$$\delta I_D(t_{D,m}) \sim s(x_{D,}t_{D,m}), \tag{14}$$

where $m = 0,\ 1,\ 2...$ is the probe pulse number. The sensitivity is expressed as

$$K(x_D) = \frac{dI_D}{ds}(t_D) = -\frac{dI_D}{d\omega_0}(t_D)\omega_0 = -4I_1\gamma^2(x_D)\omega_0 \int_{\Lambda}^{\frac{T}{2}} \Delta\tau \mathrm{Im}\{\exp\{-i2\omega_0\Delta\tau\}F(\Delta\tau,\ x_D)\}d\Delta\tau. \tag{15}$$

Although the sensitivity $K(x_D)$ of the signal modulation $\delta I_D(t_{D,m})$ to the modulation of $s(x_D t_{D,m})$ depends on the fiber point $x_D$ one can see that when an external force modulates $s(x_D t_{D,m})$ periodically, the recorded signal $I_D(t_{D,m})$ also exhibits periodic modulation. This mechanism enables the extraction of acoustic spectra applied to the fiber by analyzing the modulation patterns $I_D(t_{D,m})$. It is clear that to this effect, the external field modulation frequency should be much lower than the repetition rate

$$R = \frac{1}{W} \tag{16}$$

of the pulse interrogation. In this case processing of the signal $I_D(t_{D,m})$ for different points $x_D$ along the fiber enables distributed reconstruction of the acoustic spectra applied to the sensing fiber.

Thus, the described mechanism explains the ability of Φ-OTDR to detect weak acoustic signals over long distances, making it a promising tool for distributed acoustic sensing applications. Given its capability to detect weak acoustic signals across long distances, Φ-OTDR is an ideal technique for applications such as environmental monitoring, structural health assessment, and biological studies. In the present work, it has been applied to analyze the acoustic emissions of a single Madagascar hissing cockroach, which underscores the system's high sensitivity and resolution. This demonstrates the feasibility of distributed fiber-optic acoustic sensors in bioacoustics research, extending their applications beyond conventional industrial and security domains.

### 2.3. Experimental Setup

A laboratory-based fiber-optic distributed acoustic sensor (DAS) was used for the experiments. Its operating principle is based on Φ-OTDR with direct detection (without phase extraction). This design enables the recording of acoustic signals within the target frequency range (200–5000 Hz) with acceptable quality. Additionally, the use of a pulsed current-driven laser, along with the absence of an optical modulator and fiber Bragg gratings, reduces both setup costs and sensitivity to environmental disturbances. The implementation of a low-cost system design [23–25] is particularly important for expanding the use of distributed fiber-optic sensors in biology and agriculture.

The system utilized a GSPF 053 arbitrary waveform generator (Rudnev-Shilyaev, LLC, Moscow, Russia) to generate current pulses with a duration of 40 ns and a repetition period of 11 μs. These pulses were fed into a pulsed laser source (DL-BF12-CLS101B-S1550, DenseLight Semiconductors, Singapore), which had an output power of 10 mW, a wavelength of 1550 nm, and a bandwidth of 5 kHz. The laser converted the electrical pulses into optical pulses, which were subsequently amplified to 30.5 mW by an erbium-doped fiber amplifier (AEDFA-23-M-FA, Amonics Ltd., Hong Kong, China). The amplified pulses were then injected into the sensing element—Corning SMF-28 optical fiber—through an optical circulator (Advanced Fiber Resources, Ltd., Zhuhai, China).

Within the fiber, due to inherent refractive index inhomogeneities [26], light pulses underwent Rayleigh scattering. Given the narrow laser bandwidth, the scattered light components within each pulse length interfered with one another [27]. The time-domain amplitude of the resulting backscattered signal depended on the distribution, quantity, and size of inhomogeneities along the fiber. A portion of this backscattered light traveled back to the fiber's input end, where it was redirected via an optical circulator to an Amonics AEDFA-35-M-PA erbium-doped fiber preamplifier (Amonics Ltd., Hong Kong, China). The preamplified backscattered signal, reaching a power of 1.99 mW, was detected by a Thorlabs PDA 10D-EC photodetector (Thorlabs Inc., Newton, NJ, USA) and then digitized using a La-n1usb analog-to-digital converter (ADC) (Rudnev-Shilyaev LLC, Moscow, Russia) before being processed on a personal computer (Figure 5).

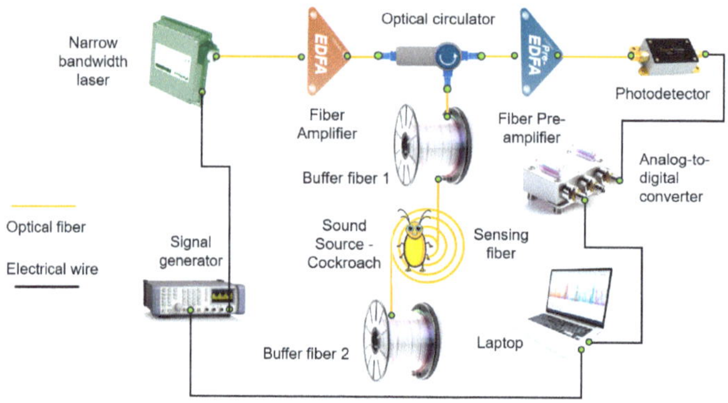


**Figure 5.** The distributed acoustic sensor (DAS) setup used in the experiment. In contrast to setup from [15], it does not require the use of acousto-optic modulator, its driver, and a fiber Bragg grating filter.

The ADC operated at a sampling frequency of 500 MHz, corresponding to a spatial resolution of approximately 0.2 m, and featured a buffer capacity that allowed continuous acquisition of up to 762 traces. The unprocessed output signal represented the dependence of scattering intensity on time for each point along the optical fiber. When an acoustic signal was present, it modulated both the position and magnitude of refractive index inhomogeneities within the optical fiber at that location, with the modulation occurring at the same frequency as the acoustic signal itself. Consequently, the Rayleigh backscattered signal from this location was also modulated at the same frequency. By utilizing knowledge of the speed of light in the given optical fiber and applying a fast Fourier transform (FFT), it was possible to determine the location, amplitude, and frequency of the acoustic signal affecting the fiber. The design of the sensing element (Figure 6) plays a crucial role in optimizing signal detection and requires a detailed description.

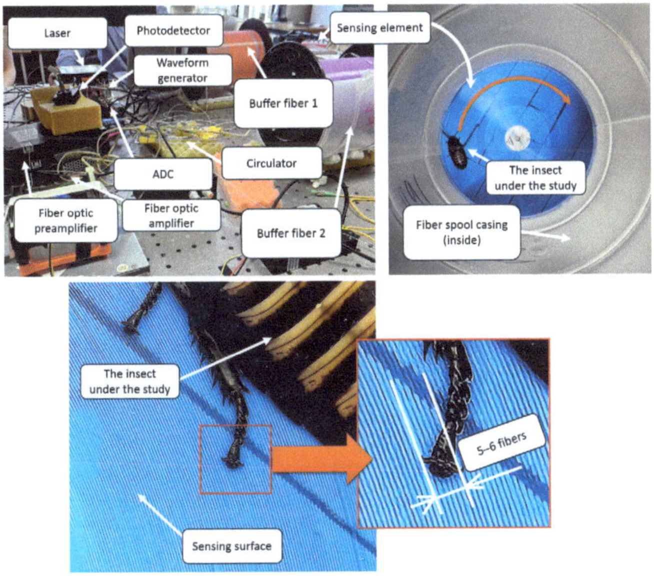


**Figure 6.** Photograph of the experimental setup, showing the sensing element and the *Gromphadorhina portentosa* (Madagascar hissing cockroach) used in the study. The fiber-optic sensor was configured to detect and analyze the insect's acoustic emissions and movements.

For high-quality signal detection using a distributed acoustic sensor (DAS), the optical fiber must be in direct physical contact with the insect under study. Since the objective was not only to record the acoustic signals emitted by the insect but also to localize its position within a defined area, it was essential to ensure that physical contact between the insect and the fiber-optic sensor could occur at any point within the sensing region. This was achieved by using a standard single-mode optical fiber with an acrylate coating, and an outer diameter of 250 μm was arranged in a spiral configuration on a flat surface. Thus, the fiber used was bare (not cabled), and the expected sensitivity of the system was relatively high. The method for forming the fiber-optic spiral was as follows. A 30 cm diameter vinyl record was used as the sensing platform. This choice allowed the setup to be installed on a music player flywheel, enabling controlled rotation of the record if needed. To securely attach the optical fiber to the disk, a double-sided adhesive tape with a polypropylene base (50 mm × 10 m) was used. This tape featured an adhesive layer on both sides, providing a strong yet manageable connection between the vinyl record and the optical fiber. The attachment process involved fully covering the surface of the vinyl record with one side of the adhesive tape. The protective layer on the exposed adhesive side was then gradually removed as the optical fiber was affixed in concentric circles, starting from the center of the record and moving outward toward the perimeter. The most labor-intensive step was ensuring a secure and precise placement of the fiber at the contact points with both the adhesive tape and the previously adhered fiber loops. A rod tool with a smooth convex end was used to press the fiber against the adhesive at a 45-degree angle to the record's surface to achieve this. This approach allowed one hand to fix the fiber's contact point while the other hand rotated the record, ensuring an even and tightly packed spiral pattern. The individual fiber loops were positioned as closely as possible to maximize the sensing area. Approximately 200 m of optical fiber were successfully placed onto the vinyl record using this method, forming a compact and efficient sensing element.

The sensing element was fusion-spliced at both ends to two buffer sections of single-mode fiber wound on transport spools, each approximately 400 m long. This was carried out to prevent dead zones in the DAS system, which are caused by strong signal reflections at both the input and output, from interfering with the studied area. Additionally, to minimize light reflections at the fiber output, the fiber end was immersed in a container filled with glycerin, effectively reducing unwanted optical feedback. The sensor's perimeter was enclosed within the casing of the optical fiber transport spool to restrict the insect's movement during the experiment. This ensured that the insect remained within the designated sensing area, allowing for accurate measurements. For precise localization of the insect relative to the sensing element, proper data processing was required. The dense spiral arrangement of the optical fiber on the vinyl record, as described earlier, follows the geometry of an Archimedean spiral (Figure 7). A key property of this spiral is that any radial line drawn from the center intersects successive turns of the spiral at equidistant points, with this spacing referred to as the spiral pitch. This property plays a crucial role in interpreting the insect's position within the sensing area.

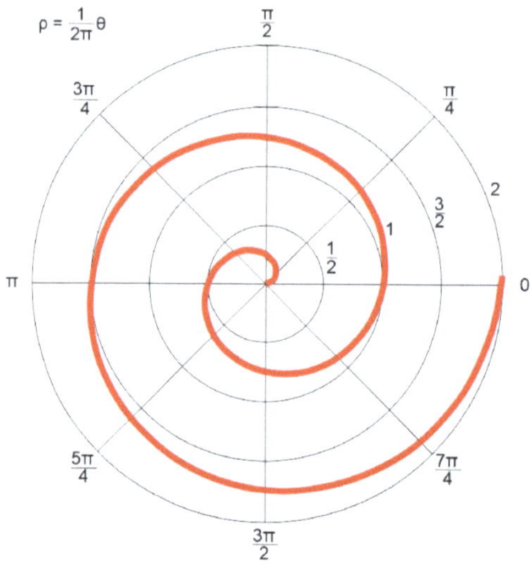


**Figure 7.** Archimedean spiral in the polar coordinate system. The spiral represents the layout of the fiber-optic sensing element used in the experiment, where each successive turn is equidistant from the previous one, enabling accurate localization of acoustic events.

The equation of an Archimedean spiral in the polar coordinate system is given by:

$$\rho = \frac{a}{2\pi}\varphi, \tag{17}$$

where $\rho$ is the radius, $\varphi$ is the angle in radians, $a$ is the spiral path, representing the radial displacement per full revolution. Since the developed data processing program outputs the image as colored pixels, it is more convenient to express Equation (17) in parametric form:

$$\begin{Bmatrix} x = \rho\cos(\varphi) \\ y = \rho\sin(\varphi) \end{Bmatrix}, \tag{18}$$

where $x$ and $y$ represent the pixel coordinates in the Cartesian plane. When constructing the sensing element model, the initial and final values of the spiral radius and its pitch are defined in the program. The spiral is generated by incrementing the angle $\varphi$ in small steps (0.01 rad), causing a gradual increase in the radius $\rho$ until it reaches its final value. Simultaneously, the arc length $l$ of the optical fiber laid in the spiral accumulates as:

$$d\rho = \frac{a}{2\pi}d\varphi. \tag{19}$$

$$dl = \frac{a}{2\pi}\sqrt{1+\varphi^2}d\varphi. \tag{20}$$

Each arc segment of the spiral is assigned an RGB color value corresponding to the normalized amplitude of the acoustic signal recorded at that specific location. The color of the $i$-th section of the spiral is expressed as:

$$\{255A[i];\, 0;\, 255(1 - A[i])\}, \tag{21}$$

where $A$ represents a set of normalized signal amplitude values recorded by the fiber sensor. Thus, when visualizing the data, the program colors sections of the fiber spiral where the acoustic signal amplitude is maximum in red, while sections with minimal amplitude are colored blue.

## 3. Results and Discussion

The insect under study, a male Madagascar hissing cockroach measuring 65 mm in length and weighing 27 g, was placed onto the sensor element. Initially, the cockroach exhibited minimal movement; however, after an adaptation period of approximately 10 min, it began to move slowly across the sensor surface. The observed movement trajectory was primarily concentrated near the outer radius of the plate. This behavior suggests that in an unfamiliar environment with an unstable surface for movement, the insect was drawn toward potential escape routes or attempted to use the inner surface of the spool casing as a shelter. The fiber trace was visualized in real time on the system display. It is important to note that slow movements (approximately up to 1 cm/s) were barely visible on the spatial scan, likely due to the lower amplitude of the resulting acoustic signal. To analyze the cockroach's response to external stimuli, hissing and rapid movements were induced multiple times during the experiment. The resulting signal, recorded and processed according to the previously described methodology, is presented in Figure 8.

The output data from distributed acoustic sensing (DAS) systems are most effectively represented in frequency-distance-amplitude coordinates. This visualization method allows a single image to encapsulate all key characteristics of the acoustic signals detected by the sensor. In the low-frequency region, it is evident that the entire sensing element exhibits an elevated noise background compared to the buffer spools. This background noise is independent of the presence of the insect on the sensing element. The most likely explanation for this phenomenon is that the current sensor design functions as a highly responsive membrane, efficiently transmitting vibrations from the optical table, including minor oscillations from auxiliary equipment. Despite this, the background noise did not interfere with obtaining the desired results. On the contrary, its localized presence in the low-frequency region allows for a preliminary estimation of the insect's position relative to the entire sensing element. Figure 8 confirms that the insect remained closer to the outer edge of the sensing element. This conclusion is supported by the fact that the first buffer spool was fusion-spliced to the center of the spiral-laid fiber, while the second spool was connected to its outermost part. Additionally, the recorded data show interference propagating along the entire fiber length, including the buffer spools, at approximately 5000 Hz. This interference is likely of electrical origin and will not be considered in further analysis. Importantly, its presence does not obstruct the detection of the insect's acoustic signature, as the useful signal level remains higher than the average noise level in the affected region. Figure 9a shows the acoustic spectrum of the studied insect hiss signal inherent in the spatial cross-section at a distance of 542.2 m, as well as the spectrum of external noise recorded at this location in the absence of an insect. Notably, the signal shape in this frequency range closely resembles previously reported data (see Figure 4). The dominant frequency peaks observed at 100, 1500, 3300, and 4500 Hz align well with the spectrogram of the insect's hiss, which was independently recorded using a smartphone (SM- A536E/DS, Samsung Electronics Co., Ltd., Suwon, Republic of Korea) during the experiment (Figure 9b).

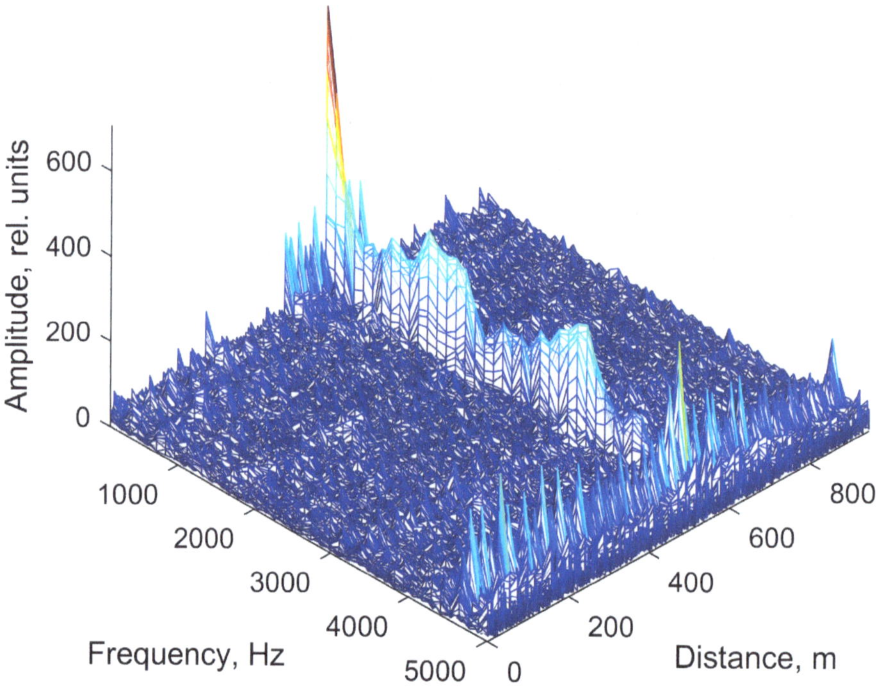


**Figure 8.** Hissing and rapid movement of a *Gromphadorhina portentosa* (Madagascar hissing cockroach) recorded using the distributed acoustic sensor (DAS). The figure illustrates the acoustic signals generated by the insect's hissing and physical interaction with the fiber-optic sensing element.

The frequency spectrum recorded by the fiber-optic sensor contains a limited number of data points. This is due to the relatively small number of available time-domain samples, which are recorded sequentially and continuously. The primary constraint stems from the limited buffer capacity of the ADC, making it a hardware limitation inherent to the current system design. This limitation could be mitigated by using an ADC with a larger buffer and improved specifications. However, it is important to consider that upgrading the ADC would increase the overall cost of the setup. Despite this constraint, the system remains capable of recording and distinguishing the most characteristic frequencies of the acoustic signals generated by the insect (Figure 9). Additionally, the time-domain representation of the insect's acoustic signal, presented in Figure 10, may provide further insights into the temporal dynamics of its sound production. Considering it, together with the spectrum analysis of this signal and background noise, one can distinguish between the frequency components generated by the insect and those resulting from environmental or equipment-related noise.

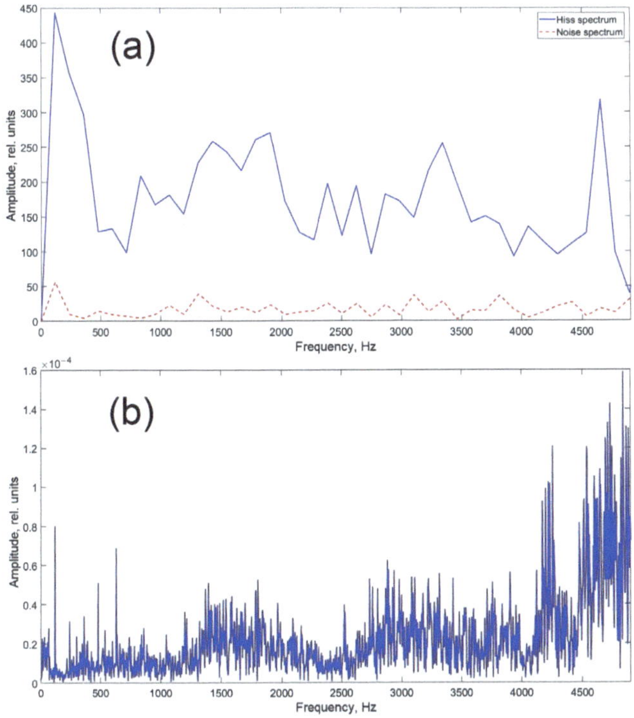


**Figure 9.** (**a**) Frequency spectra of the hiss and external noise recorded at a distance of 542.2 m along the fiber, showing the acoustic response detected by the distributed acoustic sensor (DAS); (**b**) spectrogram of the *Gromphadorhina portentosa* hiss recorded using a smartphone, serving as a reference for comparison.

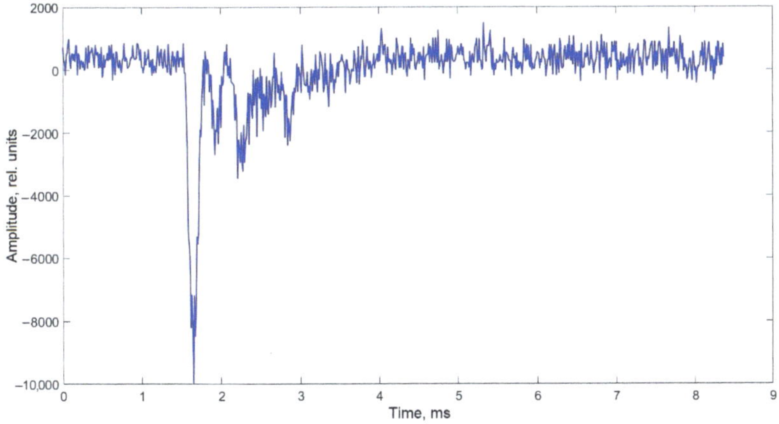


**Figure 10.** Time-domain representation of the signal recorded at 542.2 m along the fiber (corresponding frequency-domain representation can be found in Figure 9a). The plot illustrates the temporal characteristics of the acoustic signal detected by the distributed acoustic sensor (DAS), corresponding to the hissing and movement of *Gromphadorhina portentosa*.

The dip at the beginning of the signal in Figure 10, as well as the low-frequency peak around 100 Hz observed in Figures 8 and 9, is likely not directly related to the hissing sound produced by the cockroach. Instead, it is presumably associated with the insect's rapid movement, which typically coincides with hissing. This suggests that the recorded signal includes components resulting from the physical contact of the insect's legs with the optical fiber rather than purely airborne acoustic emissions. As previously mentioned, the elevated low-frequency noise level detected by the sensing element facilitated signal localization within the sensor's output data (Figure 11) and provided an approximate estimation of the cockroach's position (Figure 8).

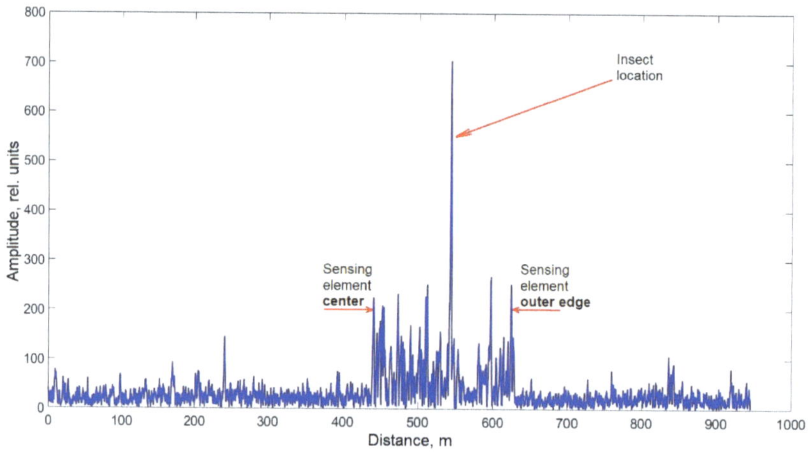


**Figure 11.** Frequency-domain cross-section of the data from Figure 8 at 119.3 Hz. Red arrows with captions indicate the location of the sensing element along the setup fiber optic part and studied insect on it.

The sensing element exhibited higher noise sensitivity compared to the buffered fiber, which not only facilitated precise detection of acoustic signals but also allowed for an accurate determination of the fiber length laid in a spiral on the vinyl record: 192.25 m. This information was crucial for calibrating the program responsible for localizing the sound source along the spiral. The output data from the localization program, which maps the detected effect relative to the sensing element, are presented in Figure 12.

The fiber transport spool casing, which restricted the insect's movement, had a diameter of approximately 23 cm. Figure 12 confirms that the system successfully detected and localized the cockroach's movements and sounds near the casing. However, the localization accuracy was inherently limited to the diameter of the circular region within which the impact was registered. The circumference of this circular region is approximately 0.6 m, which is significantly shorter than the 8 m light pulse length generated by the laser. This suggests that, in this case, the accuracy of impact localization is primarily limited by the ADC sampling rate rather than the pulse length. As previously noted, the ADC sampling rate used in this setup results in a spatial resolution of 20 cm. Consequently, the acoustic signal produced by the insect was captured in only three ADC samples, limiting the precision of its localization. The cockroach was not harmed during the experiments. After the study, it was returned to an optimal environment.

Considering the length of the fiber segment within which the cockroach's hissing was detected (0.6 m) and assuming the insect acts as a pointwise acoustic source, one can conclude that the system's localization precision is roughly 0.3 m along the fiber length. This value is in reasonable agreement with the system's spatial resolution of 0.2 m. Localization precision in polar coordinates can also be considered. Thus, the radial localization precision is significantly high—comparable to the diameter of a single spiral layer (approximately 250–300 µm)—while the angular localization error is relatively large, up to 180 degrees, due to the symmetric nature of the spiral layout.

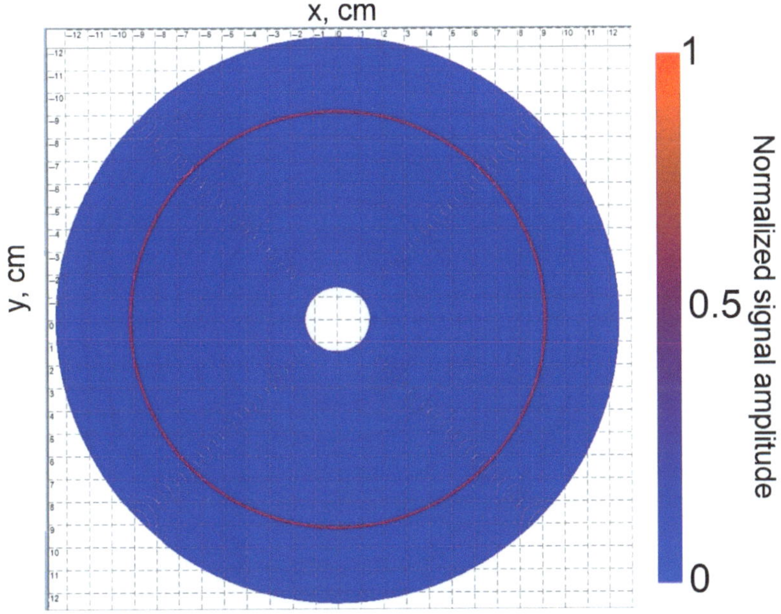


**Figure 12.** Localization of the *Gromphadorhina portentosa* (Madagascar hissing cockroach) acoustic signal relative to the sensing element. The distributed acoustic sensor (DAS) detected and mapped the insect's hissing and movement, demonstrating the system's capability to determine the sound source position.

## 4. Conclusions

Despite the aforementioned limitations, primarily related to the ADC characteristics and signal noise, the fiber-optic sensor and the proposed sensing element design successfully demonstrated the ability to record the acoustic signature of a single insect. Future research will focus on determining which parameter contributes more significantly to the inaccuracy in localizing the sound source and estimating the impact length on the fiber. The current laser source is incapable of producing light pulses shorter than 20 ns, while the ADC sampling rate could potentially be increased to 2 GHz. However, this would significantly reduce the volume of continuously recorded data, a limitation that could be mitigated by reducing the length of the optical fiber connected to the sensor. The successful recording of an acoustic signal from a single insect marks an important step in the development of new applications for fiber-optic acoustic sensors. Previous studies have either recorded entire insect colonies or individual subjects of considerably larger size and weight, which inherently emit higher-intensity acoustic vibrations. To extend this technique to even smaller organisms than the Madagascar hissing cockroach, enhancements in system sensitivity and signal processing will be necessary. International research efforts suggest that neural networks may play a key role in optimizing signal detection and analysis. Additionally, an important hardware modification would be to implement a hybrid detection scheme, enabling the extraction of optical phase information instead of merely measuring signal intensity.

The approach tested in this study, utilizing the Madagascar hissing cockroach, serves as a foundation for future research and the development of methods for species identification and invasive species detection. While previous studies have monitored insects acoustically, they have relied on conventional hardware that records sound at a single point where the insect is localized or fixed. In contrast, the fiber-optic sensor in this study not only records sound waves emitted by the insect but also localizes them on a plane. This work paves the way for distributed monitoring of insect habitats created by humans. Beyond agricultural plantations vulnerable to pest infestations [28], potential applications include beehive monitoring. The global decline in bee populations is an increasingly urgent issue, and recent studies have begun investigating bee behavior through acoustic analysis [29]. The approach demonstrated in this study could significantly enhance the functionality of such research, contributing to the broader field of bioacoustics and ecological monitoring.

Future research will also focus on signal processing and impact classification. Presumably, two main challenges must be addressed to achieve this goal. The first is the accurate classification of interference caused by equipment operation or other background sources. This interference can be identified by analyzing signal spectra using Fourier and wavelet transforms, as well as by decomposing the signal into components using empirical mode decomposition (EMD) and variational mode decomposition (VMD). To enhance the signal-to-noise ratio, dynamic filtering techniques such as Frequency Domain Dynamic Averaging (FDDA) and Activation Function Dynamic Averaging (AFDA), developed and tested by our research group, will be applied. Once a satisfactory signal-to-noise ratio is achieved, neural network-based recognition methods will be employed. As demonstrated in studies on the red palm weevil, neural networks can be trained in controlled laboratory conditions and later applied in the field to detect and classify acoustic signatures of specific biological targets in the presence of diverse ambient noise.

**Author Contributions:** Conceptualization, Y.A.K.; methodology, A.G.V.; software, A.T.T., and E.T.T.; validation, A.G.V. and A.T.T.; formal analysis, A.T.T.; investigation, A.T.T.; resources, G.F.M. and A.G.V.; data curation, A.T.T.; writing—original draft preparation, A.T.T., Y.A.K., E.E.T., G.F.M. and A.G.V.; writing—review and editing, D.A.K., A.A.F. and Y.A.K.; visualization, A.T.T. and E.T.T.; supervision, Y.A.K. and A.A.F.; project administration, Y.A.K. and A.A.F.; funding acquisition, Y.A.K., A.T.T., D.A.K. and A.A.F. All authors have read and agreed to the published version of the manuscript.

**Funding:** Sections 1 and 3 were performed as a part of State Assignment No. 122031100058-3; Sections 2.3 and 4 were carried out within the framework of State Assignment No. 124020600009-2; Section 2.2 was supported by the Russian Science Foundation (Grant No. 23-79-30017).

**Data Availability Statement:** The data reported in this manuscript are available on request from the corresponding author.

**Conflicts of Interest:** The authors declare no conflicts of interest. The funders had no role in the design of the study, in the collection, analyses, or interpretation of data, in the writing of the manuscript, or in the decision to publish the results.

## References

1. Chernutsky, A.O.; Gritsenko, T.V.; Zhirnov, A.A.; Orekhov, I.O.; Sazonkin, S.G.; Khan, R.I.; Dvoretskiy, D.A.; Pnev, A.B. Ultrashort pulse laser based Raman DTS for midrange FUT high spatial resolution measurements. In Proceedings of the 2024 International Conference Laser Optics (ICLO), Saint-Petersburg, Russia, 1–5 July 2024; p. 334.
2. Gritsenko, T.V.; Chesnokov, G.Y.; Koshelev, K.I.; Khan, R.I.; Stepanov, K.V.; Valba, O.V.; Chernutsky, A.O.; Svelto, C.; Zhirnov, A.A.; Pnev, A.B.; et al. Optical Fiber Sensor for Real-Time Monitoring of Industrial Structures and Application to Urban Telecommunication Networks. In Proceedings of the 2023 IEEE International Conference on Metrology for eXtended Reality, Artificial Intelligence and Neural Engineering (MetroXRAINE), Milano, Italy, 25–27 October 2023; pp. 383–388.
3. Tkachenko, A.Y.; Lobach, I.A.; Kablukov, S.I. Coherent Optical Frequency Reflectometry Based on a Fiber Self-Scanning Laser: Current Status and Development Prospects. *Instrum. Exp. Tech.* **2023**, *66*, 730–736. [CrossRef]

4. Taranov, M.A.; Gorshkov, B.G.; Alekseev, A.E.; Konstantinov, Y.A.; Turov, A.T.; Barkov, F.L.; Wang, Z.; Zhao, Z.; Zan, M.S.D.; Kolesnichenko, E.V. Optical Reflectometry, Metrology, and Sensing. Present and Future. *Instrum. Exp. Tech.* **2023**, *66*, 713–729. [CrossRef]
5. Ashry, I.; Mao, Y.; Wang, B.; Hveding, F.; Bukhamsin, A.Y.; Ng, T.K.; Ooi, B.S. A Review of Distributed Fiber–Optic Sensing in the Oil and Gas Industry. *J. Light Technol.* **2022**, *40*, 1407–1431.
6. Ellmauthaler, A.; LeBlanc, M.; Bush, J.; Willis, M.E.; Maida, J.L.; Wilson, G.A. Real-time DAS VSP acquisition and processing on single-and multi-mode fibers. *IEEE Sens. J.* **2020**, *21*, 14847–14852. [CrossRef]
7. Ren, L.; Jiang, T.; Jia, Z.G.; Li, D.S.; Yuan, C.L.; Li, H.N. Pipeline corrosion and leakage monitoring based on the distributed optical fiber sensing technology. *Measurement* **2018**, *122*, 57–65.
8. Liang, S.; Zhang, C.; Lin, W.; Li, L.; Li, C.; Feng, X.; Lin, B. Fiber-optic intrinsic distributed acoustic emission sensor for large structure health monitoring. *Opt. Lett.* **2009**, *34*, 1858–1860.
9. Stepanov, K.V.; Zhirnov, A.A.; Sazonkin, S.G.; Pnev, A.B.; Bobrov, A.N.; Yagodnikov, D.A. Non-invasive acoustic monitoring of gas turbine units by fiber optic sensors. *Sensors* **2022**, *22*, 4781. [CrossRef]
10. Markvart, A.A.; Liokumovich, L.B.; Ushakov, N.A. Fiber optic SMS sensor for simultaneous measurement of strain and curvature. *Tech. Phys. Lett.* **2022**, *48*, 34.
11. Anoshkin, A.N.; Voronkov, A.A.; Kosheleva, N.A.; Matveenko, V.P.; Serovaev, G.S.; Spaskova, E.M.; Shardakov, I.N.; Shipunov, G.S. Measurement of inhomogeneous strain fields by fiber optic sensors embedded in a polymer composite material. *Mech. Solids* **2016**, *51*, 542–549. [CrossRef]
12. Lu, X.; Chruscicki, S.; Schukar, M.; Münzenberger, S.; Krebber, K. Application of intensity-based coherent optical time domain reflectometry to bridge monitoring. *Sensors* **2022**, *22*, 3434. [CrossRef]
13. Juarez, J.C.; Taylor, H.F. Field test of a distributed fiber-optic intrusion sensor system for long perimeters. *Appl. Opt.* **2007**, *46*, 1968–1971. [CrossRef] [PubMed]
14. Jabidin, H.Z.; Ibrahim, S.A.; Zan, M.S.D.; Musa, S.M.A.; Mansoor, A.; Ngo, H.Y.; Aziz, N.A.A.; Choo, K.Y.; Bakar, A.A.A.; Mokhtar, M.R.B.; et al. Review on Digital Signal Processing (DSP) Algorithm for Distributed Acoustic Sensing (DAS) for Ground Disturbance Detection. *Int. J. Integr. Eng.* **2024**, *16*, 102–113.
15. Ashry, I.; Mao, Y.; Al-Fehaid, Y.; Al-Shawaf, A.; Al-Bagshi, M.; Al-Brahim, S.; Ng, T.K.; Ooi, B.S. Early detection of red palm weevil using distributed optical sensor. *Sci. Rep.* **2020**, *10*, 3155. [CrossRef] [PubMed]
16. Khait, I.; Lewin-Epstein, O.; Sharon, R.; Saban, K.; Goldstein, R.; Anikster, Y.; Zeron, Y.; Agassy, C.; Nizan, S.; Sharabi, G.; et al. Sounds emitted by plants under stress are airborne and informative. *Cell* **2023**, *186*, 1328–1336. [CrossRef]
17. Glaser, D.R.; Costley, R.D.; Hathaway, K.; Boitnott, G.; Weale, J. Snow coupled distributed acoustic sensing for intrusion detection of polar bears in arctic camps. In *Symposium on the Application of Geophysics to Engineering and Environmental Problems 2021*; Society of Exploration Geophysicists: Houston, TX, USA, 2021; p. 262.
18. Zhu, T.; Stensrud, D.J. Characterizing thunder-induced ground motions using fiber-optic distributed acoustic sensing array. *J. Geophys. Res. Atmos.* **2019**, *124*, 12810–12823. [CrossRef]
19. Morozov, O.; Tunakova, Y.; Hussein, S.M.; Shagidullin, A.; Agliullin, T.; Kuznetsov, A.; Valeev, B.; Lipatnikov, K.; Anfinogentov, V.; Sakhabutdinov, A. Addressed combined fiber-optic sensors as key element of multisensor greenhouse gas monitoring systems. *Sensors* **2022**, *22*, 4827. [CrossRef]
20. Kakei, Y.; Katayama, S.; Lee, S.; Takakuwa, M.; Furusawa, K.; Umezu, S.; Sato, H.; Fukuda, K.; Someya, T. Integration of body-mounted ultrasoft organic solar cell on cyborg insects with intact mobility. *NPJ Flex. Electron.* **2022**, *6*, 78. [CrossRef]
21. Nelson, M.C.; Fraser, J. Sound production in the cockroach, Gromphadorhina portentosa: Evidence for communication by hissing. *Behav. Ecol. Sociobiol.* **1980**, *6*, 305–314. [CrossRef]
22. Clark, D.C.; Moore, A.J. Social communication in the Madagascar hissing cockroach: Features of male courtship hisses and a comparison of courtship and agonistic hisses. *Behaviour* **1995**, *132*, 401–417. [CrossRef]
23. Escobedo, J.B.; Spirin, V.V.; López-Mercado, C.A.; Lucero, A.M.; Mégret, P.; Zolotovskii, I.O.; Fotiadi, A.A. Self-injection locking of the DFB laser through an external ring fiber cavity: Application for phase sensitive OTDR acoustic sensor. *Results Phys.* **2017**, *7*, 641–643. [CrossRef]
24. Escobedo, J.B.; Jason, J.; López-Mercado, C.A.; Spirin, V.V.; Wuilpart, M.; Mégret, P.; Korobko, D.A.; Zolotovskiy, I.O.; Fotiadi, A.A. Distributed measurements of vibration frequency using phase-OTDR with a DFB laser self-stabilized through PM fiber ring cavity. *Results Phys.* **2019**, *12*, 1840–1842.
25. Lopez-Mercado, C.A.; Korobko, D.A.; Zolotovskii, I.O.; Fotiadi, A.A. Application of dual-frequency self-injection locked DFB laser for Brillouin optical time domain analysis. *Sensors* **2021**, *21*, 6859. [CrossRef] [PubMed]
26. Lu, P.; Lalam, N.; Badar, M.; Liu, B.; Chorpening, B.T.; Buric, M.P.; Ohodnicki, P.R. Distributed optical fiber sensing: Review and perspective. *Appl. Phys. Rev.* **2019**, *6*, 041302. [CrossRef]

27. Martins, H.F.; Martin-Lopez, S.; Corredera, P.; Filograno, M.L.; Frazão, O.; González-Herráez, M. Coherent noise reduction in high visibility phase-sensitive optical time domain reflectometer for distributed sensing of ultrasonic waves. *J. Light. Technol.* **2013**, *31*, 3631–3637.
28. Wang, B.; Mao, Y.; Ashry, I.; Al-Fehaid, Y.; Al-Shawaf, A.; Ng, T.; Yu, C.; Ooi, B. Towards Detecting Red Palm Weevil Using Machine Learning and Fiber Optic Distributed Acoustic Sensing. *Sensors* **2021**, *21*, 1592. [CrossRef]
29. Abdollahi, M.; Giovenazzo, P.; Falk, T.H. Automated beehive acoustics monitoring: A comprehensive review of the literature and recommendations for future work. *Appl. Sci.* **2022**, *12*, 3920. [CrossRef]

sensors

MDPI

Article

# Detailed Determination of Delamination Parameters in a Multilayer Structure Using Asymmetric Lamb Wave Mode

**Olgirdas Tumšys *, Lina Draudvilienė and Egidijus Žukauskas**

Ultrasound Research Institute, Kaunas University of Technology, LT-51423 Kaunas, Lithuania; lina.draudviliene@ktu.lt (L.D.); e.zukauskas@ktu.lt (E.Ž.)

* Correspondence: olgirdas.tumsys@ktu.lt; Tel.: +370-616-48797

**Abstract:** A signal-processing algorithm for the detailed determination of delamination in multilayer structures is proposed in this work. The algorithm is based on calculating the phase velocity of the Lamb wave $A_0$ mode and estimating this velocity dispersion. Both simulation and experimental studies were conducted to validate the proposed technique. The delamination having a diameter of 81 mm on the segment of a wind turbine blade (WTB) was used for verification of the proposed technique. Four cases were used in the simulation study: defect-free, delamination between the first and second layers, delamination between the second and third layers, and defect (hole). The calculated phase velocity variation in the $A_0$ mode was used to determine the location and edge coordinates of the delaminations and defects. It has been found that in order to estimate the depth at which the delamination is, it is appropriate to calculate the phase velocity dispersion curves. The difference in the reconstructed phase velocity dispersion curves between the layers simulated at different depths is estimated to be about 60 m/s. The phase velocity values were compared with the delamination of the second and third layers and a hole drilled at the corresponding depth. The obtained simulation results confirmed that the drilled hole can be used as a defect corresponding to delamination. The WTB sample with a drilled hole of 81 mm was used in the experimental study. Using the proposed algorithm, detailed defect parameters were obtained. The results obtained using simulated and experimental signals indicated that the proposed new algorithm is suitable for the determination of delamination parameters in a multilayer structure.

**Keywords:** non-destructive testing; Lamb waves; delamination parametrization; phase velocity dispersion curve; finite element modelling; wind turbine blade; zero-crossing

## 1. Introduction

Recently, composite materials have been used in a wide range of structures to reduce weight and increase strength. A composite material is a combination of several materials that have different mechanical, electrical, and thermal properties [1]. Therefore, this type of material has great advantages such as low overall cost, high strength-to-weight ratio, design flexibility, directional properties, and resistance to corrosion, high temperature, and severe fatigue [2–4]. However, the combination of several materials leads to inhomogeneity and an anisotropic nature, which poses a serious problem for the effective detection of defects [5,6]. Since, in practice, these advanced composite materials are susceptible to a variety of internal damage [7,8], which can occur during operation or during manufacture, the identification of such defects becomes a complex process [4,9]. It is therefore very important to detect, diagnose, and prognose these damages at an early stage before they propagate to major damage and reach catastrophic failure [8,10]. Delamination is named as one of the most common defects contributing to the degradation of the mechanical properties of these materials, which can lead to the failure of the overall composite component [1]. One of the important aspects of delamination detection in a composite structure is to predict the size of the delamination and at what depth it exists [11]. The delamination can be inspected by techniques that require investigation through the thickness of the structure. Thus, different approaches are required for the estimation of delamination and defects in such types of structures, especially in the case of complex objects under investigation [12]. Non-destructive testing (NDT) and Structural Health Monitoring (SHM) technologies are commonly used for the evaluation of the condition status of complex objects [10,12]. The application of these two technologies generally consists of two parts. First, a measurement system is used to excite and receive waves. Second, data analysis is performed using appropriate signal processing techniques [8,10]. Thus, special techniques are continuously being developed to increase the probability of detecting a fault with a high level of confidence [1]. For the qualitative assessment of multilayer materials, the ultrasonic guided wave-based method [5,8,13,14], acoustic emission (AE) [15], electromechanical impedance (EMI) [16], computed tomography (CT), microwave, thermography, radioscopy, eddy currents, and other techniques are used [12,17–19]. Among them, the Ultrasonic Lamb Wave (ULW) method has been extensively investigated regarding the detection of delamination and has been identified as one of the most reliable and promising methods with good results [3,13,14,20–22].

Lamb waves are elastic waves created by the motion of particles between the top and bottom surfaces of the plate and propagation between these boundaries. Thus, these waves are more sensitive to various kinds of defects, can travel long distances, have a large detection area, have superior inspection sensitivity, and require minimal equipment [23]. This means that the propagation characteristics of Lamb waves in a specific medium contain information about the current state of the structure, possible defects, and their location. Therefore, various Signal Processing Methods (SPMs) have been developed to extract and parameterize Lamb wave signals, which contain a wide variety of useful information. Thus, data analysis can be implemented through multi-dimensional Fourier transform, Wavelet Transform (WT), Wigner–Ville Distribution (WVD), Hilbert–Huang transform (HHT), and other methods [24–29].

However, despite all the good features of Lamb waves, the application of such waves has some limitations due to the unusual propagation characteristics in a specific medium. The phenomenon of dispersion, the infinite number of modes, interference, and the mode conversation are main unusual wave characteristics that make defect and/or delamination detection and parameterization difficult [23,30,31]. Thus, in the general case, the propagation of Lamb waves depends on the elastic properties, density, and thickness of the test material and the frequency of the excited wave and is described by Rayleigh–Lamb equations [5,32] represented by families of dispersion curves. For isotropic plates, this propagation is straightforward and can be described in a simple way. However, if the Lamb waves propagate in a multilayer composite structure, their propagation process becomes more complicated. The propagation of waves depends on the parameters of the individual layers and the interaction between their surfaces [9]. Thus, by knowing the parameters of the individual layers and considering the position of the layers relative to each other, it is possible to calculate the dispersion curves for the propagation velocity of Lamb waves. The calculated velocity of the dispersion curve can thus parameterize the layering.

The aim of this research is to propose a signal-processing algorithm for the detailed determination of delamination parameters in a multilayer structure and to investigate the feasibility of the proposed method for determining delaminations at different depths.

In this paper, Section 2 introduces the methodology for detecting delaminations using Lamb waves, the algorithm for delamination parametrization, and the processing of finite element modelled signals. Section 3 describes the experimental study. Section 4 concludes this paper.

## 2. Methodology for Detecting Delamination's Using Lamb Waves

### 2.1. Dispersion Theory of Lamb Waves for Composite Structures with Delamination's

As mentioned in the introduction, Lamb waves are very sensitive to any changes in a structure. Thus, at the point of layering, the structure becomes thinner, which means that the geometry of the structure changes. In this case, if delamination occurs between layers, the propagation of Lamb waves should change. To confirm such a theory, an example is chosen which is a three-layer Glass Fiber-Reinforced Plate (GFRP), where the thickness of the individual layers is $d_1 = d_2 = d_3 = 5$ mm (Figure 1a). The parameters of the layers of the chosen plate are the same: density $\rho = 1828$ kg/m$^3$, Young's modulus $E_1 = 42.5$ GPa, Young's modulus $E_2 = 10$ GPa, Poisson's ratio $\nu_{12} = 0.26$, Poisson's ratio $\nu_{23} = 0.4$, and in-plane shear modulus $G_{12} = 4.3$ GPa. Two cases of delamination are considered at different depths between layers. In the first case, there is a delamination of width $dl_1$ between the first and second layers at depth $dd_1$, and in the second, there is a delamination of width $dl_2$ between the second and third layers at depth $dd_2$, as shown in Figure 1a. In order to evaluate the Lamb wave propagation in a multilayer GFRP with delaminations at different depths, the phase velocity dispersion curves were calculated using the semi-analytical finite element (SAFE) method [33]. The phase velocity dispersion curves calculated by the SAFE method in the defect-free region and in both selected delamination regions are shown in Figure 1b.

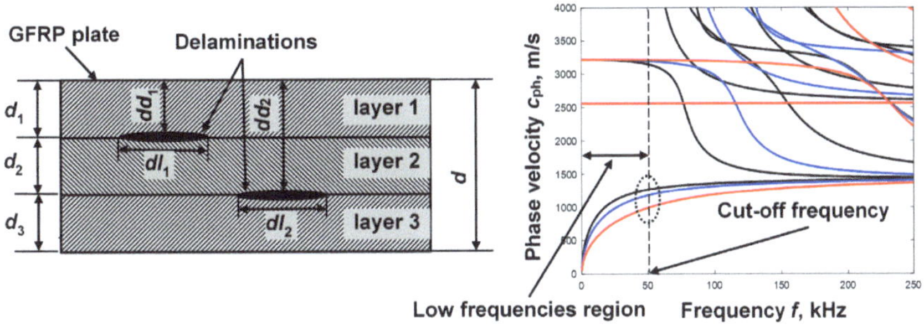


(a) (b)

**Figure 1.** Three-layer GFRP with delaminations between the different layers (**a**) and phase velocity dispersion curves in defect-free and defective regions calculated by the SAFE method (**b**); black lines are dispersion curves in the defect-free region, and red and blue lines are in the first and second regions of delaminations.

Analysis of the obtained phase velocity dispersion curves leads to several conclusions. Firstly, it is observed that only the fundamental $A_0$ and $S_0$ modes propagate in the lower frequency ranges. The cut-off frequency of the higher modes is approximately 50 kHz (Figure 1b). Secondly, the resulting dispersion curves show that changes in geometric thickness in delamination regions affect the phase velocity of the $A_0$ mode in the lower frequency region. In contrast, the $S_0$ mode is more sensitive to these changes at higher frequencies (Figure 1b). Therefore, for further study, the $A_0$ mode at the 50 kHz frequency range is chosen for the study. The higher reduction in the phase velocity of $A_0$ mode at the 50 kHz frequency range is obtained when the delamination is between the first and second layers at depth $dd_1$ (Figure 1b, red line). Meanwhile, a smaller change in phase velocity is observed when analyzing the delamination of width $dl_2$ between the second and third layers at depth $dd_2$ (Figure 1b, blue line). Thus, the calculation of the phase velocity dispersion curves could be a parameter for the determination of the depth of layering.

The change in the phase velocity of the $A_0$ mode at low frequencies was used by the authors [34] to determine the geometric dimensions (width) of the internal defect of the WTB. Thus, in order to determine, in detail, the parameters of defects and delaminations in a multilayer structure, it is necessary to know not only the area but also the depth at which they are located. A new signal-processing algorithm is therefore developed after using the already described methods [9,33] of digital signal processing, enabling the determination of the location and geometric dimensions of delamination defects. A detailed description of this algorithm is provided in the following subsection.

### 2.2. Algorithm for Delamination Parametrization

A signal-processing algorithm based on the estimation of phase velocity variations in the $A_0$ mode is proposed to determine the size, location, and depth of a delamination in a multilayer structure. The flow chart of the proposed signal-processing algorithm is presented in Figure 2.

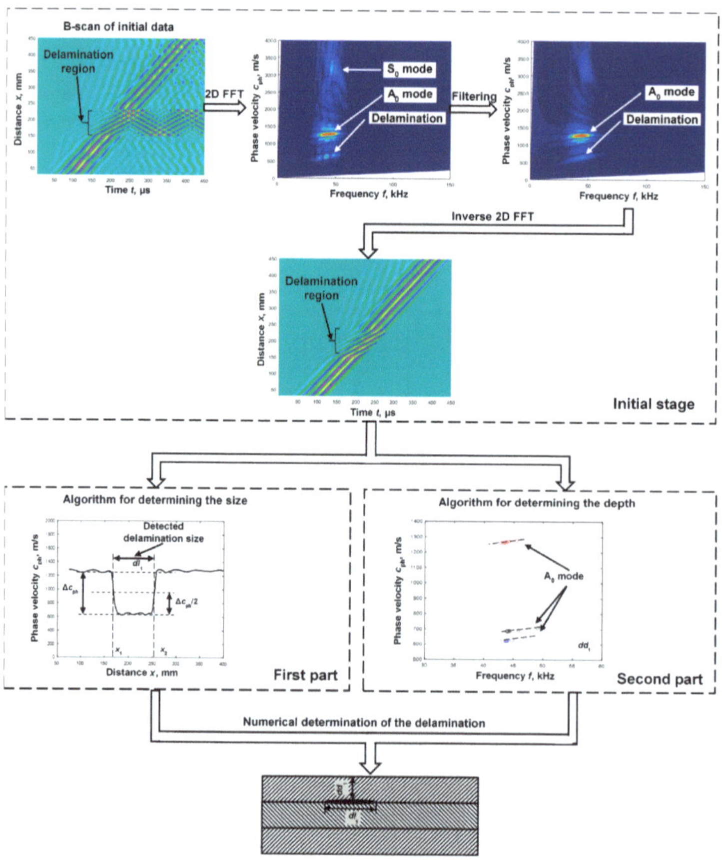


**Figure 2.** The flow chart of the proposed signal-processing algorithm for delamination parametrization.

The algorithm displays the signals emitted by the object in the initial stage in a B-scan image. Filtering is then performed to remove reflections from the delamination edges and to highlight the $A_0$ mode. Spatial filtering based on two-dimensional fast Fourier transform (2D-FFT) is used for this purpose [34]. An inverse two-dimensional fast Fourier transform is used to reconstruct a B-scan image that reflects only the propagation of the $A_0$ mode. This procedure has been described in detail in a previous paper [34]. Further processing of the received B-scan image is divided into two parts.

In the first part, a defect-sizing algorithm is applied. According to the algorithm presented, the size of the defect (delamination) is determined by measuring the change in phase velocity with respect to the distance with the moving window [34]. The phase velocity is calculated using two adjacent signals acquired at two different spatial measurement positions. In this way, the phase velocity variation is obtained in the defective region that indicates the location of the delamination. Then, the middle threshold between the maximum and minimum values (0.5 or 6 dB level) of the determined phase velocity in the defect region is determined and the distance between two points is designated as the delamination size.

In the second part, the location of the delamination depth is determined. A segment of the phase velocity dispersion curve is used to determine the more accurate depth location. The zero-crossing method is proposed to apply for the phase velocity dispersion curve calculation. This method has been previously proposed and presented in our work [35]. The method involves determining the time instants at which signals cross the zero-amplitude line to calculate phase velocity and frequency. Thus, the method consists of two main parts: obtaining phase velocity values and calculating frequency values. Phase velocity values are calculated using two adjacent signals and the instances at which both signals cross the zero-amplitude line. For the calculation of the frequency values, the determined time instances of the second signals are used. Then, the frequency values are calculated using the duration of each selected half-period.

In order to bring theoretical research closer to real experimental measurements, a fragment of a WTB was chosen as the object of theoretical modelling to verify the effectiveness of the proposed algorithm.

### 2.3. Processing of Finite Element Modelled Signals by the Proposed Algorithm

A WTB is heavily exposed to fluctuating wind or cyclic loads, making it the most potentially defective element of a wind turbine. Therefore, the regular maintenance and inspection of WTBs is necessary. A graphical representation of the multilayer structure of a WTB with delamination in one of the interlayers is presented in Figure 3. The orientations of the main layers of the GFRP in the composite structure are +45°/−45° and 0°/90°/+45°/−45°/0°, respectively. The properties of the materials of the different layers are presented in Table 1. The dispersion curves of the phase velocities of the asymmetric and symmetric modes of Lamb waves were calculated using the SAFE method; the obtained results are presented in Figure 4.

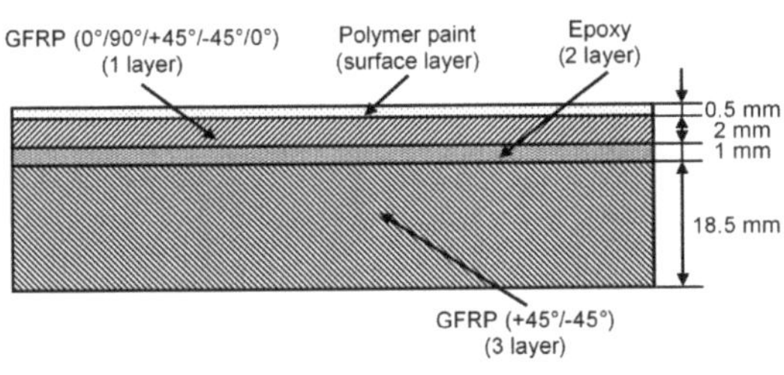


**Figure 3.** The graphical representation of the multilayer structure of a WTB.

**Table 1.** Elastic parameters of WTB composite structure layer materials.

| Parameters | Numerical Value |
|---|---|
| Paint (Surface layer): | |
| Density ($\rho$) | 1270 kg/m$^3$ |
| Young's modulus (E) | 4.2 GPa |
| Poisson's ratio ($\upsilon$) | 0.35 |
| Unidirectional GFRP layer: | |
| Density ($\rho$) | 1828 kg/m$^3$ |
| Young's modulus ($E_1$) | 42.5 GPa |
| Young's modulus ($E_2$) | 10 GPa |
| Poisson's ratio ($\upsilon_{12}$) | 0.26 |
| Poisson's ratio ($\upsilon_{23}$) | 0.4 |
| In-plane shear modulus ($G_{12}$) | 4.3 GPa |
| Epoxy: | |
| Density ($\rho$) | 1260 kg/m$^3$ |
| Young's modulus (E) | 3.6 GPa |
| Poisson's ratio ($\upsilon$) | 0.35 |

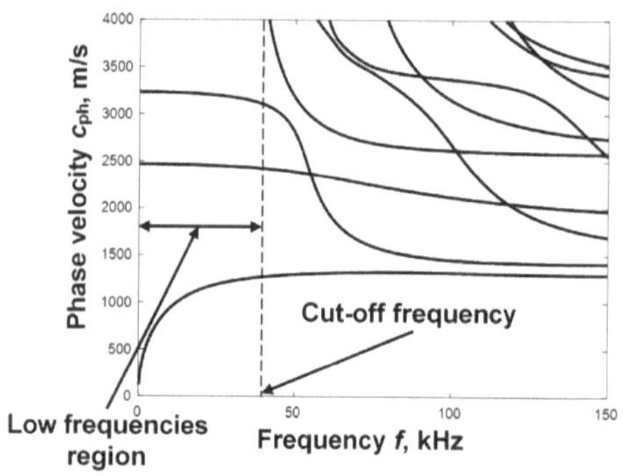


**Figure 4.** The phase velocity dispersion curves of the $A_0$ and $S_0$ modes calculated by the SAFE method in the WTB multilayer structure.

The simulation of Lamb wave propagation in the WTB composite structure was performed using Abaqus finite element software. An explicit algorithm was used to solve the transient wave equation. The structure of the WTB was simplified using a 2D plane strain model. A delamination was modelled by disconnecting relevant surfaces at desired positions. The size of the delamination was 81 mm. The model was meshed with rectangular mapped mesh. The size of the element was 0.25 mm. In order to excite the $A_0$ Lamb waves, transient excitation force was applied to the surface of the object under investigation. The width of the excitation zone was 30 mm. For the excitation of guided waves with 43 kHz frequency, five periods sine burst with Gaussian envelope was used. To selectively excite the $A_0$ mode, applied force was phased by delaying the force applied to each node by the following law:

$$\Delta t = \frac{\sin(\theta) \cdot dx}{c_{A0}} \tag{1}$$

where $\theta$ is the wave propagation angle, which in our case is 90°; $dx$ is the distance between nodes (0.25 mm); and $c_{A0}$ is the phase velocity of the $A_0$ mode at 43 kHz.

Three cases were used in the simulation study: defect-free, delamination between the first and second layers, and delamination between the second and third layers. The B-scan images obtained as a result of modelling Lamb wave propagation in a multilayer defective WTB structure are presented in Figure 5a–c. The B-scan image obtained in a defect-free region is presented in Figure 5a, in a region where delamination is between the first and second layers (Figure 5b) and in a region where delamination is between the second and third layers (Figure 5c). The filtering was performed as it is presented in Figure 2, in order to highlight the $A_0$ mode. The obtained filtered B-scan images of all cases of the study are presented in Figure 5d,e,f. According to the proposed algorithm, in Figure 2, the phase velocity of the $A_0$ mode is calculated for each case. The obtained phase velocity values in the defect free region, in the region with delamination between first and second layers, and in the region with delamination between the second and third layers are presented in Figure 6a,b,c, respectively.

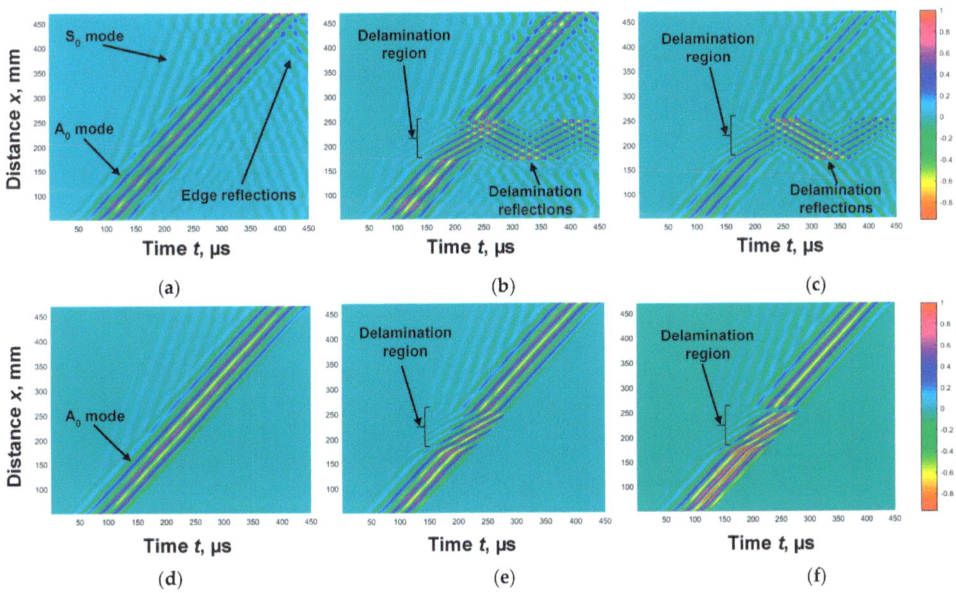


**Figure 5.** Simulated B-scan images of propagated Lamb wave in defect-free region (**a**), in region with delamination between 1st and 2nd layers (**b**), and in region with delamination between 2nd and 3rd layers (**c**); filtered B-scan images (**d**–**f**).

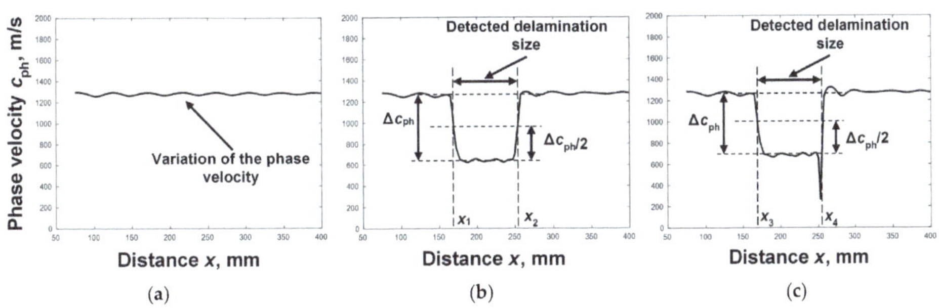


**Figure 6.** Variation in the phase velocity of the $A_0$ mode of the Lamb wave with respect to the distance when there is no defect in the multilayer structure (**a**), when delamination is between the 1st and 2nd layers (**b**), and when delamination is between the 2nd and 3rd layers (**c**).

The results obtained clearly show that in both simulated delamination cases, in the range from ~170 mm to ~250 mm, the phase velocity was reduced to 600–700 m/s. The coordinates of the first and second edges of the delamination in both cases were the same and equal to $x_1 = x_3 = 172.5$ mm and $x_2 = x_4 = 255$ mm, respectively, and the delamination size was determined to be $dl_1 = dl_2 = 82.5$ mm.

However, due to the interference phenomena, in the other mode seen in B-scan images (Figure 5d–f), scattered results were obtained in all cases of the study (Figure 6a–c). Thus, according to the calculated phase velocity values (Figure 6b,c), it is difficult to assess at which depth the delamination was and to determine between which layers it was present. Therefore, in order to obtain more accurate values, it was proposed to calculate the segments of the phase velocity dispersion curve. The zero-crossing method was applied for the calculation of the phase velocity dispersion curve segment of the $A_0$ mode [35]. The necessary parameters that needed to be selected in each case of the study were two adjusted signals, the threshold level, and the number of time instances where the signals crossed the zero-amplitude line. Two signals at distances $x_1 = 60$ and $x_2 = 200$ mm were selected in a defect-free sample and the threshold level $U_L = 0.1$ was set; $x_1 = 177$ and $x_2 = 202$ mm in the region with delamination between the first and second layers were selected and the threshold level $U_L = 0.15$ was set, and $x_1 = 175$ and $x_2 = 205$ mm were selected in the region with delamination between the second and third layers, and the threshold level $U_L = 0.2$ was set. The eight time instances were recorded in each case of the study. The obtained segments of the reconstructed phase velocity dispersion curves of the $A_0$ mode in a defect-free region, in a region with delamination between the first and second layers, and in a region with delamination between second and third layers are presented in Figure 7.

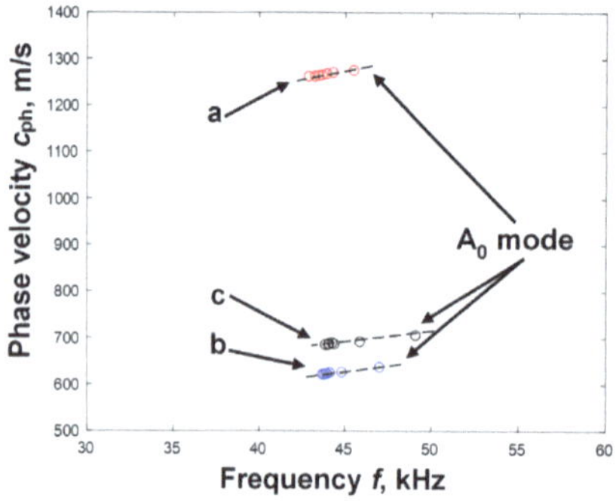


**Figure 7.** Variation in the phase velocity of the Lamb wave $A_0$ mode with respect to the frequency when there is no defect in the multilayer structure (**a**), when delamination is between the 1st and 2nd layers (**b**), and when delamination is between the 2nd and 3rd layers (**c**).

The resulting segment of the phase velocity dispersion curve of $A_0$ mode in the defect-free region was 1230–1300 m/s. This is consistent with the SAFE calculated dispersion curve presented in our previous work [34]. The segment of the phase velocity dispersion curve in the region with delamination between the first and second layers was obtained as 628–644 m/s, and in the region with delamination between the second and third layers, it was 685–708 m/s. The difference in phase velocity was about 60 m/s. The results clearly show that the phase velocity of the reconstructed dispersion curves varied with the delamination depth of the layers. Thus, the phase velocity dispersion curve can be used to determine the depth of delamination. Based on the simulation results, the proposed new algorithm is suitable for the detailed determination of the delamination parameters of a multilayer structure.

Since it was not possible to carry out the experiment with different types of delaminations in a WTB, a milled hole-type defect with a diameter of 81 mm was chosen for the study. It was assumed that a drilled hole, which has analogous characteristics with respect to Lamb wave propagation, can be used as a defect corresponding to delamination.

This assumption was tested with additional theoretical simulations. The simulation results are shown in Figure 8. Delamination between the second and third layers (Figure 8a) and a hole drilled at the corresponding depth (Figure 8b) were selected for simulation. The out-of-plane components were analyzed in the simulation. The obtained B-scan images are presented in Figure 8c,d. The obtain phase velocity variation in the $A_0$ mode at the defect location is presented in Figure 8e.

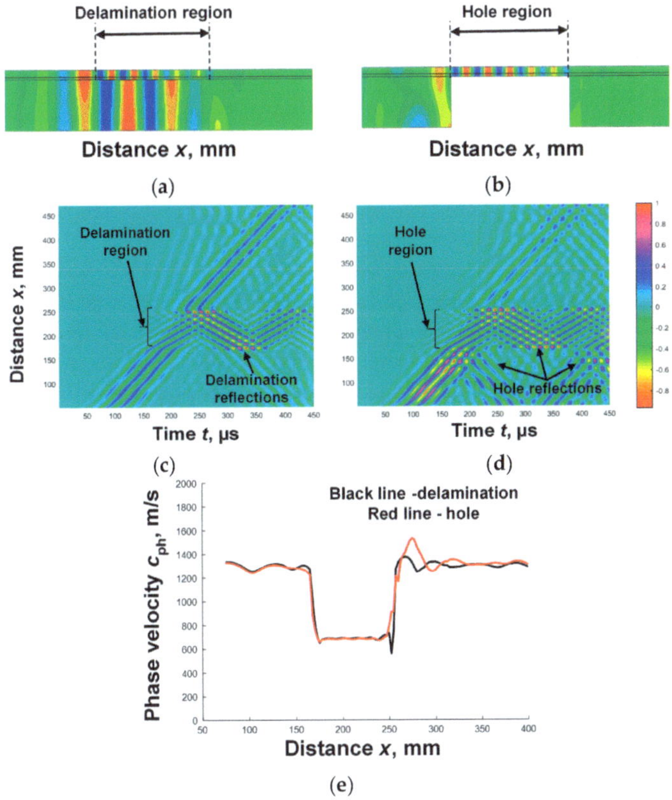


**Figure 8.** Simulation results of Lamb wave propagation in a WTB specimen with a delamination between the 2nd and 3rd layers (**a**), a hole drilled at the corresponding depth (**b**); the B-scan images are presented in (**c**,**d**) accordingly, and phase velocity variation in the $A_0$ mode at the defect location is shown in (**e**).

According to the algorithm proposed in Figure 2, the phase velocity variation in the $A_0$ mode at the defect location was calculated for each case. The obtained simulation results (Figure 8e) confirmed the assumption that a drilled hole can be used as a defect corresponding to delamination. Thus, the experiment verification was performed using a milled hole-type defect.

## 3. Experimental Verification

The performance of the proposed new algorithm was experimentally verified on the WTB sample. A milled hole-type defect with a diameter of 81 mm was used, which had analogous characteristics with respect to Lamb wave propagation. The real WTB sample is presented in Figure 9.

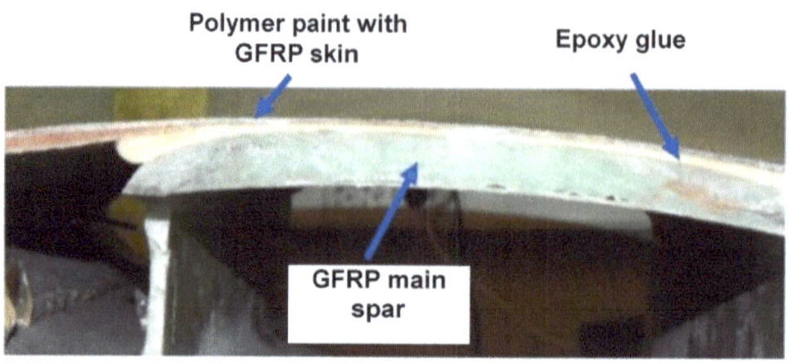


**Figure 9.** The real WTB sample.

The low-frequency ultrasonic system developed by the Ultrasound Research Institute of Kaunas University of Technology [36] was used for the experimental study. The macro-fibre composite (MFC) transducer of a P1-type MFC-P1-2814 (S.n. 08J100791) transducer manufactured by Smart Materials was used for the excitation of ultrasonic guided waves. The characteristics of this MFC transducer have been studied in detail [37]. The wideband contact-type piezoceramic transducer was used as a receiver. The experimental setup of WTB inspection is presented in Figure 10a. In previous work [34], the resonant frequency of the MFC transducer was found to be 43 kHz, and therefore, this frequency was chosen as the excitation frequency. The transducer was excited by a rectangular pulse with a duration of 11.6 μs. The wideband contact-type ultrasonic receiving transducer was scanned up to the distance of $x_2$ = 160 mm away from the transmitter with a step of 0.1 mm. The initial distance between the transmitter and receiver was $x_1$ = 30 mm. The experimentally obtained B-scan images as a result of Lamb wave propagation in a multilayer defective WTB structure are presented in Figure 10b. According to the proposed algorithm, filtering was performed. The obtained filtered B-scan image is presented in Figure 10c.

According to the algorithm presented in Figure 2, the phase velocity was obtained in the defect-free and defect regions (Figure 11). The obtained phase velocity variation clearly indicated the defect region (Figure 11a). According to the calculated threshold level between the minimum and maximum values of the phase velocity, the size of the defect was determined, which was $dl_1$ = 83.2 mm. The phase velocity of the $A_0$ mode over the defect-free region was obtained as about 1260–1300 m/s, and over the defect region, it was about 780–810 m/s (Figure 11b). Two signals at distances $x_1$ = 45 mm and $x_2$ = 55 mm were selected in a defect-free sample and $x_1$ = 90 mm and $x_2$ = 110 mm in the defect region. The threshold level $U_L$ = 0.1 was set and six time instances were recorded in each case of the study. Applying the zero-crossing method, the segments of the phase velocity dispersion curves in the defect-free and defect regions were calculated (Figure 11b).

Based on the obtained results, the location, size, and depth of the defect could be evaluated using the proposed algorithm. The results obtained with experimental signals indicated that the proposed signal-processing algorithm is a suitable tool for the detailed determination of delamination parameters in a multilayer structure.

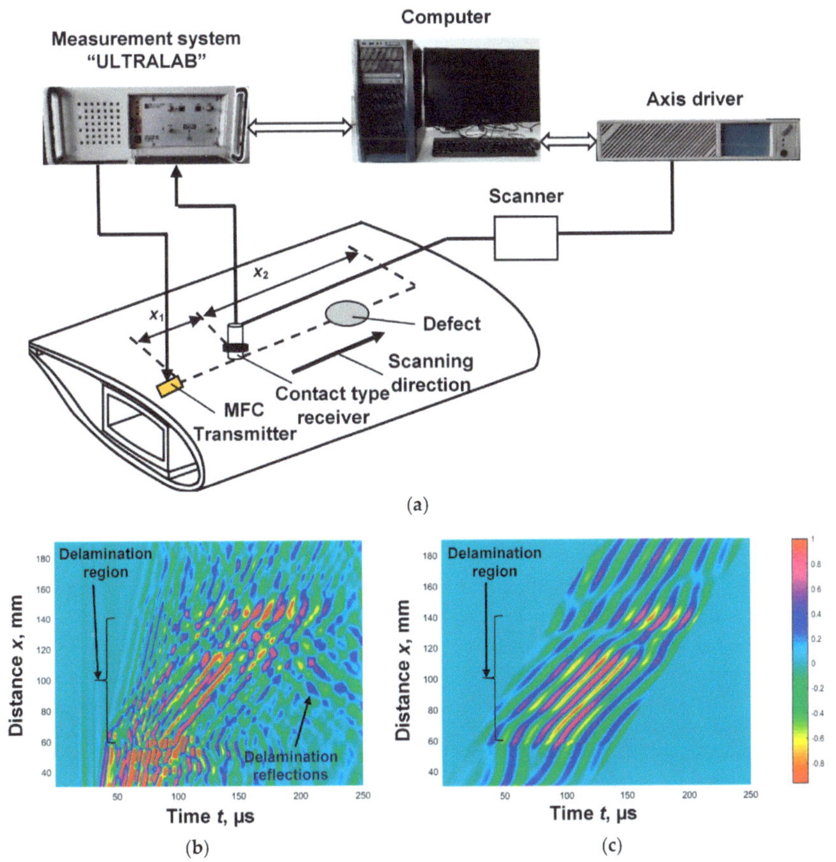


**Figure 10.** Experimental setup of WTB inspection (**a**), B-scan image of propagated Lamb wave in region (**b**) and filtered experimental B-scan image (**c**).

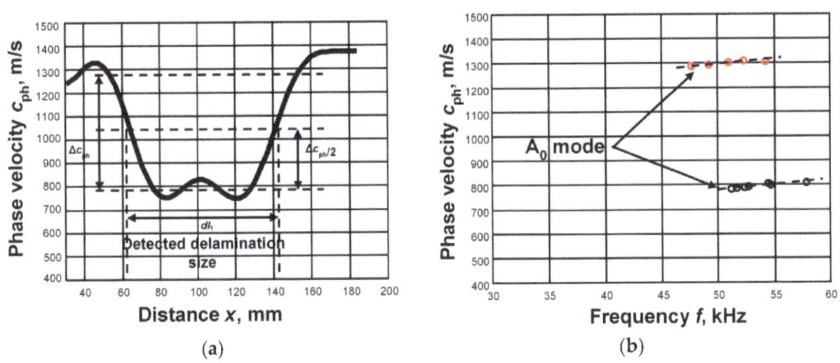


**Figure 11.** Variation in the phase velocity of the Lamb wave $A_0$ mode in defect-free and defect regions (**a**); the segments of the phase velocity dispersion curves in defect-free and defect regions (**b**).

## 4. Conclusions

A new signal-processing algorithm is proposed for the detailed determination of delamination parameters of multilayer structures using the $A_0$ mode of Lamb waves. The algorithm consists of two main parts: the calculation of the phase velocity variation and estimation of the dispersion of this velocity. Applying the first part of the proposed algorithm, the coordinates of the delamination edge are determined, which indicate the size of delamination. Applying the second part of the algorithm, information is obtained regarding at what depth or between which layers the delamination is. The simulation and experimental studies were conducted to validate the proposed technique. The WTB sample was chosen and the $A_0$-mode signals at 43 kHz range were used in both studies. The proposed method was verified by the use of an 81 mm diameter WTB segment delamination in the simulation study. Four cases were used: defect-free, delamination between the first and second layers, delamination between the second and third layers, and defect (hole). The obtained phase velocity values showed the location and width of both delaminations. The delamination size in both cases was determined to be $dl_1 = dl_2 = 82.5$ mm. The reconstructed phase velocity dispersion curve of the $A_0$ mode was obtained at 628–644 m/s in the region with delamination between the first and second layers and in the region with delamination between the second and third layers was obtained at 685–708 m/s. A simulation study was carried out to confirm that a drilled hole can be used as a defect corresponding to delamination. Then, the experimental study was carried out using the WTB sample with a drilled hole of 81 mm to validate the proposed algorithm. The phase velocity variations in the $A_0$ mode, calculated according to the proposed algorithm, showed a defect size of 83.2 mm. The calculated $A_0$-mode dispersion curve segments were obtained for the defect-free region at about 1260–1300 m/s and for the defect region at about 780–810 m/s. The simulated and experimental results indicated that the proposed signal-processing algorithm is a suitable tool for the parameterization of delamination in a multilayer structure.

**Author Contributions:** O.T. and L.D.: conceptualization, methodology, investigation, writing—original draft preparation, and visualization. E.Ž.: validation, software, data curation, visualization, and writing—review and editing. All authors have read and agreed to the published version of the manuscript.

**Funding:** This research was funded by the Research Foundation of the Research Council of Lithuania under the project "COMMET". A combined signal processing method for determining the location and size of defects using higher-order Ultrasound Guided Waves, No. MIP-23-119.

**Institutional Review Board Statement:** Not applicable.

**Informed Consent Statement:** Not applicable.

**Data Availability Statement:** Data are contained within the article.

**Conflicts of Interest:** The authors declare no conflicts of interest.

## References

1. Cheng, L.; Tian, G.Y. Comparison of Nondestructive Testing Methods on Detection of Delaminations in Composites. *J. Sens.* **2012**, *2012*, 408437. [CrossRef]
2. Cawley, P.; Adams, R.D. Defect types and non-destructive testing techniques for composites and bonded joints. *Mater. Sci. Technol.* **1989**, *5*, 413–425. [CrossRef]
3. Yu, L.; Xiao, W.; Mei, H.; Giurgiutiu, V. Delamination imaging in composites using cross-correlation method by non-contact air-coupled Lamb waves. *Smart Mater. Struct.* **2023**, *32*, 105013. [CrossRef]
4. Wang, B.; Wang, B.; Shi, W.; Shi, W.; Zhao, B.; Zhao, B.; Tan, J.; Tan, J. A modal decomposition imaging algorithm for ultrasonic detection of delamination defects in carbon fiber composite plates using air-coupled Lamb waves. *Measurement* **2022**, *195*, 111165. [CrossRef]
5. Ben, B.; Vikram, K.; Yang, S. Damage identification in composite materials using ultrasonic based Lamb wave method. *Measurement* **2013**, *46*, 904–912. [CrossRef]
6. Mei, H.; Giurgiutiu, V. Characterization of multilayer delaminations in composites using wavenumber analysis: Numerical and experimental studies. *Struct. Health Monit.* **2021**, *20*, 1004–1029. [CrossRef]

7. Roundi, W.; El Mahi, A.; El Gharad, A.; Rebière, J.-L. Experimental and numerical investigation of the effects of stacking sequence and stress ratio on fatigue damage of glass/epoxy composites. *Compos. B Eng.* **2017**, *109*, 64–71. [CrossRef]
8. Yang, B.; Xuan, F.-Z.; Chen, S.; Zhou, S.; Gao, Y.; Xiao, B. Damage localization and identification in WGF/epoxy composite laminates by using Lamb waves: Experiment and simulation. *Compos. Struct.* **2017**, *165*, 138–147. [CrossRef]
9. Papanaboina, M.R.; Jasiuniene, E.; Žukauskas, E.; Mažeika, L. Numerical Analysis of Guided Waves to Improve Damage Detection and Localization in Multilayered CFRP Panel. *Materials* **2022**, *15*, 3466. [CrossRef]
10. Badihi, H.; Zhang, Y.; Jiang, B.; Pillay, P.; Rakheja, S. A Comprehensive Review on Signal-Based and Model-Based Condition Monitoring of Wind Turbines: Fault Diagnosis and Lifetime Prognosis. *Proc. IEEE* **2022**, *110*, 754–806. [CrossRef]
11. Munian, R.K.; Mahapatra, D.R.; Gopalakrishnan, S. Lamb wave interaction with composite delamination. *Compos. Struct.* **2018**, *206*, 484–498. [CrossRef]
12. Ding, S.; Yang, C.; Zhang, S. Acoustic-Signal-Based Damage Detection of Wind Turbine Blades—A Review. *Sensors* **2023**, *23*, 4987. [CrossRef]
13. Radzieński, M.; Kudela, P.; Marzani, A.; De Marchi, L.; Ostachowicz, W. Damage identification in various types of composite plates using guided waves excited by a piezoelectric transducer and measured by a laser vibrometer. *Sensors* **2019**, *19*, 1958. [CrossRef]
14. Nzouatchoua, C.B.; Bentahar, M.; Montresor, S.; Colin, N.; Le Cam, V.; Trottier, C.; Terrien, N. Damage Localization on Composite Structures Based on the Delay-and-Sum Algorithm Using Simulation and Experimental Methods. *Sensors* **2023**, *23*, 4368. [CrossRef]
15. Nikbakht, M.; Yousefi, J.; Hosseini-Toudeshky, H.; Minak, G. Delamination evaluation of composite laminates with different interface fiber orientations using acoustic emission features and micro visualization. *Compos. B Eng.* **2017**, *113*, 185–196. [CrossRef]
16. Gresil, M.; Yu, L.; Giurgiutiu, V.; Sutton, M. Predictive modeling of electromechanical impedance spectroscopy for composite materials. *Struct. Health Monit.* **2012**, *11*, 671–683. [CrossRef]
17. Wang, P.; Li, Z.; Zhou, L.; Pei, Y. Microwave nondestructive detection and quantitative evaluation of kissing defects in GFRP laminates. *Compos. Sci. Technol.* **2018**, *162*, 117–122. [CrossRef]
18. Zappalorto, M.; Panozzo, F.; Carraro, P.A.; Quaresimin, M. Electrical response of a laminate with a delamination: Modelling and experiments. *Compos. Sci. Technol.* **2017**, *143*, 31–45. [CrossRef]
19. Su, Z.; Cheng, L.; Wang, X.; Yu, L.; Zhou, C. Predicting delamination of composite laminates using an imaging approach. *Smart Mater. Struct.* **2009**, *18*, 074002. [CrossRef]
20. Gangadharan, R.; Mahapatra, D.R.; Gopalakrishnan, S.; Murthy, C.; Bhat, M. On the sensitivity of elastic waves due to structural damages: Time–frequency based indexing method. *J. Sound Vib.* **2009**, *320*, 915–941. [CrossRef]
21. Chia, C.C.; Lee, S.Y.; Harmin, M.Y.; Choi, Y.; Lee, J.-R. Guided ultrasonic waves propagation imaging: A review. *Meas. Sci. Technol.* **2023**, *34*, 052001. [CrossRef]
22. Yang, S.; Guo, Z.; He, J.; Cheng, Q.; Xiong, X. Detection of delamination using laser-generated Lamb waves with improved wavenumber analysis. *Meas. Sci. Technol.* **2025**, *36*, 015412. [CrossRef]
23. Zhang, H.; Hua, J.; Tong, T.; Zhang, T.; Lin, J. Dispersion compensation of Lamb waves based on a convolutional auto-encoder. *Mech. Syst. Signal Process.* **2023**, *198*, 110432. [CrossRef]
24. Su, Z.; Ye, L.; Lu, Y. Guided Lamb waves for identification of damage in composite structures: A review. *J. Sound Vib.* **2006**, *295*, 753–780. [CrossRef]
25. Jia, H.; Zhang, Z.; Liu, H.; Dai, F.; Liu, Y.; Leng, J. An approach based on expectation-maximization algorithm for parameter estimation of Lamb wave signals. *Mech. Syst. Signal Process.* **2019**, *120*, 341–355. [CrossRef]
26. Dai, D.; He, Q. Structure damage localization with ultrasonic guided waves based on a time–frequency method. *Signal Process.* **2014**, *96*, 21–28. [CrossRef]
27. Golub, M.V.; Doroshenko, O.V.; Arsenov, M.A.; Bareiko, I.A.; Eremin, A.A. Identification of Material Properties of Elastic Plate Using Guided Waves Based on the Matrix Pencil Method and Laser Doppler Vibrometry. *Symmetry* **2022**, *14*, 1077. [CrossRef]
28. Ghose, B.; Panda, R.S.; Balasubramaniam, K. Phase velocity measurement of dispersive wave modes by Gaussian peak-tracing in the f-k transform domain. *Meas. Sci. Technol.* **2021**, *32*, 124006. [CrossRef]
29. Pai, P.F.; Deng, H.; Sundaresan, M.J. Time-frequency characterization of lamb waves for material evaluation and damage inspection of plates. *Mech. Syst. Signal Process.* **2015**, *62–63*, 183–206. [CrossRef]
30. Gao, X.; Gao, X.; Tian, Y.; Tian, Y.; Jiao, J.; Jiao, J.; Gao, J.; Gao, J.; Li, C.; Li, C. An accurate measurement method of lamb wave phase velocity based on clustering algorithms. *Measurement* **2022**, *195*, 111178. [CrossRef]
31. Vu, A.; Bond, L.J.; Chakrapani, S.K. Numerical Study of Mode Conversion between Fundamental Lamb and Rayleigh Waves. *Appl. Sci.* **2023**, *13*, 5613. [CrossRef]
32. Rose, J.L. (Ed.) *Ultrasonic Guided Waves in Solid Media*; Cambridge University Press: Cambridge, UK, 2014. [CrossRef]
33. Hayashi, T.; Song, W.-J.; Rose, J.L. Guided wave dispersion curves for a bar with an arbitrary cross-section, a rod and rail example. *Ultrasonics* **2003**, *41*, 175–183. [CrossRef]

34. Raišutis, R.; Tiwari, K.A.; Žukauskas, E.; Tumšys, O.; Draudvilienė, L. A novel defect estimation approach in wind turbine blades based on phase velocity variation of ultrasonic guided waves. *Sensors* **2021**, *21*, 4879. [CrossRef] [PubMed]
35. Draudvilienė, L.; Raišutis, R. Estimation of Lamb Wave Anti-Symmetric Mode Phase Velocity in Various Dispersion Ranges Using Only Two Signals. *Symmetry* **2023**, *15*, 1236. [CrossRef]
36. Tiwari, K.A.; Raisutis, R.; Mazeika, L.; Samaitis, V. 2D Analytical Model for the Directivity Prediction of Ultrasonic Contact Type Transducers in the Generation of Guided Waves. *Sensors* **2018**, *18*, 987. [CrossRef] [PubMed]
37. Tiwari, K.A.; Raisutis, R. Investigation of the 3D displacement characteristics for a macro-fiber composite transducer (MFC-P1). *Mater. Tehnol.* **2018**, *52*, 235–239. [CrossRef]

sensors

Article

# An Experimental Study of the Acoustic Signal Characteristics of Locked-Segment Damage Evolution in a Landslide Model

**Xing Zhu [1,2,3], Hui Chen [1,4,*], Zhanglei Wu [5], Shumei Yang [1,2], Xiaopeng Li [1,4] and Tiantao Li [1]**

1 State Key Laboratory of Geohazard Prevention and Geoenvironment Protection, Chengdu University of Technology, Chengdu 610059, China; zhuxing15@cdut.edu.cn (X.Z.)
2 College of Computers and Cyber Security, Chengdu University of Technology, Chengdu 610059, China
3 Sichuan Engineering Technology Research Center of Industrial Internet Intelligent Monitoring and Application, Chengdu University of Technology, Chengdu 610059, China
4 The Geomathematics Key Laboratory of Sichuan Province, Chengdu University of Technology, Chengdu 610059, China
5 Powerchina Chengdu Engineering Corporation Limited, Chengdu 610050, China
* Correspondence: huichencdut@cdut.edu.cn

**Abstract:** Three-section landslides are renowned for their immense size, concealed development process, and devastating impact. This study conducted physical model tests to simulate one special geological structure called a three-section-within landslide. The failure process and precursory characteristics of the tested samples were meticulously analyzed using video imagery, micro-seismic (MS) signals, and acoustic emission (AE) signals, with a focus on event activity, intensity, and frequency. A novel classification method based on AE waveform characteristics was proposed, categorizing AE signals into burst signals and continuous signals. The findings reveal distinct differences in the evolution of these signals. Burst signals appeared exclusively during the crack propagation and failure stages. During these stages, the cumulative AE hits of burst signals increased gradually, with amplitude rising and then declining. High-amplitude burst signals were predominantly distributed in the middle- and high-frequency bands. In contrast, cumulative AE hits of continuous signals escalated rapidly, with amplitude monotonously increasing, and high-amplitude continuous signals were primarily distributed in the low-frequency band. The emergence of burst signals and high-frequency AE signals indicated the generation of microcracks, serving as early-warning indicators. Notably, the early-warning points of AE signals were detected earlier than those of video imagery and MS signals. Furthermore, the early-warning point of burst signals occurred earlier than those of continuous signals, and the early-warning point of the classification method preceded that of overall AE signals.

**Keywords:** three-section landslide; locking section; video image; micro-seismic signal; acoustic emission

## 1. Introduction

The majority of rock landslides are highly sudden, destructive, and widespread geological disasters. The overall stability of most large rock landslides is controlled by the key rock masses in the locking section on weak structural planes or shear zones. Such landslides are referred to as locking-section landslides [1]. Among them, the three-section landslide is a typical type of locking-section landslide, which is common in the western regions of China [2]. The generic mechanism of the three-section landslide involves sliding, tension cracking, and shearing. In essence, creep occurs on the low inclined cataclinal structural planes, followed by the appearance of tensile cracks at the slope's rear edge, ultimately resulting in damage to the central locking section [2]. Globally, the economic losses and casualties caused by rock landslides, particularly three-section landslides, are increasing [3]. The locking section is a crucial load-bearing component of the slope, and its strength and deformation determine the overall stability of the slope [4–7]. Therefore, a deeper understanding of the failure mechanisms and early-warning indicators of the locking section is imperative.

In recent years, researchers have conducted numerous studies on rock landslides involving the locking section, focusing on failure mechanisms and mechanical properties of rock. Huang et al. [8] and Pan et al. [9] classified rock landslides with a locking section by studying essential characteristics of multiple large-scale catastrophic landslides and elucidated the sliding mechanisms of the locking section. From the perspective of rock's mechanical properties, many scholars have discussed the influence of different angles of joints [10,11] and varying bridge lengths [12] on the failure mechanism of the locking section. Although these studies have shed light on the failure mechanism of three-section landslides, significant challenges in early warning persist due to the inconspicuous macroscopic failure deformation.

The deformation or failure of rock involves the generation and propagation of cracks under external loads, releasing elastic waves known as AE activity in the laboratory tests [13–17]. Acoustic emission (AE) signals are widely utilized for their capability to monitor failure behavior [18]. Currently, AE technology has become an innovative and powerful tool for monitoring rocks, slopes and tunnels, especially in the study of the correlation between slope instability and AE behavior [19–21]. Kumar et al. [22] conducted physical experiments on artificial soil slopes and demonstrated that the threshold deformation rate can be achieved in the engineering design of a landslide early-warning system based on AE technology. Deng et al. [23] developed a complete AE monitoring system and verified its practical performance at 22 landslide sites. These studies provide a reliable basis for the landslide early-warning system. Further investigation into AE phenomena during the fracture process of the locking section is warranted. Moradian et al. [24] established the relationship between cracking levels in brittle rocks and cumulative AE hits and the energy of AE signals. Additionally, Moradian et al. [24] analyze the relationship between shear cracking behavior and AE by examining count and energy parameters, highlighting the AE monitoring's effectiveness in interpreting the shear cracking process of in situ discontinuities. While these studies provide robust support for understanding rock damage mechanisms, AE characteristic parameters offer only a simplistic statistical description of the waveform, lacking in-depth information [25].

Numerous scholars have delved into the AE waveform to reveal deeper insights into the rock's damage process. Schiavi et al. [26] and Patricia Rodríguez et al. [27] interpreted the behavior of the main frequency components of AE signals. Wang et al. [28] analyzed the frequency–amplitude distributions of AE signals throughout the failure process. They observed a noticeable increase in the amplitude of AE signals as stress increments, further accompanied by low-frequency components. More recently, researchers have delved into the sub-frequency characteristics. Mei et al. [29] classified AE signals based on both main frequencies and sub-frequencies characteristics, exploring their variations. However, focusing solely on the overall features of AE may overlook crucial information regarding rock fracture. Therefore, some scholars have categorized AE signals into burst signals and continuous signals based on their waveform characteristics, investigating their behavior [30]. Nevertheless, comprehensive analyses of these two types of AE signals during the process of rock fracture remain limited.

The AE characteristics of rock deformation and failure have been extensively studied in laboratory tests, while micro-seismic (MS) signals monitoring is commonly employed to predict the rock failure modes in engineering applications. Leśniak [31] and Pastén et al. [32] have analyzed rock failure mechanisms and stability by examining the spatial and temporal evolution of MS events. Calder and Semadeni [33], as well as Xiao et al. [34], have studied the frequency characteristics of MS events and proposed precursor indicators for rock failure. Furthermore, rock failure typically initiates with the formation of microcracks, followed by macroscopic crack propagation [35]. With advancements in digital imaging technology, techniques such as high-speed photography [36], scanning electron microscope [37], infrared thermography [38], and computerized tomography [39,40] have been introduced for monitoring and early warning of rock mass failure. Consequently, digital image processing has become an essential technical method for studying crack propagation modes and failure patterns.

This research, based on the video image and characteristics data obtained from laboratory tests designed to replicate the geological structure of the three-section landslide, aims to explore the evolutionary patterns and precursory characteristics of MS and AE signals. The investigation focuses on event activity, event intensity, and frequency. Furthermore, a novel classification method is proposed to categorize AE signals into burst signals and continuous signals. The evolution characteristics of burst signals and continuous AE signals are further examined through analysis of AE characteristic parameters (AE hits) and waveform attributes (amplitude and frequency). The method is more suitable as a precursor warning for sudden landslide disasters than deformation monitoring indicators.

## 2. Experimental Methodology

### *2.1. Sample Preparation*

The geo-mechanical model tests of three-section landslides with different angles of rock bridge were designed and prepared. The tested samples were fabricated by casting concrete using pre-prepared molds. The angle of the rock bridge of the locking section, denoted as α in Figure 1b, is defined as the angle between the line extending from the end of the tensile crack to the end of the creep section and the horizontal direction. In Figure 2, the angles of the rock bridge are 70°, 90°, and 110°, respectively. The sample material consisted of cement, lime, sand, and water, with a mass ratio of 5:1:13:4. The physical model of the specimen, as depicted in Figure 1a, was divided into four parts. The size and shape of the slope model are illustrated in Figure 1b. After the mold was filled tightly, the concrete was cured and dried under natural conditions to form the sample.

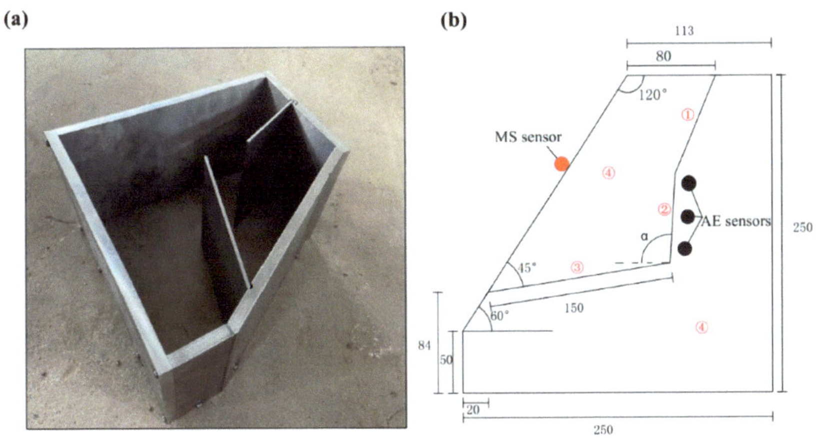


**Figure 1.** (**a**) Preprepared mold of tested samples, and (**b**) conceptual model of the locking section (unit: mm). 1—Tensile crack of the back; 2—Locking section in the middle; 3—Creep segment of the front; 4—Rock slope; α—The angles of the rock bridge.

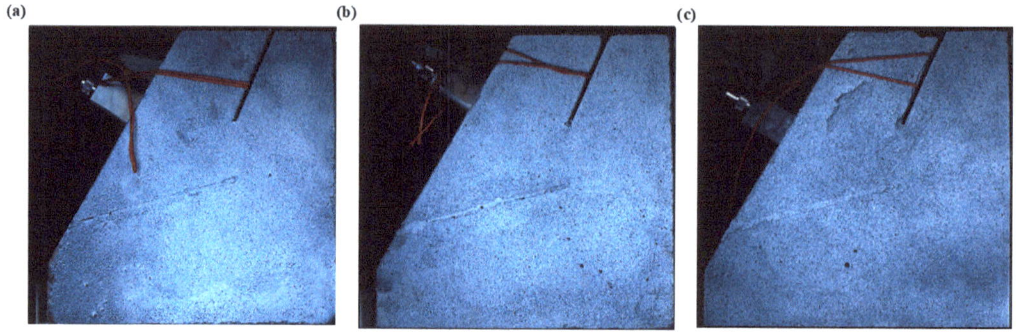


**Figure 2.** Photos of test samples: (**a**) sample #1; (**b**) sample #2; and (**c**) sample #3.

### 2.2. Test Apparatus

In this study, a WA08 strain acceleration sensor was utilized to collect MS signal, with an acquisition frequency set as 1 kHz. Concurrently, AE signals were automatically detected and collected using a Micro-II type AE system by Physical Acoustics from Princeton in America. The main amplifier of the AE test system was set to a gain of 40 dB, with a threshold set at 40 dB. The sampling frequency was configured as 1 MHz, and each waveform was sampled with 1024 points per second. To ensure the accuracy of the test results, a layer of petroleum jelly was applied between each rock specimens and the AE sensors to maintain good connections. The test employed a YH1000 microcomputer-controlled pressure testing machine from Chengdu in China. The loading method employed was displacement control mode, with loading rate of 0.2 mm/min, and preloading was set at 1 kN. Finally, the failure termination criterion for the sample was set as a displacement of 5 mm.

### 2.3. Test Schemes

A high-speed camera was employed to record the fracture process of the tested samples. Additionally, an MS sensor and three AE sensors were installed on each sample, as illustrated in Figure 1b. Once the test commenced, the lower pressing plate of the specimen machine was gradually raised to allow the cushion block to contact with the upper pressing plate, initiating stress on the locking-section model. Upon reaching a preload of 1 kN, the corresponding MS and AE signals were synchronously recorded. Simultaneously, the loading system recorded the stress data until the stress decreased rapidly upon instability and the specimen destruction. Statistical analysis of the time of AE signals received from three sensors in each test was conducted in this study. Effective time points were selected from the numerous AE signals for reduced computational burden. Each sensor exhibited good consistency in AE data across each test. Therefore, the sensor data with the largest amplitude were selected for analysis at the effective time point. Due to data similarity across the three tests and space limitations, detailed analysis in this paper focused primarily on sample #1. The results of sample #2 and sample #3 were briefly discussed.

## 3. Results and Analysis

The relationship between stress and time for sample #1 is depicted in Figure 3. The time–stress characteristics are similar to the axial deformation characteristics of rocks under uniaxial compression conditions. Therefore, with reference to the crack propagation stages commonly used in rock mechanics tests, the loading process of sample #1 was segmented into the following stages [41]:

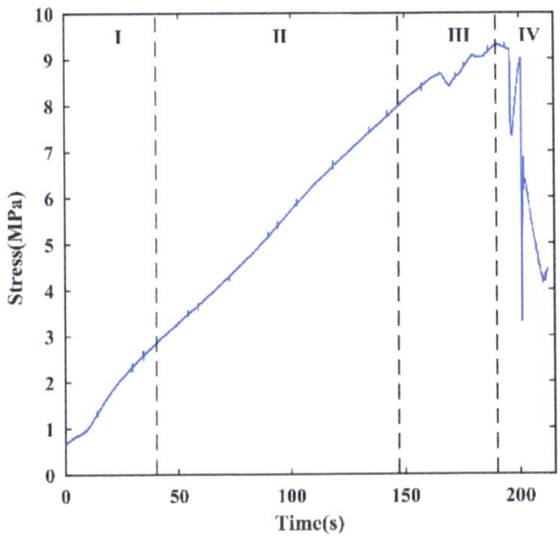


**Figure 3.** The relational curve between stress and time of sample #1.

Stage I: Compaction stage (0 to 40 s). This stage involves the mutual friction and compression of particles in the model under the applied load, resulting in an accelerated increase in the stress curve over time.

Stage II: Elastic deformation stage (40 to 147 s). During this stage, the time–stress curve that approximates a straight line, indicating a well-recoverable shear resistance within the locked section, is shown.

Stage III: Crack propagation stage (147 to 190 s). This stage shows an overall upward trend in the stress curve with multiple sudden drops, indicating the occurrence of large-scale microcracking events within the locked section due to shear compression from the upper load.

Stage IV: Rock instability and failure stage (190 to 212 s). Characterized by a significant drop in the stress curve, this stage signifies the complete loss of locking ability as the fractures in the locked section become fully connected, resulting in the complete release of energy.

For clarity, stages III and IV are collectively referred to as the crack propagation and failure stage.

### 3.1. Analysis of Video Image

Video imaging serves as an effective method to observe the specimen's failure process. Figure 4 illustrates the crack growth of sample #1. During stages I and II, the video imagery did not reveal any significant information [23]. Subsequently, three macrocracks, denoted as A, B, and C, respectively, emerged in the sample. The lower end of crack A appeared in the front part of the locking section at 167 s (0:02:47), with its propagation direction perpendicular to the creep segment. The upper end of crack B was generated at 169 s (0:02:49). With increasing load, crack A progressed upward while crack B extended downward, indicating alternating shear failure in the front and back of the locking segment. The primary crack C emerged at 194 s (0:03:14), propagating along the direction of the locked section. At 198 s, the main crack C caused relative slippage, resulting in instability and failure of the locking segment. The failure of the slope is delineated by cracks connection in the locking section. Considering the occurrence of macrocracks as the early-warning point of rock instability, this key point precedes the time of unstable failure by 27 s.

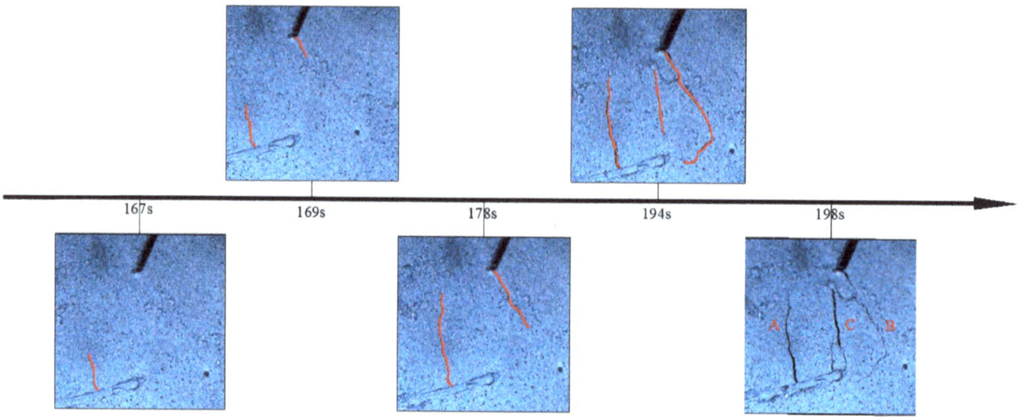


**Figure 4.** Video images of macrocracks on sample #1. The red lines in the figures indicated the nascent macroscopic cracks.

### 3.2. Analysis of MS Signal

#### 3.2.1. Activity and Intensity

In this test, the sampling frequency of the MS signal is set at 1 kHz. Firstly, the MS signal is prefiltered using the wavelet-based de-noising method to mitigate noise interference in the experimental environment [42]. The density and amplitude of the MS signal waveform directly reflect the activity and intensity of the event, respectively. The time-domain waveform of the MS signal is presented in Figure 5. The typical evolutionary process of the MS signal and the corresponding failure process can be described in detail as follows.

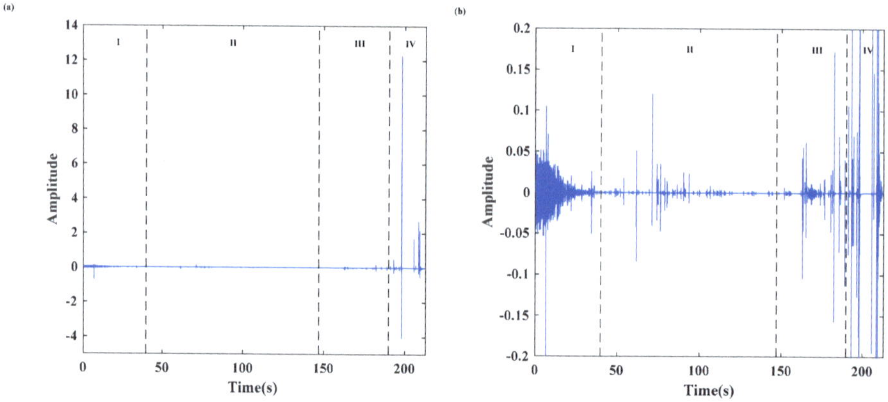


**Figure 5.** (**a**) The MS waveform, and (**b**) a partial enlargement of (**a**).

(1) In the stage of compaction due to initial damage (stage I), the density and amplitude of the MS signal are small as the external pressure gradually compresses the pre-existing cracks, voids, and/or other defects.

(2) During the elastic deformation stage (stage II), the density and amplitude of the MS signal are lower than those in the previous stage (stage I), indicating insufficient force to induce the formation of new microcracks in the locking section under the current load level. Notably, a significant locking effect and stress accumulation along the locking section are observed due to the creeping segment at the front and the tensile crack at the back of the slope.

In the stage of cracks propagation and failure (stages III and IV), the density and amplitude of the MS signal increase, signifying intense activity within the specimen and gradual failure of the locked segment. Particularly after the main crack is generated, the high-amplitude MS signal emerges continuously. The maximum amplitude was 12.25 V at 198 s when the main crack caused relative slippage, indicating sudden brittle shear failure in the locking section and an instantaneous release of tremendous energy.

3.2.2. Frequency

The continuous MS signal, filtered and sampled at 1024 points per second, underwent Fast Fourier Transform (FFT) analysis to extract the main frequencies from the sampled MS events. The corresponding main frequencies were identified based on their definition, and Figure 6 illustrates the evolutionary process of the main frequencies during the failure process of the tested specimen. This process can be described in detail as follows.

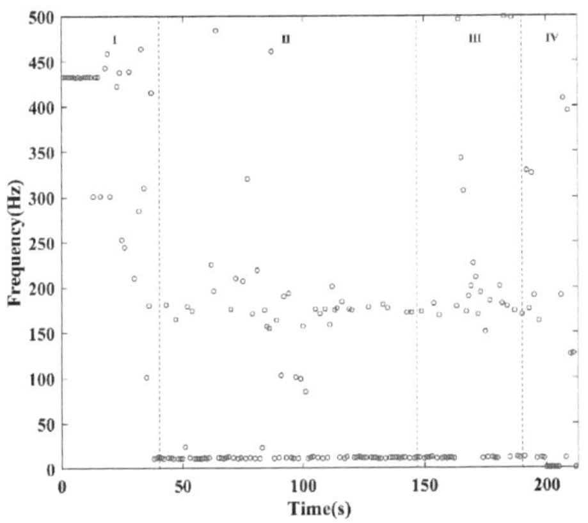


**Figure 6.** The variation in main frequencies of the MS.

(1) In the compaction stage due to initial damage (stage I), the main frequencies ranged from 250 to 500 Hz, gradually transitioning towards the low-frequency band.
(2) During the elastic deformation stage (stage II), the main frequencies of the intercepted MS signals were predominantly distributed below 250 Hz.
(3) In the crack propagation and failure stage (stages III and IV), high-frequency MS signals reappeared. At 163 s, the main frequency was 487.3 Hz. The resurgence of the high-frequency component signifies shear failure in the locking section and the emergence of microcracks appeared in the locking section. With increasing load, numerous microcracks in the locking section will expand and form macrocracks on the free face of the locking section. Thus, the reappearance of the high-frequency component serves as an early-warning point for rock instability. Notably, the first macrocrack appeared on the free face of the locking section under loading conditions at 167 s, while the time of the early-warning point of the MS signal was 163 s. Therefore, the time of the early-warning point of the MS signal precedes that of the video image.

### 3.3. Classification Method of AE Signals

AE signals are the elastic waves released in the deformation or failure process. Although both AE and MS sensors utilize the piezoelectric principle to convert vibration signals into electrical signals, the MS signals are mainly generated by stress release resulting from rock fractures or sliding. These signals are typically associated with macroscopic geological activities, rock deformations, and rock layer ruptures. On the other hand, AE signals are generated within small, localized areas of the material and are usually caused by internal stress, crack propagation, particle displacement, and other factors. Based on its time-domain morphology, AE signals can be classified into burst signals and continuous signals [43]. Burst signals are AE signals whose waveform is pulse waveform and can be separated in the time domain. In contrast, continuous signals do not have clear separations in time and typically indicate the simultaneous occurrence of numerous AE events [44]. The blue line in Figure 7a,b shows the typical burst signal waveform and continuous signal waveform, respectively. Tens of thousands of AE signals are released during the rock failure process. Efficiently classifying and extracting key information from these massive waveform data are crucial. The traditional classification methods rely on manual screening, which is time-consuming and prone to errors. Therefore, a scientific and efficient calculation method is urgently needed for automatic classification.

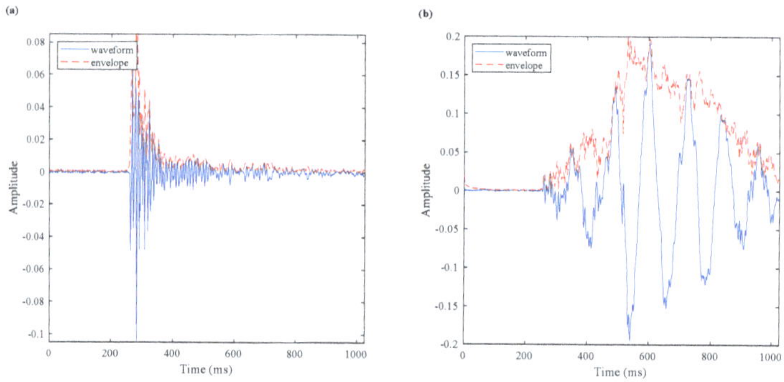


**Figure 7.** The waveform and its envelope: (**a**) typical burst signal, and (**b**) typical continuous signal.

In this paper, a new classification method is proposed based on the characteristics of the AE waveform. The flow of the algorithm is as follows: (1) To fit the waveform and the red line in Figure 7a,b is the fitting envelope; (2) To find the peak of the envelope, that is, to find the peak (amplitude) and peak position (time); (3) Count the number of peaks whose peak value is greater than 50% of the maximum peak value, and the number is denoted by $N$; (4) If $1 < N \leq 3$, and these peaks are all adjacent, go to the next step. If $N = 1$, it is a burst signal (algorithm end). Otherwise, it is a continuous signal (algorithm end); (5) Calculate the distance between its adjacent peaks, denoted by $L_i$, respectively. If $\forall L_i \leq 100, i = 1, \ldots, N-1$ (the distance is less than 0.1 ms), it is a burst signal. Otherwise, it is a continuous signal.

Using this algorithm, AE signals were classified into burst signals and continuous signals. The distribution of AE signals across stages during the failure process is presented in Table 1. The table shows the total proportion of continuous signals is as high as 80.75%, indicating a scarcity of burst signals during the test. Importantly, burst signals were exclusively observed during the stages of crack propagation and failure (stages III and IV). Therefore, subsequent analyses will focus solely on burst signals starting from stage III.

**Table 1.** Proportions of the number of AE signals in each stage.

| Proportions (%) | Stage I | Stage II | Stage III | Stage IV | Whole Process |
|---|---|---|---|---|---|
| Burst signals | 0.00 | 0.00 | 15.32 | 3.93 | 19.25 |
| Continuous signals | 1.86 | 5.18 | 37.27 | 36.44 | 80.75 |
| AE signals | 1.86 | 5.18 | 52.59 | 40.37 | 100.00 |

## 4. Analysis of Classified AE Signals

### *4.1. AE Hits (Activity)*

During the failure process of the specimens, multiple AE signals are released every second. To explore the internal relationship between AE hits and the damage of the locking section, the evolution characteristics of AE hits and cumulative AE hits of the two types of signals are analyzed. AE hits represent the number of AE signals released per second, related to the number of cracks [24]. Cumulative AE hits denote the cumulative sum of signals up to the current moment, reflecting its growth trend [45]. Figure 8 illustrates the variations in AE hits over time in the test, depicting both instantaneous AE hits and cumulative AE hits, which can be described in detail as follows.

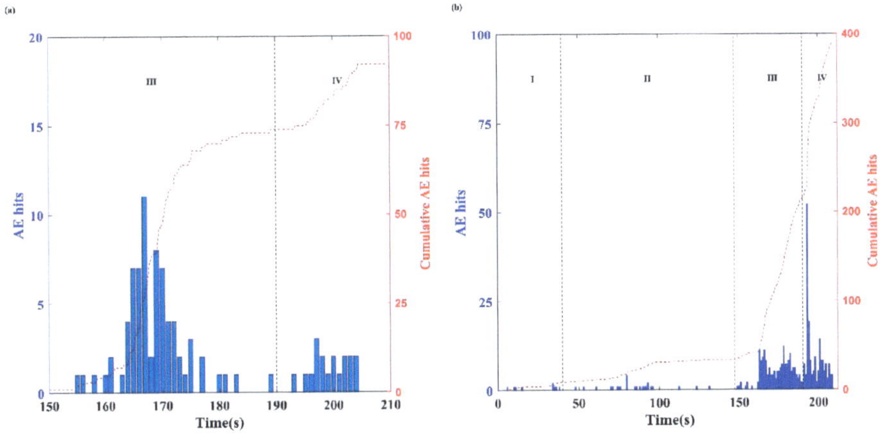


**Figure 8.** The variations in AE hits and cumulative AE hits of two types of the AE signals: (**a**) burst signals, and (**b**) continuous signals.

(1) In the stage of compaction by initial damage (stage I), the cumulative AE hits of the continuous signal gradually increased, reflecting the closure of pores and cracks in the rock caused by the stress.

(2) In the elastic deformation stage (stage II), AE hits from the continuous signal remained at a low level, and the increase rate of the cumulative AE hits gradually stabilized. AE inactivity implies the rock grains reached a state of balance and associated with accumulating energy inside the locking section.

(3) In the crack propagation and failure stages (stages III and IV), a burst signal first appeared at 147.8 s, indicating the formation of microcracks within the locking section. Therefore, in this study, the initial appearance of the burst signal was also used as the first early-warning point obtained from the AE hits. The emergence of the burst signal provides critical information obtained exclusively from the classification results. Accordingly, the time of the first early-warning point obtained from the AE hits (147.8 s) is 15.2 s earlier than that obtained from the MS signal (163 s). From 147.8 to 163 s, the AE hits increased slowly, indicating steady initiation and propagation of microcracks. Then, the cumulative AE hits of burst and continuous signals increased sharply induced by the shear failure of the locking section and microcracks were coalesced and propagated unstably. Therefore, the rapid increase in cumulative AE hits was regarded as the second early-warning point obtained from the AE hits, and the time of the second early-warning key point obtained from the AE hits is consistent with that of the MS signal, which is 163 s. After the shear failure occurred in the front and back of the locking segment alternately, the number of AE hits and cumulative AE hits increased again at 193 s, indicating that the main crack was forming as well as the sudden brittle shear failure of the locking section. It is worth noting that the number of AE hits of burst signal shows a decreasing trend, while the number of AE hits of continuous signal continues to remain at a high level.

### 4.2. Maximum Amplitude of Waveform (Intensity)

The amplitude of the AE waveform can reflect the intensity of the AE event. The process of microcrack initiation, propagation, and gradual convergence in the rock led to low amplitude. The increasing amplitude means that more and more microcracks develop and fuse to form large-scale cracks [46]. In this study, by calculating the maximum value of each waveform data, the maximum amplitude of the two types of AE signals during the failure process was obtained, as shown in Figure 9. The evolutionary process of the maximum amplitude is described as follows.

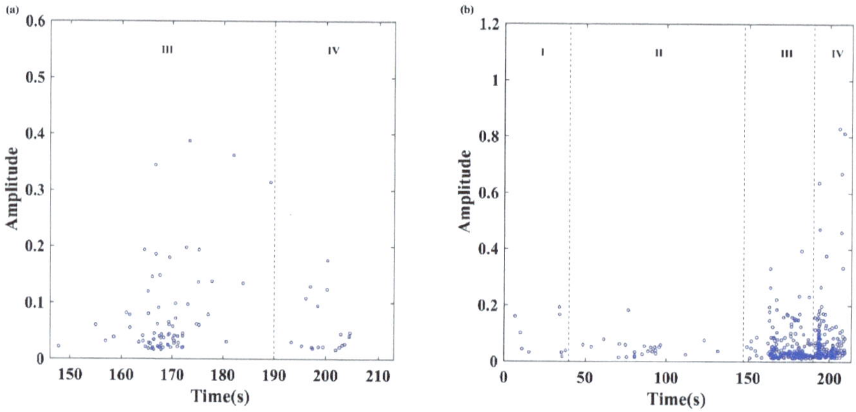


**Figure 9.** The distribution of the maximum amplitude of AE waveform: (**a**) burst signals, and (**b**) continuous signals.

(1) In the stage of compaction by initial damage (stage I), some continuous signals with low amplitude can be observed due to the compaction of internal pore.
(2) In the elastic deformation stage (stage II), the load level is insufficient to induce new microcracks inside the locking section. AE events were caused by the occlusal failure of rough surfaces when cracks caused a relative slip. Therefore, a small number of continuous signals with low energy were observed.
(3) In the stage of cracks propagation and failure (stages III and IV), the amplitude of the burst signals gradually increased and then decreased, while the amplitude of the continuous signals increased monotonously. The above phenomenon implies that several microcracks appeared with low energy after the main crack appeared, and continuous signals with higher energy were caused by the occlusal failure of rough surfaces when the cracks caused relative slippage.

### 4.3. Frequency

AE signals are characterized as non-stationary signals, and FFT is a classical spectrum analysis method to analyze the non-stationary signal. FFT was carried out for each of the AE data to extract the main frequency and sub-frequency of the corresponding waveform [40]. In this section, we analyze the frequency characteristics of burst signal and continuous signal, respectively.

#### 4.3.1. Frequency Characteristics of Burst Signals

Figure 10 shows the frequency distribution of burst signals, including the main frequency and sub-frequency. The frequencies were in the range from 0 to 180 kHz. Based on the frequency distribution, the entire frequency band was divided into three bands: 0–20 kHz (low-frequency band), 20–70 kHz (mid-frequency band), and 70–180 kHz (high-frequency band). The main frequency–amplitude characteristics of burst signals are presented in Figure 10a. It is observed that the main frequencies of burst signals were all distributed in the high- and middle-frequency bands, and there were relatively few low-frequency burst signals. Furthermore, high-amplitude burst signals were predominantly concentrated in the high- and middle-frequency bands.

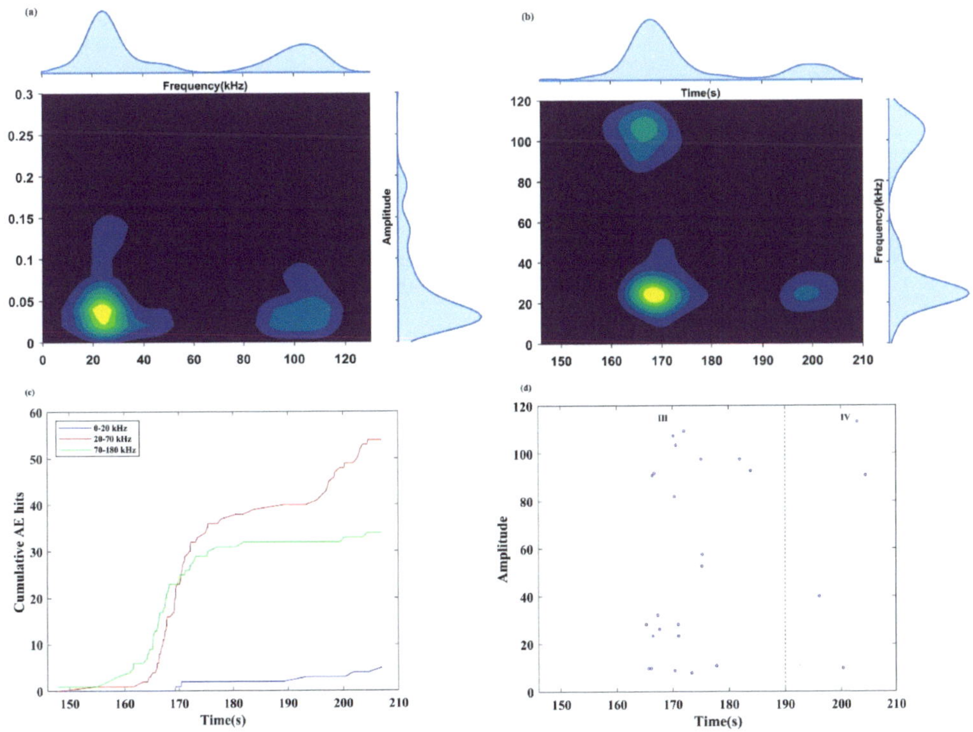


**Figure 10.** The characteristics of burst signals: (**a**) the main frequency–amplitude distribution; (**b**) the distribution of the main frequency; (**c**) the distribution of the cumulative AE hits in each main frequency band; and (**d**) the sub-frequency distribution.

The evolution characteristics of the main frequency and the cumulative AE hits in each of the main frequency bands of burst signals are shown in Figure 10b and c, respectively.

Before 147.8 s, the signal has no significant change in characteristics. From 147.8 to 163 s, the cumulative AE hits in the high-frequency band of the burst signals increased from 1 to 7, while the cumulative AE hits in the other frequency bands remained relatively stable. According to Cai et al. [47], high-frequency AE signals correspond to the initiation of microcracks, and the low-frequency signals correspond to the formation of large cracks. Therefore, the appearance of high-frequency AE signals indicates the occurrence of microcracks inside the locking segment and is the early-warning point of rock instability. The high-frequency burst signal was detected at 147.8 s, so the time of the early-warning point obtained from the frequency of burst signals (147.8 s) is 15.2 s earlier than that obtained from the MS signal (163 s). The microcracks were coalesced and propagated stably until 163 s.

From 163 to 180 s, burst signals began to appear in large numbers and were concentrated in the middle and high frequency; the cumulative AE hits in the middle- and high-frequency bands increased rapidly. The main frequency showed a penetrating trend in the whole frequency band, indicating that microcracks were rapidly generated, and the microcracks began to propagate to a few large cracks, to coalescence, and to form macrocracks.

From 180 to 190 s, the increase rate of the cumulative AE hits of burst signals in the middle- and high-frequency bands decreased gradually. After the main crack occurred, the increased rate of the cumulative AE hits in high frequency decreased obviously, indicating there were fewer microcracks in the specimen.

Figure 10d shows the sub-frequency characteristics of the burst signals. Burst signals in each frequency band have scattering characteristics in crack propagation and fracture stage (stages III and IV).

4.3.2. Frequency Characteristics of Continuous Signals

Figure 11 shows the frequency distribution of continuous signals. Like the burst signal, the frequency band is divided into three parts. Figure 11a presents the main frequency–amplitude distribution of continuous signals. The main frequency–amplitude distribution of continuous signals is significantly different from that of burst signals. Continuous signals were mainly distributed in the low- and middle-frequency bands, with high-amplitude continuous signals predominantly found in the low-frequency band.

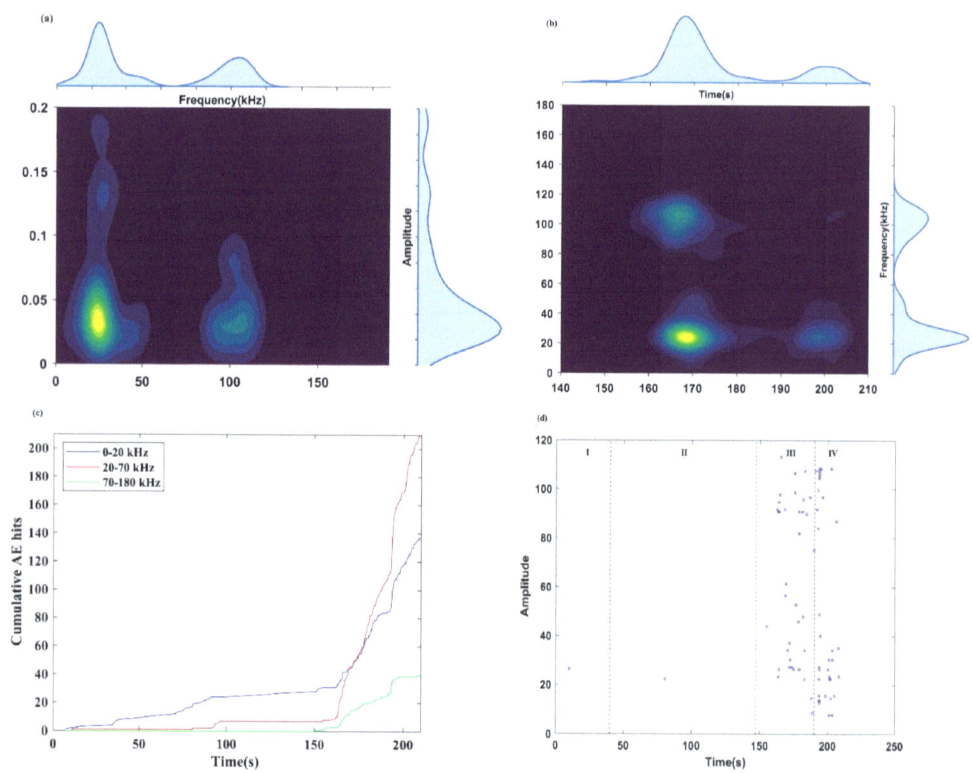


**Figure 11.** The characteristics of continuous signals: (**a**) the main frequency–amplitude distribution; (**b**) the distribution of the main frequency; (**c**) the distribution of the cumulative AE hits in each main frequency band; and (**d**) the sub–frequency distribution.

Figure 11b and c, respectively, show the evolution process of the main frequency and the cumulative AE hits in each main frequency band of the continuous signals.

Only a few continuous signals with low frequency were received during the initial compaction stage and elastic stage (stages I and II). In the stage of cracks propagation (stage III), a high-frequency continuous signal appeared for the first time at 150.9 s, marking the early-warning point of rock instability. And the time of the early-warning point obtained from the frequency of continuous signals (150.9 s) is 12.1 s earlier than that obtained from the MS signal (163 s) but 3.1 s later than that obtained from the frequency of burst signals (147.8 s). The microcracks coalesced and propagated stably from 150.9 to 163 s.

From 163 to 193 s, the cumulative AE hits of continuous signals in the low- and middle-frequency bands increased at a high rate. In contrast, the cumulative AE hits of high-frequency continuous signals increased slowly, indicating that the microcracks spread and penetrated to form large-scale ruptures at this time. The growth rate of the cumulative AE hits in 193 s was the highest, and the cumulative AE hits in each frequency band shows a sudden increase trend again, indicating that the main crack was about to form.

The main crack caused relative slippage at 198 s, and the cumulative AE hits of the continuous signals in low- and medium-frequency bands increased steadily, but the cumulative AE hits of the continuous signals in the high-frequency band remained almost stable, indicating that a small amount of small-scale fracture occurred at this time.

According to the evolution law of the sub-frequency in Figure 11d, the sub-frequency mainly appeared in the stage of cracks propagation and failure (stages III and IV), and the sub-frequencies in the low-frequency band were only detected in the stage of rock instability and failure (stage IV), indicating that the sub-frequency migrated to a low frequency.

## 5. Discussion

### *5.1. Evolution Characteristics of Sample #1*

In the stage of compaction by initial damage (stage I), the internal cracks and voids in the locking section of the three-section model were compacted and closed. The amplitude of the MS signal decreased gradually. Frequency components of the MS signal were in the range from 250 to 500 Hz, and then migrated to a low frequency. Meanwhile, a small number of AE hits occurred, primarily continuous signals with low amplitude and low frequency.

In the elastic deformation stage (stage II), the amplitude of the MS signal remained at a low level. The main frequencies of the MS signals were almost all distributed below 250 Hz. Relatively few continuous signals with low amplitude and low frequency were detected at the same time.

In the stage of cracks propagation and failure (stages III and IV), the high-frequency burst signal appeared for the first time at 147.8 s, indicating that the locking section was sheared and microcracks inside the locking section occurred, and the high-frequency continuous signal was detected for the first time at 150.9 s. The cumulative AE hits of the AE signals increased rapidly at 163 s, indicating the rapid expansion of the microcracks in the locking section. Concurrently, the high-frequency components of the MS signal appeared again. All of these are indicators as early-warning points. From 163 to 190 s, numerous burst signals with higher amplitudes were noted in the middle- and high-frequency bands. Meanwhile, a substantial number of low-amplitude continuous signals appeared in low- and middle-frequency bands. From 190 to 212 s (stage IV), low-amplitude burst signals in the mid-frequency band occurred, and several high-amplitude continuous signals in the low-frequency band were detected.

In summary, the above analysis reveals the time axis of the early-warning key points and crack propagation. As shown in Figure 12, the MS signal, burst signal, and continuous signal all exhibited one early-warning point at 163 s, which was 4 s earlier than the appearance of a macroscopic crack (derived from video image). Comparative analysis indicates that the early-warning points of AE signals preceded those of the MS signal, with the MS signal showing only one early-warning point. The appearance of a burst signal is the key warning information of sample rupture and instability, and its time was earlier than the time of the early-warning point of the overall AE signals. The first appearance of high-frequency AE signals indicates the occurrence of microcracks, so it can be used as beneficial information for prediction. Notably, the high-frequency burst signal first appeared at 147.8 s, 3.1 s earlier than the continuous signal.

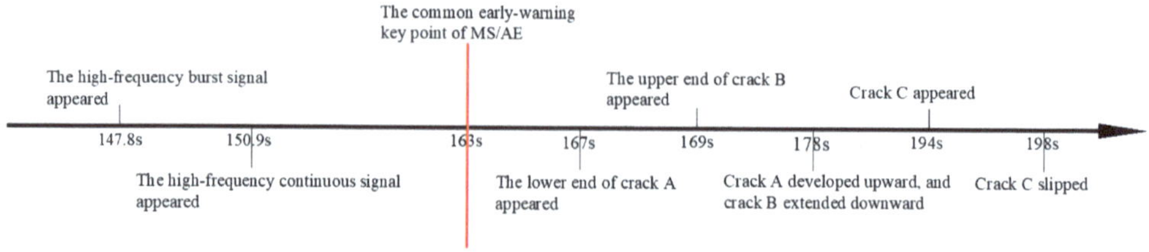


**Figure 12.** The time axis of the early-warning points and macrocracks propagation of sample #1.

## 5.2. Evolution Characteristics of Other Samples

Based on the above processing procedure, the test data of samples #2 and #3 were processed. The time axis of the early-warning points and crack propagation are shown in Figures 13 and 14. It is evident from the figures that the MS signal and AE signals respond to the early-warning point earlier than the video image. Importantly, the early-warning points of AE signals preceded those of the MS signals. It is worth noting that the time of the early-warning point of the AE signals was earlier than the MS signals. Among AE signals, the time of the early-warning point of the burst signal was earlier than the continuous signal. The time of the early-warning point of the classification results was earlier than the time of the overall AE signals. These findings align consistently with those observed in sample #1.

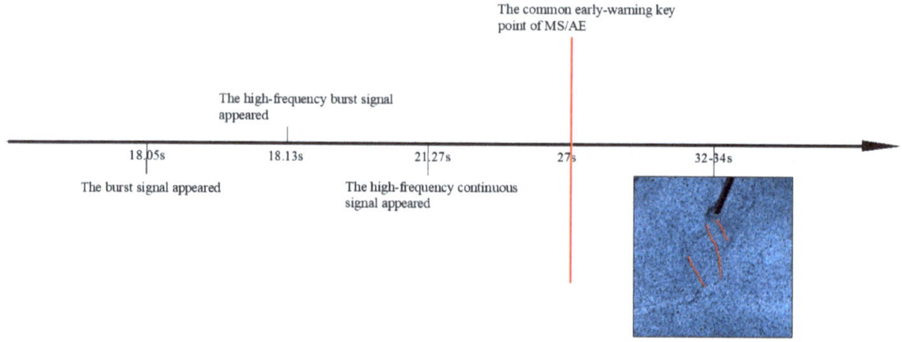


**Figure 13.** The time axis of the early-warning points and macrocracks propagation of sample #2. The red lines in picture indicates the macrocracks.

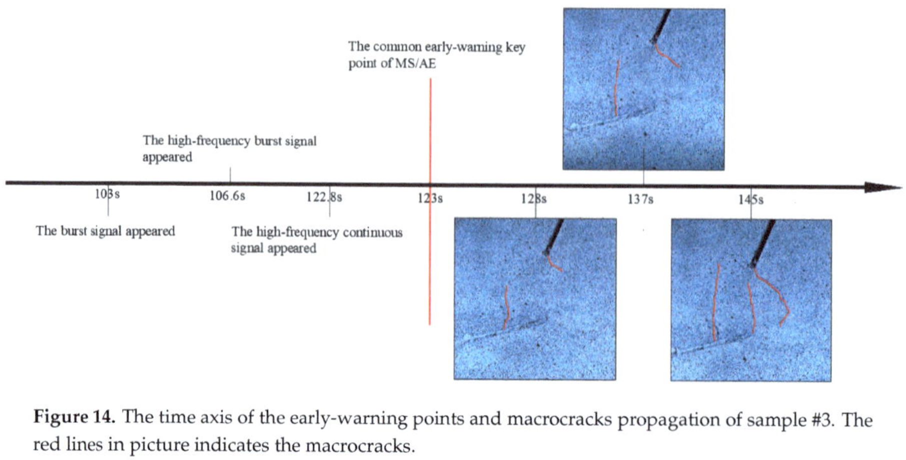


**Figure 14.** The time axis of the early-warning points and macrocracks propagation of sample #3. The red lines in picture indicates the macrocracks.

## 6. Conclusions

This study utilizes video imaging, MS signal, and AE signal to analyze the evolution characteristics and early-warning indicators of three-section landslides. We proposed a novel method to classify AE signals into burst signals and continuous signals, studying the features of the two types, respectively. The findings of this study are summarized as follows:

(1) Pronounced distinctions are evident between burst signals and continuous signals concerning event frequency, intensity, and activity. Continuous signals exhibited significantly higher total AE hits compared to burst signals, which were exclusively observed during the crack propagation and failure stage. High-amplitude burst signals were predominantly distributed in the middle- and high-frequency bands, while high-amplitude continuous signals were primarily distributed in the low-frequency band. During the crack propagation stage, both burst signals and continuous signals with high amplitude occurred. In the stage of rock instability and failure, the AE hits of burst signals exhibited a decreasing trend, with their amplitude gradually declining, while the AE hits of the continuous signal remained at a high level, with further amplitude increases.

(2) The emergence of a burst signal indicates the occurrence of microcracks in the rock mass, serving as an early-warning point of rock instability. Notably, this critical information is only obtained from the classification results. The generation of high-frequency AE signals indicates the presence of microcracks in the specimen, which can propagate and converge into macrofractures. Therefore, the occurrence of high-frequency AE signals is employed as an early-warning point for rock instability.

(3) Comparison of the timing of the early-warning points derived from the video imaging, MS signals, and AE signals revealed that both MS signals and AE signals preceded video imaging, with AE signals leading MS signals. The occurrence of early-warning points was more frequent with AE signals compared to MS signals. Importantly, the timing of the early-warning point for burst signals preceded that of continuous signals, and the timing of the early-warning point for the classification results preceded that of the overall AE signals. The research findings indicate that AE and MS are more suitable as precursor warnings for sudden landslide disasters than deformation monitoring indicators.

**Author Contributions:** Conceptualization, X.Z.; methodology, H.C.; validation, X.Z. and H.C.; formal analysis, X.L.; investigation, Z.W.; resources, T.L.; data curation, X.L. and T.L.; writing—original draft preparation, X.L. and S.Y.; writing—review and editing, X.Z.; visualization, S.Y.; supervision, H.C.; project administration, H.C. and X.Z.; funding acquisition, H.C. and T.L. All authors have read and agreed to the published version of the manuscript.

**Funding:** The research was funded in part by the State Key Laboratory of Geohazard Prevention and Geoenvironment Protection Independent Research Project under Grants SKLGP2022Z011; in part by the Technology Innovation Research and Development Project of Chengdu Science and Technology Bureau under Grant 2024-YF05-01584-SN; in part by the Sichuan Province Central Government Guides Local Science and Technology Development Special Project (2023ZYD0151); Sichuan Science and Technology Program (grant number 2023NSFSC0282); in part by the Creative Research Groups of the Natural Science Foundation of Sichuan under Grants 2023NSFSC1984; in part by the Research Project of Chunhui Plan, Ministry of Education.

**Institutional Review Board Statement:** Not applicable.

**Informed Consent Statement:** Not applicable.

**Data Availability Statement:** The data that support the findings of this study are available from the corresponding author upon reasonable request.

**Acknowledgments:** The authors sincerely thank the editors and reviewers for their constructive comments, which greatly help improve the quality of this paper. Many thanks to Jie Fan for his contribution in revising the paper.

**Conflicts of Interest:** Author Zhanglei Wu was employed by the company Powerchina Chengdu Engineering Corporation Limited. The remaining authors declare that the research was conducted in the absence of any commercial or financial relationships that could be construed as a potential conflict of interest.

## References

1. Huang, R.; Chen, G.; Guo, F.; Zhang, G.; Zhang, Y. Experimental Study on the Brittle Failure of the Locking Section in a Large-Scale Rock Slide. *Landslides* **2016**, *13*, 583–588. [CrossRef]
2. Huang, R.Q. Mechanisms of Large-Scale Landslides in China. *Bull. Eng. Geol. Environ.* **2012**, *71*, 161–170. [CrossRef]
3. Amare, K.; Mebrahtu, G.; Gebremiceal, A. Landslide Characterization and Distribution in Northern Ethiopia: A Case Study from the Adishu Area. *Bull. Eng. Geol. Environ.* **2018**, *77*, 581–593. [CrossRef]
4. Eberhardt, E.; Stead, D.; Coggan, J.S. Numerical Analysis of Initiation and Progressive Failure in Natural Rock Slopes—The 1991 Randa Rockslide. *Int. J. Rock Mech. Min. Sci.* **2004**, *41*, 69–87. [CrossRef]
5. Elmo, D.; Donati, D.; Stead, D. Challenges in the Characterisation of Intact Rock Bridges in Rock Slopes. *Eng. Geol.* **2018**, *245*, 81–96. [CrossRef]
6. Brideau, M.A.; Yan, M.; Stead, D. The Role of Tectonic Damage and Brittle Rock Fracture in the Development of Large Rock Slope Failures. *Geomorphology* **2009**, *103*, 30–49. [CrossRef]
7. Romer, C.; Ferentinou, M. Numerical Investigations of Rock Bridge Effect on Open Pit Slope Stability. *J. Rock Mech. Geotech. Eng.* **2019**, *11*, 1184–1200. [CrossRef]
8. Huang, R.Q.; Chen, G.Q.; Tang, P. Precursor Information of Locking Segment Landslides Based on Transient Characteristics. *Chinese J. Rock Mech. Eng.* **2017**, *36*, 521–533. [CrossRef]
9. Pan, X.H.; Xue, L.; Qin, S.Q.; Li, G.L.; Li, P.; Wang, M.M. Types, formation Conditions and Pre-Decision Method for Large Landslides with Potential Locked Patches. *J. Eng. Geol.* **2014**, *22*, 1159–1167. [CrossRef]
10. Gehle, C.; Kutter, H.K. Breakage and Shear Behaviour of Intermittent Rock Joints. *Int. J. Rock Mech. Min. Sci.* **2003**, *40*, 687–700. [CrossRef]
11. Sarfarazi, V.; Ghazvinian, A.; Schubert, W.; Blumel, M.; Nejati, H.R. Numerical Simulation of the Process of Fracture of Echelon Rock Joints. *Rock Mech. Rock Eng.* **2014**, *47*, 1355–1371. [CrossRef]
12. Ghazvinian, A.; Sarfarazi, V.; Schubert, W.; Blumel, M. A Study of the Failure Mechanism of Planar Non-Persistent Open Joints Using PFC2D. *Rock Mech. Rock Eng.* **2012**, *45*, 677–693. [CrossRef]
13. Labuz, J.F.; Cattaneo, S.; Chen, L.H. Acoustic Emission at Failure in Quasi-Brittle Materials. *Constr. Build. Mater.* **2001**, *15*, 225–233. [CrossRef]
14. Kasai, N.; Utatsu, K.; Park, S.; Kitsukawa, S.; Sekine, K. Correlation between Corrosion Rate and AE Signal in an Acidic Environment for Mild Steel. *Corros. Sci.* **2009**, *51*, 1679–1684. [CrossRef]
15. Tan, C.K.; Mba, D. Identification of the Acoustic Emission Source during a Comparative Study on Diagnosis of a Spur Gearbox. *Tribol. Int.* **2005**, *38*, 469–480. [CrossRef]
16. Pieri, M.; Burlini, L.; Kunze, K.; Stretton, I.; Olgaard, D.L. Rheological and Microstructural Evolution of Carrara Marble with High Shear Strain: Results from High Temperature Torsion Experiments. *J. Struct. Geol.* **2001**, *23*, 1393–1413. [CrossRef]
17. Zhang, Y.; Jiang, Y.; Asahina, D.; Wang, Z. Shear Behavior and Acoustic Emission Characteristics of En-Echelon Joints under Constant Normal Stiffness Conditions. *Theor. Appl. Fract. Mech.* **2020**, *109*, 102772. [CrossRef]
18. Moradian, Z.A.; Ballivy, G.; Rivard, P.; Gravel, C.; Rousseau, B. Evaluating Damage during Shear Tests of Rock Joints Using Acoustic Emissions. *Int. J. Rock Mech. Min. Sci.* **2010**, *47*, 590–598. [CrossRef]
19. Smith, A.; Dixon, N.; Meldrum, P.; Haslam, E.; Chambers, J. Acoustic Emission Monitoring of a Soil Slope: Comparisons with Continuous Deformation Measurements. *Proc. Inst. Civ. Eng. Struct. Build.* **2014**, *4*, 255–261. [CrossRef]
20. Dixon, N.; Smith, A.; Flint, J.A.; Khanna, R.; Clark, B.; Andjelkovic, M. An Acoustic Emission Landslide Early Warning System for Communities in Low-Income and Middle-Income Countries. *Landslides* **2018**, *15*, 1631–1644. [CrossRef]
21. Berg, N.; Smith, A.; Russell, S.; Dixon, N.; Proudfoot, D.; Andy Take, W. Correlation of Acoustic Emissions with Patterns of Movement in an Extremely Slow-Moving Landslide at Peace River, Alberta, Canada. *Can. Geotech. J.* **2018**, *55*, 1475–1488. [CrossRef]
22. Kumar, D.; Mahapatro, A.K.; Singh, S.K. Active Waveguide Deformation Dynamics Using Acoustic Emission Technology for Landslide Early Warning System. *Bull. Eng. Geol. Environ.* **2024**, *83*, 1–13. [CrossRef]
23. Deng, L.; Yuan, H.; Chen, J.; Zhang, M.; Su, G.; Zhou, Y.; Chen, Y. Experimental Investigation and Field Application of Acoustic Emission Array for Landslide Monitoring. *Landslides* **2024**, *21*, 71–81. [CrossRef]
24. Moradian, Z.; Einstein, H.H.; Ballivy, G. Detection of Cracking Levels in Brittle Rocks by Parametric Analysis of the Acoustic Emission Signals. *Rock Mech. Rock Eng.* **2016**, *49*, 785–800. [CrossRef]
25. Behnia, A.; Chai, H.K.; Shiotani, T. Advanced Structural Health Monitoring of Concrete Structures with the Aid of Acoustic Emission. *Constr. Build. Mater.* **2014**, *65*, 282–302. [CrossRef]
26. Schiavi, A.; Niccolini, G.; Tarizzo, P.; Lacidogna, G.; Manuello, A.; Carpinteri, A. Analysis of Acoustic Emissions at Low Frequencies in Brittle Materials under Compression. In Proceedings of the Society for Experimental Mechanics—SEM Annual Conference and Exposition on Experimental and Applied Mechanics, Albuquerque, NM, USA, 1–4 June 2009.

27. Rodríguez, P.; Celestino, T.B. Application of Acoustic Emission Monitoring and Signal Analysis to the Qualitative and Quantitative Characterization of the Fracturing Process in Rocks. *Eng. Fract. Mech.* **2019**, *210*, 54–69. [CrossRef]
28. Wang, C.; Chang, X.; Liu, Y.; Chen, S. Mechanistic Characteristics of Double Dominant Frequencies of Acoustic Emission Signals in the Entire Fracture Process of Fine Sandstone. *Energies* **2019**, *12*, 3959. [CrossRef]
29. Mei, F.; Hu, C.; Li, P.; Zhang, J. Study on Main Frequency Precursor Characteristics of Acoustic Emission from Deep Buried Dali Rock Explosion. *Arab. J. Geosci.* **2019**, *12*, 645. [CrossRef]
30. Liu, G.H. *The Research on the Key Technologies in Acoustic Emission's Signal Processing*; Zhejiang University: Hangzhou, China, 2008.
31. Leśniak, A.; Isakow, Z. Space-Time Clustering of Seismic Events and Hazard Assessment in the Zabrze-Bielszowice Coal Mine, Poland. *Int. J. Rock Mech. Min. Sci.* **2009**, *46*, 918–928. [CrossRef]
32. Pastén, D.; Estay, R.; Comte, D.; Vallejos, J. Multifractal Analysis in Mining Microseismicity and Its Application to Seismic Hazard in Mine. *Int. J. Rock Mech. Min. Sci.* **2015**, *78*, 74–78. [CrossRef]
33. Calder, P.N.; Semadeni, T.J. High Frequency Microseismic Monitoring Trial near a Rapidly Changing Stress Field in a Hard Rock Mine. *Gerlands Beitrage Geophys.* **1990**, *99*, 21–34.
34. Xiao, P.; Qian, B.; Jiang, P.; Xu, N.; Li, B. Deformation Forecasting of Surrounding Rock Mass Based on Correlation between Frequency and Fracture Scale of Microseismicity. *Adv. Civ. Eng.* **2018**, *2018*, 1–13. [CrossRef]
35. Liu, B.; Zhou, Y.; Gao, Y.T.; Xu, C. Experimental and Numerical Study on Crack Development Characteristics between Two Cavities in Rock-like Material under Uniaxial Compression. *Theor. Appl. Fract. Mech.* **2020**, *109*, 102755. [CrossRef]
36. Lotidis, M.A.; Nomikos, P.P.; Sofianos, A.I. Numerical Study of the Fracturing Process in Marble and Plaster Hollow Plate Specimens Subjected to Uniaxial Compression. *Rock Mech. Rock Eng.* **2019**, *52*, 4361–4386. [CrossRef]
37. Zuo, J.; Wei, X.; Pei, J.; Zhao, X. Investigation of Meso-Failure Behaviors of Jinping Marble Using SEM with Bending Loading System. *J. Rock Mech. Geotech. Eng.* **2015**, *7*, 593–599. [CrossRef]
38. Pappalardo, G.; Mineo, S.; Angrisani, A.C.; Di Martire, D.; Calcaterra, D. Combining Field Data with Infrared Thermography and DInSAR Surveys to Evaluate the Activity of Landslides: The Case Study of Randazzo Landslide (NE Sicily). *Landslides* **2018**, *15*, 2173–2193. [CrossRef]
39. Gao, R.; Du, X.; Zeng, Y.; Li, Y.; Yan, J. A New Method to Simulate Irregular Particles by Discrete Element Method. *J. Rock Mech. Geotech. Eng.* **2012**, *4*, 276–281. [CrossRef]
40. Kou, M.; Liu, X.; Tang, S.; Wang, Y. 3-D X-Ray Computed Tomography on Failure Characteristics of Rock-like Materials under Coupled Hydro-Mechanical Loading. *Theor. Appl. Fract. Mech.* **2019**, *104*, 102396. [CrossRef]
41. Zhao, K.; Yang, D.; Gong, C.; Zhuo, Y.; Wang, X.; Zhong, W. Evaluation of Internal Microcrack Evolution in Red Sandstone Based on Time–Frequency Domain Characteristics of Acoustic Emission Signals. *Constr. Build. Mater.* **2020**, *260*, 120435. [CrossRef]
42. Zhu, X.; Xu, Q.; Zhou, J.; Tang, M. International Journal of Rock Mechanics & Mining Sciences Experimental Study of Infrasonic Signal Generation during Rock Fracture under Uniaxial Compression. *Int. J. Rock Mech. Min. Sci.* **2013**, *60*, 37–46. [CrossRef]
43. Zhao, J.R. *Acoustic Emission Signal Processing System and Source Recognition Methods*; Jilin University: Changchun, China, 2010.
44. Liu, H.D.; Zhang, Y.B.; Lu, L.P. Types of the Locked Section Landslide in the Western Henan Province. *J. North China Univ. Water Resources Electr. Power* **2018**, *39*, 2018. [CrossRef]
45. Wang, C.Y.; Chang, X.K.; Liu, Y.L.; Guo, W. Bin Spectrum Evolution Characteristics of Acoustic Emission during the Rupture Process of Marble under Uniaxial Compression Condition. *Rock Soil Mech.* **2020**, *41*, 51–62. [CrossRef]
46. Zhang, G.K.; Li, H.B.; Wang, M.Y.; Li, J.; Deng, S.X. Comparative Study on Damage Characterization and Damage Evolution of Rock under Uniaxial Compression. *Chin. J. Geotech. Eng.* **2019**, *41*, 1074–1082.
47. Cai, M.; Kaiser, P.K.; Morioka, H.; Minami, M.; Maejima, T.; Tasaka, Y.; Kurose, H. FLAC/PFC Coupled Numerical Simulation of AE in Large-Scale Underground Excavations. *Int. J. Rock Mech. Min. Sci.* **2007**, *44*, 550–564. [CrossRef]

*sensors*

MDPI

*Article*

# Three-Dimensional Reconstruction and Visualization of Underwater Bridge Piers Using Sonar Imaging

**Jianbin Luo [1], Shaofei Jiang [1,*], Yamian Zeng [2] and Changqin Lai [1]**

[1] College of Civil Engineering, Fuzhou University, Fuzhou 350108, China; ljb312@fzu.edu.cn (J.L.)
[2] College of Civil Engineering, Quanzhou University of Information Engineering, Quanzhou 362008, China
* Correspondence: cejsf@fzu.edu.cn

**Abstract:** The quality of underwater bridge piers significantly impacts bridge safety and long-term usability. To address limitations in conventional inspection methods, this paper presents a sonar-based technique for the three-dimensional (3D) reconstruction and visualization of underwater bridge piers. Advanced MS1000 scanning sonar is employed to detect and image bridge piers. Automated image preprocessing, including filtering, denoising, binarization, filling, and morphological operations, introduces an enhanced wavelet denoising method to accurately extract the foundation contour coordinates of bridge piers from sonar images. Using these coordinates, along with undamaged pier dimensions and sonar distances, a model-driven approach for a 3D pier reconstruction algorithm is developed. This algorithm leverages multiple sonar data points to reconstruct damaged piers through multiplication. The Visualization Toolkit (VTK) and surface contour methodology are utilized for 3D visualization, enabling interactive manipulation for enhanced observation and analysis. Experimental results indicate a relative error of 13.56% for the hole volume and 10.65% for the spalling volume, demonstrating accurate replication of bridge pier defect volumes by the reconstructed models. Experimental validation confirms the method's accuracy and effectiveness in reconstructing underwater bridge piers in three dimensions, providing robust support for safety assessments and contributing significantly to bridge stability and long-term safety assurance.

**Keywords:** underwater bridge pier; sonar imaging; three-dimensional reconstruction

## 1. Introduction

To ensure the safety and reliability of underwater bridge piers, regular physical examinations and monitoring are crucial. The harsh hydrogeological environment and long-term exposure to water erosion and human activities make these foundations susceptible to various issues that can compromise the safety of bridge operations. Therefore, it is imperative for the highway department to prioritize research on underwater bridge piers' detection technology. This research enables the timely detection and assessment of problems in underwater bridge piers, allowing for necessary repairs and reinforcements. Inspection techniques such as sonar and diving equipment can be utilized to conduct comprehensive evaluations of structural integrity, load-bearing capacity, and durability. The results of these inspections inform repair and reinforcement efforts, thereby ensuring the normal operation of the bridge and traffic safety.

Due to the challenging environment of underwater bridge piers, conventional methods for bridge pile detection, such as ultrasonic pulse detection, drilling coring, and low-strain methods, are difficult to apply effectively. Instead, various methods based on detecting the geometric characteristics of surface defects are commonly used. These methods include artificial diving/underwater optical imaging technology [1,2], underwater laser imaging technology [3], and underwater sonar imaging technology [4,5].

In the field of underwater detection, the underwater sonar imaging technology has significant advantages compared to other techniques [6]. The underwater sonar imaging technology employs a sonar array to emit sound pulses into the water. When the sound pulse encounters a detection object, it generates a reflected echo. By receiving and analyzing the information from the reflected echo, one can determine the distance between the detection object and the sonar and generate corresponding images. Compared to optical imaging technology, underwater sonar imaging technology has a longer sound wave wavelength (in the megahertz range), weaker scattering, slower propagation, and attenuation in water. It provides a broad imaging range and is unaffected by illumination [7]. These advantages make underwater sonar imaging technology a critical method for underwater bridge detection.

Currently, underwater sonar imaging technology is widely used in various fields, including engineering structure inspection [4,5], underwater pipelines [8,9], seabed topography [10,11], and riverbed exploration. In the context of underwater bridge inspection, relevant research and applications have been conducted. De et al. [12] conducted experiments on bridge scour conditions to study riverbed evolution. Topczewski et al. [13] performed the sonar scanning and imaging of underwater bridge piers and nearby riverbeds, obtaining information on pier outlines and adjacent riverbeds through acoustic image processing. Clubley et al. [14] conducted sonar imaging scans of elevated bridge piers and identified multiple defects that divers had not detected, indicating a higher reliability of sonar detection compared to diver inspections. Hunt et al. [15] analyzed the impact of scour on bridges and developed scour monitoring systems using sonar devices.

Meanwhile, three-dimensional (3D) reconstruction technology uses modern measurement techniques to acquire 2D data and, through computer image processing, reconstructs 3D models that include detailed surface information [16]. Three-dimensional imaging sonar technology offers more detailed descriptions and stereoscopic positioning, providing significant advantages over traditional 2D sonar. Typically, this technology involves a series of processes such as data acquisition, image preprocessing, registration and fusion, and the generation of three-dimensional surfaces. It has a wide range of applications in fields like cultural heritage preservation [17], clinical medicine [18], architectural modeling [19], game development [20], and film special effects [21]. Davis et al. [22] used the Coda Echoscope 3D sonar for high-speed underwater inspections in ports and harbor security, generating real-time 3D images of underwater scenes. Other studies include sonar scanning and imaging detections of railway bridge substructures, dock piles, and bridge foundation scour. Regarding the 3D reconstruction of underwater bridge piers, it can be broadly classified into two categories of methods: model-driven and data-driven. Model-driven approaches involve matching known 3D model information with data obtained from scanning to generate authentic 3D models. In contrast, data-driven methods directly generate planar images from the scanned data and subsequently combine these planar images to construct 3D models. Model-driven methods are suitable for relatively simple structural models, exhibiting generally moderate effectiveness for complex structural models. Conversely, data-driven methods demand a higher data quality but can achieve 3D reconstruction of complex structures.

In conclusion, in the field of underwater bridge pier detection, while many scholars have devoted considerable effort and made significant progress, there are still several pressing challenges. For example, interpreting sonar images depicting surface defects on underwater bridge piers can prove to be quite daunting, especially for individuals lacking a background in sonar technology. Relying solely on a single 2D sonar image for diagnosing these surface defects is often insufficient. Instead, a comprehensive diagnosis necessitates the integration of multiple 2D sonar images captured from various angles. Moreover, it requires inspection personnel to possess a fundamental understanding of sonar imaging principles and to draw upon their expert experience to render accurate diagnoses. This situation bears some resemblance to the way medical doctors must grasp the fundamental principles of CT imaging and rely on their professional expertise to make precise diagnoses based on CT scans of patients.

The sonar image data obtained from single-beam scanning is relatively sparse and of lower accuracy. Additionally, practical engineering constraints limit the range of observation angles, and underwater bridge piers are relatively simple. Therefore, to quickly achieve a 3D model with sufficient engineering accuracy, it seems more appropriate to combine known pier dimensions and use the model-driven approach in this situation. In this approach, we propose a methodology that involves matching the sonar images acquired through scanning with established bridge pier models. We collaboratively reconstruct the underwater bridge piers by combining information sourced from multiple measurement points within the sonar images. Compared to Reference [4], this approach capitalizes on the correspondence between 2D sonar image features of surface defects and 3D shapes, facilitating the precise reconstruction of 3D models depicting surface defects on underwater bridge piers. Subsequently, we use triangular mesh techniques to reconstruct the topology of underwater bridge piers, enabling an approximation of the representation of underwater bridge piers' surface information. Additionally, we develop specialized digital reconstruction system software tailored for this purpose.

This paper is organized as follows: Section 2 presents an introduction to the mechanism underlying the 3D reconstruction of sonar images of underwater bridge piers. Section 3 introduces an automated recognition method specifically tailored for extracting surface profiles from 2D sonar images. Section 4 focuses on the algorithm employed for the 3D reconstruction of underwater bridge piers. It revolves around the reconstruction of cross-sections of the piers and the generation of a 3D model of the undamaged pier using known information such as dimensions and sonar distance. The identified contour coordinates serve as control points, significantly enhancing the accuracy of the reconstruction process. Section 5 facilitates digital analysis and the observation of surface defects in the 3D models, enabling actions such as 360° rotation, multi-angle viewing, and digital management. This section also discusses how the 3D reconstruction algorithm contributes to the foundational data necessary for 3D model visualization. Section 6 provides a detailed description of the underwater sonar detection experiments conducted on a pier model with man-made surface defects. The reconstructed pier model is subsequently compared with actual measurements to validate the proposed reconstruction and visualization techniques.

## 2. Mechanism of 3D Reconstruction of Sonar Images of Underwater Bridge Piers

To achieve the 3D reconstruction of surface defects on underwater bridge piers using sonar images, it is evident that relying solely on the 2D information derived from individual sonar images of pier contours is insufficient. This limitation becomes particularly pronounced when dealing with underwater bridge piers that exhibit surface defects. Therefore, it becomes imperative to integrate data obtained from multiple sonar image acquisition points and combine them to facilitate recognition and reconstruction. During the fusion and reconstruction process, it is crucial to delve deeper into understanding how information gathered from multiple sonar image acquisition points synergistically contributes to the generation of a 3D model of the underwater bridge piers. This section will begin with an introduction to the imaging principles underlying the MS1000 sonar. Subsequently, it will provide a comprehensive explanation of the mechanism involved in fusing information originating from multiple sonar image acquisition points. It will delve into the analysis of the correspondence between the 2D characteristics of various surface defects within sonar images and their corresponding 3D shapes, presenting potential scenarios. By harnessing the 3D reconstruction mechanism that relies on multiple sonar image acquisition points, we can achieve a more precise determination of potential scenarios, thereby enabling a more accurate reconstruction of the 3D model representing surface defects.

### 2.1. *The MS1000 Mechanical Scanning Imaging Sonar*

The MS1000 mechanical scanning imaging sonar, proudly crafted by the Canadian company Kongsberg, stands out as a global leader in sonar technology due to exceptional performance characteristics, characterized by precision and resolution.

As a representative of active sonar systems, the MS1000 sonar possesses the capability to generate sonar images by analyzing echo data. At its core lies the transducer, a vital element within the sonar system. This transducer emits sonar beam pulses into the submerged target area following mechanical rotation to a specific angle. These pulses interact with underwater targets, subsequently reflecting back to the sonar system, thereby generating echo signals. The MS1000 sonar effectively processes and analyzes these echo signals to produce high-quality sonar images, capturing the characteristics and structures of the underwater environment. Notably, the standout feature of the transducer is its ability to achieve a maximum scanning range of 360°, as depicted in Figure 1's schematic diagram of single-beam scanning sonar.

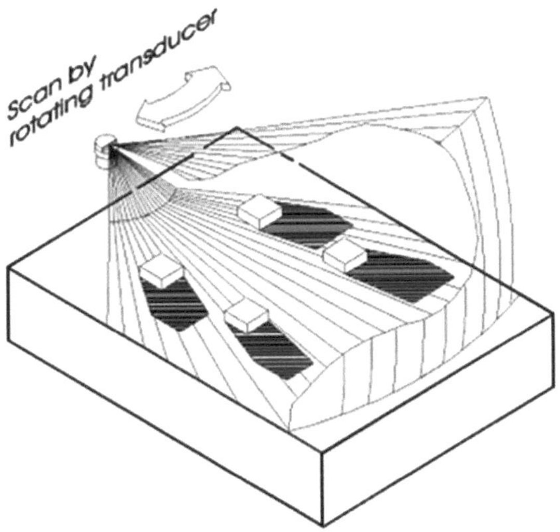


**Figure 1.** Schematic diagram of acoustic imaging.

Figure 1 elucidates the fundamental principles of sonar imaging, where the sonar transducer, when rotated to a specific angle, begins to receive and process reflected waves from identified targets. The collected information correlates with the intensity of the reflected waves, ultimately manifesting as color variations within the sonar image. Regions with higher reflected wave intensities display vibrant colors, while those with lower intensities appear darker. Through a 360° scanning rotation, the sonar transducer constructs a comprehensive sonar image.

Single-beam scanning imaging sonar operates in distance, azimuth, and intensity modes, generating 2D planar images of the underwater environment. While this sonar device is adept at detecting 3D objects, its primary function remains the production of planar images, and it does not provide complete 3D information.

## *2.2. Mechanism of 3D Reconstruction Using Multi-Point Sonar Imaging*

To achieve precise 3D reconstructions of bridge piers and explore the connection between 2D sonar image features and 3D structures, we initiate by establishing a mapping relationship between the 2D sonar image data and the actual scanning space. As illustrated in Figure 2, we define a spherical coordinate system, denoted as ρθφ, within the spatial domain, with the sonar's position serving as the origin. Key parameters encompass the sonar's detection range, denoted as "Range", the present sonar azimuth angle, represented as $\theta$, the current sonar elevation angle, designated as $\varphi$, and the sample count for each beam, referred to as "NumBins." The equation characterizing the spatial region associated with the $n$th sample point $b$ in spherical coordinates is succinctly described as Equation (1).

$$\begin{cases} \theta - 0.45^\circ \leq \theta_b \leq \theta + 0.45^\circ \\ \varphi - 15^\circ \leq \varphi_b \leq \varphi + 15^\circ \\ \text{Range} \times \frac{\left[\rho \times \frac{\text{NumBins}}{\text{Range}}\right]}{\text{NumBins}} \leq \rho_b \leq Range \times \left[\rho \times \frac{\text{NumBins}}{\text{Range}} + 1\right] / NumBins \end{cases} \quad (1)$$

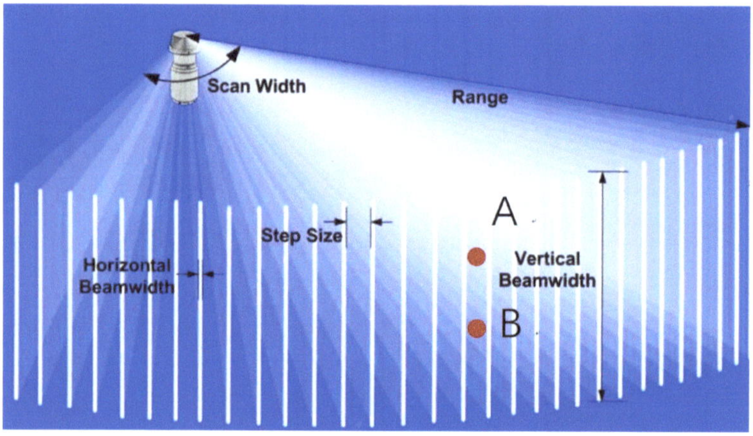


**Figure 2.** Schematic diagram of single-beam scanning imaging sonar scan.

Sonar images encapsulate temporal, angular, and intensity data. It is noteworthy that for data points where different angles correspond to identical time-of-arrival instances, such as points A and B (See Figure 2), the sonar is incapable of distinguishing between them.

Figure 3 illustrates that when the sonar is positioned at measurement point $C_1$, the transducer emits sonar beam pulses towards the underwater bridge pier region. Points A and B reside on the same beam arc, and at the peak of reverse-scattered echo intensity along the direct linear distance, they concurrently reach the sonar. In this scenario, the sonar cannot differentiate between points A and B, leading to their coalescence in the sonar image, manifesting as a single pixel. Nonetheless, when the sonar scans from an alternative measurement point, $C_2$, points A and B occupy disparate beam pulse arcs, causing variations in their arrival times at the sonar. Consequently, the sonar distinguishes between points A and B, representing them as distinct pixels in the sonar image. Analogously, points that are indistinguishable when the sonar is positioned at measurement point $C_2$ can typically be discerned when positioned at measurement point $C_1$. Leveraging this complementary relationship between measurements at points $C_1$ and $C_2$ facilitates the derivation of 3D information concerning the underwater bridge piers.

When conducting sonar scanning imaging on cylindrical bridge piers, a multi-point approach is utilized to generate multiple 2D sonar images, as illustrated in Figure 4. Eight designated measurement points, labeled $C_1, C_2, \ldots, C_8$, are strategically positioned. The dashed lines in the figure represent the emitted beam pulses originating from measurement points $C_1, C_2, \ldots, C_8$. By performing the multiplication of intensity data corresponding to pixels at the intersection of these two beam pulses, a dataset is generated. If a specific region is covered by multiple beam pulses and exhibits a heightened reflection intensity, the result of the multiplication will increase, indicating the presence of an underwater bridge pier in that region. Conversely, if a region displays a substantial reflection intensity on only one beam pulse while the other beam pulse indicates no significant reflection intensity, the result of the multiplication will decrease, implying the absence of an underwater bridge pier in that area. Consequently, the determination of underwater bridge pier presence or absence can be made based on these multiplication values.

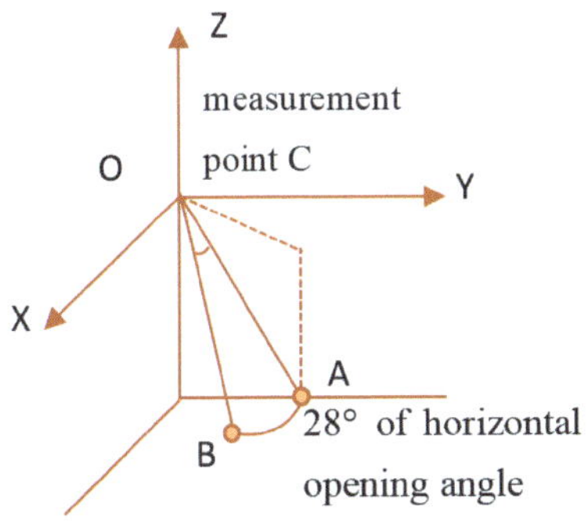


**Figure 3.** Schematic diagram of imaging principle.

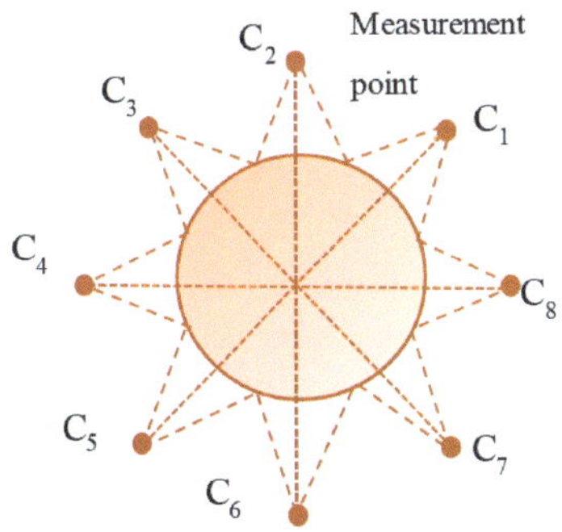


**Figure 4.** Schematic of measurement point beam coverage area.

In simpler terms, this means that if the sonar image data acquired from one measurement point are converted into 3D spatial data and then multiplied with 3D spatial data derived from sonar image data obtained from another measurement point, and both results demonstrate a high reflection intensity, they will form spatial data with peak regions. This facilitates the accurate spatial localization of underwater bridge piers. By adopting this method, the actual direction of the beam pulse can be ascertained using data obtained from two sonar images, as depicted in Figure 5.

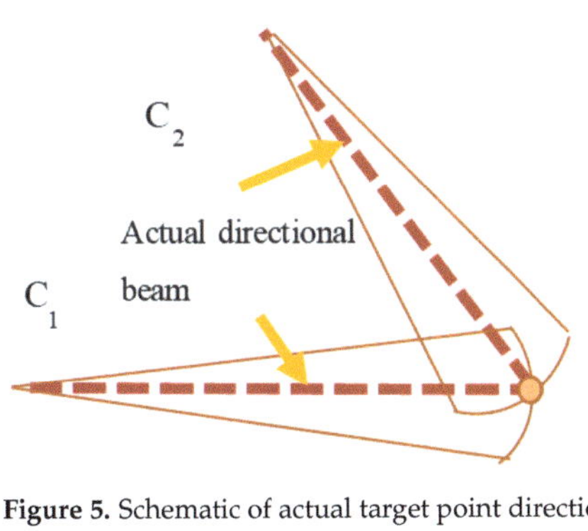


**Figure 5.** Schematic of actual target point directional beam.

Hence, by multiplying sonar image data from multiple measurement points, along with utilizing pre-existing knowledge of undamaged bridge pier models and the established correspondence between specific surface defect features in 2D sonar images and their 3D morphologies, it becomes feasible to achieve the 3D reconstruction of underwater bridge piers afflicted with surface defects.

## 3. Automated Recognition of 2D Sonar Images for Underwater Bridge Piers Surface Profile

The recognition of underwater bridge piers' surface profiles plays a pivotal role in the field of 3D digital reconstruction, as it directly influences key factors such as efficiency, complexity, and precision in the 3D digital reconstruction process. In this section, we introduce an automated recognition method designed to extract underwater bridge pier surface profiles from 2D sonar images. The workflow for this method is visually depicted in Figure 6, and it encompasses the following primary steps:

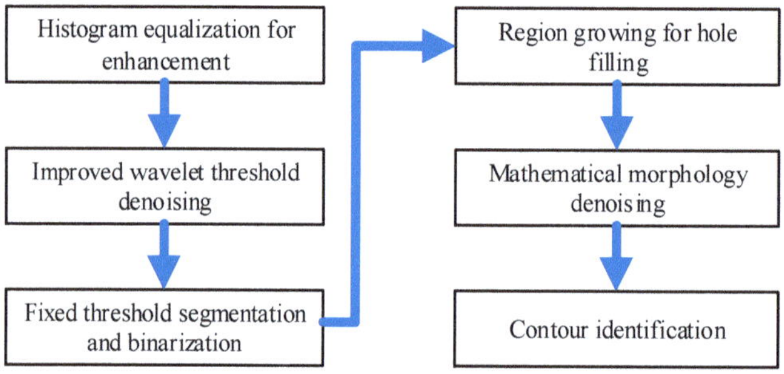


**Figure 6.** Flowchart of automated recognition of 2D sonar images for underwater bridge piers surface profile.

(1) Histogram equalization for enhancement: As an initial step, we employ the histogram equalization method to broaden the grayscale range of sonar images with the goal of enhancing the data. This step is undertaken to bolster image contrast and quality.

(2) Improved wavelet threshold denoising: Subsequently, we apply an enhanced wavelet threshold denoising function to the sonar images. This step serves to mitigate noise in the images and elevate the accuracy of subsequent processing stages.

(3) Fixed threshold segmentation and binarization: We employ a fixed thresholding approach to segment underwater bridge piers information and subsequently convert it into binary images, thus facilitating superior differentiation between targets and the background.

(4) Region growing for hole filling: To ensure the integrity of the bridge piers' information, we implement a region-growing method to autonomously fill gaps within the sonar images.

(5) Mathematical morphology denoising: Further refinement of image quality is achieved through the use of mathematical morphology operations designed to eliminate protrusions and indentations along the edges of bridge piers.

(6) Contour identification: Finally, we leverage the findContours function to identify the surface profile of underwater bridge piers, extracting both the inner and outer contour information of the piers and the upper and lower contour data of the riverbed.

In summary, we have successfully integrated this methodology into digital detection software, resulting in the automated recognition of 2D sonar images for underwater bridge piers' surface profiles.

In this section, we illustrate the automated recognition process of the 2D sonar image for the surface profile of underwater bridge piers using an example of original pseudocolor sonar images of underwater bridge piers. The dimensions of the original pseudocolor sonar image of underwater bridge piers are 450 pixels by 900 pixels, with the sonar positioned 500 mm away from the center of the pier, and the underwater portion of the pier measuring 1085 mm in length and 245 mm in diameter, as depicted in Figure 7.

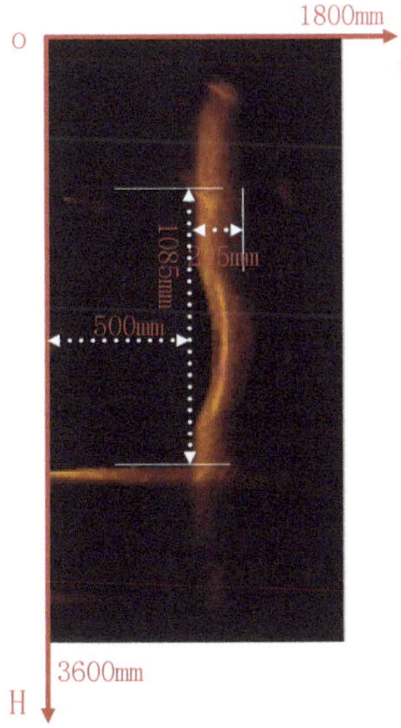


**Figure 7.** Original pseudo-color sonar image of an underwater bridge pier.

### 3.1. Histogram Equalization for Enhancement

Sonar images are typically grayscale images, and their grayscale information plays a crucial role in the intelligent recognition of surface anomalies and 3D digital reconstruction. Histogram equalization is a process that involves statistically analyzing grayscale values in sonar images to remap them into a uniform distribution. This process stretches and compresses the grayscale levels of the image, thereby expanding the dynamic range of grayscale values to achieve data enhancement [23].

Let f and g represent the original image and the image after histogram equalization, respectively, with image dimensions of $m \times n$ and a grayscale value range of 0 to 255, i.e., L = 256. The steps for histogram equalization are as follows:

(1) Calculate the grayscale histogram of the sonar image f, denoted as h.

(2) Calculate the cumulative grayscale distribution frequency $v(\mu)$, defined by the following formula:

$$v(\mu) = \sum_{k=0}^{\mu} \frac{h(k)}{m \times n} \tag{2}$$

Here, h(k) represents the number of pixels with the k-th grayscale level, and $m \times n$ represents the image size.

(3) Determine the minimum cumulative grayscale distribution frequency vmin.

(4) Calculate the grayscale values $\omega$ for g using the following formula:

$$\omega = \left\lfloor (L-1)\left(\frac{v(\mu) - v_{min}}{1 - v_{min}}\right) \right\rfloor \tag{3}$$

Here, $\lfloor \; \rfloor$ represents floor rounding.

Insufficient contrast in an image is often manifested by a concentration of the pixel values in the histogram. Images with lower illumination tend to have their histogram components concentrated in the low grayscale levels, while images with higher illumination and distinct grayscale tend to have their histogram components concentrated in the high grayscale levels. Figure 8 illustrates an example of grayscale histogram statistics after grayscale processing for some original pseudocolor sonar images. From Figure 8, it is evident that the grayscale distribution of underwater bridge piers sonar images is uneven. Hence, it is necessary to apply local histogram equalization.

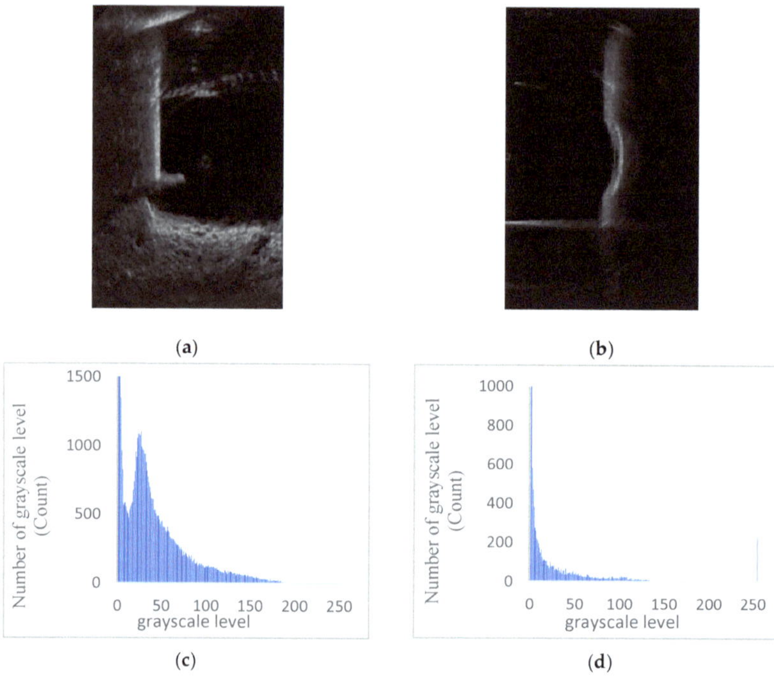


**Figure 8.** Histogram statistical examples of partial bridge pier sonar grayscale images. (**a**) Grayscale sonar image of actual bridge pier. (**b**) Grayscale sonar image of bridge pier experimental model. (**c**) Actual bridge pier sonar grayscale image histogram. (**d**) Bridge pier experimental model sonar grayscale image histogram.

Figure 9 displays the results after applying histogram equalization. Through this process, the contrast of sonar images is significantly enhanced, the grayscale distribution range is widened, and the bridge piers' contour information becomes clearer.

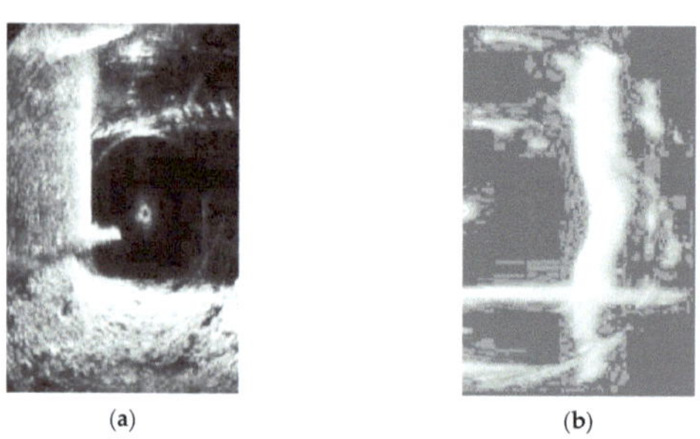


**Figure 9.** *Cont.*

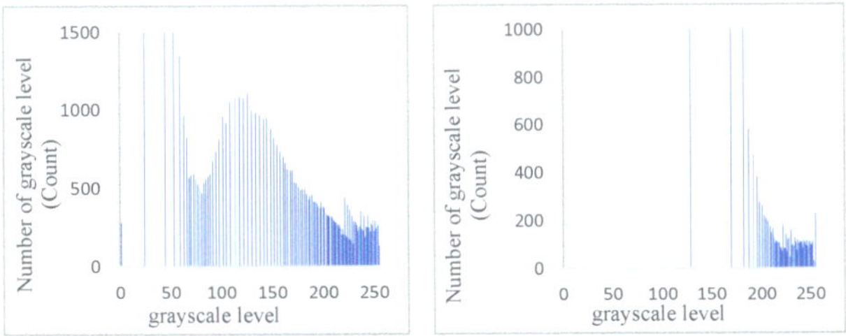


(c) (d)

**Figure 9.** Results after histogram equalization. (**a**) Processed grayscale image of actual bridge pier. (**b**) Processed grayscale image of bridge pier experimental model. (**c**) Processed histogram of actual bridge pier. (**d**) Processed histogram of bridge pier experimental model.

*3.2. Improved Wavelet Threshold Denoising*

Sonar images are a type of vector signal, and thus, spatial domain filtering methods like mean filtering and median filtering have limited effectiveness in removing noise from such stationary random vector signals [24].

In this section, we have improved and combined the Garrote function and soft threshold function to create an enhanced wavelet threshold denoising function, which can be mathematically expressed as follows:

$$\hat{d_{j,k}} = \begin{cases} d_{j,k} - \mathrm{m}\lambda sgn\left(d_{j,k}\right) - (1-m)\frac{\lambda^2}{d_{j,k}}, \ \left|d_{j,k}\right| \geq \lambda, \ m \in [0,1] \\ 0, \ \left|d_{j,k}\right| < \lambda \end{cases} \tag{4}$$

Here, the parameter "m" is used for adjustment. The enhanced function possesses the characteristics of both the soft threshold function and the Garrote function. When $m = 1$, it takes the form of a soft threshold function, and when $m = 0$, it resembles the Garrote function. When $m$ falls within the range of (0, 0.5), the enhanced function tends to exhibit the characteristics of the Garrote function, resulting in image smoothing while preserving edge features. When $m$ falls within the range of (0.5, 1), the enhanced function leans more towards the soft threshold function, with a more continuous function curve, leading to smoother images. The improved wavelet threshold denoising function combines the advantages of the Garrote function and the soft threshold function, allowing for flexible selection of the parameter "m" based on the denoising requirements.

According to research findings, the use of the improved wavelet threshold denoising function in filtering, compared to basic mean filtering, basic median filtering, and traditional wavelet threshold denoising methods, results in the smallest mean squared error and the highest signal-to-noise ratio and peak signal-to-noise ratio. Therefore, this study employs this filtering method for denoising. As shown in Figure 10, the denoised results are presented, clearly illustrating that the denoised image is smoother compared to the original image.

To validate the denoising effectiveness of the improved wavelet threshold function, this paper objectively compares and analyzes various denoising methods using empirical data. Initially, Gaussian white noise is added to sonar images of underwater bridge piers. Subsequently, denoising techniques are applied to these noisy sonar images, followed by numerical evaluation using the mean square error (MSE), peak signal-to-noise ratio (PSNR), and signal-to-noise ratio (SNR). The denoising methods evaluated include basic mean filtering, basic median filtering, as well as hard thresholding, soft thresholding, Garrote function, and the improved wavelet threshold function for wavelet denoising. Evaluation metrics are summarized in Table 1. Analysis of Table 1 reveals that the improved wavelet threshold function achieves the lowest MSE and highest SNR and PSNR values, demonstrating a superior denoising performance. Consequently, this paper adopts the improved wavelet threshold function for preprocessing image denoising.

**Figure 10.** Filtering of sonar image.

**Table 1.** Evaluation metrics of different denoising methods.

| Denoising Methods | MSE | PSNR | SNR |
|---|---|---|---|
| Noisy Sonar Images | 0.0071 | 69.6331 | 1.8188 |
| Basic Mean Filtering | 0.0042 | 72.968 | 5.1538 |
| Basic Median Filtering | 0.002 | 75.138 | 7.3237 |
| Hard Thresholding Function | 0.0032 | 74.968 | 6.1538 |
| Soft Thresholding Function | 0.0018 | 76.7792 | 8.9649 |
| Garrote Function | 0.0036 | 73.5295 | 5.7152 |
| Proposed Algorithm | 0.0017 | 76.9467 | 9.0899 |

### *3.3. Fixed Threshold Segmentation and Binarization*

Following the previous histogram equalization process, the histogram exhibits a bimodal distribution, representing two distinct object classes with significant grayscale differences. This corresponds to an image consisting of a darker background and brighter targets, as depicted in Figure 9d. By carefully selecting an appropriate threshold value "T" at the trough point between the two histogram peaks, it becomes possible to effectively separate the bridge pier from the background. Consequently, for image segmentation in this section, we have opted for the fixed threshold method, configuring "T" to 220. Simultaneously, the image undergoes binarization, where pixels less than or equal to the threshold are assigned a grayscale value of zero, while pixels exceeding the threshold are assigned a grayscale value of 255. Figure 11 illustrates the results of applying the fixed threshold segmentation to Figure 10.

By comparing Figures 10 and 11, it becomes evident that after the process of image segmentation and binarization, the image presents a distinct black-and-white effect. This effectively accentuates the contour information of the underwater bridge pier while faithfully representing the mirrored reflection of the underwater pier.

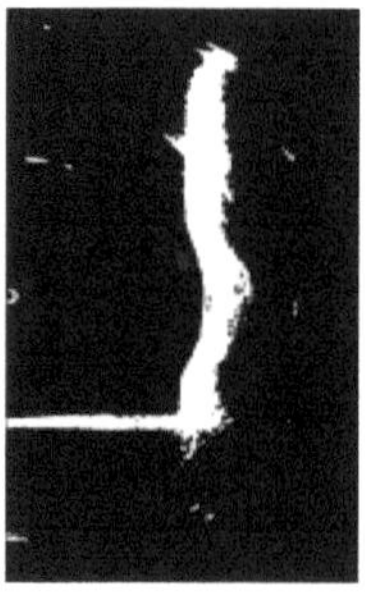

**Figure 11.** Binarization of sonar image.

### 3.4. Region Growing for Hole Filling

Due to the uneven surface of the bridge pier, non-reflective areas exist on its surface. Within the bridge pier region, dark areas with grayscale values lower than the ideal threshold are present. Furthermore, due to noise and the reflection and refraction of the water surface, white regions appear outside the bridge pier area, as depicted in Figure 11. Achieving precise recognition necessitates filling these data gaps and removing both the dark and white regions. This task is accomplished using a region-growing method [25].

Specifically, for the removal of dark regions in the underwater bridge piers' sonar image, the entire image is scanned. White pixels are considered valid, and a 4-neighbor detection is utilized to calculate the number of pixels in each region. If the total count falls below a predetermined threshold, the region is cleared, effectively filling the dark regions, as illustrated in Figure 12a. When eliminating white regions, a subsequent scan of the entire image is conducted, with black background pixels being deemed valid. An 8-neighbor detection scheme is employed to compute the pixel count in each region. If the total count falls below a predefined threshold, the region is cleared, resulting in the filling of white regions, as shown in Figure 12b.

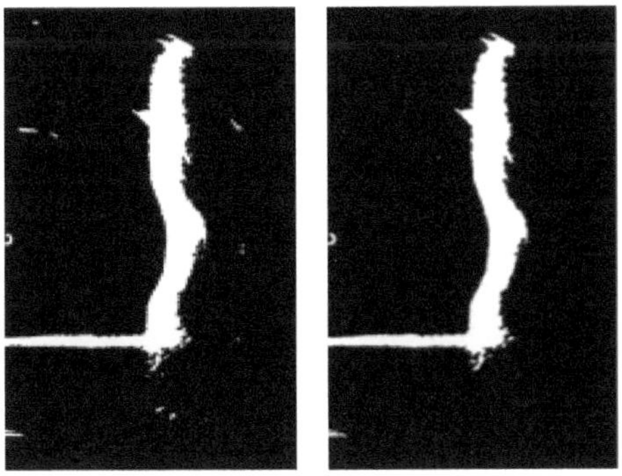

(a) (b)

**Figure 12.** Removal of black and white regions in sonar image. (**a**) Sonar image after black region removal. (**b**) Sonar image after white region removal.

Upon comparing Figure 12a,b, it becomes evident that by removing the dark and white regions from the sonar image, extraneous noise information is further eliminated. The sonar image now exclusively retains the underwater bridge pier area, consequently reducing computation time for subsequent defect extraction and 3D digital reconstruction processes.

### 3.5. Mathematical Morphology Denoising

Despite the removal of dark and white regions, the edges of the bridge pier still exhibit numerous jagged protrusions, leading to elongated edge contours. These protrusions increase the computational complexity during subsequent contour recognition and can potentially cause misidentification. To address this issue, mathematical morphology operations, such as erosion and dilation, are commonly employed to smooth out these jagged features along the bridge pier's edges.

Mathematical morphology serves as a valuable tool for nonlinear image (signal) processing and analysis, with its primary focus on the geometric structure and interrelationships within images. By employing various "structuring elements", it is possible to measure or extract corresponding shapes and features from the image. Morphological operations encompass erosion and dilation, and a variety of practical morphological algorithms can be developed and combined based on these two fundamental operations. Erosion and dilation operations in binary images are dual operations. The erosion of X by S, denoted as X ⊖ S, is defined as the set of all points x for which the translation of S by x is entirely contained within X:

$$X \ominus S = \{x | S + x \subseteq X\} \tag{5}$$

Dilation of X by S, denoted as X⊕S, is defined as the set of all points x for which the translation of S by x has a non-empty intersection with X:

$$A \oplus S = \{x | S + x \cup x \neq \varnothing\} \tag{6}$$

In this study, a 3 × 3 square structuring element is employed for further morphological erosion and dilation operations on the binary image, as illustrated in Figure 13. Figure 13a showcases the outcome following the erosion operation. When compared to the binary image in Figure 12b, it is evident that the white isolated noise points along the underwater bridge pier's edge have been significantly reduced in the sonar binary image after the erosion operation. Figure 13b presents the image following the application of the dilation operation to the image from Figure 13a. In contrast to Figure 13a, the white pixels appear more concentrated, indicating stronger connectivity within the underwater bridge pier's edge region. Post-processing, when compared to the original image, results in smoother edges with fewer jagged features, thus facilitating subsequent data extraction. Furthermore, a substantial portion of the internal and external noise is eliminated, thereby reducing the potential for misjudgment.

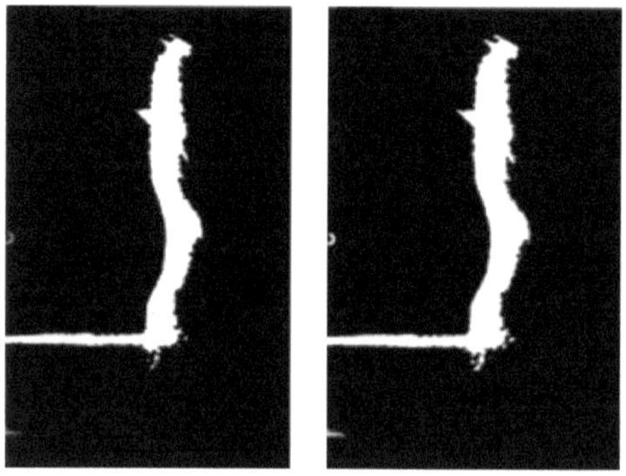

(**a**) (**b**)

**Figure 13.** Morphological erosion and dilation. (**a**) Erosion treatment to eliminate edge spikes. (**b**) Dilation treatment to preserve spikes.

*3.6. Contour Identification*

To meet the requirements for subsequent 3D reconstruction, it is necessary to further extract the contours of the underwater bridge pier and riverbed. This involves identifying both the inner and outer contours of the bridge pier in the sonar image and the upper and lower contours of the riverbed. The automated steps for sonar image contour recognition are as follows:

(1) In the sonar image, the underwater bridge pier and riverbed form a connected structure and represent the largest target contours in the sonar image. Therefore, the "findContours" function provided by OpenCV is utilized to identify the contours of image regions. The area of each contour is recorded, and the contour with the largest area is identified as the target, which is the bridge pier, as shown in Figure 14a.

(2) Determine the coordinates of the uppermost contour and the lowermost contour.

(3) For the outer contour, trace the next contour coordinate point to the left from the uppermost contour coordinate, continuing until the lowermost contour coordinate is reached.

(4) For the inner contour, trace the next contour coordinate point to the right from the uppermost contour coordinate, continuing until the lowermost contour coordinate is reached.

(5) Repeat the above steps for another sonar image until all sonar images have been processed.

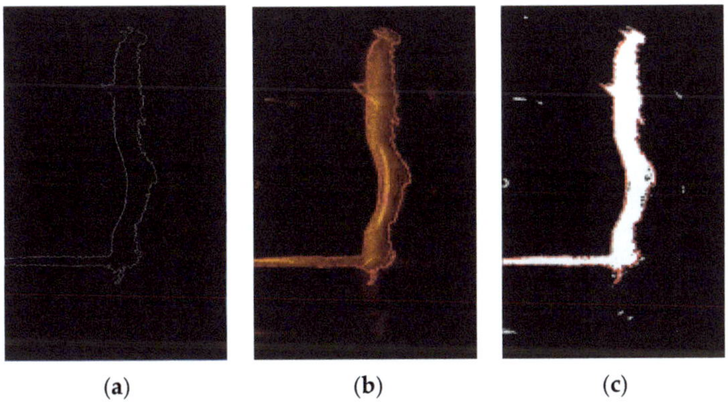

**Figure 14.** Bridge pier contour recognition. (**a**) Extracted contour information. (**b**) Comparison with the original sonar image. (**c**) Comparison with the binarized sonar image.

Finally, the extracted underwater bridge pier contour information is overlaid onto the original sonar image and the binary sonar image. To emphasize the contours, they are marked in red, as shown in Figure 14a,b. From Figure 14a,b, it can be observed that the extracted underwater bridge pier contour information closely aligns with the actual contour information, demonstrating the accuracy of this method in extracting underwater bridge pier contours. In the end, all the inner and outer contours of the bridge piers in the sonar images and the upper and lower contours of the riverbed are obtained for subsequent 3D digital reconstruction purposes.

## 4. Underwater Bridge Piers 3D Reconstruction Algorithm

This algorithm is centered around the reconstruction of underwater bridge pier cross-sections. It generates a 3D model of the bridge pier in its undamaged state, leveraging known information like the pier's dimensions, the distance between the sonar device and the pier, and more. Furthermore, it utilizes the identified coordinates of the bridge pier's contours as control points and combines them with the multiplication of multi-point sonar data to further refine potential scenarios. Ultimately, it culminates in the reconstruction of the authentic 3D model of the underwater bridge pier with surface defects. Here are the detailed algorithm steps:

Step 1: Based on actual measurements of the bridge pier's dimensions, create a three-dimensional array V to represent the undamaged bridge pier's 3D model. The elements of the array $V_{xyz}$ are defined according to Formula (7), with a visual representation as depicted in Figure 15.

$$\begin{cases} V_{xyz} = 255 \text{ when} \sqrt{x^2+y^2} \leq R \text{ and } 0 \leq R \leq H, \text{ inside the bridge pier} \\ V_{xyz} = 0 \text{ when} \sqrt{x^2+y^2} \geq R \text{ or } \; R < 0 \text{ or } R > H, \text{ outside the bridge pier} \end{cases} \tag{7}$$

Step 2: Set Z to a value m and select $V_{xyz}$ to represent a cross-section of the bridge pier. Proceed to intersect the coordinates of n measurement points with the current cross-section. As illustrated in Figure 16, the aim of this step is to derive potential scenarios for the 3D morphological analysis of continuous sonar images gathered from multiple measurement points, each of which contains information that cannot be ascertained solely from a single-point sonar image. Op is the center of the bridge pier cross-section; *R* is the radius of the bridge pier; Os is the position of the sonar transducer; possible bridge pier profile shapes such as ABC, a′cb, ac′b, acb′ are obtained based on the distance of the profile.

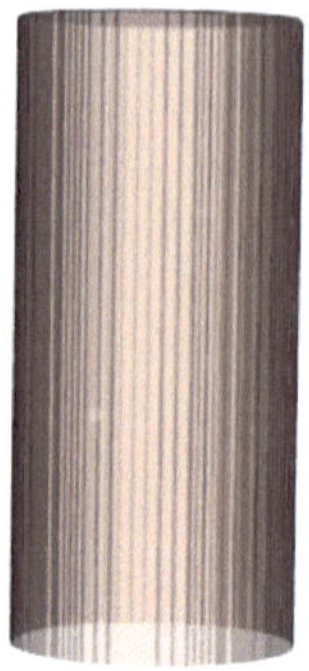

**Figure 15.** Undamaged bridge pier model.

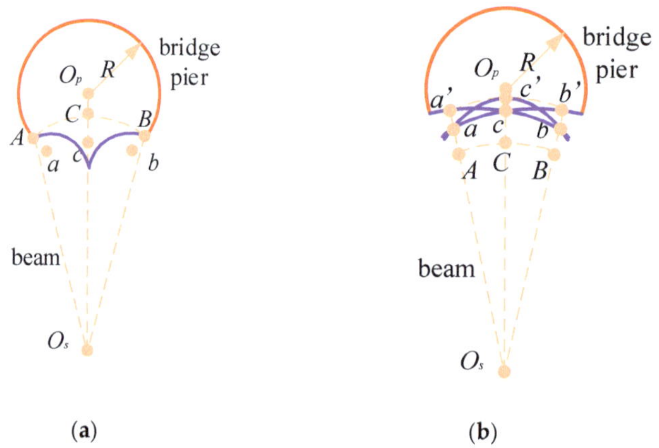


**Figure 16.** Schematic representation of possible 3D shapes of continuous measurement points. (**a**) Measurement Point 1. (**b**) Measurement Point 2.

Step 3: Within the current cross-section, perform a multiplicative assessment of the 3D morphological scenarios obtained in Step 2, using data from multiple measurement points. The scenario yielding the highest multiplication result is chosen as the final outcome, and the cross-sectional contour of the current bridge pier is adjusted accordingly. For instance, hole contour reconstruction can be accomplished by utilizing sonar images from neighboring measurement points, as depicted in Figure 16. All key points from Figure 16a and Figure 16b are combined, as displayed in Figure 17a. Then, based on the relative positions of the measurement points, select the surface defect contour on the side close to $O_{s2}$ and the one close to $O_{s1}$, resulting in a contour shape that best matches the key points, as shown in Figure 17b. Ultimately, the constructed surface defect contour is presented in Figure 17c.

Step 4: Set Z to values 0, 1, . . ., H, and then repeat Step 2 and Step 3 to obtain the bridge pier contours for all cross-sections. In the end, these contours will be employed for the reconstruction of the 3D model of the bridge pier, as depicted in Figure 18.

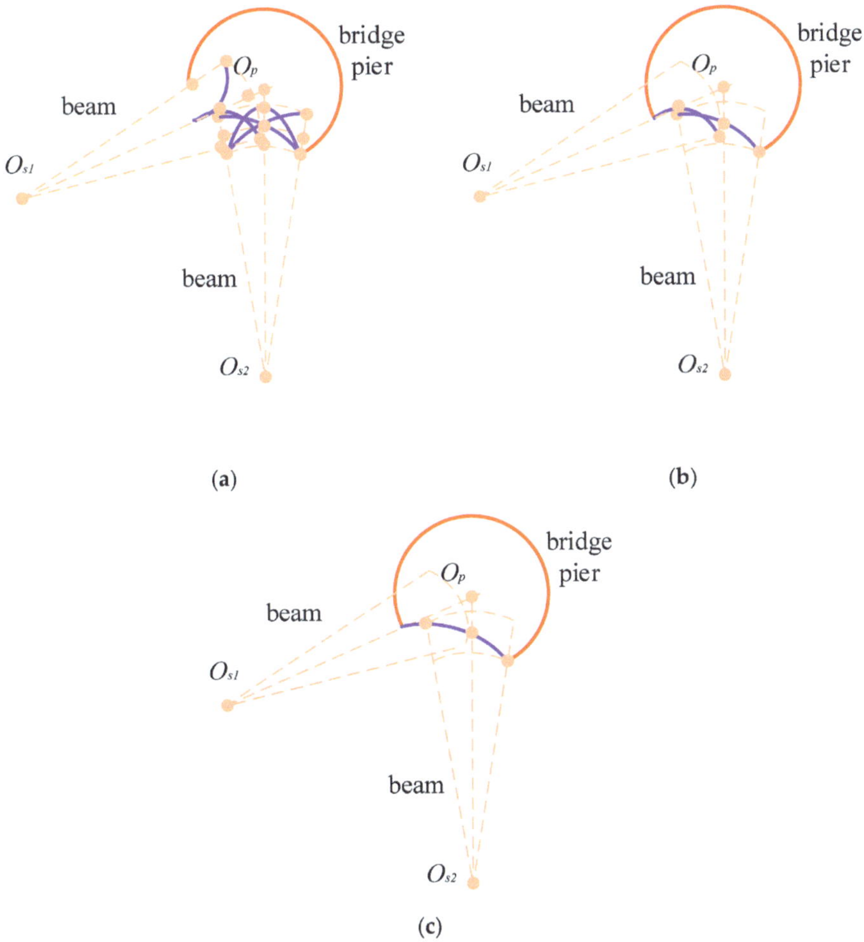


**Figure 17.** Schematic representation of surface damage reconstruction for adjacent measurement points. (**a**) All control points corresponding to the measurement point. (**b**) Excluded contour. (**c**) Final surface damage contour.

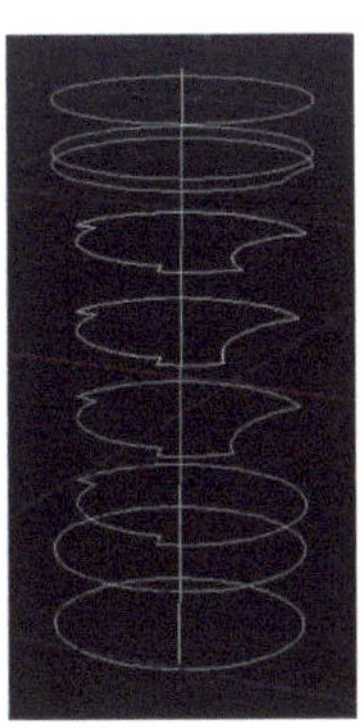

**Figure 18.** Fitting the bridge piers contour for the current cross-section.

## 5. Three-Dimensional Visualization of Underwater Bridge Piers Sonar Images Using VTK

Three-dimensional visualization provides an intuitive and dynamic representation of underwater bridge pier models on a computer, aiding researchers in digitally analyzing surface defects. However, the 3D reconstruction algorithm described earlier focuses primarily on fitting cross-sectional contours and obtaining raw data. These data serve as foundational for 3D model visualization but do not convey detailed 3D model information.

To achieve computer-based 3D visualization of underwater bridge pier models, we utilize the Visualization Toolkit (VTK). This section compares different 3D visualization algorithms, focusing on surface rendering methods for displaying surface defects in the bridge pier model. The visualization program is integrated into digital inspection software.

There are various 3D reconstruction algorithms, broadly categorized into volume rendering and surface rendering. Surface rendering, chosen here for its suitability in handling underwater bridge pier surface defects with reduced computational complexity, employs VTK's Marching Cubes (MC) algorithm. This approach involves extracting isocontours, reducing triangles, smoothing data, generating normals, and triangulating isocontours to achieve 3D visualization.

Furthermore, leveraging sonar image data processing and 3D reconstruction algorithms, we developed the Digital Inspection System Software for Underwater Bridge Pier Surface Defects. This software, developed in C++ with an interface in QT and using VTK for visualization, supports parameter setting, image processing, 3D visualization, feature extraction, result analysis, and model export (Figure 19).

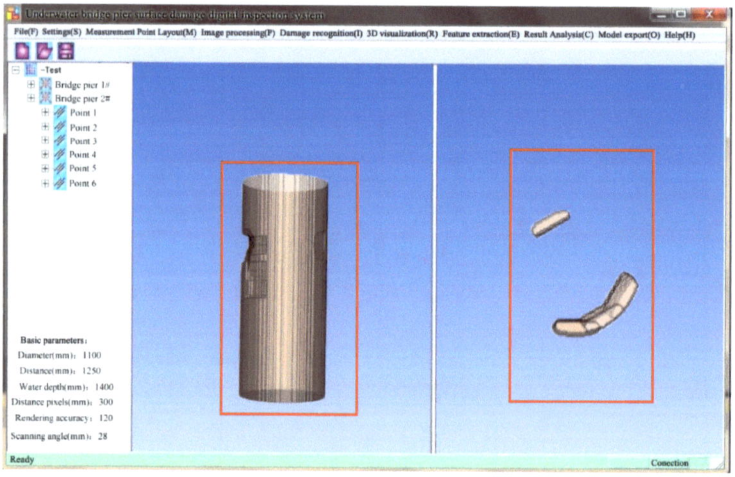


**Figure 19.** The visualization software.

As shown in Figure 19, this software serves as a platform for processing sonar images and 3D reconstruction of underwater bridge piers. It enables 3D visualization of underwater bridge pier models and offers interactive functions such as translation, scaling, and rotation, facilitating the analysis and observation of surface defects in the 3D models of underwater bridge piers by researchers.

## 6. Underwater Bridge Piers' 3D Reconstruction Experiment

To validate the feasibility and accuracy of the 3D reconstruction technology, this section conducts underwater bridge pier sonar scanning experiments in a controlled environment. The sonar images obtained from the scans are used as input for the 3D reconstruction process. The previously described 3D reconstruction algorithm and visualization methods are employed to reconstruct the 3D model of the underwater bridge pier. Finally, the dimensions of the reconstructed 3D model are measured and compared with the actual model dimensions for analysis and validation.

### 6.1. Experimental Environment

A laboratory environment was set up to simulate underwater inspection conditions. A water tank measuring 7.1 m × 5.1 m × 1.5 m was constructed and filled with water, as depicted in Figure 20. To facilitate hoisting large concrete column models, we installed a miniature overhead crane with horizontal movement capabilities and a rotatable lifting rod, as shown in the water tank hoisting system illustrated in Figure 21.

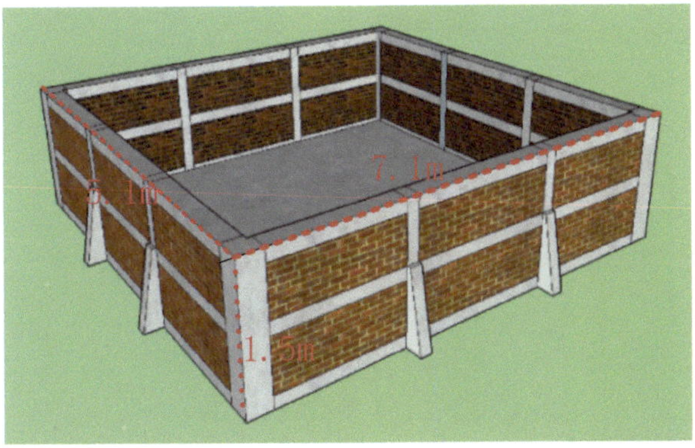


**Figure 20.** Schematic diagram of the 7.1 m × 5.1 m × 1.5 m water tank structure.

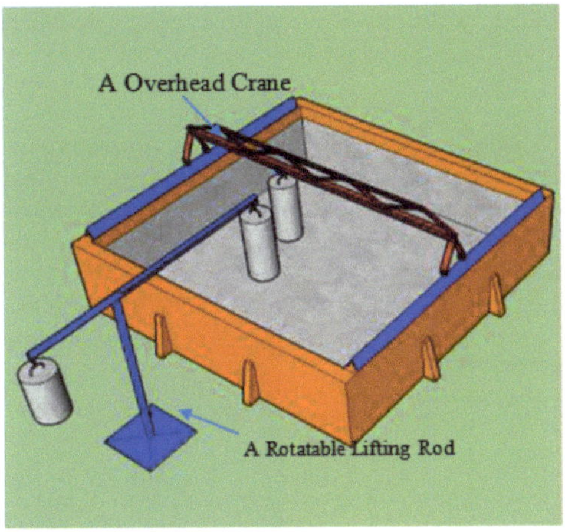


**Figure 21.** The water tank hoisting system.

For the stability and safety of the sonar equipment during inspections, a set of auxiliary devices, depicted in Figure 22, were designed and constructed. This testing auxiliary device includes an angle steel frame, a movable operating platform, and a turntable. The angle steel frame enables adjustments in the sonar angle and height within the water environment, as seen in Figure 22a. The movable operating platform, positioned above the water tank, supports the angle steel frame, allowing its free movement, as depicted in Figure 22b. At the bottom of the tank, a turntable holds the underwater bridge pier model, allowing us to alter the positions of the sonar measurement points by rotating the turntable, as shown in Figure 22c.

(a)

(b) (c)

**Figure 22.** Experimental auxiliary device diagram. (**a**) Schematic of sonar fixed on angle steel frame. (**b**) Test tank and mobile operating platform. (**c**) Underwater turntable.

### 6.2. Concrete Column Bridge Pier Models

Concrete column bridge pier models were designed in the laboratory, simulating two types of surface defects: holes and spalling. A concrete column bridge pier model was cast using C30 concrete with a diameter of 490 mm and a height of 1500 mm. The pier model was reinforced with longitudinal bars and hoop bars. The longitudinal bars consisted of 6 B16 steel bars, and the hoop bars consisted of 5 A8 steel bars. Additionally, the bridge pier model was designed to incorporate two types of surface defects: holes and spalling. Photos of the bridge pier model are shown in Figure 23, and detailed dimensions and the sizes of the surface defects are illustrated in Figure 24.

(a) (b)

**Figure 23.** Concrete column with surface defects. (**a**) Hole. (**b**) Spalling.

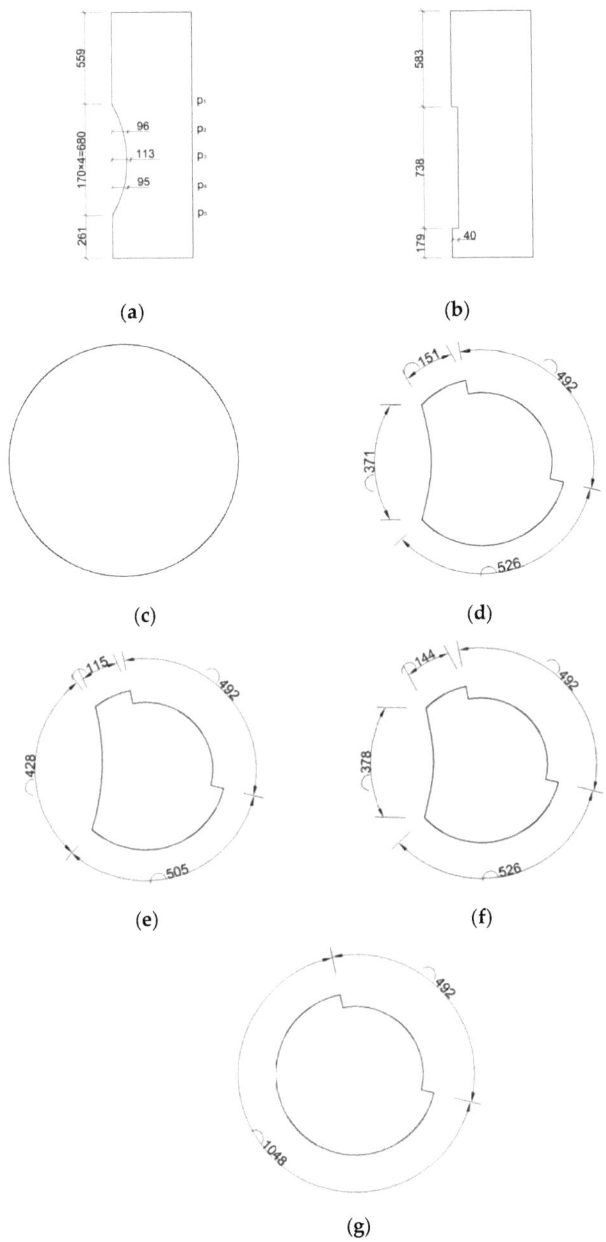


**Figure 24.** Surface defect dimensions (Unit: mm). (**a**) Hole dimensions. (**b**) Spalling dimensions. (**c**) Section p1. (**d**) Section p2. (**e**) Section p3. (**f**) Section p4. (**g**) Section p5.

### 6.3. Layout of Measurement Points

The layout of measurement points involves determining the sonar scanning parameters, scanning distances, and the number and positions of measurement points, and calculating the azimuth and elevation of points.

(1) Azimuthal layout of measurement points

To obtain multi-point data for the underwater bridge pier model, the azimuthal layout of measurement points is calculated based on the requirements [4], known bridge pier model information, and the set sonar scanning parameters. Generally, the higher the sonar scanning frequency, the better the quality of the sonar image, but the scanning speed is relatively slower and the scanning range becomes smaller. To obtain clear sonar images, the scanning frequency is set to the maximum of 1.2 MHz, with a scanning speed of one sonar image every 5 s, which is acceptable. Due to the close scanning distance, the beamwidth at this frequency is 28° × 0.6°, which is also an acceptable scanning range. According to the previous model design, the radius of the pier is known to be r = 0.245 m.

Assuming that the horizontal beam angle's edge line is tangent to the bridge piers surface, we can calculate $l_0$ as follows:

$$l_0 = r\left(\frac{1}{\sin\left(\frac{\alpha_0}{2}\right)} - 1\right) = 0.245 \times \left(\frac{1}{\sin\left(\frac{28^\circ}{2}\right)} - 1\right) = 0.768 \text{ m}$$

To obtain detailed apparent information from the pier's surface, the distance $l$ between the sonar measurement point and the pier's surface is set to 0.5 m. Consequently, we can calculate $\beta$ as follows:

$$\beta = 2\arcsin\left[\left(1 + \frac{l}{r}\right)\sin\left(\frac{\alpha_0}{2}\right)\right] - \alpha_0 = 66.7^\circ$$

Considering that the scanning range of the measurement points must completely cover the entire pier, and adding 2 more points for full coverage, we can determine the number of scanning points, $n$, as follows:

$$n = \left\lfloor \frac{360^\circ}{\beta} + 0.5 \right\rfloor + 2 = \left\lfloor \frac{360^\circ}{66.7^\circ} + 0.5 \right\rfloor + 2 = 7$$

Here, $\lfloor\ \rfloor$ represents the floor function. Of course, to increase the overlapping area, n can be further increased.

Once $\alpha$, $\beta$, $l$, and $n$ are determined, the points are evenly distributed along the circular cross-section, with each point at the same distance from the pier's edge. The angular distance between adjacent measurement points is equal, resulting in the azimuthal layout of measurement points for the bridge pier model, as shown in Figure 25.

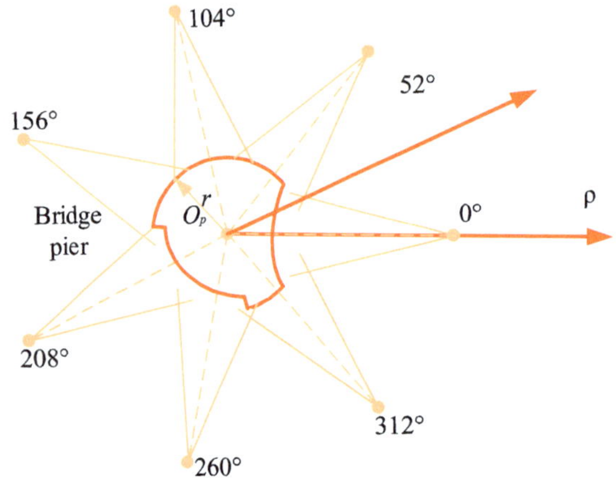


**Figure 25.** Azimuthal layout of measurement points.

(2) Vertical layout of measurement points

Given that the height of the bridge pier model is 1.088 m, the water depth is h = 1.400 m, and the sonar scanning radius is R = 1m, and considering the small size of the pier in the vertical direction, all measurement points are placed just 0.8 m below the water surface to ensure effective imaging.

### 6.4. Experimental Procedure

The experimental procedure involved several steps:

(1) Firstly, the turntable device was installed at the designated location. Next, the bridge pier model was suspended onto the turntable device using a hoisting bar. During the experiment, the turntable was rotated to change the position of the measurement points.

(2) Setting up the sonar system: Afterward, the sonar system was connected, and the sonar was mounted on the angle steel frame. The angle steel frame was securely fixed to the movable platform. Scanning of the bridge pier model was performed using horizontal and lateral scanning methods.

(3) Configuration of sonar parameters: The relevant parameters of the sonar system were configured using computer control software. The 1.2 MHz frequency, which provided better imaging results, was selected for scanning.

(4) Imaging scans: The imaging scans began at the 0° position, and the bridge pier model was scanned. After completing the scan at this point, the bridge pier model was rotated to 52°, and scanning at this measurement point commenced. This process continued, with imaging scans carried out at each measurement point, until all points were scanned.

The experimental setup and process are depicted in Figure 26 for reference.

**Figure 26.** Test scene image.

### 6.5. Experimental Results

The bridge pier model was subjected to imaging scans following the scanning plan, resulting in sonar images corresponding to various measurement points. These sonar images are shown in Figure 27.

When the bridge pier model at a measurement point did not exhibit any defects, the inner and outer contours of the bridge target in the sonar image appeared as straight lines, as seen in Figure 27f. This indicates that the bridge pier in that area has no surface defects.

However, when there were surface defects in the bridge pier model at a measurement point, both the inner and outer contours of the bridge target in the sonar image showed varying degrees of concavity and convexity, indicating the presence of surface defects in the bridge pier in that area.

Based on the sonar images, preliminary conclusions can be drawn as follows:

(1) Sonar images at 0°, 52°, and 312° correspond to areas with cavity surface defects.

(2) Sonar images at 52°, 104°, 156°, and 208° correspond to areas with spalling surface defects.

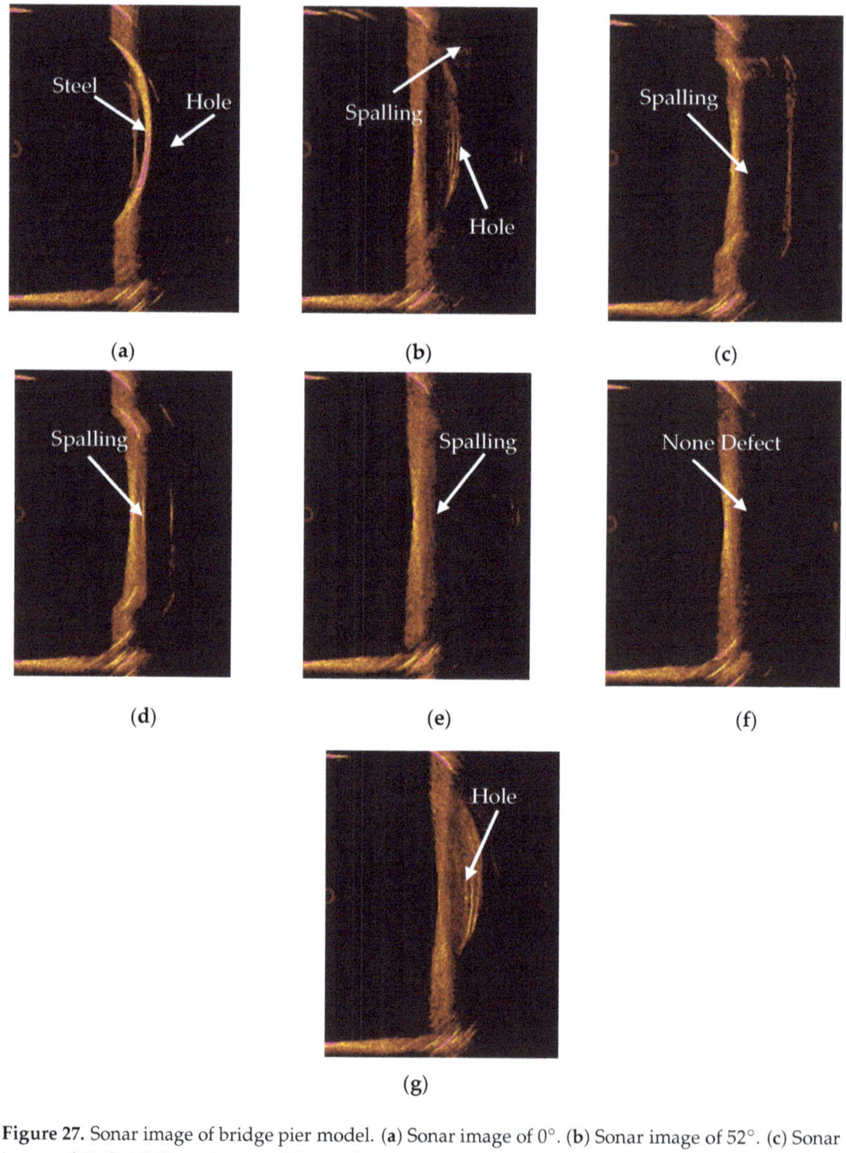


**Figure 27.** Sonar image of bridge pier model. (**a**) Sonar image of 0°. (**b**) Sonar image of 52°. (**c**) Sonar image of 104°. (**d**) Sonar image of 156°. (**e**) Sonar image of 208°. (**f**) Sonar image of 260°. (**g**) Sonar image of 312°.

(3) The sonar image at 260° corresponds to an area with no surface defects.

These observations help to identify the presence or absence of surface defects in different regions of the bridge pier model.

By obtaining this set of sonar images, they can serve as input for the 3D reconstruction system. After undergoing image preprocessing in Section 3, sonar image 3D reconstruction in Section 4, and visualization in Section 5, the 3D model representing the surface defects in the underwater bridge pier has been reconstructed, as depicted in Figure 28.

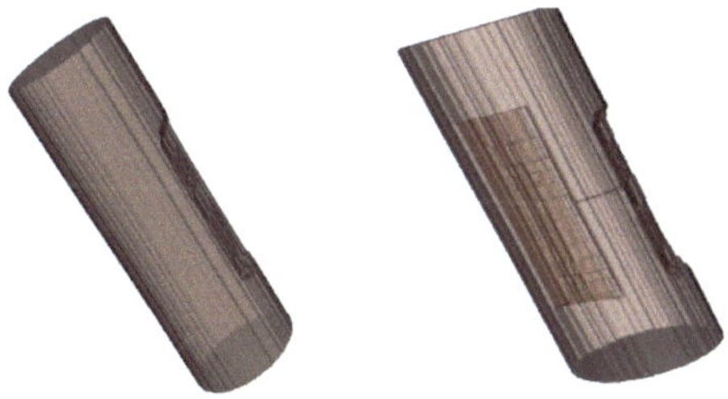

**Figure 28.** Reconstructed 3D model of the bridge pier.

## 6.6. Three-Dimensional Model Measurements

Using the generated underwater bridge pier 3D model, measurements of the surface defect portions were conducted to obtain positional information, 1D (distance measurement), 2D (area measurement), and 3D (volume measurement) data.

### 6.6.1. Distance Measurement

As shown in Figure 28, the pixels on the reconstructed 3D model were converted into distances (with a conversion factor of 2.07 mm per pixel) and compared to the actual measurements from Figure 23, as presented in Table 2. It can be observed that there is a significant difference (15.04% difference) in identifying the exposed reinforcement as having no surface defects in the 0° sonar image. However, for the other geometric dimensions, the error does not exceed 12.00%. The reconstructed 3D model accurately represents the size of the actual bridge pier in most cases.

**Table 2.** Distance measurement results.

| | Actual Distance A (mm) | 3D Model Distance B (mm) | Relative Error (%) (1 − A/B) × 100 |
|---|---|---|---|
| Hole maximum height | 680 | 682 | 0.29 |
| Hole maximum arc height | 428 | 395 | −8.35 |
| Hole maximum depth | 113 | 133 | 15.04 |
| Spalling maximum height | 738 | 751 | 1.73 |
| Spalling maximum arc height | 680 | 682 | 0.29 |
| Spalling maximum depth | 428 | 395 | −8.35 |

### 6.6.2. Area Measurement

Using the scanned arc length of holes and spalls in the cross-section on the cylinder, the surface area can be calculated as follows:

$$S = \sum_{i=1}^{x} l_i \cdot \Delta x \tag{8}$$

where S is the surface area. $l_i$ is the arc length on the cylinder when $x = i$. $\Delta x = 2.07$ mm is the distance between two pixels in the x-direction.

A comparison of the calculated surface area of the model's surface defects with the surface area obtained from actual measurements is shown in Table 3. The relative error for the hole area is −8.08%, and for the spalling area, it is −0.52%. The reconstructed model accurately reproduces the surface area of the bridge pier's surface defects.

Table 3. Comparison of defects surface area between reconstructed model and actual pier.

| | Actual Pier Surface Area $S_a(m^2)$ | 3D model Surface Area $S_b/(m^2)$ | Relative Error (%) $(1 - S_a/S_b) \times 100$ |
|---|---|---|---|
| Hole | 0.20009 | 0.18513 | −8.08 |
| Spalling | 0.36310 | 0.36123 | −0.52 |

6.6.3. Volume Measurement

Based on the surface area calculation Formula (8), we can calculate the volume using the following formula:

$$V = \sum_{j=1}^{z} S_j \cdot \Delta z \tag{9}$$

where V is the volume. $\Delta z$ = 2.07 mm is the distance between two pixels in the z-direction. $S_j$ is the arc surface area when $z = j$.

A comparison of the calculated volume of the model's defects with the volume obtained from actual direct measurements is shown in Table 4. The relative error for the hole volume is 13.56%, and for the spalling volume, it is 10.65%. The reconstructed model accurately reproduces the volume of the bridge pier's defects. In practical engineering applications, when the defect size is small, it has a relatively minor impact on the pier's load-bearing capacity and safety. When the defect size is large, despite the higher relative error, it still remains within acceptable limits. This is because, based on the load capacity experimental calculations, it is below the 15% safety assessment requirement [4]. This method has been successfully applied to actual engineering projects, including the Fuzhou Wulongjiang Bridge and the Fuqing Xizixiang Bridge sonar inspections, where the actual measurement results are consistent with the laboratory model test results.

Table 4. Comparison of defects volumes between the reconstructed model and the actual pier.

| | Actual Pier Volumes $V_{cal}$ $(m^3)$ | 3D Model Volumes $V_{meas}$ $(m^3)$ | Relative Error (%) $(1 - V_{cal}/V_{meas}) \times 100$ |
|---|---|---|---|
| Hole | 0.20009 | 0.18513 | −8.08 |
| Spalling | 0.36310 | 0.36123 | −0.52 |

The piers used in the experiment are scaled-down models of real bridge piers, and the accuracy of the 3D reconstruction algorithm depends on precise data. With sufficient accuracy, the research findings can be extrapolated beyond the confines of the specific experimental setup. The accuracy of sonar images depends on the frequency of the sonar; higher frequencies provide higher precision. In our experiment, we selected the highest frequency available for this particular sonar model.

## 7. Conclusions

This study has summarized the research on the 3D digital reconstruction and visualization of sonar images depicting surface defects on underwater bridge piers. The following key conclusions have been drawn from this research:

(1) We have introduced an automated method for recognizing the surface contours of underwater bridge piers. This method incorporates a series of steps in sonar image processing, including histogram equalization, wavelet threshold denoising, image segmentation, binarization, hole filling, mathematical morphology operations, and contour extraction. Importantly, this automated recognition method has been seamlessly integrated into digital inspection software, facilitating the accurate and automated extraction of underwater bridge pier surface contours.

(2) The study has presented a 3D reconstruction algorithm tailored for underwater bridge piers' surface defects. Leveraging contour information extracted from sonar images

as control points and combining it with existing non-destructive bridge pier model data, this algorithm effectively reconstructs various 3D models representing different types of surface defects.

(3) We have developed 3D visualization software for underwater bridge piers based on sonar images, utilizing the Visualization Toolkit (VTK). This software adopts surface rendering techniques and leverages VTK's visualization capabilities to display 3D models of underwater bridge piers. Furthermore, it offers interactive functionalities such as translation, scaling, and rotation, enhancing the convenience of researchers in analyzing and observing surface defects within underwater bridge pier 3D models.

(4) The effectiveness of the proposed 3D model reconstruction technology has been successfully validated through experiments. The experimental results unequivocally demonstrate the capacity of the reconstructed models to faithfully represent the 3D surface defects on underwater bridge piers. Importantly, the relative errors in surface defect measurements remain within acceptable limits, affirming the reliability and efficacy of this technology.

In conclusion, this research has furnished robust tools and methodologies for digitally detecting and managing surface defects in underwater bridge piers. The practical engineering applications of this research hold substantial promise for the field. In conclusion, this study has equipped us with robust tools and methodologies for digitally detecting and managing surface defects in underwater bridge piers. The practical engineering applications of this research show substantial promise for the field. Nonetheless, there remain unresolved issues, such as integrating optical and acoustic detection technologies to develop integrated opto-acoustic devices capable of operating effectively in complex underwater environments, thereby accurately locating and identifying surface defects on underwater bridge piers. The 2D sonar image-based 3D reconstruction method proposed in this paper relies on recognizing bridge pier cross-sectional profiles. It was initially designed to accurately reconstruct various types of basic defects with minimal image data. However, compared to 3D reconstruction methods using 3D sonar, the resulting 3D models are still relatively coarse.

**Author Contributions:** Conceptualization, J.L. and S.J.; Methodology, J.L.; Software, J.L.; Validation, J.L., Y.Z., and C.L.; Formal analysis, J.L.; Investigation, J.L. and Y.Z.; Resources, S.J.; Data curation, Y.Z.; Writing—original draft preparation, J.L.; Writing—review and editing, J.L. and S.J.; Visualization, J.L.; Supervision, S.J.; Project administration, S.J.; Funding acquisition, S.J. All authors have read and agreed to the published version of the manuscript.

**Funding:** This research was funded by the National Natural Science Foundation of China (No. 52278295); the Natural Science Foundation of Fujian Province, China (No. 2022J02016); and the Talent Start-up Fund Project of Fuzhou University (No. XRC-23030).

**Institutional Review Board Statement:** Not applicable.

**Informed Consent Statement:** Not applicable.

**Data Availability Statement:** The data are available from the National Natural Science Foundation of China (No. 52278295); the Natural Science Foundation of Fujian Province, China (No. 2022J02016); the Talent Start-up Fund Project of Fuzhou University (No. XRC-23030).

**Conflicts of Interest:** The authors declare that they have no conflicts of interest.

## References

1. Wu, T.; Hou, S.; Sun, W.; Shi, J.; Yang, F.; Zhang, J.; Wu, G.; He, X. Visual measurement method for three-dimensional shape of underwater bridge piers considering multirefraction correction. *Autom. Constr.* **2023**, *146*, 104706. [CrossRef]
2. Kondo, H.; Kobayashi, S.; Tashiro, T.; Saigo, N.; Hiraike, T.; Kuroki, K. Development of a hybrid underwater vehicle for visual inspection of bridge piers. In Proceedings of the 2023 IEEE Underwater Technology (UT), Tokyo, Japan, 6–9 March 2023; pp. 1–5.
3. Walcarius, K.; Rosin, J.; Hespel, L.; Chami, M.; Dartigalongue, T. Impact of blur on 3D laser imaging: Monte-Carlo modelling for underwater applications. *Opt. Express* **2023**, *31*, 26194–26207. [CrossRef] [PubMed]
4. Shen, S.; Zeng, Y.; Lai, C.; Jiang, S.; Wu, S.; Ma, S. Rapid three-dimensional reconstruction of underwater defective pile based on two-dimensional images obtained using mechanically scanned imaging sonar. *Struct. Control Health Monit.* **2023**, 3647434. [CrossRef]

5. Luo, J.; Jiang, S.; Shen, S.; Chen, J. Deep learning and intelligent detection of apparent defects on underwater foundations of bridges with sonar imaging. *China Civ. Eng. J.* **2021**, *54*, 90–100.
6. Park, C.; Kim, Y.; Lee, H.; Choi, S.; Jung, H. Development of a 2 MHz Sonar Sensor for Inspection of Bridge Substructures. *Sensors* **2018**, *18*, 1222. [CrossRef] [PubMed]
7. Sun, K.; Cui, W.; Chen, C. Review of Underwater Sensing Technologies and Applications. *Sensors* **2021**, *21*, 7849. [CrossRef] [PubMed]
8. Chen, J.; Gong, Z.; Li, H.; Xie, S. A detection method based on sonar image for underwater pipeline tracker. In Proceedings of the 2011 Second International Conference on Mechanic Automation and Control Engineering, IEEE, Inner Mongolia, China, 15–17 July 2011; pp. 3766–3769.
9. Zhang, W.; Zhou, T.; Peng, D.; Shen, J. Underwater pipeline leakage detection via multibeam sonar imagery. *J. Acoust. Soc. Am.* **2017**, *141*, 3917. [CrossRef]
10. Okino, M.; Higashi, Y. Measurement of seabed topography by multibeam sonar using CFFT. *IEEE J. Ocean. Eng.* **1986**, *11*, 474–479. [CrossRef]
11. Shang, X.; Zhao, J.; Zhang, H. Obtaining high-resolution seabed topography and surface details by co-registration of side-scan sonar and multibeam echo sounder images. *Remote Sens.* **2019**, *11*, 1496. [CrossRef]
12. De Falco, F.; Mele, R. The monitoring of bridges for scour by sonar and sedimetri. *NDT E Int.* **2002**, *35*, 117–123. [CrossRef]
13. Topczewski, Ł.; Cieśla, J.; Mikołajewski, P.; Adamski, P.; Markowski, Z. Monitoring of scour around bridge piers and abutments. *Transp. Res. Procedia* **2016**, *14*, 3963–3971. [CrossRef]
14. Clubley, S.; Manes, C.; Richards, D. High-resolution sonars set to revolutionise bridge scour inspections. *Proc. Inst. Civ. Eng.-Civ. Eng.* **2015**, *168*, 35–42. [CrossRef]
15. Hunt, B.E. Scour monitoring programs for bridge health. In Proceedings of the 6th International Bridge Engineering Conference: Reliability, Security, and Sustainability in Bridge Engineering, Boston, MA, USA, 17–20 July 2005.
16. Dai, F.; Rashidi, A.; Brilakis, I.; Vela, P. Comparison of image-based and time-of-flight-based technologies for three-dimensional reconstruction of infrastructure. *J. Constr. Eng. Manag.* **2013**, *139*, 69–79. [CrossRef]
17. Pan, Y.; Dong, Y.; Wang, D.; Chen, A.; Ye, Z. Three-dimensional reconstruction of structural surface model of heritage bridges using UAV-based photogrammetric point clouds. *Remote Sens.* **2019**, *11*, 1204. [CrossRef]
18. Zhang, X.; Xu, Z.; Tan, L.; Li, Y.; Liu, L.; Chen, N.; Zhang, S.; Lamers, W.H.; Wu, C.; Wu, Y. Application of three-dimensional reconstruction and printing as an elective course for undergraduate medical students: An exploratory trial. *Surg. Radiol. Anat.* **2019**, *41*, 1193–1204. [CrossRef] [PubMed]
19. Zhu, Q.; Hu, M.; Zhang, Y.; Du, Z. Research and practice in three-dimensional city modeling. *Geo-Spatial Inf. Sci.* **2009**, *12*, 18–24. [CrossRef]
20. Huang, Z.; Bai, X.; Yu, Z.; Chen, Z.; Feng, N.; Ai, Y.; Song, S.; Xue, L. A novel three-dimensional reconstruction technology for the defect inspection of tubing and casing. *Processes* **2023**, *11*, 2168. [CrossRef]
21. Wang, H.; Wang, H.; Zhu, X.; Song, L.; Guo, Q.; Dong, F. Three-dimensional reconstruction of dilute bubbly flow field with light-field images based on deep learning method. *IEEE Sens. J.* **2021**, *21*, 13417–13429. [CrossRef]
22. Davis, A.; Lugsdin, A. High-speed underwater inspection for port and harbour security using Coda Echoscope 3D sonar. In Proceedings of the OCEANS 2005 MTS/IEEE, IEEE, Washington, DC, USA, 17–23 September 2005; pp. 2006–2011.
23. Li, P. Research on Image Matching Method for the Side-Scan Sonar Image. Master's Thesis, Harbin Engineering University, Harbin, China, 2016. (In Chinese).
24. Sheng, H.; Meng, F.; Li, Q.; Ma, G.; Cao, Y. Enhancement algorithm of side-scan sonar image in curvelet transform domain. *Hydrogr. Surv. Chart.* **2012**, *24*, 8–10. (In Chinese)
25. Zhang, X.; Bao, T.; Gao, X. Crack detection method of concrete dams based on computer vision. *Adv. Sci. Technol. Water Resour.* **2021**, *41*, 83–88. (In Chinese)

sensors

MDPI

Article

# Assessment of Anisotropic Acoustic Properties in Additively Manufactured Materials: Experimental, Computational, and Deep Learning Approaches

Ivan Malashin [1,*], Vadim Tynchenko [1,*], Dmitry Martysyuk [1], Nikita Shchipakov [1], Nikolay Krysko [1], Maxim Degtyarev [1], Vladimir Nelyub [1,2], Andrei Gantimurov [1], Aleksei Borodulin [1] and Andrey Galinovsky [1]

1 Bauman Moscow State Technical University, Moscow 105005, Russia
2 Far Eastern Federal University, Vladivostok 690922, Russia
* Correspondence: ivan.p.malashin@gmail.com (I.M.); vadimond@mail.ru (V.T.); Tel.: +7-926-875-7128 (I.M.)

**Abstract:** The influence of acoustic anisotropy on ultrasonic testing reliability poses a challenge in evaluating products from additive technologies (AT). This study investigates how elasticity constants of anisotropic materials affect defect signal amplitudes in AT products. Experimental measurements on AT samples were conducted to determine elasticity constants. Using Computational Modeling and Simulation Software (CIVA), simulations explored echo signal changes across ultrasound propagation directions. The parameters $A_{13}$ (the ratio between the velocities of ultrasonic transverse waves with vertical and horizontal polarizations at a 45-degree angle to the growth direction), $A_3$ (the ratio for waves at a 90-degree angle), and $A_g$ (the modulus of the difference between $A_{13}$ and $A_3$) were derived from wave velocity relationships and used to characterize acoustic anisotropy. Comparative analysis revealed a strong correlation (0.97) between the proposed anisotropy coefficient $A_g$ and the amplitude changes. Threshold values of $A_g$ were introduced to classify anisotropic materials based on observed amplitude changes in defect echo signals. In addition, a method leveraging deep learning to predict $A_g$ based on data from other anisotropy constants through genetic algorithm (GA)-optimized neural network (NN) architectures is proposed, offering an approach that can reduce the computational costs associated with calculating such constants.

**Keywords:** additive technologies; selective laser melting; ultrasonic inspection; anisotropy; deep learning

## 1. Introduction

Additive manufacturing (AM) is a modern technology for creating complex three-dimensional objects. AM is becoming increasingly common in various industries [1–3]. One reason is that additive manufacturing allows for the creation of parts with complex geometric shapes that may be labor-intensive or even impossible to create by traditional manufacturing methods. Materials produced by additive technologies have fundamental differences in terms of their morphology [4] compared to materials produced by traditional methods such as casting or forging. In all cases, it is important to ensure that manufactured parts meet the required quality standards.

In order to reliably detect internal flaws in products produced by additive manufacturing, non-destructive testing (NDT) methods must be used. Depending on the stage of application, inspection of additive manufacturing products can be divided into two main types: direct inspection during the manufacturing process of the product [5], and inspection after production of the product [6]. Among the main NDT methods used to inspect the AM parts are X-ray computed tomography (XCT), eddy current testing, infrared thermography, acoustic emission, and ultrasonic testing (UT). In these methods, the most promising are XCT and ultrasonic testing [6–8]. These inspection methods are not interchangeable; in certain cases, combinations of methods are used for more effectively flaw detection. In this article, we study the use of ultrasonic testing for the inspection of finished AM products.

Metal AM is a rapidly advancing technology poised to transform product design across the biomedical, aerospace, automotive, marine, and offshore industries. While early adopters have demonstrated significant performance gains, a comprehensive understanding of the microstructure and mechanical properties of additively manufactured metals is lacking. In order to fully harness the design potential of metal AM, especially for structural components, it is vital to address the anisotropic and heterogeneous nature of the microstructures and mechanical properties of metal AM parts. Kok et al. [9] reviewed various metal AM technologies and discussed factors contributing to anisotropy and heterogeneity, offering insights into overcoming these challenges for improved performance and reliability.

Ultrasonic automation offers precise control over particle size and shape for metal powders used in additive manufacturing. Sridharan et al. [10] observed cavity formation and droplet generation using high-speed imaging techniques. Cavitation events were found to play a crucial role in the atomization process. Experiments with liquid aluminum produced spherical particles, with particle size influenced by vibration amplitude. Their research provides insights into ways of optimizing powder production for additive manufacturing applications.

Du et al. [11] examined the anisotropic microstructure of a nickel IN718 sample fabricated using additive manufacturing (AM) and direct laser deposition (DLD). Ultrasonic scattering measurements employing longitudinal-to-transverse (L-T) mode conversion and the continuous wavelet transform (CWT) revealed increasing anisotropy towards the AM-forged interface. These results demonstrate the method's stability in characterizing microstructural anisotropy, suggesting its suitability for additive manufacturing quality control.

Sol et al. [12] utilized pulse-echo ultrasonic testing to investigate anisotropy in $AlSi_{10}Mg$ samples produced by selective laser melting (SLM) additive manufacturing. Various ultrasonic analyses were conducted, revealing anisotropy in both transverse wave velocity and attenuation relative to the build direction. This anisotropy was symmetric around the build direction and persisted after heat treatments. These findings suggest that transverse wave velocity and frequency-dependent attenuation are sensitive tools for detecting subtle changes in additive manufacturing products.

Thevet et al. [13] investigated the anisotropic elastic properties of a $Ti_6Al_4V$ alloy under various manufacturing methods, including additive manufacturing (AM) and wrought alloy. Using dynamic pulse-echo ultrasonic techniques, the study measured sound wave velocities to derive elastic constants and Young's module maps. Minor anisotropy was found in AM materials, while the wrought alloy exhibited higher density and ultrasonic anisotropy due to a thin phase layer along the grain boundaries. Understanding these variations is crucial for product design.

Gou et al. [14] proposed an ultrasonic peening treatment (UPT) in three directions to refine large columnar prior-$\beta$ grains and secondary $\alpha$ grains in the cold-metal transfer additive manufacturing process of $Ti_6Al_4V$ thin-walled structures, thereby enhancing the tensile property anisotropy. Their experimental results revealed significant grain refinement post-UPT, along with minimal surface deformation. Microstructural changes and dislocations were observed, which were attributed to the mechanical effects of ultrasonic treatment within the $\alpha'$ dissolution temperature range. Specimens treated with UPT exhibited improved properties, including higher load capacity in nano-indentation tests along with increased ultimate tensile strength and reduced anisotropic percentage in tensile tests.

Inter-layer ultrasonic impact (UI) strengthening in wire and arc additive manufacturing (WAAM) has been found to effectively reduce anisotropy in both microstructure and mechanical properties [15]. In this study, columnar microstructures were transformed into fine and uniform equiaxial ones, leading to improved stress distribution during tensile testing. Anisotropy in the tensile and yield strength decreased significantly, from 4.2% to 1.6%

and from 10.1 to 2.3%, respectively. UI treatment promoted dislocation movement, leading to substructure formation and grain refinement. It also reduced local misorientations and refined the grain size while weakening the texture orientation strength.

Several studies [6,16,17] have shown the presence of anisotropy in AM materials. The properties of anisotropic materials depends on the direction. In [2,18], the authors showed that the amplitude of a signal reflected from artificial reflectors during ultrasonic testing of TC18 titanium alloy using a phased array depends on the sound beam's direction relative to the direction of sample growth. In other works [16,19], it was found that the velocities of compression and shear waves in Inconel 718 and AlSi10Mg materials obtained by selective laser melting (SLM) changed symmetrically when changing the direction of ultrasonic wave propagation relative to the direction of samples growth. In [16], the dependence of the ultrasonic wave speed on the direction was shown to be preserved even after heat treatment. The directional dependence of acoustic properties with respect to the growth direction may be caused by the characteristics of the crystallographic texture of additive materials. Several works [20,21] have shown that grains have a predominant orientation along the growth direction when using SLM. The crystallographic texture is noted for many materials manufactured by additive methods, for example Inconel 718, $AlSi_{10}Mg$, 316L, etc. [20,22,23].

Table 1 provides an overview of articles related to defect detection using ultrasonic and other related methods along with their limitations in the context of additive technologies (AT).

**Table 1.** Summary and limitations of approaches to studying anisotropy and quality control in additive manufacturing.

| Source | Proposed Approach | Limitations |
|---|---|---|
| Sridharan et al. [10] | Investigation of ultrasonic atomization of metals using high-speed imaging to control particle size and shape. | Research is limited to aluminum, and the influence of cavitation events needs further study for other metals. |
| Du et al. [11] | Use of ultrasonic methods to measure anisotropy in nickel alloy IN718 fabricated by AM and DLD. | Methodology focuses on a specific material (IN718) and needs validation for other materials and additive manufacturing methods. |
| Sol et al. [12] | Application of pulse-echo ultrasonic testing to study anisotropy in $AlSi_{10}Mg$ samples produced by selective laser melting (SLM). | Limited to one type of material ($AlSi_{10}Mg$) and requires additional studies for other additive materials. |
| Thevet et al. [13] | Study of anisotropic elastic properties of $Ti_6Al_4V$ alloy using dynamic pulse-echo ultrasonic techniques. | Focuses on $Ti_6Al_4V$ alloy and needs validation for other materials and manufacturing methods. |
| Gou et al. [14] | Use of ultrasonic peening treatment to enhance anisotropy of mechanical properties and microstructure in cold metal transfer process of $Ti_6Al_4V$. | Primarily addresses one process (cold metal transfer) and material ($Ti_6Al_4V$), requiring further research for other processes and materials. |
| Sun et al. [15] | Inter-layer ultrasonic impact strengthening in wire and arc additive manufacturing (WAAM) to reduce anisotropy. | Limited to one manufacturing process (WAAM) and needs validation on other additive manufacturing methods and materials. |
| Markanday et al. [16], Aleshin et al. [6,17], Li et al. [18], Kransutskaya et al. [2] | Investigation of anisotropic material properties dependency on direction using ultrasonic methods. | Combines different studies and materials but requires a systematic approach for standardizing methods and generalizing results. |
| Simonelli et al. [20], Sufiiarov et al. [21] | Study of crystallographic texture and its impact on anisotropy of materials produced by additive manufacturing methods. | Textural studies focus on a few materials and methods, requiring broader validation. |

A symmetrical change in acoustic properties depending on the direction is common for anisotropic single crystals [24]. The additive materials discussed in this work are polycrystalline. Anisotropy of acoustic properties in polycrystalline materials is largely determined by the presence of a crystallographic texture. Because of the influence of the growth direction on the structure, AM materials are usually considered orthotropic [25]. The propagation of an acoustic wave in a three-dimensional anisotropic medium is generally described by the following Equation (1):

$$C_{ijkl}\frac{\partial^2 u_j}{\partial x_k \partial x_l} = \rho \frac{\partial^2 u_i}{\partial t^2} \tag{1}$$

where $\rho$ represents the material's density and $C_{ijkl}$ denotes its elasticity modulus tensor.

As per Equation (1), the acoustic properties of the material are contingent upon its elastic properties. From the standpoint of the elastic properties, an orthotropic material behaves in a manner akin to the crystals of an orthorhombic system [16]. Crystals in an orthorhombic system possess three mutually perpendicular axes with second-order symmetry. The tensor $C_{ijkl}$ from Equation (1) exhibits symmetry in the first and second pair of indices, and its elements can be represented as a 6 × 6 matrix. In the case of orthotropy or the rhomboid system, the elastic moduli matrix consists solely of the nine independent constants $C_{11}, C_{22}, C_{33}, C_{44}, C_{55}, C_{66}, C_{12}, C_{13}, C_{23}$, and can be delineated as shown below.

$$[C] = \begin{bmatrix} C_{11} & C_{12} & C_{13} & 0 & 0 & 0 \\ C_{12} & C_{22} & C_{23} & 0 & 0 & 0 \\ C_{13} & C_{23} & C_{33} & 0 & 0 & 0 \\ 0 & 0 & 0 & C_{44} & 0 & 0 \\ 0 & 0 & 0 & 0 & C_{55} & 0 \\ 0 & 0 & 0 & 0 & 0 & C_{66} \end{bmatrix} \tag{2}$$

Hence, in certain instances the nature of acoustic anisotropy in additive materials can parallel the anisotropy observed in single crystals. Ultrasonic testing of anisotropic single crystals differs significantly from ultrasonic testing of isotropic materials [6,19]. The propagation of ultrasonic waves is heavily reliant on the orientation of the sound beam relative to the crystal's axis of symmetry. When an ultrasonic wave is incident obliquely to the sample boundary, three elastic waves typically arise, each propagating at its distinct speed. Waves propagating in an anisotropic material are not purely compressional or shear. Due to the aforementioned characteristics, when ultrasonic oscillations propagate in an anisotropic material, deviations in the acoustic field's propagation direction relative to the principal acoustic axis may ensue, along with alterations in the focusing and energy distribution of the ultrasonic beam [17].

For additive materials, a direct correlation exists between the parameters of part growth and microstructure. Factors such as scanning speed, laser beam power, and atmospheric conditions during the AM process, among others, can impact the thermal processes occurring during layer-by-layer material deposition, consequently affecting the resultant microstructure [26,27] and the anisotropy of acoustic properties. Consequently, the degree of anisotropy among samples fabricated from the same material but under different production conditions can vary significantly.

During the development of ultrasonic testing procedures, accounting for anisotropy of the material's properties is imperative. In most instances, the amplitude of the ultrasonic signal serves as the primary informative feature for ultrasonic testing. However, when inspected from different angles relative to the symmetry axis of properties, the impact of anisotropy on the signal amplitude reflected from flaws is contingent upon numerous factors, and is rather intricate. An approach for testing anisotropic single-crystalline materials is elucidated in [26], involving the generation of ultrasonic waves normal to the surface of the test object (TO) followed by inspection along the symmetry axis of the property. While effective for single-crystalline materials, this approach often proves impractical for ultrasonic testing of additive manufacturing products due to their geometric complexity. In such scenarios, the undesirable effects of anisotropy may impede reliable ultrasonic testing. In this study, deep learning is employed to predict $A_g$ based on other acoustic characteristics of the material. This approach is advantageous because deep learning models can capture complex nonlinear relationships within high-dimensional data, leading to accurate and generalizable predictions.

This article delves into the challenge of determining an effective method for evaluating the influence of property anisotropy on the characteristics of an ultrasonic wave in AM product materials. An expedient method for assessing the anisotropy of acoustic properties could facilitate preliminary evaluations for conducting reliable ultrasonic testing on additive materials.

## 2. Materials and Methods

### 2.1. Experimental Study

#### 2.1.1. Sample Description

Samples were fabricated from the heat-resistant nickel alloys Inconel 718, VG159, and EP648 using the selective laser melting (SLM) method on a Concept Laser M2 Cusing device. The growth parameters for SLM samples are presented in Table 2. Additionally, Sample No. 2 underwent further processing by hot isostatic pressing (HIP).

**Table 2.** Growth parameters for SLM samples.

| Sample No | Growth Technology | Material | Laser Parameters | | |
|---|---|---|---|---|---|
| | | | Power, V | Scan Speed, mm/s | Hatch Type |
| 1 | SLM | EP648 | 170 | 800 | Solid |
| 2 | SLM + HIP | EP648 | 180 | 800 | Staggered |
| 3 | SLM | Inconel 718 | 180 | 700 | Staggered |
| 4 | SLM | VG159 | 300 | 805 | Staggered |

Samples were also fabricated from Inconel 718 and AISI 321 using direct laser melting deposition (DLMD) technology. The DLMD setup utilized an IPG Photonics LS-3 fiber laser, and the process was conducted in an argon protective environment. The growth parameters for the DLMD samples are detailed in Table 3.

**Table 3.** Growth parameters for the DLMD samples.

| Sample No | Growth Technology | Material | Power, V | Process Speed, mm/c | Spot Diameter, mm | Layer Step, mm |
|---|---|---|---|---|---|---|
| 5 | DLMD | AISI 321 | 2200 | 25 | 3.2 | 0.6 |
| 6 | DLMD | Inconel 718 | 1570 | 25 | 2.2 | 0.6 |

The samples were designed as cubes with side lengths of 25 mm. Before the experiment, the surfaces of the samples were ground to achieve a roughness level of Rz40. The cube shape was chosen to facilitate the measurement of ultrasonic wave velocities in various directions, which is essential for calculating the coefficients of the elasticity matrix ($C_{11}$ to $C_{66}$). After measuring the necessary ultrasonic speed values, chamfers were removed at a 45° angle to create additional plane-parallel surfaces. Figure 1 shows the general view of the samples before and after processing.

Flat surfaces oriented at a 45° angle relative to directions 1–3 were also necessary for measuring the velocities of ultrasonic waves in order to calculate the elasticity matrix coefficients $C_{12}$, $C_{13}$, and $C_{23}$.

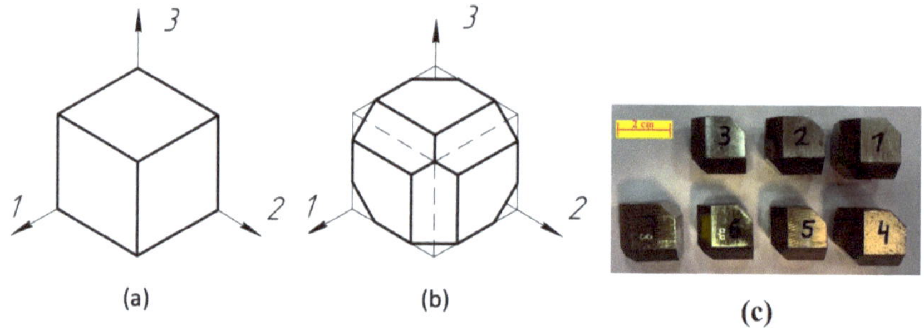


**Figure 1.** General view of the printed samples: (**a**) sample before machining; (**b**) sample after mechanical processing; (**c**) photos of sample after processing.

### 2.1.2. Acoustic Characteristics Measurement

To measure the acoustic properties of additive samples, A-scans were obtained using an Omniscan MX flaw detector. Figure 2 shows an example scan for Inconel 718 SLM with a transcript showing the pulse-echo numbers. Time-of-flight measurements of echo signals were performed using the contact method with a C543SM 5 MHz direct compression-wave transducer and a V155RB direct shear-wave transducer. The time of flight of the echo signal was measured using the maximum amplitude technique. The error in measuring the speeds of ultrasonic waves corresponded to about 0.5%.

Measurements of longitudinal wave attenuation coefficients were performed using an immersion contact on cubic samples prior to chamfer removal. The measurement technique corresponded to that described in [28], method 2.

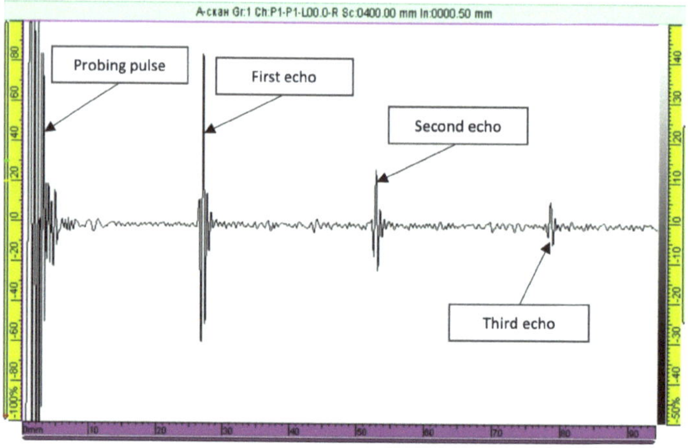


**Figure 2.** A-scan obtained for an Inconel 718 SLM sample at frequency $f = 5$ MHz with the C543SM longitudinal wave transducer.

### 2.1.3. Calculation of Elasticity Matrix Coefficients $C_{ij}$

Nine independent coefficients of the elasticity matrix $C_{ij}$ for the orthotropic case were determined for the six studied samples. These coefficients were calculated after measuring the speed of ultrasonic waves using the dependencies shown below.

$$C_{11} = \rho \cdot V^2_{1/1} \tag{3}$$

$$C_{22} = \rho \cdot V^2_{2/2} \tag{4}$$

$$C_{33} = \rho \cdot V^2_{3/3} \tag{5}$$

$$C_{44} = \rho \cdot V_{2/3}^2 \quad (6)$$

$$C_{55} = \rho \cdot V_{1/3}^2 \quad (7)$$

$$C_{66} = \rho \cdot V_{1/2}^2 \quad (8)$$

$$C_{23} = \sqrt{(C_{22} + C_{44} - 2\rho \cdot V_{23/23}^2) \cdot (C_{33} + C_{44} - 2\rho \cdot V_{23/23}^2)} - C_{44} \quad (9)$$

$$C_{13} = \sqrt{(C_{11} + C_{55} - 2\rho \cdot V_{12/12}^2) \cdot (C_{33} + C_{55} - 2\rho \cdot V_{12/12}^2)} - C_{55} \quad (10)$$

Here, $V_{i/i}$ represents the speed of a longitudinal wave in direction $i$, $V_{i/j}$ represents the speed of a transverse wave propagating in direction $i$ with polarization in direction $j$, and $V_{ij/ij}$ represents the speed of a quasi-longitudinal or quasi-transverse wave propagating and polarized in the $ij$ plane, while $\rho$ denotes the density of the material. A detailed description of the methodology used to calculate the coefficients $C_{ij}$ is provided in [13]. Sample density measurements were not conducted in this study; therefore, the $\rho$ values of materials provided by the manufacturer of samples manufactured under these conditions were utilized. The values of $\rho$ are presented in Table 4.

**Table 4.** Density values used to calculate $C_{ij}$ coefficients.

| Material | $\rho$, g/cm$^3$ |
|---|---|
| EP648 | 8.0 |
| Inconel 718 | 8.2 |
| VG159 | 8.2 |
| AISI 321 | 7.9 |

2.1.4. Porosity Analysis

X-ray tomography (X-ray CT) was conducted using a General Electric v | tome | x m300 tomograph equipped with a tube to achieve micrometer resolution tomography. The parameters of the X-ray CT performance modes are outlined in Table 5. The results were processed using specialized VGSTUDIO MAX software employing algorithms for visualizing pores and solid inclusions. Based on the sample material and voxel size, the minimum reliably detectable pore size in all samples was approximately 40 µm.

**Table 5.** X-ray CT performance modes.

| Sample Number | Cathode Voltage, kV | Current, µA | Focal Spot Size, µm | Number of Projections | Voxel Size, µm |
|---|---|---|---|---|---|
| VG159 SLM | 220 | 70 | 15.4 | 1700 | 15.38 |
| AISI321 DLMD | 220 | 70 | 15.4 | 1700 | 15.22 |
| EP648 SLM | 220 | 70 | 15.4 | 1700 | 15.45 |
| EP648 SLM+HIP | 220 | 70 | 15.4 | 1700 | 15.22 |
| Incinel718 DLMD | 220 | 70 | 15.4 | 1700 | 15.22 |
| Inconel718 SLM | 220 | 70 | 15.4 | 1700 | 15.22 |

*2.2. Computer Modelling*

2.2.1. Assessment of Changes in the Amplitude of the Echo Signal Reflected from the Flaw

As previously noted, the anisotropy of acoustic properties can significantly impact test results. When the difference in signal amplitudes from identical reflectors in different ultrasound propagation directions exceeds permissible error limits, concerns arise regarding the reliability of ultrasonic testing due to uneven sensitivity. This issue has been extensively documented for Austenitic materials and welded joints [17,27,29,30]. In developing inspection techniques for Austenitic materials, a preliminary assessment of testability is conducted to determine the feasibility of reliable ultrasonic testing. Testability is typically evaluated based on several criteria, including the signal-to-noise ratio and the deviation from rectilinear ultrasonic beam propagation.

This study proposes assessing the change in the amplitude of the reflected signal from an artificial reflector when inspected in different directions as an indirect indicator of the feasibility of reliable ultrasonic testing of anisotropic additive materials. In a previous study [31], it was demonstrated using Inconel 718 SLM samples that the amplitude change of the reflected signal from a side drill hole (SDH) at a depth of 13.5 mm when inspected from different angles can reach 4 dB. Experimentally determined elasticity matrix coefficients $C_{ij}$ of Inconel 718 were used as input data for modeling in the CIVA UT Simulation Tool. A quantitative comparison of the simulation results with experimental data showed that the maximum deviation between the results did not exceed 1 dB, indicating high accuracy of the amplitude change estimates obtained from CIVA simulations using these elastic constants.

Based on these findings, it can be argued that for the samples studied in [19], the crystallographic texture significantly contributes to the change in the echo signal amplitude from the SDH. The directional dependence of the amplitude change was determined by modeling using only experimentally derived coefficients of the Inconel 718 elasticity matrix. Using the CIVA modeling package and known elasticity matrix coefficients as input data, the influence of the material's anisotropy on the ultrasonic testing results can be preliminarily assessed [32]. The experimental setup of the approach is shown in Figure 3.

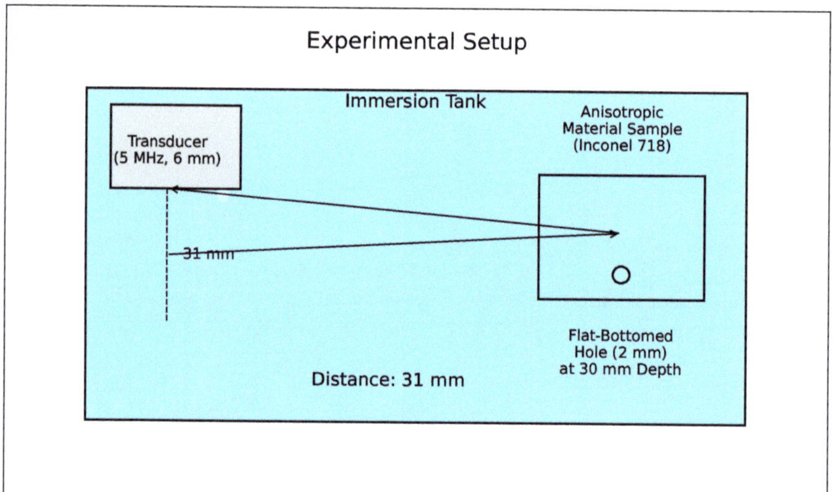


**Figure 3.** Schematic diagram of the experimental setup for assessing changes in echo signal amplitude.

### 2.2.2. Description of CIVA Algorithms

The ultrasonic propagation simulation tools in the CIVA UT Simulation Tool are based on semi-analytical solutions used to calculate the ultrasonic field within a material and its interaction with flaws. CIVA utilizes ray tube theory [33] to model the acoustic field and the ray tracing method for visualization. The calculation of the field emitted by the transducer occurs by representing the field as a sum of fields created by elementary point sources on the surface of the piezoelectric plate. Using this method, it is possible to determine the time of flight, wave type transformations, and ray amplitudes. When calculating the amplitude and phase of the field reflected from defects, CIVA employs the Kirchhoff approximation method, variable separation method, and the geometric theory of diffraction [34].

2.2.3. Model Description

To calculate the change in the signal amplitude from a defect for different directions of ultrasound propagation in an anisotropic material, an immersion ultrasonic model was employed. A flat-bottomed hole (FBH) with a 2 mm diameter located at a depth of 30 mm was used as a defect model for calculations. The general diagram of the model is presented in Figure 4. The amplitude of the reflected signal from the FBH was calculated for different positions of the material properties' symmetry axes, simulating various directions of ultrasound propagation with respect to the properties' symmetry axes. The calculation of the echo signal amplitude from the FBH was carried out by rotating the coordinate system (1, 2, 3) with a step of 15° separately relative to each of the three axes, as shown in Figure 4a. A model of an unfocused piezoelectric transducer was used as a source of ultrasonic vibrations with a frequency of 5 MHz and a piezoelectric element diameter of 6 mm. The distance between the probe and the sample surface was 31 mm, ensuring that the near zone was completely in the immersion layer.

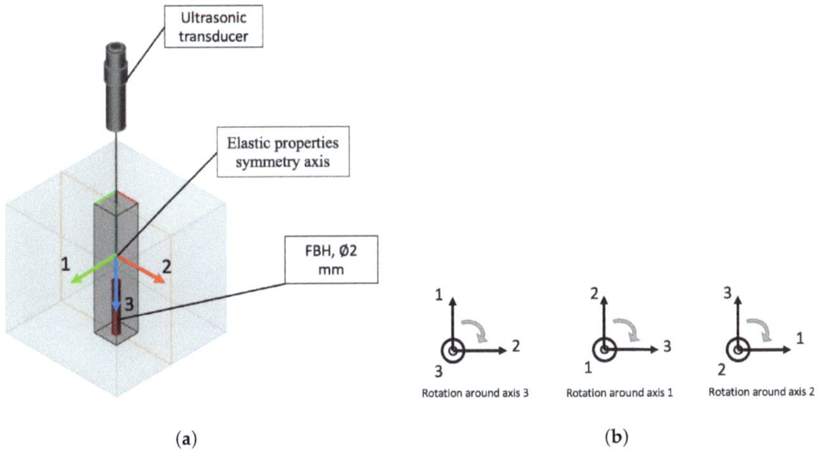


**Figure 4.** (**a**) Model diagram for calculation of amplitude changes and (**b**) rotation of property symmetry axes.

For each calculated value of the reflected FBH echo signal amplitude, the change in amplitude $\Delta A$ was determined relative to the amplitude value for the initial position of the axes shown in Figure 4. This calculation was performed for all studied materials using the measured single-crystal coefficients of the elasticity matrix as input data.

## 2.3. Assessment of Acoustic Anisotropy

### 2.3.1. Methods Used to Assess Acoustic and Elastic Anisotropy

To conduct a comparative analysis of the effectiveness of different anisotropy coefficients for assessing amplitude changes obtained as a result of CIVA simulations, the following anisotropy coefficients were considered and calculated for the studied materials.

Many studies [35–38] have been devoted to assessing the anisotropy of the elastic properties of rocks. Similar to additive materials, rocks are generally orthotropic media; therefore, methods for assessing elastic anisotropy using acoustic methods can in some cases be applied to polycrystalline materials.

For transversely isotropic media, the following coefficients which allow for estimating the elastic anisotropy were proposed in [38]:

$$\delta = \frac{(C_{13} + C_{44})^2 - (C_{33} - C_{44})^2}{2C_{33}(C_{33} - C_{44})} \approx 4\left(\frac{\nu_p(\frac{\pi}{4})}{\nu_p(0)} - 1\right) - \left(\frac{\nu_p(\frac{\pi}{2})}{\nu_p(0)} - 1\right) \quad (11)$$

$$\epsilon = \frac{C_{11} - C_{33}}{2C_{33}} \approx \frac{\nu_p(\frac{\pi}{2}) - \nu_p(0)}{\nu_p(0)} \quad (12)$$

$$\gamma = \frac{C_{66} - C_{44}}{2C_{44}} \approx \frac{\nu_{SH}(\frac{\pi}{2}) - \nu_{SH}(0)}{\nu_{SH}(0)} \quad (13)$$

where $C_{ij}$ are the coefficients of the elasticity matrix; $\nu_p(0)$, $\nu_p(\frac{\pi}{4})$, and $\nu_p(\frac{\pi}{2})$ are the velocities of longitudinal waves, measured in the direction of the property's symmetry axis at an angle of 45° relative to the symmetry axis and perpendicular to the property's symmetry axis, respectively; and $\nu_{SH}(0)$ and $\nu_{SH}(\frac{\pi}{2})$ are the velocities of the transverse wave with horizontal polarization, measured in the direction of the property's symmetry axis and perpendicular to its symmetry axis, respectively.

For orthotropic rock media, the following coefficients are used to estimate anisotropy based on the birefringence factor of shear waves [35]:

$$B_1 = \frac{2(V_{1/2} - V_{1/3})}{V_{1/2} + V_{1/3}} = \frac{2(\sqrt{C_{55}} - \sqrt{C_{66}})}{\sqrt{C_{55}} + \sqrt{C_{66}}} \quad (14)$$

$$B_2 = \frac{2(V_{2/1} - V_{2/3})}{V_{2/1} + V_{2/3}} = \frac{2(\sqrt{C_{66}} - \sqrt{C_{44}})}{\sqrt{C_{66}} + \sqrt{C_{44}}} \quad (15)$$

$$B_3 = \frac{2(V_{3/1} - V_{3/2})}{V_{3/1} + V_{3/2}} = \frac{2(\sqrt{C_{55}} - \sqrt{C_{44}})}{\sqrt{C_{55}} + \sqrt{C_{44}}} \quad (16)$$

$$B_S = \sqrt{B_1^2 + B_2^2 + B_3^2} \quad (17)$$

where $V_{i/j}$ is the velocity of the transverse wave in direction $i$ with polarization $j$.

There is also a method for estimating the anisotropy of crystals [39] through the coefficient $G$:

$$G = \frac{2C_{44}}{C_{11} - C_{12}} = \left(\frac{\nu_2[100]}{\nu_2[110]}\right)^2 \quad (18)$$

where $\nu_2[100]$ and $\nu_2[110]$ are transverse wave velocities propagating along the [100] and [110] directions, respectively, with polarization perpendicular to [001].

### 2.3.2. Proposed Method for Estimating Acoustic Anisotropy

In this work, we propose evaluating the anisotropy of additive samples using the birefringence factor, similar to the method described in [35], but focused on two directions of ultrasonic vibration propagation, i.e., 45° and 90° relative to the direction of the additive sample growth. Specifically, we introduce the following coefficients:

- $A_{13}$: The percentage ratio between the velocities of ultrasonic transverse waves with vertical (SV) and horizontal (SH) polarizations, propagating at a 45-degree angle to the growth direction.
- $A_3$: The similarity ratio for ultrasonic waves propagating at a 90-degree angle relative to the growth direction.
- $A_g$: The modulus of the difference between $A_{13}$ and $A_3$, indicating the extent of variation in the deviation between the velocities of transverse waves with mutually perpendicular polarizations in the two directions, i.e., at a 45-degree angle and normal to the growth direction.

These coefficients are determined with the following formulas:

$$A_{13} = \frac{(\nu_{SH}(\frac{\pi}{4}) - \nu_{SV}(\frac{\pi}{4}))}{\nu_{SH}(\frac{\pi}{4})} \times 100\% =$$

$$= \frac{\sqrt{C_{44}+C_{66}} - \sqrt{C_{44}+\frac{1}{2}(C_{33}+C_{11}) - \sqrt{\frac{1}{4}(C_{11}-C_{33})^2+(C_{13}+C_{44})^2}}}{\sqrt{C_{44}+C_{66}}} \times 100\% \quad (19)$$

$$A_3 = \frac{(v_{SH}(\frac{\pi}{2}) - v_{SV}(\frac{\pi}{2}))}{v_{SH}(\frac{\pi}{2})} \times 100\% = \frac{\sqrt{C_{44}} - \sqrt{C_{66}}}{\sqrt{C_{44}}} \times 100\% \quad (20)$$

$$A_g = |A_3 - A_{13}| \quad (21)$$

where $v_{SH}$ and $v_{SV}$ are the velocities of transverse waves with horizontal and vertical polarization, respectively. We propose using the coefficient $A_g$ as the general indicator of a material's anisotropy. To calculate the coefficients, use a special cube-shaped sample with removed chamfers at 45° relative to the growth direction, as shown in Figure 5.

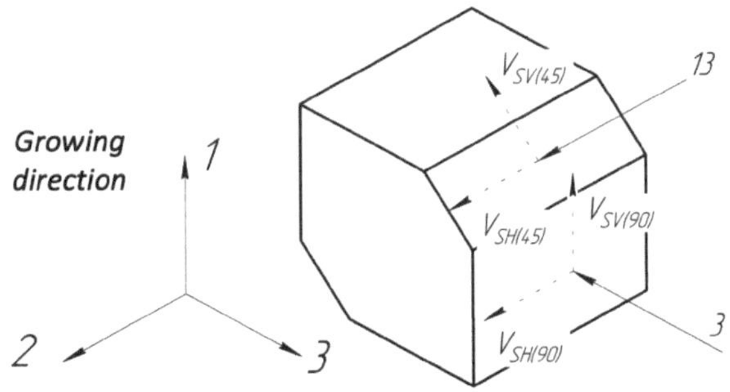


**Figure 5.** Sample for determining anisotropy coefficients using Formulas (19)–(21).

## 3. Results

### *3.1. Ultrasonic Speed and Attenuation*

The values of the velocities of longitudinal and transverse ultrasonic waves measured for various directions on the samples described in Section 2.1.1 are presented in Table 6. The values of the measured attenuation coefficients of longitudinal waves measured in three mutually perpendicular directions are presented in Table 7. The values of the calculated elasticity matrix coefficients $C_{ij}$ are provided in Table 8. It should be noted that in all the studied materials a difference of approximately 1–3% was noted between the measured velocities of longitudinal waves in the growth direction and normal to it.

**Table 6.** Ultrasound wave velocity values.

| Measured Velocity, km/s | EP648 SLM | EP648 SLM+HIP | Inconel718 DLMD | Inconel718 SLM | VG159 SLM | AISI 321 DLMD |
|---|---|---|---|---|---|---|
| $v_{(1/1)}$ | 5.69 | 5.82 | 5.40 | 5.62 | 5.53 | 5.66 |
| $v_{(2/2)}$ | 5.80 | 5.95 | 5.60 | 5.72 | 5.48 | 5.75 |
| $v_{(3/3)}$ | 5.80 | 5.92 | 5.57 | 5.72 | 5.48 | 5.73 |
| $v_{(3/2)}$ | 2.64 | 2.84 | 2.92 | 2.96 | 3.42 | 2.96 |
| $v_{(1/3)}$ | 3.08 | 3.14 | 3.10 | 3.25 | 3.48 | 3.18 |
| $v_{(1/2)}$ | 3.07 | 3.17 | 3.04 | 3.25 | 3.44 | 3.20 |
| $v_{(13/13)}$ | 5.78 | 5.89 | 5.60 | 5.85 | 6.03 | 5.77 |
| $v_{(12/12)}$ | 5.82 | 5.92 | 5.56 | 5.85 | 6.03 | 5.79 |
| $v_{(23/23)}$ | 5.54 | 5.76 | 5.57 | 5.72 | 5.99 | 5.62 |

**Table 7.** Values of the longitudinal wave attenuation coefficients $f = 5$ MHz, dB/cm.

| Material | EP648 SLM | EP648 SLM+HIP | Inconel718 DLMD | Inconel718 SLM | VG159 SLM | AISI 321 DLMD |
|---|---|---|---|---|---|---|
| **Direction** | | | | | | |
| 1 | 0.17 | 0.5 | 1.66 | 0.11 | 0.63 | 0.72 |
| 2 | 0.12 | 0.3 | 1.59 | 0.32 | 0.83 | 0.59 |
| 3 | 0.43 | 0.14 | 1.61 | 0.25 | 1.18 | 0.69 |

**Table 8.** Values of the elasticity matrix coefficients $C_{ij}$, GPa.

| Coefficients $C_{ij}$ | EP648 SLM | EP648 SLM+HIP | Inconel718 DLMD | Inconel718 SLM | VG159 SLM | AISI 321 DLMD |
|---|---|---|---|---|---|---|
| $C_{11}$ | 259 | 271 | 259 | 257 | 251 | 253 |
| $C_{22}$ | 269 | 283 | 269 | 255 | 246 | 261 |
| $C_{33}$ | 269 | 281 | 269 | 239 | 246 | 259 |
| $C_{44}$ | 56 | 66 | 72 | 76 | 96 | 71 |
| $C_{55}$ | 75 | 79 | 87 | 75 | 98 | 79 |
| $C_{66}$ | 75 | 79 | 87 | 70 | 97 | 81 |
| $C_{23}$ | 110 | 119 | 125 | 115 | 151 | 97 |
| $C_{13}$ | 121 | 121 | 125 | 108 | 152 | 113 |
| $C_{12}$ | 127 | 125 | 125 | 113 | 154 | 110 |

The most significant velocity differences are noted for transverse waves. For instance, in EP648 SLM the velocity difference between waves $\nu_{1/3}$ and $\nu_{1/2}$ from Table 6 is about 17%. The largest deviation for longitudinal waves is observed in VG159 SLM, with a 10% difference between $\nu_{1/1}$ and $\nu_{13/13}$. The attenuation coefficients measured in three perpendicular directions (Table 7) are relatively low compared to heat-resistant alloys produced by traditional methods, where they are twice as high [29]. An F-test showed a statistically significant relationship between direction and attenuation coefficients for Inconel718 SLM, EP648 SLM, EP648 SLM+HIP, and VG159 SLM at a 5% significance level. This relationship can affect the amplitude of the reflected echo signal, which is important for ultrasonic testing of these materials.

### 3.2. Porosity Analysis

A small number of pores were found in almost every sample. The corresponding results are shown in Table 9.

**Table 9.** Values of porosity analysis.

| Sample | Volume Porosity (%) | Presence of Cracks | Other Defects | Maximum Defect Volume (mm$^3$) | Quantity of Founded Defects |
|---|---|---|---|---|---|
| VG59 SLM | 0.00087 | No | No | 0.000419 | 47 |
| AISI 321 DLMD | 0.00061 | No | No | 0.001291 | 25 |
| EP648 SLM | 0.00113 | No | No | 0.000872 | 78 |
| EP648 SLM+HIP | 0 | No | No | - | - |
| INCONEL718 DLMD | 4.97 | No | No | 0.03789 | 51,638 |
| Inconel 718 SLM | 0.00002 | No | No | 0.000158 | 2 |

In the Inconel 718 DLMD sample, a significant number of large-volume voids were observed; these were almost completely spherical in shape and were distributed evenly throughout the entire volume of the sample. The high level of porosity in Inconel 718 DLMD is likely the reason for the longitudinal wave velocities measured in three mutually perpendicular directions being lower than in the SLM sample of the same alloy. In addition, it should be noted that the attenuation coefficients of longitudinal waves in the Inconel 718 SLM sample are significantly lower than in the Inconel 718 DLMD sample. These results are consistent with studies of the relationship between the level of porosity of additive samples and the values of acoustic characteristics [4,5].

### 3.3. Maximum Change in Echo Signal Amplitude and Degree of Anisotropy

Using the method described in Section 2.2.3, changes in the amplitude $\Delta A$ of the echo signal from the FBH were calculated for the six samples and for varying values of the angle between the direction of propagation of vibrations and the properties' axes of symmetry. Among the values of $\Delta A$ obtained by modeling, the values of the maximum changes in the amplitude $\Delta A_{\max}$ were selected for each material under study. To increase the set of data under study, we used the coefficients $C_{ij}$ from [13] for the Ti6Al4V alloy produced using SLM technology and metal forming (MF), as well as the elastic coefficients determined experimentally in [16] for the Inconel 718 and Inconel718NbC materials produced using DLMD technology. Calculations performed in CIVA showed that the largest changes in the amplitude of the echo signal occur when the angle between the propagation direction of the vibrations and the properties' axis of symmetry co-directed with the direction of the material growth is approximately 45° or 90°. For all samples, anisotropy coefficients and maximum changes in amplitude $\Delta A_{\max}$ were calculated using the methods described in Section 2.3. The values of the anisotropy coefficients and maximum changes in amplitude $\Delta A_{\max}$ are provided in Table 10 along with the maximum changes in amplitude and degree of anisotropy for each material.

**Table 10.** Maximum changes in amplitude and degree of anisotropy.

| Material | $\Delta A_{\max}$ (dB) | $B_S$ | $\epsilon$ | $\delta$ | $\gamma$ | $G$ | $A_g$ (%) |
|---|---|---|---|---|---|---|---|
| EP648 SLM | 2.9 | 0.21 | 0.02 | 0.071 | −0.126 | 0.94 | 11.2 |
| EP648 SLM+HIP | 2 | 0.14 | −0.004 | −0.023 | −0.001 | 0.99 | 6 |
| Inconel718 SLM | 4.3 | 0.14 | 0.018 | 0.168 | −0.084 | 1.2 | 16.1 |
| Inconel718 DLMD | 2.1 | 0.06 | 0.036 | 0.097 | −0.043 | 1.06 | 9.2 |
| VG159 SLM | 9.5 | 0.02 | −0.01 | 0.518 | −0.01 | 2.06 | 30 |
| AISI 321 DLMD | 2.5 | 0.09 | −0.011 | −0.035 | 0.085 | 0.97 | 9.6 |
| Ti6Al4V SLM [13] | 1 | 0.1 | 0.026 | −0.019 | 0.075 | 0.86 | 4.11 |
| Ti6Al4V MF [13] | 0.4 | 0 | 0 | −0.019 | 0 | 1.02 | 1.4 |
| Inconel 718 DLMD [16] | 6.1 | 0.21 | 0.057 | 0.066 | −0.131 | 0.9 | 20.1 |
| Inconel718NbC DLMD [16] | 8 | 0.26 | 0.074 | 0.058 | −0.159 | 0.88 | 25.4 |
| AlSi10Mg SLM [40–42] | 3.4 | 0.19 | 0.012 | 0.049 | −0.097 | 1.1 | 12.5 |
| 316L SLM [43–45] | 2.2 | 0.15 | 0.008 | 0.35 | −0.065 | 1.03 | 9.0 |
| 17-4 PH SLM [46,47] | 2.7 | 0.13 | 0.02 | 0.067 | −0.089 | 0.96 | 10.4 |
| CoCrW SLM [48,49] | 3.9 | 0.18 | 0.015 | 0.056 | −0.105 | 1.08 | 14.3 |
| Hastelloy X SLM [50–52] | 4.6 | 0.22 | 0.023 | 0.074 | −0.111 | 1.2 | 17.0 |
| 18Ni-300 MS [53–55] | 3.1 | 0.17 | 0.019 | 0.061 | −0.092 | 1.05 | 13.2 |
| AlSi12 SLM [41,56] | 2.8 | 0.16 | 0.011 | 0.048 | −0.086 | 1.09 | 11.7 |
| CuSn10 SLM [57–59] | 3.5 | 0.20 | 0.014 | 0.055 | −0.095 | 1.12 | 13.6 |
| Al7075 SLM [60,61] | 4.0 | 0.21 | 0.018 | 0.062 | −0.108 | 1.15 | 15.2 |
| Al2024 SLM [62,63] | 3.6 | 0.19 | 0.016 | 0.057 | −0.102 | 1.11 | 13.8 |
| NiTi SLM [64,65] | 4.4 | 0.23 | 0.025 | 0.073 | −0.114 | 1.17 | 16.8 |

To compare the effectiveness of various coefficients for estimating the value of $\Delta A_{\max}$, the correlation coefficients $R$ between the degree of anisotropy and the change in amplitude were calculated. The values of the calculated correlation coefficients are provided in Table 11.

**Table 11.** Correlation coefficients between the degree of anisotropy and change in amplitude.

| | $\epsilon$ | $\delta$ | $\gamma$ | $G$ | $B_S$ | $A_g$ |
|---|---|---|---|---|---|---|
| $\Delta A_{\max}$ | 0.31 | 0.74 | −0.50 | 0.58 | 0.31 | 0.97 |

These values were calculated using the Pearson correlation coefficient $r$ using the formula

$$r = \frac{\sum(x_i - \bar{x})(y_i - \bar{y})}{n\sigma_x\sigma_y}, \tag{22}$$

where $x_i$ and $y_i$ represent individual sample points, $\bar{x}$ and $\bar{y}$ are the means of the samples, $\sigma_x$ and $\sigma_y$ are the standard deviations of the samples, and $n$ is the number of sample points.

This analysis allows for the determination of which anisotropy parameter most effectively predicts $\Delta A_{\max}$. By evaluating the Pearson correlation coefficient for each anisotropy parameter, it is possible to identify the parameter that exhibits the strongest linear relationship with the change in amplitude. As shown in Table 11, the results highlight the comparative effectiveness of these coefficients in estimating $\Delta A_{\max}$, providing insights for optimizing the evaluation of anisotropy in additive manufacturing materials.

It can be seen from Table 11 that the maximum value of 0.97 for the correlation coefficient was found for our proposed anisotropy coefficient $A_g$, which indicates that the coefficient $A_g$ has the strongest correlation with the change in amplitude.

### 3.4. Criteria for Classifying Anisotropic Additive Materials by Assessing the Magnitude of the Amplitude Change

To develop a criterion based on the $A_g$ coefficient which can allow for the classification of anisotropic materials depending on the magnitude of the change in amplitude, more data were needed. To solve this problem, data obtained as a result of numerically calculating the coefficients $C_{ij}$ from the system of Equations (13) and (14) for given values of the anisotropy coefficients $A_{13}$ and $A_3$, varying in the range from −15% to 15%, were added to the array of data in Table 7. The use of data obtained by modeling the parameters of anisotropic media was necessary in order to obtain estimated information about the relationship between the coefficient $A_g$ and the magnitude of the possible change in amplitude $\Delta A_{\max}$ in materials where the anisotropy coefficients $A_{13}$ and $A_3$ differ significantly from those calculated on experimental samples. The total set of data obtained as a result of modeling and experimental samples included 59 points. A scatterplot of the data between the anisotropy parameter $A_g$ and the magnitude of the maximum amplitude change $\Delta A_{\max}$ for the overall dataset is shown in Figure 6.

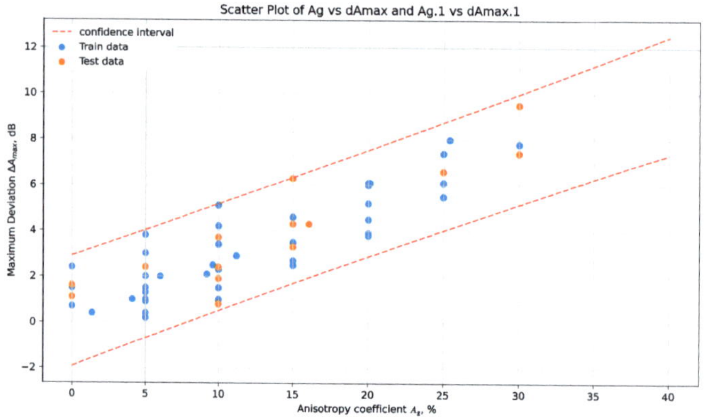


**Figure 6.** Scatterplot of the data between the anisotropy parameter $A_g$ and the magnitude of the maximum amplitude change $\Delta A_{\max}$.

The scatterplot in Figure 6 shows that the data are distributed along a conditional inclined line (trend line), indicating a strong positive relationship between $\Delta A_{\max}$ and $A_g$. The regression equation has the following form:

$$\Delta A_{\max} = 0.24 A_g + 0.48 \tag{23}$$

while the $R^2$ coefficient equals 0.74. Significance testing for linear regression was performed using Fisher's test of significance. At a 5% significance level, the $p$-value equals $2.4 \times 10^{-13} < 0.05$.

Based on Equation (23), the following criteria are proposed:

- If $A_g < 6.5\%$, the estimated change in the amplitude of the echo signal reflected from the FBH with a 2 mm diameter and located at a depth of 30 mm does not exceed 2 dB when the propagation direction of the vibrations changes relative to the axis of the properties' symmetry.
- If $6.5\% \leq A_g < 23\%$, the estimated change in the amplitude of the echo signal reflected from the FBH with a 2 mm diameter and located at a depth of 30 mm can take values from 2 to 6 dB when the vibrations' propagation direction changes with respect to the axis of the properties' symmetry.
- If $A_g \geq 23\%$, the estimated change in the amplitude of the echo signal reflected from the FBH with a 2 mm diameter and located at a depth of 30 mm can take values in excess of 6 dB when the vibrations' propagation direction changes relative to the axis of the properties' symmetry. In this case, it is permissible to inspect the material by exciting ultrasonic waves only normally to the surface, with propagation along the axis of the material properties' symmetry.

### 3.5. Deep Learning Approach for Predicting the Anisotropy Indicator $A_g$

Based on the data presented in Table 10, our objective in this section is to develop a machine learning model capable of accurately predicting the anisotropy indicator $A_g$. To achieve this, we employ an approach based on optimizing TensorFlow Keras neural network (NN) hyperparameters using a genetic algorithm (GA) [66,67].

A possible architecture of the suggested approach is shown in Figure 7. This architecture is designed to effectively capture the relationships between the input features and the target output. The input layer comprises nodes representing acoustic features. The architecture includes multiple hidden layers with varying numbers of neurons which are optimized via the GA. These hidden layers are responsible for learning complex patterns and interactions between the input features through nonlinear transformations. The output layer [68] contains a single node representing $A_g$. This output is the primary target of the model, providing insights into the effectiveness of ultrasonic testing under varying conditions. The architecture leverages the strengths of each input feature [69], transforming them through a series of hidden layers to generate an accurate prediction of the output. The specific design and depth of the hidden layers are determined through hyperparameter optimization, ensuring that the model captures the underlying relationships within the data.

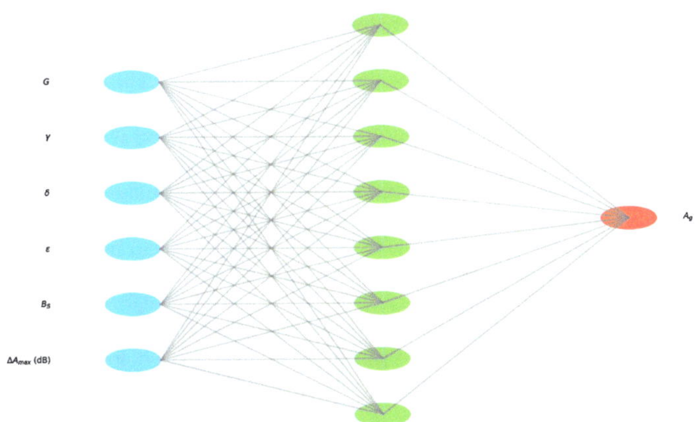


**Figure 7.** Example showing possible optimized NN architecture.

The methodology involves defining the the hyperparameter space to consider various configurations, including different numbers of layers ranging from one to five, neurons per layer ranging from four to sixty-four in increments of four, and a diverse set of activation functions, including softmax [70], elu [71], selu [72], softplus [73], softsign [74], relu [75], tanh [76], sigmoid [77], hard sigmoid [78], exponential [79], and linear [80]. This defines the space in which the GA searches for the optimal hyperparameters.

The fitness function [81] used to build and trained the neural network is based on the following specified hyperparameters: layers, neurons per layer, and activation functions. The model is initialized with a specified input shape and trained using the Adam optimizer and mean squared error loss function for 100 epochs. The model's performance can then be evaluated based on maximization of the R-squared ($R^2$) score, calculated by comparing the true and predicted values on a test dataset comprising 20% of the overall sample.

The genetic algorithm (GA) optimizes the neural network's hyperparameters through a series of structured operations, namely, population initialization, crossover, mutation, and parent selection.

The process begins by initializing a population of chromosomes, each of which represents a unique set of hyperparameters randomly selected from predefined search spaces. This initial population serves as the foundation for the evolutionary search process.

During the crossover phase, new chromosomes are generated by combining the hyperparameters of two parent chromosomes. This operation fosters genetic diversity within the population, as offspring inherit characteristics from both parents, potentially combining advantageous traits that enhance performance.

Mutation introduces variability into the population by randomly altering one or more hyperparameters within a chromosome [82]. This stochastic modification ensures a comprehensive exploration of the search space, preventing premature convergence on suboptimal solutions and enhancing the algorithm's ability to discover globally optimal hyperparameters.

Parents are selected based on their fitness scores [83–85], which reflect the performance of the neural network using the corresponding set of hyperparameters. Several strategies exist for parent selection, such as roulette wheel selection, tournament selection, and rank-based selection, which ensure that higher-performing chromosomes have a greater probability of contributing to the next generation.

This iterative process of selection, crossover, and mutation is repeated over multiple generations. Each cycle refines the population to gradually optimize the hyperparameters and enhance the neural network's predictive performance.

Through these iterative evolutionary operations, the genetic algorithm effectively navigates the hyperparameter space, converging on configurations that yield superior predictive accuracy and generalization capabilities for the neural network.

To optimize the parameters of the neural network for determining the anisotropy indicator, we ran several iterations of the GA with different parameters. For instance, Figure 8a shows the results of the first two generations, where the population size was 10, the number of generations was 3, and the number of parents for selection was 5. Figure 8b depicts the results of the third generation, where the parameters were altered; the population size remained 10, but the number of generations increased to 5 and the number of parents for selection was set to 10. Subsequently, Figure 8b displays the results for the fourth generation, where the population size was increased to 20, the number of generations to 10, and the number of parents for selection to 15. It can be observed that as the number of iterations increases, the $R^2$ score becomes more stable, indicating enhancement of the model's performance.

The advantage of this method lies in its ability to automatically optimize the parameters of the neural network to achieve maximum prediction accuracy. This helps to reduce the time and resources spent on manual parameter tuning and ensures more stable and reliable results.

In order to validate the effectiveness of our method, we performed spatial cross-validation [86] on the neural networks with the best $R^2$ scores (>0.9).

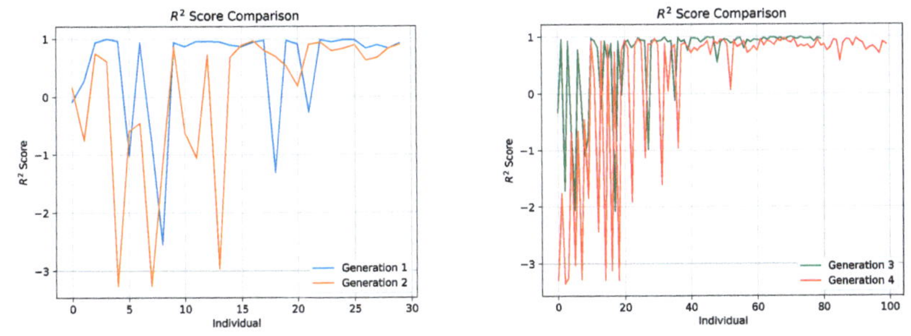


(**a**) (**b**)

**Figure 8.** Change in $R^2$ score as a function of the number of GA generations during fine-tuning of the neural network hyperparameters: (**a**) the orange and blue curves represent experiments with a population size of 10, three generations, and five parents selected per generation; (**b**) the red curve corresponds to experiments with a population size of 20, ten generations, and fifteen parents selected per generation, while the green curve indicates experiments with a population size of 10, eight generations, and ten parents selected per generation.

Figure 9 depicts the $R^2$ scores obtained from spatial cross-validation, reflecting the performance of the neural network architecture. Each point in the plot corresponds to a specific combination of neurons and layers within the network. The plot captures the relationship between the number of neurons, the number of layers, and the resulting $R^2$ score, providing insights into the model's predictive capability. Each data point represents the $R^2$ score obtained from spatial cross-validation for the respective neural network architecture, highlighting the influence of architectural choices on model performance.

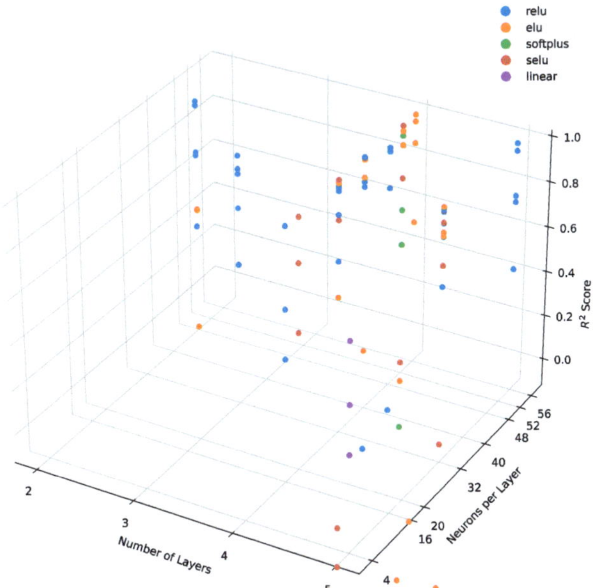


**Figure 9.** Cumulative $R^2$ scores obtained from spatial cross-validation for various neural network architectures, represented by different combinations of neurons and layers.

Table 12 summarizes the key statistics of the $R^2$ scores obtained from different configurations. These configurations varied in terms of the number of layers, neurons per layer, and activation functions used. The median $R^2$ score, along with the standard deviation, minimum, and maximum $R^2$ values, provides insights into the overall predictive performance and variability across different network architectures.

**Table 12.** Summary statistics of $R^2$ scores for different neural network configurations.

| Number of Layers | Neurons per Layer | Activation Function | Median $R^2$ | Standard Deviation of $R^2$ | Minimum $R^2$ | Maximum $R^2$ |
|---|---|---|---|---|---|---|
| 2 | 48 | relu | 0.699986 | 0.205056 | 0.360193 | 0.927981 |
| 5 | 32 | relu | 0.928194 | 0.120986 | 0.657188 | 0.986920 |
| 4 | 32 | relu | 0.949115 | 0.123375 | 0.643705 | 0.968744 |
| 3 | 32 | relu | 0.906979 | 0.172669 | 0.506309 | 0.985874 |
| 5 | 56 | relu | 0.757946 | 0.199017 | 0.430448 | 0.987120 |

## 4. Discussion

Traditional methods for evaluating the acoustic properties of materials often involve direct measurements of the velocity ($V$) and attenuation coefficient ($\alpha$) of ultrasound waves propagating through the material [87,88]. These measurements are fundamental in calculating acoustic parameters such as elastic moduli and anisotropy coefficients. The velocity $V$ and attenuation coefficient $\alpha$ are related to the material's properties through equations such as

$$V = \frac{\omega}{k} \quad \text{and} \quad \alpha = \frac{\omega}{2Q},$$

where $\omega$ is the angular frequency of the wave, $k$ is the wave number, and $Q$ is the quality factor.

Another common approach involves employing standard elasticity models tailored for anisotropic materials [89,90]. These models often utilize elasticity tensors or compliance matrices to describe the material's response to stress and strain. The elastic constants can be derived from these models using equations such as

$$\sigma_{ij} = C_{ijkl}\epsilon_{kl},$$

where $\sigma_{ij}$ and $\epsilon_{kl}$ represent the stress and strain components, respectively, and $C_{ijkl}$ denotes the elasticity tensor components.

Some methods rely on theoretical models such as elasticity theory or wave propagation models. These theoretical frameworks provide insights into how ultrasound signals interact with the material [91,92], allowing for prediction of wave behaviors based on mathematical and physical models.

In contrast to traditional methods, the approach proposed in this study introduces the $A_g$ coefficient, which is based on evaluating the ratio of the velocities of transverse ultrasound waves with perpendicular polarizations at specific angles relative to the material's growth direction [93]. The proposed coefficient $A_g$ demonstrates a strong linear correlation with changes in defect echo signal amplitudes. This method allows for more precise identification and classification of anisotropic properties in materials produced by additive manufacturing technologies.

To ensure the robustness and reliability of the proposed method using the $A_g$ coefficient for assessing acoustic anisotropy in additive manufacturing materials, it is crucial to thoroughly analyze potential error sources that could influence the experimental results. Minor deviations in measurement angles or velocity measurements can significantly impact the accuracy of ultrasound wave velocity and polarization angle measurements, thereby affecting the calculated $A_g$ coefficient. Techniques for improving measurement accuracy include precise angle alignment and calibration methods.

Variations in sample preparation [94], such as surface roughness [95], thickness uniformity [96], and material porosity [97], can introduce discrepancies in ultrasound wave propagation characteristics. These variations may lead to inconsistent $A_g$ values between different samples. Moreover, the assumption of material homogeneity [98] across the sample volume may not always hold true, especially in complex additive manufacturing structures. Variations in material composition and microstructure could influence ultrasound wave propagation and subsequently impact $A_g$ measurements. Advanced imaging techniques and microstructural analysis can aid in accurately assessing materials' homogeneity.

Additionally, external environmental conditions [99] such as temperature and humidity fluctuations during measurements can alter the material's properties and affect ultrasound wave behavior. These factors must be carefully controlled or accounted for in order to minimize their influence on $A_g$ calculations and ensure reliable results.

Understanding potential error sources is necessary in order to assess their influence on the accuracy and reliability of $A_g$ coefficients and subsequent analyses. These errors may introduce uncertainties into the correlation between $A_g$ and defect echo signal amplitudes, potentially affecting the predictive capability of the method.

For instance, inaccuracies in velocity measurements could lead to misinterpretation of anisotropy levels, while inconsistencies in sample preparation might obscure the true anisotropic effects [100]. By acknowledging and mitigating these potential error sources through rigorous experimental protocols and calibration procedures, the reliability of $A_g$ as an anisotropy indicator can be enhanced.

## 5. Conclusions

Our study investigated the acoustic properties of samples made from Inconel 718, EP648, AISI 321, and VG159 produced using additive technologies. Ultrasound wave velocities and attenuation coefficients were measured for six samples, and the elasticity coefficient matrices were obtained.

A review of existing methods for assessing acoustic anisotropy was carried out, on the basis of which we propose a new anisotropy indicator, the $A_g$ coefficient. The method for estimating anisotropy through the $A_g$ coefficient is based on determining the ratio between the velocities of transverse waves of mutually perpendicular polarization measured in two directions: at a 45-degree angle, and normal to the growth direction.

A comparison of methods for assessing acoustic anisotropy was carried out for the problem of establishing a relationship between anisotropy coefficients and the magnitude of the change in the echo signal amplitude from a defect when the direction of ultrasonic wave propagation changes relative to the axis of the properties' symmetry. The results showed that the proposed $A_g$ anisotropy index had the strongest linear relationship with the change in amplitude. The value of the correlation coefficient was equal to 0.97.

We propose that threshold values of the $A_g$ anisotropy index can make it possible to distinguish between anisotropic additive materials depending on the magnitude of the estimated change in the echo signal amplitude from the defect when the inspection angle changes relative to the axis of the properties' symmetry.

Additionally, we propose an approach for predicting $A_g$ using deep learning. Specifically, based on the optimization of neural network architecture hyperparameters using a genetic algorithm, it is possible to identify the architecture that can most accurately predict the anisotropy. This approach leverages the power of machine learning to efficiently analyze complex relationships between material properties and ultrasonic inspection outcomes, offering a promising avenue for enhanced defect detection in additive manufacturing processes.

**Author Contributions:** Data curation, D.M., A.B. and A.G. (Andrey Galinovsky); Formal analysis, I.M., V.T., D.M., N.K., M.D. and A.G. (Andrei Gantimurov); Funding acquisition, V.N., A.G. (Andrei Gantimurov) and A.B.; Investigation, V.T., N.S., N.K., M.D., V.N. and A.G. (Andrey Galinovsky); Methodology, D.M. and N.S.; Project administration, V.T., V.N., A.G. (Andrei Gantimurov) and A.B.; Resources, I.M., N.S., V.N., A.B. and A.G. (Andrey Galinovsky); Software, I.M., V.T., D.M., N.S., N.K.,

M.D. and A.G. (Andrey Galinovsky); Supervision, N.K., V.N., A.G. (Andrei Gantimurov) and A.B.; Validation, I.M., N.S., N.K., A.G. (Andrei Gantimurov) and A.G. (Andrey Galinovsky); Visualization, I.M., N.K. and M.D.; Writing—original draft, I.M., N.S. and N.K.; Writing—review and editing, V.T. and A.G. (Andrey Galinovsky). All authors have read and agreed with the published version of the manuscript.

**Funding:** This research received no external funding.

**Institutional Review Board Statement:** Not applicable.

**Informed Consent Statement:** Not applicable.

**Data Availability Statement:** The original contributions presented in the study are included in the article, further inquiries can be directed to the corresponding authors.

**Conflicts of Interest:** The authors declare no conflicts of interest.

## References


1. Uss, A.; Chernyshev, A.; Pugachuk, A. Application of additive technologies in the development of a vortex valve to control the thrust of a gas-jet drive. *AIP Conf. Proc.* **2021**, *2412*, 030041. [CrossRef]
2. Kransutskaya, A.; Funtikov, V.; Syritskii, A.; Komshin, A. Modern Trends in Measurement Assurance of Additive Manufacturing. *AIP Conf. Proc.* **2022**, *2503*, 070008. [CrossRef]
3. Barzov, A.; Galinovskiy, A.; Golubev, E.; Filimonov, A. Ultrajet Diagnostics of the Structural and Anisotropic Features of the Parts Fabricated by Selective Laser Melting. *Russ. Metall. (Met.)* **2021**, *2021*, 794–800. [CrossRef]
4. Khlybov, A.; Uglov, A.; Demchenko, A. On Spectral-Acoustic Method for Estimating Porosity of Metals Produced by Hot Isostatic Pressing. *Russ. J. Nondestruct. Test.* **2022**, *58*, 1051–1063. [CrossRef]
5. Aleshin, N.; Grigor'ev, M.; Shchipakov, N.; Prilutskii, M.A.; Murashov, V.V. Using nondestructive testing methods for in-production quality control of additive manufactured parts. *Russ. J. Nondestr. Test.* **2016**, *52*, 532–537. [CrossRef]
6. Aleshin, N.; Grigor'ev, M.; Shchipakov, N.; Prilutskii, M.A.; Murashov, V.V. Applying nondestructive testing to quality control of additive manufactured part. *Russ. J. Nondestr. Test.* **2016**, *52*, 600–609. [CrossRef]
7. Taheri, H.; Shoaib, M.M.; Koester, L.; Bigelow, T.; Collins, P.; Bond, L. Powder-based additive manufacturing–a review of types of defects, generation mechanisms, detection, property evaluation and metrology. *Int. J. Addit. Subtractive Mater. Manuf.* **2017**, *1*, 172–209. [CrossRef]
8. Honarvar, F.; Varvani-Farahani, A. A review of ultrasonic testing applications in additive manufacturing: Defect evaluation, material characterization, and process control. *Ultrasonics* **2020**, *108*, 106227. [CrossRef]
9. Kok, Y.; Tan, X.P.; Wang, P.; Nai, M.; Loh, N.H.; Liu, E.; Tor, S.B. Anisotropy and heterogeneity of microstructure and mechanical properties in metal additive manufacturing: A critical review. *Mater. Des.* **2018**, *139*, 565–586. [CrossRef]
10. Sridharan, N.; Gussev, M.; Seibert, R.; Parish, C.; Norfolk, M.; Terrani, K.; Babu, S.S. Rationalization of anisotropic mechanical properties of Al-6061 fabricated using ultrasonic additive manufacturing. *Acta Mater.* **2016**, *117*, 228–237. [CrossRef]
11. Du, H.; Yao, D. Ultrasonic Measurement of Microstructural Anisotropy in a Ni-Based Superalloy Fabricated by Direct Laser Deposition. *J. Nondestruct. Eval.* **2023**, *42*, 60. [CrossRef]
12. Sol, T.; Hayun, S.; Noiman, D.; Tiferet, E.; Yeheskel, O.; Tevet, O. Nondestructive ultrasonic evaluation of additively manufactured AlSi10Mg samples. *Addit. Manuf.* **2018**, *22*, 700–707. [CrossRef]
13. Tevet, O.; Svetlizky, D.; Harel, D.; Barkay, Z.; Geva, D.; Eliaz, N. Measurement of the Anisotropic Dynamic Elastic Constants of Additive Manufactured and Wrought Ti6Al4V Alloys. *Materials* **2022**, *15*, 638. [CrossRef] [PubMed]
14. Gou, J.; Wang, Z.; Hu, S.; Shen, J.; Tian, Y.; Zhao, G.; Chen, Y. Effects of ultrasonic peening treatment in three directions on grain refinement and anisotropy of cold metal transfer additive manufactured Ti-6Al-4V thin wall structure. *J. Manuf. Process.* **2020**, *54*, 148–157. [CrossRef]
15. Sun, L.; Guo, C.; Huang, L.; Jiang, F.; Xu, K.; Huang, R. Effect and mechanism of inter-layer ultrasonic impact strengthening on the anisotropy of low carbon steel components fabricated by wire and arc additive manufacturing. *Mater. Sci. Eng. A* **2022**, *848*, 143382. [CrossRef]
16. Markanday, J.; Carpenter, M.; Thompson, R.; Jones, N.; Christofidou, K.; Fairclough, S.; Heason, C.; Stone, H. Effect of NbC inoculants on the elastic properties and microstructure of additively manufactured IN718. *Materialia* **2023**, *27*, 101701. [CrossRef]
17. Aleshin, N.; Bely, V.; Vopilkin, A.; Voshchanov, A.; Ermolov, I.; Gurvich, A. *Methods of Acoustic Control of Metals*; Mechanical Engineering: Moscow, Russia, 1989; p. 456.
18. Li, W.; Zhou, Z.; Li, Y. Application of Ultrasonic Array Method for the Inspection of TC18 Addictive Manufacturing Titanium Alloy. *Sensors* **2019**, *19*, 4371. [CrossRef]
19. Aleshin, N.; Shchipakov, N.; Degtyarev, M. Modeling propagation of ultrasonic waves in anisotropic material of additive manufactured products. *Russ. J. Nondestr. Test.* **2022**, *58*, 425–432. [CrossRef]
20. Simonelli, M.; Tse, Y.; Tuck, C. Effect of the build orientation on the Mechanical Properties and Fracture Modes of SLM Ti–6Al–4V. *Mater. Sci. Eng. A* **2014**, *616*, 1–11. [CrossRef]

21. Sufiiarov, V.; Popovich, A.; Borisov, E.; Polozov, I. Evolution of structure and properties of heatresistant nickel alloy afterselective laser melting, hot isostatic pressing and heat treatment. *Tsvetnye Met.* **2017**, *1*, 77–82. [CrossRef]
22. Popovich, A.; Sufiiarov, V.; Borisov, E.; Polozov, I.; Masaylo, D.; Grigoriev, A. Anisotropy of mechanical properties of products manufactured using selective laser melting of powdered materials. *Russ. J. Non-Ferr. Met.* **2017**, *58*, 389–395. [CrossRef]
23. Im, Y.; Kim, K.; Jung, K.; Lee, Y.; Song, K. Anisotropic mechanical behavior of additive manufactured AISI 316L steel. *Metall. Mater. Trans. A Phys. Metall. Mater. Sci.* **2019**, *50*, 2014–2021. [CrossRef]
24. Pao, Y.-H. Elastic waves in solids. *ASME J. Appl. Mech.* **1983**, *50*, 1152–1164.
25. Vlasov, A.; Juravleva, L.; Ismagilov, K. The concept of using additive technologies for digital prototyping of assembly devices. *J. Phys. Conf. Ser.* **2022**, *2373*, 072035. [CrossRef]
26. Rinkevich, A.; Smorodinskii, Y.; Volkova, H.; Zagrebin, B. Group velocity of ultrasound in a transversally isotropic medium. *Defektoskopiya* **1994**, *1*, 58–63.
27. Smorodinskii, Y. Calculation of polarization of elastic waves in transversally isotropic austenite. *Defektoskopiya* **1999**, *1*, 9–12.
28. Ermolov, I.N. Progress in the theory of ultrasonic flaw detection. Problems and prospects. *Russ. J. Nondestruct. Test.* **2004**, *40*, 655–678. [CrossRef]
29. Rinkevich, A.; Stepanova, N.; Rodionov, D.; Perov, D.V. Ultrasonic testing of single- and polycrystal articles made of nickel-based heat-resistant alloys. *Russ. J. Nondestruct. Test.* **2009**, *45*, 745–759. [CrossRef]
30. Rinkevich, A.; Smorodinsky, Y. Elastic energy distribution in aluminum single crystal. *Nondestruct. Test. Eval.* **2002**, *18*, 159–170. [CrossRef]
31. Spies, M.; Müller, S.; Rieder, H.; Bamberg, J.; Henkel, B. On- und Offline Materialcharakterisierung bei der Additiven Fertigung. In Proceedings of the DGZfP-Symposium Zerstörungsfreie Materialcharakterisierung, Berlin, Germany, 28 November 2017. [CrossRef]
32. Mahaut, S.; Darmon, M.C.; Sylvain, C.; Jenson, F.J.; Calmon, P. Recent advances and current trends of ultrasonic modelling in CIVA. *Insight-Non-Destr. Test. Cond. Monit.* **2009**, *51*, 78–81. [CrossRef]
33. Gengembre, N. Pencil method for ultrasonic beam computation. In Proceedings of the 5th World Congress on Ultrasonics, Paris, France, 7–10 September 2003.
34. Darmon, M.; Dorval, V.; Chatillon, S.; Fradkin, L. Generic GTD-Kirchhoff scattering model for the ultrasonic response of planar defects. In Proceedings of the 10th International Conference on NDE in Relation to Structural Integrity for Nuclear and Pressurized Components, Cannes, France, 1–3 October 2013.
35. Gorbatsevich, F.F. On the Problem of Elastic Anisotropy Estimation in the Rocks with Quasi Orthotropic Symmetry. *Izv. Phys. Solid Earth* **2019**, *55*, 927–935. [CrossRef]
36. Aleksandrov, K.; Prodayvoda, G. *Anisotropy of Elastic Properties of Minerals and Rocks*; Publishing House of the SB RAS: Novosibirsk, Russia, 2000; p. 354.
37. Batugin, S.; Nirenburg, R. Approximate relation between the elastic constants of anisotropic rocks and the anisotropy parameters. *Sov. Min.* **1972**, *8*, 5–9. [CrossRef]
38. Thomsen, L. Week elastic anisotropy. *Geophysics* **1986**, *51*, 1–37. [CrossRef]
39. Dieulesaint, E.; Royer, D. *Elastic Waves in Solids: Applications to Signal Processing*; Bastin, A.; Motz, M., Translators; John Wiley and Sons: New York, NY, USA, 1980; p. 511.
40. Maconachie, T.; Leary, M.; Zhang, J.; Medvedev, A.; Sarker, A.; Ruan, D.; Lu, G.; Faruque, O.; Brandt, M. Effect of build orientation on the quasi-static and dynamic response of SLM AlSi10Mg. *Mater. Sci. Eng. A* **2020**, *788*, 139445. [CrossRef]
41. Awd, M.; Stern, F.; Kampmann, A.; Kotzem, D.; Tenkamp, J.; Walther, F. Microstructural characterization of the anisotropy and cyclic deformation behavior of selective laser melted AlSi10Mg structures. *Metals* **2018**, *8*, 825. [CrossRef]
42. Song, L.; Yuan, S.; Zhao, L.; Zhu, Y.; Liang, S.; Huang, M.; Simar, A.; Li, Z. Deciphering phase stress partition and its correlation to mechanical anisotropy of laser powder bed fusion AlSi10Mg. *Addit. Manuf.* **2023**, *71*, 103594. [CrossRef]
43. Deev, A.A.; Kuznetcov, P.A.; Petrov, S. Anisotropy of mechanical properties and its correlation with the structure of the stainless steel 316L produced by the SLM method. *Phys. Procedia* **2016**, *83*, 789–796. [CrossRef]
44. Zhou, B.; Xu, P.; Li, W.; Liang, Y.; Liang, Y. Microstructure and anisotropy of the mechanical properties of 316L stainless steel fabricated by selective laser melting. *Metals* **2021**, *11*, 775. [CrossRef]
45. Wang, H.; Jiang, P.; Yang, G.; Yan, Y. An Investigation of the Anisotropic Mechanical Properties of Additive-Manufactured 316L SS with SLM. *Materials* **2024**, *17*, 2017. [CrossRef] [PubMed]
46. Sarila, V.K.; Moinuddin, S.Q.; Cheepu, M.; Rajendran, H.; Kantumuchu, V.C. Characterization of microstructural anisotropy in 17–4 PH stainless steel fabricated by DMLS additive manufacturing and laser shot peening. *Trans. Indian Inst. Met.* **2023**, *76*, 403–410. [CrossRef]
47. Lebrun, T.; Tanigaki, K.; Horikawa, K.; Kobayashi, H. Strain rate sensitivity and mechanical anisotropy of selective laser melted 17-4 PH stainless steel. *Mech. Eng. J.* **2014**, *1*, SMM0049. [CrossRef]
48. Lee, H.W.; Jung, K.H.; Hwang, S.K.; Kang, S.H.; Kim, D.K. Microstructure and mechanical anisotropy of CoCrW alloy processed by selective laser melting. *Mater. Sci. Eng. A* **2019**, *749*, 65–73. [CrossRef]
49. Zou, S.; Zhao, Z.; Xu, W.; Ni, X.; Zhang, L.; Wu, W.; Kong, D.; He, X.; Wang, L.; Dong, C. Effects of scanning speeds on the wear behavior of CoCrW alloy fabricated by selective laser melting. *Opt. Laser Technol.* **2022**, *147*, 107652. [CrossRef]

50. Kong, D.; Ni, X.; Dong, C.; Zhang, L.; Yao, J.; Man, C.; Wang, L.; Xiao, K.; Li, X. Anisotropic response in mechanical and corrosion properties of hastelloy X fabricated by selective laser melting. *Constr. Build. Mater.* **2019**, *221*, 720–729. [CrossRef]
51. Tomus, D.; Tian, Y.; Rometsch, P.A.; Heilmaier, M.; Wu, X. Influence of post heat treatments on anisotropy of mechanical behaviour and microstructure of Hastelloy-X parts produced by selective laser melting. *Mater. Sci. Eng. A* **2016**, *667*, 42–53. [CrossRef]
52. Etter, T.; Kunze, K.; Geiger, F.; Meidani, H. Reduction in mechanical anisotropy through high temperature heat treatment of Hastelloy X processed by Selective Laser Melting (SLM). *IOP Conf. Ser. Mater. Sci. Eng.* **2015**, *82*, 012097. [CrossRef]
53. Kim, D.; Kim, T.; Ha, K.; Oak, J.J.; Jeon, J.B.; Park, Y.; Lee, W. Effect of heat treatment condition on microstructural and mechanical anisotropies of selective laser melted maraging 18Ni-300 steel. *Metals* **2020**, *10*, 410. [CrossRef]
54. Yao, Y.; Wang, K.; Wang, X.; Li, L.; Cai, W.; Kelly, S.; Esparragoza, N.; Rosser, M.; Yan, F. Microstructural heterogeneity and mechanical anisotropy of 18Ni-330 maraging steel fabricated by selective laser melting: The effect of build orientation and height. *J. Mater. Res.* **2020**, *35*, 2065–2076. [CrossRef]
55. Karlapudy, S.P.; Nancharaiah, T.; Rao, V.S. Influence of different build orientation and laser scan strategies on surface quality, Mechanical and Material Characteristics of 18 Ni-300 maraging steel processed through DMLS. *Aust. J. Mech. Eng.* **2023**, *21*, 1381–1395. [CrossRef]
56. Rashid, R.; Masood, S.; Ruan, D.; Palanisamy, S.; Rashid, R.R.; Elambasseril, J.; Brandt, M. Effect of energy per layer on the anisotropy of selective laser melted AlSi12 aluminium alloy. *Addit. Manuf.* **2018**, *22*, 426–439. [CrossRef]
57. Song, C.; Hu, Z.; Xiao, Y.; Li, Y.; Yang, Y. Study on interfacial bonding properties of NiTi/CuSn10 dissimilar materials by selective laser melting. *Micromachines* **2022**, *13*, 494. [CrossRef]
58. Kremer, R.; Khani, S.; Appel, T.; Palkowski, H.; Foadian, F. Selective laser melting of CuSn10: Simulation of mechanical properties, microstructure, and residual stresses. *Materials* **2022**, *15*, 3902. [CrossRef] [PubMed]
59. Chen, J.; Yang, Y.; Song, C.; Wang, D.; Wu, S.; Zhang, M. Influence mechanism of process parameters on the interfacial characterization of selective laser melting 316L/CuSn10. *Mater. Sci. Eng. A* **2020**, *792*, 139316. [CrossRef]
60. Yi, J.; Chang, C.; Yan, X.; Xie, Y.; Liu, Y.; Liu, M.; Zhou, K. A novel hierarchical manufacturing method of the selective laser melted Al 7075 alloy. *Mater. Charact.* **2022**, *191*, 112124. [CrossRef]
61. Sun, S.; Liu, P.; Hu, J.; Hong, C.; Qiao, X.; Liu, S.; Zhang, R.; Wu, C. Effect of solid solution plus double aging on microstructural characterization of 7075 Al alloys fabricated by selective laser melting (SLM). *Opt. Laser Technol.* **2019**, *114*, 158–163. [CrossRef]
62. Konecna, R.; Varmus, T.; Nicoletto, G.; Jambor, M. Influence of Build Orientation on Surface Roughness and Fatigue Life of the Al2024-RAM2 Alloy Produced by Laser Powder Bed Fusion (L-PBF). *Metals* **2023**, *13*, 1615. [CrossRef]
63. Varmus, T.; Konecna, R.; Nicoletto, G.; Uriati, F. Microstructure and Fatigue Properties of Al2024-RAM2 Aluminum Alloy Obtained by Laser Powder Bed Fusion. *Procedia Struct. Integr.* **2023**, *43*, 184–189. [CrossRef]
64. Dadbakhsh, S.; Vrancken, B.; Kruth, J.P.; Luyten, J.; Van Humbeeck, J. Texture and anisotropy in selective laser melting of NiTi alloy. *Mater. Sci. Eng. A* **2016**, *650*, 225–232. [CrossRef]
65. Moghaddam, N.S.; Saghaian, S.E.; Amerinatanzi, A.; Ibrahim, H.; Li, P.; Toker, G.P.; Karaca, H.E.; Elahinia, M. Anisotropic tensile and actuation properties of NiTi fabricated with selective laser melting. *Mater. Sci. Eng. A* **2018**, *724*, 220–230. [CrossRef]
66. Nikbakht, S.; Anitescu, C.; Rabczuk, T. Optimizing the neural network hyperparameters utilizing genetic algorithm. *J. Zhejiang Univ. Sci. A* **2021**, *22*, 407–426. [CrossRef]
67. Aszemi, N.M.; Dominic, P. Hyperparameter optimization in convolutional neural network using genetic algorithms. *Int. J. Adv. Comput. Sci. Appl.* **2019**, *10*, 269–278 [CrossRef]
68. Zhao, X.; Wang, L.; Zhang, Y.; Han, X.; Deveci, M.; Parmar, M. A review of convolutional neural networks in computer vision. *Artif. Intell. Rev.* **2024**, *57*, 99. [CrossRef]
69. Salehin, I.; Islam, M.S.; Saha, P.; Noman, S.; Tuni, A.; Hasan, M.M.; Baten, M.A. AutoML: A systematic review on automated machine learning with neural architecture search. *J. Inf. Intell.* **2024**, *2*, 52–81. [CrossRef]
70. Dunne, R.A.; Campbell, N.A. On the pairing of the softmax activation and cross-entropy penalty functions and the derivation of the softmax activation function. In Proceedings of the 8th Australian Conference on the Neural Networks, Melbourne, Australia, 1–13 June 1997; Citeseer: Camp Hill, PA, USA, 1997; Volume 181, p. 185.
71. Cococcioni, M.; Rossi, F.; Ruffaldi, E.; Saponara, S. A novel posit-based fast approximation of elu activation function for deep neural networks. In Proceedings of the 2020 IEEE International Conference on Smart Computing (SMARTCOMP), Bologna, Italy, 14–17 September 2020; IEEE: Piscataway, NJ, USA, 2020; pp. 244–246.
72. Sakketou, F.; Ampazis, N. On the invariance of the SELU activation function on algorithm and hyperparameter selection in neural network recommenders. In Proceedings of the Artificial Intelligence Applications and Innovations: 15th IFIP WG 12.5 International Conference, AIAI 2019, Hersonissos, Crete, Greece, 24–26 May 2019; Proceedings 15; Springer: Berlin/Heidelberg, Germany, 2019; pp. 673–685.
73. Sun, K.; Yu, J.; Zhang, L.; Dong, Z. A convolutional neural network model based on improved softplus activation function. In Proceedings of the International Conference on Applications and Techniques in Cyber Intelligence ATCI 2019: Applications and Techniques in Cyber Intelligence 7, Huainan, China, 22–24 June 2019; Springer: Berlin/Heidelberg, Germany, 2020; pp. 1326–1335.
74. Zhou, Y.; Li, D.; Huo, S.; Kung, S.Y. Soft-root-sign activation function. *arXiv* **2020**, arXiv:2003.00547.

75. Eckle, K.; Schmidt-Hieber, J. A comparison of deep networks with ReLU activation function and linear spline-type methods. *Neural Netw.* **2019**, *110*, 232–242. [CrossRef] [PubMed]
76. Abdelouahab, K.; Pelcat, M.; Berry, F. Why TanH is a hardware friendly activation function for CNNs. In Proceedings of the 11th International Conference on Distributed Smart Cameras, Stanford, CA, USA, 5–7 September 2017; pp. 199–201.
77. Pratiwi, H.; Windarto, A.P.; Susliansyah, S.; Aria, R.R.; Susilowati, S.; Rahayu, L.K.; Fitriani, Y.; Merdekawati, A.; Rahadjeng, I.R. Sigmoid activation function in selecting the best model of artificial neural networks. *J. Phys. Conf. Ser.* **2020**, *1471*, 012010. [CrossRef]
78. Gulcehre, C.; Moczulski, M.; Denil, M.; Bengio, Y. Noisy activation functions. In Proceedings of the International Conference on Machine Learning, New York City, NY, USA, 19–24 June 2016; PMLR: London, UK, 2016; pp. 3059–3068.
79. Savchenko, A.V. Probabilistic neural network with complex exponential activation functions in image recognition. *IEEE Trans. Neural Netw. Learn. Syst.* **2019**, *31*, 651–660. [CrossRef] [PubMed]
80. Sharma, S.; Sharma, S.; Athaiya, A. Activation functions in neural networks. *Towards Data Sci.* **2017**, *6*, 310–316. [CrossRef]
81. Bottaci, L. A genetic algorithm fitness function for mutation testing. In Proceedings of the SEMINALL-Workshop at the 23rd International Conference on Software Engineering, Toronto, ON, Canada, 14 May 2001; Citeseer: Camp Hill, PA, USA, 2001.
82. Jaderberg, M.; Dalibard, V.; Osindero, S.; Czarnecki, W.M.; Donahue, J.; Razavi, A.; Vinyals, O.; Green, T.; Dunning, I.; Simonyan, K.; et al. Population based training of neural networks. *arXiv* **2017**, arXiv:1711.09846
83. Culberson, J.C. Mutation-crossover isomorphisms and the construction of discriminating functions. *Evol. Comput.* **1994**, *2*, 279–311. [CrossRef]
84. Lin, W.Y.; Lee, W.Y.; Hong, T.P. Adapting crossover and mutation rates in genetic algorithms. *J. Inf. Sci. Eng.* **2003**, *19*, 889–903.
85. Islam, S.M.; Das, S.; Ghosh, S.; Roy, S.; Suganthan, P.N. An adaptive differential evolution algorithm with novel mutation and crossover strategies for global numerical optimization. *IEEE Trans. Syst. Man Cybern. B (Cybern.)* **2011**, *42*, 482–500. [CrossRef]
86. Pohjankukka, J.; Pahikkala, T.; Nevalainen, P.; Heikkonen, J. Estimating the prediction performance of spatial models via spatial k-fold cross validation. *Int. J. Geogr. Inf. Sci.* **2017**, *31*, 2001–2019. [CrossRef]
87. Bouda, A.B.; Lebaili, S.; Benchaala, A. Grain size influence on ultrasonic velocities and attenuation. *NDT E Int.* **2003**, *36*, 1–5. [CrossRef]
88. Raišutis, R.; Kažys, R.; Mažeika, L. Application of the ultrasonic characterization methods for highly attenuating plastic materials. *NDT E Int.* **2007**, *40*, 324–332. [CrossRef]
89. Zou, R.; Xia, Y.; Liu, S.; Hu, P.; Hou, W.; Hu, Q.; Shan, C. Isotropic and anisotropic elasticity and yielding of 3D printed material. *Compos. B Eng.* **2016**, *99*, 506–513. [CrossRef]
90. Fernández, M.; Fritzen, F.; Weeger, O. Material modeling for parametric, anisotropic finite strain hyperelasticity based on machine learning with application in optimization of metamaterials. *Int. J. Numer. Methods Eng.* **2022**, *123*, 577–609. [CrossRef]
91. Wu, Y.; Yang, L. Modeling and analysis of material anisotropy-topology effects of 3D cellular structures fabricated by powder bed fusion additive manufacturing. *Int. J. Mech. Sci.* **2021**, *197*, 106325. [CrossRef]
92. Zohdi, N.; Yang, R. Material anisotropy in additively manufactured polymers and polymer composites: A review. *Polymers* **2021**, *13*, 3368. [CrossRef] [PubMed]
93. Wang, Y.; Gao, T.; Liu, D.; Sun, H.; Miao, B.; Qing, X. Propagation characteristics of ultrasonic weld-guided waves in friction stir welding joint of same material. *Ultrasonics* **2020**, *102*, 106058. [CrossRef] [PubMed]
94. Paul, K.; Ladani, L. Relationship between peak density and acoustic scattering in high-frequency ultrasound wave propagation. *SN Appl. Sci.* **2020**, *2*, 1–12. [CrossRef]
95. Wang, Z.; Cui, X.; Ma, H.; Kang, Y.; Deng, Z. Effect of surface roughness on ultrasonic testing of back-surface micro-cracks. *Appl. Sci.* **2018**, *8*, 1233. [CrossRef]
96. Xu, C.; He, L.; Xiao, D.; Ma, P.; Wang, Q. A novel high-frequency ultrasonic approach for evaluation of homogeneity and measurement of sprayed coating thickness. *Coatings* **2020**, *10*, 676. [CrossRef]
97. Hosokawa, A. Effect of porosity distribution in the propagation direction on ultrasound waves through cancellous bone. *IEEE Trans. Ultrason. Ferroelectr. Freq. Control* **2010**, *57*, 1320–1328. [CrossRef] [PubMed]
98. Haïat, G.; Naili, S.; Grimal, Q.; Talmant, M.; Desceliers, C.; Soize, C. Influence of a gradient of material properties on ultrasonic wave propagation in cortical bone: Application to axial transmission. *J. Acoust. Soc. Am.* **2009**, *125*, 4043–4052. [CrossRef] [PubMed]
99. Moyano, D.B.; Paraiso, D.A.; González-Lezcano, R.A. Possible effects on health of ultrasound exposure, risk factors in the work environment and occupational safety review. *Healthcare* **2022**, *10*, 423. [CrossRef]
100. De Souza, A.E.C.M.; Sarout, J.; Luzin, V.; Sari, M.; Vialle, S. Laboratory measurements of ultrasonic wave velocities and anisotropy across a gold-hosting structure: A case study of the Thunderbox Gold Mine, Western Australia. *Ore Geol. Rev.* **2022**, *146*, 104928. [CrossRef]

sensors

MDPI

*Article*

# In Situ Non-Destructive Stiffness Assessment of Fiber Reinforced Composite Plates Using Ultrasonic Guided Waves

**Maarten Adams [1,*], Arnaud Huijer [1], Christos Kassapoglou [2], Johannes A. A. Vaders [3] and Lotfollah Pahlavan [1]**

[1] Department of Maritime and Transport Technology, Delft University of Technology, Mekelweg 2, 2628 CD Delft, The Netherlands; a.j.huijer@tudelft.nl (A.H.); l.pahlavan@tudelft.nl (L.P.)
[2] Department of Aerospace Structures and Materials, Delft University of Technology, Kluyverweg 1, 2629 HS Delft, The Netherlands; c.kassapoglou@tudelft.nl
[3] Materiel and IT Command, The Netherlands Ministry of Defence, Kromhout Kazerne, Herculeslaan 1, 3584 AB Utrecht, The Netherlands; jaa.vaders@mindef.nl
* Correspondence: m.adams@sintegsystems.com

**Abstract:** The multimodal and dispersive character of ultrasonic guided waves (UGW) offers the potential for non-destructive evaluation of fiber-reinforced composite (FRC) materials. In this study, a methodology for in situ stiffness assessment of FRCs using UGWs is introduced. The proposed methodology involves a comparison between measured wave speeds of the fundamental symmetric and antisymmetric guided wave modes with a pre-established dataset of UGW speeds and translation of them to corresponding stiffness properties, i.e., $ABD$-components, in an inverse manner. The dispersion relations of guided waves have been calculated using the semi-analytical finite element method. First, the performance of the proposed methodology has been assessed numerically. It has been demonstrated that each of the independent $ABD$-components of the considered laminate can be approximated with an error lower than 10.4% compared to its actual value. The extensional and bending stiffness properties can be approximated within an average error of 3.6% and 9.0%, respectively. Secondly, the performance of the proposed methodology has been assessed experimentally. This experimental assessment has been performed on a glass fiber-reinforced composite plate and the results were compared to mechanical tensile and four-point bending tests on coupons cut from the plate. Larger differences between the estimated $ABD$-components according to UGW and mechanical testing were observed. These differences were partly attributed to the variation in material properties across the test plate and the averaging of properties over the measurement area.

**Keywords:** ultrasonic guided waves; structural stiffness assessment; glass fiber-reinforced composites; semi-analytical finite element method

## 1. Introduction

Fiber-reinforced composite (FRC) materials have been gaining great popularity in marine structures over the past couple of decades because of their excellent strength-to-weight ratio [1,2], low density [3], corrosion resistance [1], and additional degrees of freedom in the design process [4]. However, the non-isotropic material properties in combination with rather complex manufacturing procedures provoke uncertainties in material properties and structural integrity of FRC materials after production and during their use [5,6]. Process-induced defects, such as voids, fiber misalignment, and delamination are common problems encountered during composite manufacturing [7–9]. The formation of these irregularities can significantly affect the mechanical performance of FRCs [9–12]. Moreover, structural degradation arises during service of the structure due to (cyclic) loading, the operating environment, and/or human errors.

To fully exploit the advantages of FRC materials, non-destructive evaluation (NDE) techniques have been proposed to analyze structural properties and identify damage [13]. These techniques utilize mechanical, chemical, or electromagnetic forces to disrupt the structure and measure the response. By anticipating that any internal irregularity will alter the returned signal, this signal offers insight into the material properties or structural damage [14]. Commonly used techniques for damage detection and structural integrity assessment are visual inspection and tap testing [15], radiographic testing [16,17], electromagnetic testing [18–20], shearography [21], vibration-based testing [22], acoustic emission testing [23,24], and ultrasonic testing [25–30]. However, these inspection methods are often limited by their insensitivity to small forms of damage and/or by their inability to detect damage that is not in close proximity to the inspection point. Additionally, these methods may not accurately assess the severity of the detected damage, making it difficult to determine the appropriate course of action for repairs or mitigation [31].

Over the past decade, ultrasonic guided wave (UGW) inspection methods have emerged as a promising technique for NDE due to their notable advantages. These methods offer a cost-effective, rapid, and repeatable means of inspecting large areas in a short amount of time without requiring the motion of transducers. UGWs are sensitive to small-size damage and can quantitatively evaluate both surface and internal damages that have a size greater than half its wavelength. Additionally, UGW devices can have a low power consumption, making them well-suited for use in remote or hard-to-reach locations [31–33].

The multimodal and dispersive character of UGW propagation is sensitive to the structural properties and has therefore been the basis of multiple studies on damage detection [29,32,34–37] and elastic properties characterization [38–40] of FRCs. Combining these features with their non-destructive nature shows the high potential of UGWs in the field of NDE [41,42]. Several studies have been conducted on the stiffness determination of FRC materials using UGWs [39,43]. Most of these methods involve a computationally intensive optimization process between the results obtained from experiments and the predictions generated by a forward numerical model [44].

This study introduces an enhanced non-destructive method for in situ stiffness assessment of FRCs using UGWs. The proposed methodology utilizes a computationally efficient inversion algorithm to evaluate the structural stiffness of FRCs by comparing experimentally measured UGW speeds with a pre-established dataset of UGW speeds and corresponding stiffness properties. The approach offers potential for future rapid in situ assessment of large-scale composite structures.

The performance of the proposed methodology was initially evaluated through a numerical evaluation. To demonstrate its practical feasibility for in situ applications, a glass fiber reinforced sample plate was fabricated and subjected to the assessment methodology. Subsequently, the plate was cut into coupons and mechanical tests were conducted to evaluate the stiffness properties obtained from this new assessment methodology.

The proposed methodology and evaluation procedure are described in Section 2. The experiments consisting of UGW testing and mechanical testing are discussed in Section 3. The results and discussion are presented in Section 4. Lastly, Section 5 presents the conclusions.

## 2. Methodology

The zero-order symmetric ($S_0$) and antisymmetric ($A_0$) wave modes are most often used in guided wave NDE techniques [45,46]. The main reason for that is their sensitivity to structural damage and strong correlation to mechanical stiffness [35,36,47]. Next to that, these wave modes are more straightforward to excite and measure than the higher-order guided wave modes.

The classical laminate theory (CLT) is commonly used to describe the behavior of composite materials under different types of loading conditions through use of the *ABD*-matrix, as described in Equation (1) [4].

$$\left\{ \begin{array}{c} \{N\} \\ \{M\} \end{array} \right\} = \left[ \begin{array}{cc} [A] & [B] \\ [B] & [D] \end{array} \right] \left\{ \begin{array}{c} \{\epsilon^0\} \\ \{\kappa^0\} \end{array} \right\} \quad (1)$$

Here, $\{N\}$ and $\{M\}$ are the external forces and moments applied on the structure, respectively, $\{\epsilon^0\}$ and $\{\kappa^0\}$ denote the internal strains and curvatures. $A_{ij}$ represents the in-plane stiffnesses, $B_{ij}$ captures the coupling between in-plane forces and out-of-plane deformations, and $D_{ij}$ signifies the out-of-plane bending stiffnesses. Using these $ABD$ stiffness components, a generic expression for group wave propagation in anisotropic media is formulated as:

$$c_g = G(m, \omega, A_{ij}, B_{ij}, D_{ij}, I_0, I_1, I_2) \tag{2}$$

Here, $m$ denotes the guided wave mode, $\omega$ denotes the wave frequency, and $I_0$, $I_1$, and $I_2$ denote the first, second, and third mass moments of inertia, describing the total mass, center of mass, and moment of inertia, respectively. Generally, it can be expected that symmetric wave modes are predominantly influenced by the extensional stiffness, while antisymmetric wave modes are dominated by bending stiffness.

Establishing an analytical solution for Equation (2) is not deemed feasible due to the complexity of the governing equations for guided waves in anisotropic media. This research investigates the possibility of utilizing an approximate description of $c_g$ in terms of the $ABD$-components using a set of coupling coefficients ($c_i$). For an arbitrary guided wave mode $m$ propagating at frequency $\omega$, this relationship is given as

$$c_{m,1}A_{11} + c_{m,2}A_{12} + c_{m,3}A_{16} + \cdots + c_{m,S}D_{66} = c_{g,m}^2 + e_m \tag{3}$$

Here, subscript $S$ denotes the total number of unknown $ABD$-components and $e_m$ denotes the approximation error. This error is dependent on the wave mode, material properties, and wave frequency, and may not be considered generally negligible. When dealing with symmetric wave modes, the error is expected to be fairly small for laminates with weak axial-bending coupling. However, for antisymmetric wave modes, a larger error may be expected, as the relationship between material stiffness and $c_g$ is generally more complex and involves higher-order terms [48,49]. Increased axial-bending coupling is expected to further increase the approximation error. Equation (3) can also be expressed in matrix-vector format:

$$[\boldsymbol{C}]\{\boldsymbol{\Psi}\}^{\mathrm{T}} = \{\boldsymbol{c_g^2}\} + \{\boldsymbol{e}\} \tag{4}$$

In this system of equations, $[\boldsymbol{C}]$ represents the matrix of coupling coefficients, while $\{\boldsymbol{c_g^2}\}$ and $\{\boldsymbol{e}\}$ denote the vectors containing the squared group speeds and approximation errors, respectively. At sufficiently low frequencies where only the $S_0$ and $A_0$ wave modes are involved, the vector $\{\boldsymbol{c_g^2}\}$ reduces to

$$\{\boldsymbol{c_g^2}\} = \{\ c^2_{g,S_0,1}\ \ c^2_{g,S_0,2}\ \ c^2_{g,S_0,j}\ \ \cdots\ \ c^2_{g,S_0,W}\ \mid\ c^2_{g,A_0,1}\ \ c^2_{g,A_0,2}\ \ c^2_{g,A_0,j}\ \ \cdots\ \ c^2_{g,A_0,W}\ \}^T \tag{5}$$

Here, subscript $W$ indicates the total number of $S_0$ and $A_0$ wave velocities included. Vector $\{\boldsymbol{\Psi}\}$ (size $[1 \times S]$) in Equation (4) represents the unknown stiffness properties of the FRC plate under analysis, structured as

$$\{\boldsymbol{\Psi}\} = \{\ A_{11}\ \ A_{12}\ \ \cdots\ \ A_{66}\ \mid\ B_{11}\ \ B_{12}\ \ \cdots\ \ B_{66}\ \mid\ D_{11}\ \ D_{12}\ \ \cdots\ \ D_{66}\ \} \tag{6}$$

Based on this system, it would be possible to estimate $\{\boldsymbol{\Psi}\}$ using an inverse procedure when matrix $[\boldsymbol{C}]$ (size $[2W \times S]$) is known and the squared group speed vector $\{\boldsymbol{c_g^2}\}$ (size $[2W \times 1]$) is obtained through measurements.

### 2.1. Calculation of the Coupling Coefficients

To calculate the coupling coefficients ($c_{m,i}$) in matrix $[\boldsymbol{C}]$, a specific composite plate of interest is considered. The design process for composite laminates allows for a wide range of possible stiffness properties resulting from design properties, such as material type, stacking sequence, and plate/ply thickness. By utilizing prior information (for example, a known stacking sequence and/or $E_1$ ply stiffness) of the plate of interest, this wide range of possible stiffness properties can be narrowed down to a reduced range of stiffness possibilities. The proposed method captures this range of stiffness possibilities in the coupling coefficients. To achieve this, the coefficients are numerically determined by analyzing a set of $R$ reference laminates $p_r$, where $1 \leq r \leq R$. These reference laminates are chosen so that their stiffness properties fall within the range of stiffness possibilities. By using a sufficient number of reference laminates to sufficiently cover the range of stiffness possibilities, it is expected that a converged stiffness approximation can be obtained. Determination of the set of coupling coefficients $\{c_m\}$ related to wave mode $m$ (Equation (3)) is described as follows:

$$\boldsymbol{\Psi}_{ref}\{c_m\} = \{c^2_{g,m,ref}\} \tag{7}$$

where

$$\boldsymbol{\Psi}_{ref} = \begin{bmatrix} A_{11,p_1} & A_{12,p_1} & A_{16,p_1} & \cdots & D_{66,p_1} \\ A_{11,p_2} & A_{12,p_2} & A_{16,p_2} & \cdots & D_{66,p_2} \\ A_{11,p_r} & A_{12,p_r} & A_{16,p_r} & \cdots & D_{66,p_r} \\ \vdots & \vdots & \vdots & & \vdots \\ A_{11,p_R} & A_{12,p_R} & A_{16,p_R} & \cdots & D_{66,p_R} \end{bmatrix} \tag{8}$$

$$\{c_m\} = \{\ c_{m,1} \quad c_{m,2} \quad \cdots c_{m,S}\ \}z^{\mathrm{T}} \tag{9}$$

$$\{c^2_{g,m,ref}\} = \{\ c^2_{g,m.p_1} \quad c^2_{g,m,p_2} \quad c^2_{g,m,p_r} \quad \cdots \quad c^2_{g,m,p_R}\ \}^{\mathrm{T}} \tag{10}$$

Here, each row of matrix $\boldsymbol{\Psi}_{ref}$ (size $[R \times S]$) consists of the $ABD$-components of a single reference laminate $p_r$, which is calculated using the CLT. Similarly, each element of vector $\{c^2_{g,m,ref}\}$ (size $[R \times 1]$) consists of the squared group speed of wave mode $m$ of reference laminate $p_r$. Equation (7) is solved in a least-squares sense.

There is generally a large difference in magnitude of the extensional stiffness components $A_{ij}$, coupling stiffness components $B_{ij}$, and bending stiffness components $D_{ij}$. To improve the condition of the numerical operations, matrix $\boldsymbol{\Psi}_{ref}$ is column-wise normalized by the absolute maximum component included in the column. This matrix scaling can be expressed as

$$\overline{\boldsymbol{\Psi}}_{ref} = \boldsymbol{\Psi}_{ref} \odot (1/\boldsymbol{\Psi}_{ref,max}) \tag{11}$$

Here, vector $\boldsymbol{\Psi}_{ref,max}$ contains the absolute maximum stiffness component of each column of matrix $\boldsymbol{\Psi}_{ref}$ and $\odot$ indicates the element-wise matrix multiplication. This results in the following modified version of Equation (7):

$$\overline{\boldsymbol{\Psi}}_{ref}\{c_m\} = \{c^2_{g,m,ref}\} \tag{12}$$

Consequently, vector $\{\boldsymbol{\Psi}\}$ in Equation (4) is column-wise normalized, resulting in the following modification of Equation (4):

$$[C]\{\overline{\boldsymbol{\Psi}}\}^{\mathrm{T}} = \{c^2_g\} \tag{13}$$

where

$$\{\overline{\boldsymbol{\Psi}}\} = \{\boldsymbol{\Psi}\} \odot 1/(\boldsymbol{\Psi}_{ref,max}) \tag{14}$$

Dispersion Analysis Using the Semi-Analytical Finite Element Method

The reference velocities in $\{c^2_{g,m,ref}\}$ (Equation (12)) are calculated from $\boldsymbol{\Psi}_{ref}$ by using the semi-analytical finite element method (SAFEM). SAFEM is a particularly efficient tool for calculating phase and group speed dispersion curves of guided waves in multilayered composite laminates and is commonly used as forward numerical model in NDE [50–55].

SAFEM operates under the assumption of plane strain behavior, employing finite element discretization along the thickness direction or cross section of the waveguide. The displacement in the direction of wave propagation is analytically described using harmonic exponential functions. This makes it more computationally efficient than conventional 3D FEM [56]. Figure 1 shows a discretization of wave propagation in the $x$-direction used in 1D SAFEM, assuming an infinitely wide plate and three-node elements. The equations of motion are expressed by Hamilton's equation [57] and the SAFEM solutions are obtained in a stable manner from an eigenvalue problem. A detailed description of SAFEM is provided by Barazanchy [50] and Bartoli [51].

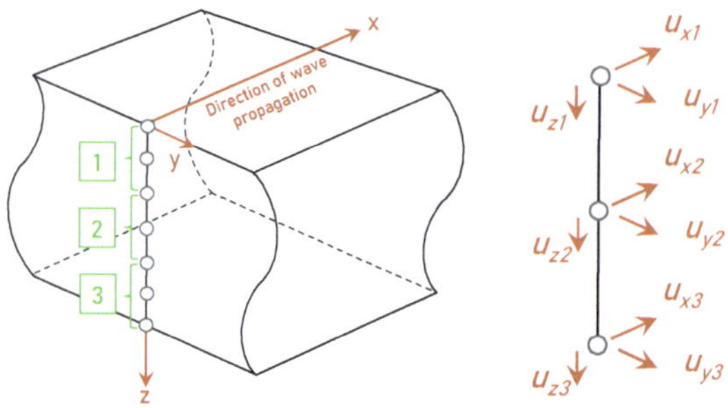


**Figure 1.** (**a**) Schematic representation of SAFEM for wave propagation in $x$-direction. (**b**) Degrees of freedom of the $i$th node element.

## 2.2. Robustness of the Algorithm

When applied in practice, the measured squared group speed vector $\{c_g^2\}$ (used in Equation (13)) may be affected by environmental conditions and/or measurement errors. To study the robustness of the algorithm with respect to imperfect input data, a numerical sensitivity study is performed. In this sensitivity study, different system configurations are considered. These configurations vary in the number of unknown $ABD$-components ($S$) included in Equation (13) and are discussed in Section 2.3.3. For each configuration the effect of the presence of measurement errors on the approximation of $\{\Psi\}$ is studied. The group speed vector, including measurement errors $\{c_{g,ME}\}$, is defined as follows:

$$\{c_{g,ME}\} = \{c_g\} + \{\Delta c_g\} \tag{15}$$

Here, vector $\{c_g\}$ is the original group speed vector as defined in Equation (13). Vector $\{\Delta c_g\}$ includes the measurement errors and is calculated as:

$$\{\Delta c_g\} = \{c_g\} \odot \{f_{ME}\} \tag{16}$$

Here, the original velocity vector is element-wise multiplied by error vector $\{f_{ME}\}$ defined as:

$$\{f_{ME}\} = \{\; e_1 \quad e_2 \quad e_i \quad \cdots \quad e_{2W} \;\}^T \qquad \text{where,} \quad -E_{max} \leq e_i \leq E_{max} \tag{17}$$

Here, $e_i$ denotes an arbitrary value between $-E_{max}$ and $+E_{max}$, which defines the maximum possible measurement error included in the error vector.

## 2.3. Evaluation Procedure

The potential of the proposed methodology is demonstrated in a numerical and experimental evaluation. For this evaluation procedure, a stiffness approximation is performed on a glass fiber-reinforced plate that is manufactured by vacuum infusion processing. First, a numerical evaluation is performed, in which conclusions are drawn on (i) the convergence of the $ABD$-approximation as function of the number of reference laminates ($R$) and (ii) the robustness of the algorithm as function of the number of unknown $ABD$-components ($S$). The findings of this numerical evaluation are used in the experimental evaluation in which the stiffness of the manufactured panel is assessed by measuring the UGW velocities. Afterwards, the test panel will be cut into test coupons and subjected to bending and tensile tests to obtain the stiffness properties according to mechanical testing.

### 2.3.1. Plate Specifications

The plate in this investigation is a cross-ply laminate consisting of transversely isotropic plies made of glass fibers and vinyl ester resin with the specifications given in Table 1. The general properties of the panel are given in Table 2. The expected ply properties are provided by manufacturing and are given in Table 3. Based on these expected plate properties, the dispersion curves are derived using SAFEM and presented in Figure 2.

**Table 1.** Material components used.

| Component | Name | $V_f/V_m$ |
|---|---|---|
| Fiber | Seartex U-E-640 g/m$^2$ | 48% |
| Resin | Atlac E-Nova MA 6215 | 52% |
| Hardener | Curox CM-75 | - |

**Table 2.** General plate properties.

| Width [mm] | Length [mm] | $\rho_{resin}$ [kg/m$^2$] | $\rho_{fiber}$ [kg/m$^2$] | $\rho_{overall}$ [kg/m$^2$] | Fiber Type [-] | $t_{total}$ [mm] |
|---|---|---|---|---|---|---|
| 600 | 600 | 1200 | 2600 | 1872 | UD 600 | 9.30 |

**Table 3.** The expected ply properties of the test panel according to manufacturing.

| $E_1$ [GPa] | $E_2, E_3$ [GPa] | $G_{12}, G_{13}$ [GPa] | $G_{23}$ [GPa] | $\nu_{12}, \nu_{13}$ [-] | $\nu_{23}$ [-] | Layup [-] | $t_{ply}$ [mm] | $\rho$ [kg/m$^3$] |
|---|---|---|---|---|---|---|---|---|
| 46.2 | 13.1 | 4.1 | 5.1 | 0.29 | 0.28 | $[0_5/90_5]_S$ | 0.465 | 1872 |

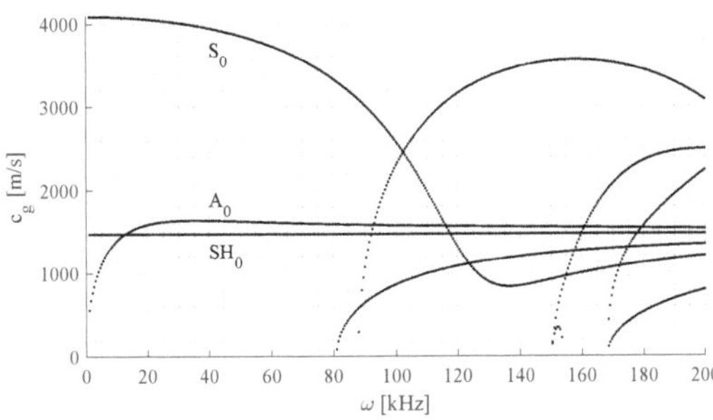


**Figure 2.** Group speed dispersion curves for the produced glass fiber-reinforced laminate with wave propagation in the 0°-direction. The $S_0$, $A_0$, and $SH_0$ wave modes are labeled, higher order waves modes emerge above 80 kHz.

### 2.3.2. System Configuration

Equation (3) is established for waves propagating at three different frequencies and along five different directions, resulting in a total of 30 equations included in the system of equations of Equation (13) ($2W = 30$). Given the symmetric and balanced cross-ply layup, it can be inferred that there is no coupling stiffness ($B_{ij} = 0$), no stretching–shearing coupling ($A_{16} = A_{26} = 0$), and no bending–twisting coupling ($D_{16} = D_{26} = 0$). As a result, coefficient matrix $[C]$ in Equation (13) reduces in dimensions to $[2W \times S] = [30 \times 8]$.

Reference Laminates

For this evaluation procedure, the stacking sequence, ply thickness ($t_{ply}$), and density ($\rho$) of the test panel are assumed to be known and the panel is defect-free. Furthermore, it is assumed that the actual properties of the plies ($E_1$, $E_2 = E_3$, $G_{12} = G_{13}$, $G_{23}$, $\nu_{12} = \nu_{13}$, and $\nu_{23}$) are unknown but located within a range of ±20% with respect to a set of expected ply properties (the so-called baseline laminate). The unknown ply stiffness properties are arbitrarily generated within the range of expected ply properties. This arbitrary process is for $E_1$ described as

$$E_{1,p_r} = E_{1,BL} + \Delta E_1 \qquad \text{where,} \quad \Delta E_1 = \alpha_{E_1} \cdot E_{1,BL} \tag{18}$$

Here, $E_{1,p_r}$ is the randomly generated value of $E_1$ for the reference laminate $p_r$, $E_{1,BL}$ is the value of $E_1$ of the baseline laminate, and $\alpha_{E_1}$ is a randomly generated value between −20% and +20%. The same approach is used for all unknown ply properties. The ply properties of each reference laminate $p_r$ are randomly generated, independently of the other reference laminates. In this manner a set of 3100 reference laminates is generated. It is expected that this amount sufficiently covers the range of stiffness possibilities.

For this evaluation procedure, two sets of reference laminates are generated, each using a different baseline laminate. The first set of reference laminates (the manufacturer's set) uses the expected ply properties provided by the manufacturer (Table 3). The second set (the mechanical testing set) uses the ply properties according to the mechanical tensile and four-point bending tests, which will be discussed in Section 4.2

2.3.3. Robustness of the Algorithm

The robustness of the algorithm with respect to imperfect input data is investigated in the numerical evaluation. Three system configurations are considered, as defined in Table 4. Configuration 1 includes all the unknown stiffness components of the panel under investigation. In configurations 2 and 3, a subset of these components has been selected to shed light on the possibilities of reducing the system size based on the expected relationship between stiffness components and wave modes.

**Table 4.** The system configurations considered for the numerical sensitivity study.

| Configuration | $\{\Upsilon\}$ |
|---|---|
| 1 | $\{ A_{11} \ A_{12} \ A_{22} \ A_{66} \mid D_{11} \ D_{12} \ D_{22} \ D_{66} \}$ |
| 2 | $\{ A_{11} \ A_{22} \ A_{66} \mid D_{11} \ D_{22} \ D_{66} \}$ |
| 3 | $\{ A_{11} \ A_{12} \ A_{22} \mid D_{11} \ D_{12} \ D_{22} \}$ |

## 3. Experiments

The procedure for measuring the ultrasonic guided waves on the test panel and the performance of mechanical tests afterwards are described in Sections 3.1 and 3.2, respectively.

### 3.1. Ultrasonic Guided Wave Testing

An overview of the experimental setup for measurement of the ultrasonic guided waves is provided in Figure 3a. A close-up of the measurement device is given in Figure 3b. The waveform generator creates an input wave signal that is amplified and emitted through a piezoelectric transducer, indicated as the actuator in Figure 3a. In total, three different wave signals are used. These signals are narrow-banded Hann-windowed sinusoidal pulses with a center frequency of 70, 80, and 90 kHz. The signals are recorded in five directions (0°, 30°, 45°, 60°, and 90° with respect to the reference axis of the laminate) by two dry point contact transducers. Each measurement set consists of a total of 30 input signals that are emitted and recorded one after another and then averaged. This procedure helps to improve the signal-to-noise ratio and mitigates the presence of background noise components in the signal. Moreover, all experiments are conducted at room temperature (20 °C), aligning with the conditions under which the reference velocities are calculated. Temperature variation was below 1 °C during the measurement period, making the effect on wave speed insignificant [58].

(a) (b)

**Figure 3.** (**a**) Overview of the experimental setup including the (1) RS PRO RSDG1032X wave generator, (2) Falco Systems WMA-300 high-voltage amplifier, (3) Vallen Systeme VS600-Z1 actuator, (4) ACS Group S2803 dry-point contact transducers, (5) Vallen Systeme AEPH5 pre-amplifiers, (6) Vallen Systeme AMSY-6 data acquisition system chassis type MB6, and (7) Vallen Systeme AE-Suite software version R2023.1218.2. (**b**) Topview of the measurement device.

An example of an averaged wave signal recorded by the two transducers is given in Figure 4. This signal corresponds to an 80 kHz wave propagating in the 0°-direction. In this figure, the dispersion effect of the $S_0$ wave, propagating faster than the $A_0$ wave, is clearly visible. Based on the arrival time of the wave at both transducers, the group wave speed of the $A_0$ and $S_0$ wave modes can be calculated using Equations (19) and (20), respectively.

$$c_{g,A_0} = \frac{d_{12}}{t_{A_0 2} - t_{A_0 1}} \tag{19}$$

$$c_{g,S_0} = \frac{1}{\frac{1}{c_{g,A_0}} - \frac{t_{S_0 2} - t_{A_0 2}}{d_{02}}} \tag{20}$$

Here, $t_{A_0 1}$ and $t_{A_0 2}$ represent the arrival time of the $A_0$ wave mode at the first and second transducer, respectively; $t_{S_0 2}$ represent the arrival time of the $S_0$ wave mode at the second transducer. Lastly, $d_{12}$ and $d_{02}$ denote the distance between the first and second transducer and between the actuator and the second transducer; respectively, see Figure 3b.

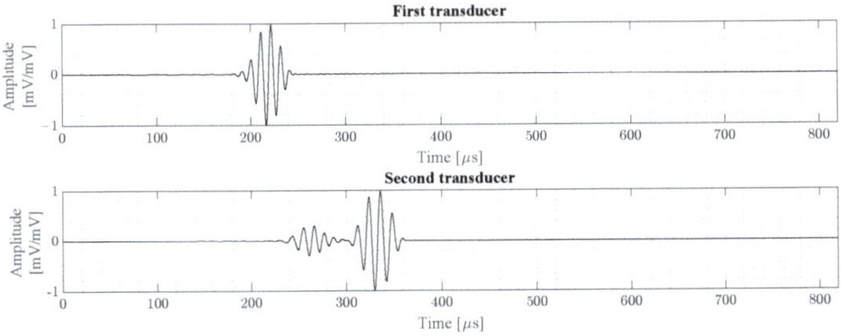


**Figure 4.** Example of an averaged wave signal recorded by the two transducers. The signal corresponds to a 80 kHz wave propagating in the 0°-direction.

### 3.2. Mechanical Testing

To assess the obtained stiffness properties using UGW testing, the test panel is subjected to mechanical testing. Test coupons are cut from the panel and subjected to both tensile and four-point bending tests to determine the stiffness components $A_{11}$, $A_{22}$, $D_{11}$, and $D_{22}$. The tests were performed in accordance with ASTM D3039/D3039M [59] and ASTM D6272 [60] with a minor deviation in the coupon dimensions driven by the limitations of the available test facilities. The coupon dimensions for the tensile and four-point bending tests were 150 × 50 mm and 130 × 25 mm, respectively. Nevertheless, these dimensions are considered reasonable and not expected to have influenced the outcomes.

The cutting plan of the panel is depicted in Figure 5. Here, orange coupons represent those used for tensile testing and green coupons are utilized for four-point bending testing. The mechanical tests are carried out on a test bench (Figure 6) that has a maximum tensile capacity of 250 kN. In Figures 7 and 8, pictures of a tensile and bending test coupon are provided, respectively. During the tensile tests, axial strain is measured using an extensometer, while transverse strain is measured using strain gauges. During the four-point bending tests, only strain gauges are used to measure longitudinal and transverse strain. Unidirectional strain gauges are employed, requiring the transverse strain gauges to be placed slightly off-center, as depicted in Figure 8. Nevertheless, it is considered that the measured strain at these positions is indicative of the strain at the center of the coupon. Figures 9 and 10 illustrate a tensile and bending coupon, respectively, during the measurement.

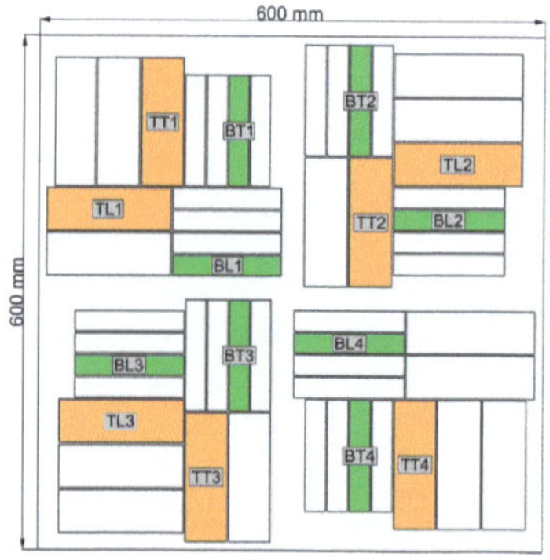


**Figure 5.** Cutting plan mechanical test coupons.

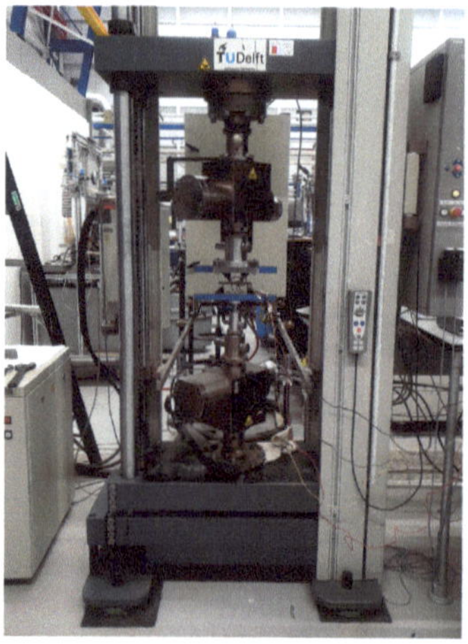


**Figure 6.** The Zwick 250 kN test bench.

**Figure 7.** Tensile test coupon.

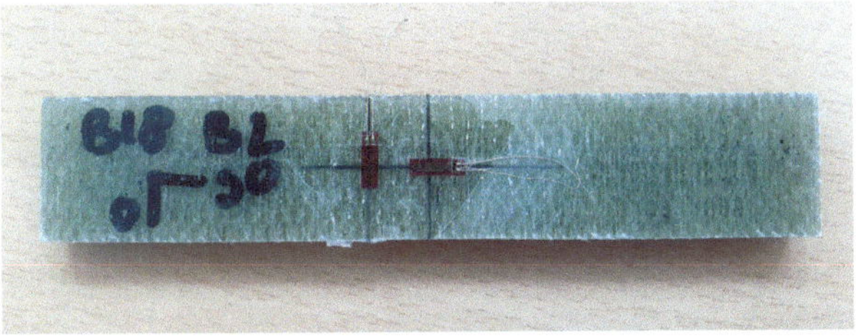


**Figure 8.** Four-point bending test coupon.

**Figure 9.** Tensile test coupon clamped in the test bench.

**Figure 10.** Four-point bending test coupon positioned in the test setup.

## 4. Results and Discussion

### 4.1. Numerical Evaluation

In the numerical evaluation, the convergence of the stiffness approximation as function of the number of reference laminates ($R$) and the robustness of the algorithm as function of the number of unknown $ABD$-components ($S$) are investigated. The manufacturer's set of reference laminates is used for the numerical evaluation.

#### 4.1.1. Convergence Study

For the convergence study, system configuration 1 of the robustness study (Table 4) is used. Each reference laminate $p_r$ included in the manufacturer's set is used as test case for the convergence study. Velocity-squared vector $\{c_g{}^2\}$, which is calculated using SAFEM, is input for the methodology; see Equation (4). For each test case, the size of $R$ is increased from 2 up to 3100 reference laminates to conclude what set size is sufficient to obtain a converged $ABD$ approximation. For each value of $R$, the reference laminates $p_r$ are arbitrarily selected from the manufacturer's set. The approximated $ABD$-components ($\boldsymbol{\Psi}_r$) are compared to the results according to the CLT ($\boldsymbol{\Psi}_{ref_n}$). Eventually, the mean absolute percentage error (MAPE) of the total of test cases ($TC$) is calculated as follows:

$$\text{MAPE} = \frac{\sum_{r=2}^{TC} \left| \frac{\boldsymbol{\Psi}_r}{\boldsymbol{\Psi}_{ref_r}} \right| \cdot 100\%}{TC} \qquad \text{where,} \quad r \in [2, 3100] \tag{21}$$

The results of the convergence study are shown in Figure 11. It can be observed that the approximation of all $ABD$-components is converged around 2000 reference laminates. The approximation of components $A_{11}$ and $A_{22}$ shows the fastest convergence around 1500 reference laminates. In the figure, significant peaks are observed in the range of $R$ lower than 1000 reference laminates. These peaks are mainly the result of the arbitrary selection procedures of the reference laminates $p_r$ included in $R$ and indicate insufficient coverage of the range of stiffness possibilities. Repeating this convergence study with again an arbitrary $p_r$ selection will lead to a shift in the peak locations with respect to $R$. The generated set of 3100 reference laminates is considered sufficient to obtain converged results. In Figure 12, the error distributions of the test cases are shown as well as the MAPE value for each $ABD$-component. These results are obtained using the complete set of 3100 reference laminates. It is shown that each $ABD$-component can be approximated within a MAPE of 10.4%. The pure extensional and bending stiffness properties can be approximated with an average MAPE of 3.6% and 9.1%, respectively. This difference in MAPE between the extensional and bending stiffness properties may be related to the quality of the approximation in Equation (3), leading to a larger error for the antisymmetric wave modes than for the symmetric wave modes.

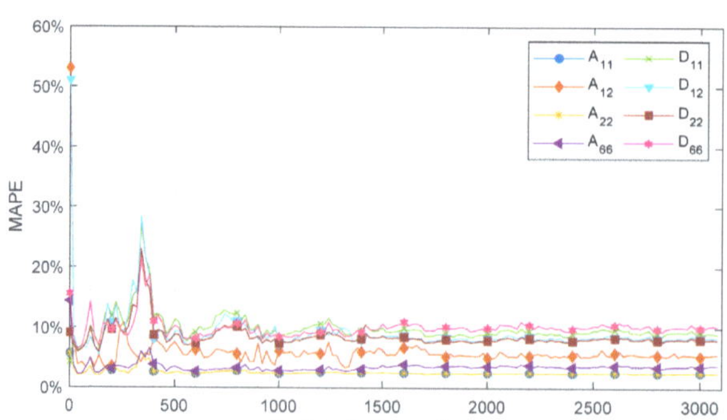


**Figure 11.** Convergence study with respect to the size ($R$) of the manufacturer's set of reference laminates. The MAPE of 3100 test cases compared to CLT is calculated using Equation (21).

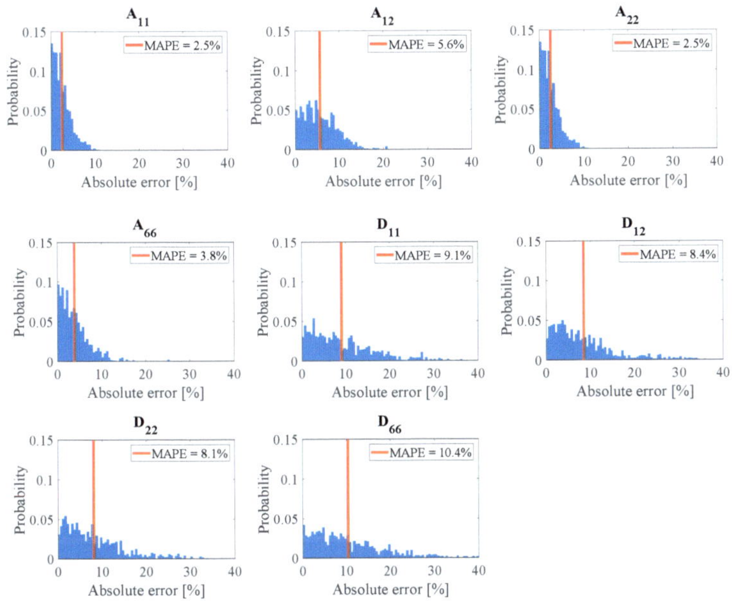


**Figure 12.** The distribution of the absolute error for each ABD-components for the 3100 test cases as well as the MAPE value.

#### 4.1.2. Robustness of the Algorithm

The sensitivity of the algorithm on measurement errors in the UGW input data is evaluated by analyzing the MAPE (Equation (21)) of the approximated $A_{11}$, $A_{22}$, $D_{11}$, and $D_{22}$ stiffness components for increasing $E_{max}$. For this study the complete set of 3100 reference laminates is used and the range of $E_{max}$ is set from 0% to 10%.

The results are presented in Figure 13. The results indicate that system configuration 3 (Table 4) is least sensitive to measurement errors and can offer the best robustness in practical environments. Configuration 3, like configuration 1, provides a converged stiffness approximation for a set size of 3100 reference laminates, as illustrated in Figure 14. Therefore, this configuration is employed in the experiments, implying that $[C]$ and $\overline{\boldsymbol{\Psi}}$ of Equation (13) are of dimensions $[2W \times S] = [30 \times 6]$ and $[1 \times S] = [1 \times 6]$, respectively.

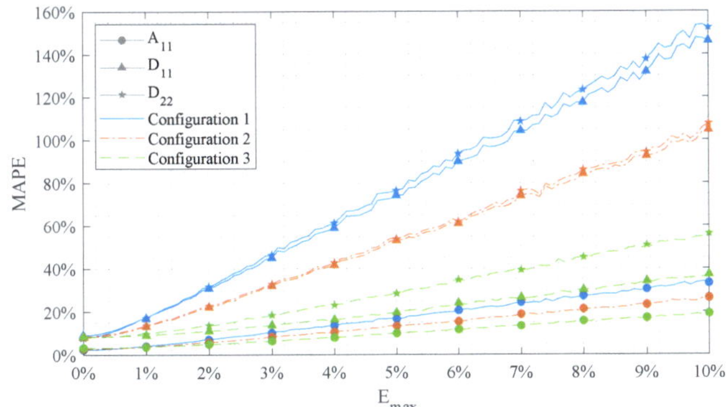


**Figure 13.** Sensitivity study on the robustness of the algorithm on measurement errors in the input data. The MAPE of 3100 test cases compared to CLT is calculated using Equation (21).

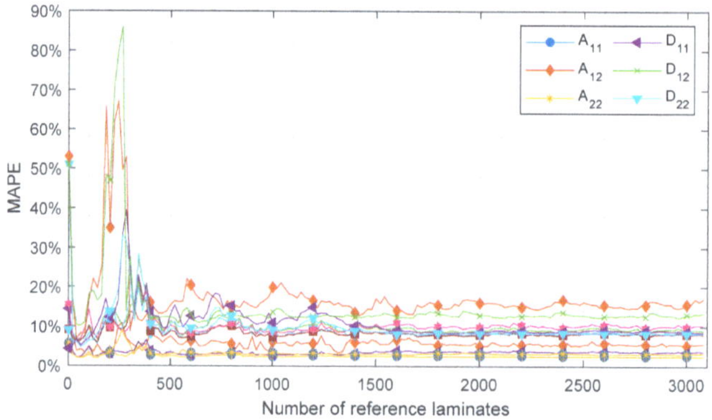


**Figure 14.** Convergence study with respect to the size of the manufacturer's set of reference laminates of system configuration 3. The MAPE of 3100 test cases compared to CLT is calculated using Equation (21).

## 4.2. Experimental Evaluation

### 4.2.1. Stiffness Assessment by Mechanical Testing

The results of the tensile tests and four-point bending tests are presented in Tables 5 and 6, respectively.

**Table 5.** Results of tensile tests.

| | | Coupons | Mean | Std |
|---|---|---|---|---|
| $E_{1m}$ | [GPa] | 3 | 25.81 | 0.69 |
| $E_{2m}$ | [GPa] | 4 | 29.78 | 6.47 |
| $\nu_{12m}$ | [-] | 3 | 0.14 | 0.01 |
| $\nu_{21m}$ | [-] | 4 | 0.16 | 0.03 |

**Table 6.** Results of four-point bending tests.

| | | Coupons | Mean | Std |
|---|---|---|---|---|
| $E_{1b}$ | [GPa] | 3 | 34.21 | 0.58 |
| $E_{2b}$ | [GPa] | 4 | 15.68 | 1.31 |
| $\nu_{12b}$ | [-] | 3 | 0.18 | 0.01 |
| $\nu_{21b}$ | [-] | 4 | 0.10 | 0.01 |

Figure 15 displays the results of the mechanical tests, along with the expected laminate stiffness properties of the plate of interest according to manufacturing (Table 3) represented as red lines. These expected properties according to manufacturing are calculated using the following equations [61]:

$$E_{1m} = \frac{1}{ha_{11}} \qquad E_{2m} = \frac{1}{ha_{22}} \qquad \nu_{12m} = -\frac{a_{12}}{a_{22}} \qquad \nu_{21m} = -\frac{a_{12}}{a_{11}} \tag{22}$$

$$E_{1b} = \frac{12}{h^3 d_{11}} \qquad E_{2b} = \frac{12}{h^3 d_{22}} \qquad \nu_{12b} = -\frac{d_{12}}{d_{22}} \qquad \nu_{21b} = -\frac{d_{12}}{d_{11}} \tag{23}$$

Here, $h$ denotes the thickness of the plate, and $a_{ij}$ and $d_{ij}$ denote the elements of the inverse $ABD$-matrix which is calculated using CLT and the properties in Table 3.

The tensile properties in the 0°-direction ($E_{1m}$) and the bending properties in both the 0°- and 90°-directions ($E_{1b}$ and $E_{2b}$, respectively) show a reasonable variation. However, there is a larger variation in the tensile properties of the 90°-coupons ($E_{2m}$). Furthermore, the extensional stiffness $E_{1m}$ is 13% lower than $E_{2m}$, indicating that the actual laminate does not behave as a balanced laminate as initially assumed. Lastly, the higher stiffness properties of the laminate of interest (red lines) suggest that the overall stiffness of the sample plate is lower than expected according to the manufacturer's data.

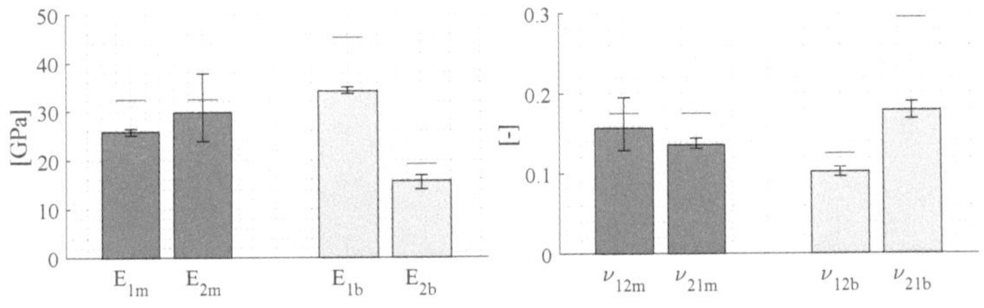


(a) Young's Modulus (b) Poisson Ratio

**Figure 15.** Results of mechanical testing. The expected laminate stiffness properties of the plate of interest according to manufacturing are indicated by the red lines.

Mechanical Testing Set of Reference Laminates

Using the findings from the mechanical tests, the mechanical testing set of reference laminates is constructed. The assumption for this set is that the lower overall stiffness properties are caused by lower $E_1$ and $E_2$ values for all plies. Additionally, it is assumed that the stiffness imbalance is caused by a difference in $E_1$ between the 0°- and 90°-plies, while all other known and unknown properties remain the same as those in the manufacturer's set (Table 3). The resulting ply stiffness properties for the laminate of interest, used as baseline laminate, in the mechanical testing set of reference laminates are provided in Table 7.

**Table 7.** Ply stiffness properties for the baseline laminate in the mechanical testing set of reference laminates.

| $E_1$ [GPa] | $E_2$,$E_3$ [GPa] | $G_{12}$,$G_{13}$ [GPa] | $G_{23}$ [GPa] | $\nu_{12}$, $\nu_{13}$ [-] | $\nu_{23}$ [-] | Layup [-] | $t_{ply}$ [mm] | $\rho$ [kg/m$^3$] |
|---|---|---|---|---|---|---|---|---|
| 29.6 (0°)<br>39.6 (90°) | 10.5 | 4.1 | 5.1 | 0.29 | 0.28 | $[0_5/90_5]_S$ | 0.465 | 1872 |

4.2.2. Stiffness Assessment by Ultrasonic Guided Waves Testing

The stiffness assessment using UGW is performed at four distinct locations on the panel, as depicted in Figure 16. These locations are selected with the consideration to minimize interference of the emitted wave signals with reflections from the boundaries of the structure. At each location, the average group wave speed is determined of 30 emitted wave signals.

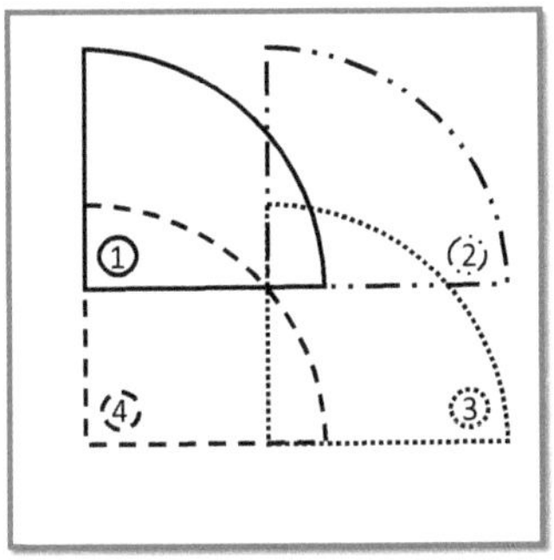


**Figure 16.** The four assessed locations on the plate labeled as 1-4 and distinguished by different line patterns.

The resulting mean group wave speeds and the range over the four locations are presented in Figure 17. The figure indicates that there are minimal variations in wave speed of the $A_0$ mode across the panel. The wave speed of $S_0$ exhibits greater variations across the panel.

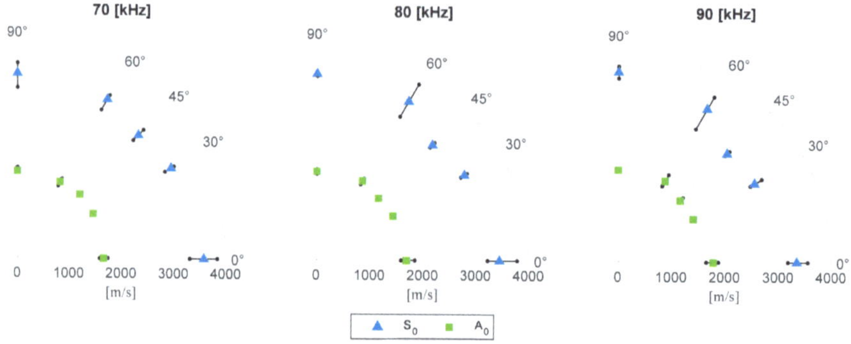


**Figure 17.** The mean $A_0$ and $S_0$ group wave speeds along with their range, measured at the four locations on the panel.

Results Using the Manufacturer's Set of Reference Laminates

Figure 18 displays the stiffness properties approximated by implementing the algorithm using the manufacturer's set of reference laminates (labeled as UGW). The figure also shows the results of mechanical testing (labeled as Tensile and Bending) and the range of stiffness properties included in the set of reference laminates (labeled as Set). The extensional stiffness $A_{22}$ as well as the bending stiffness $D_{11}$ show reasonable agreement with the mechanical tests, with an average deviation of +2% and +7%, respectively. However, extensional stiffness $A_{11}$ and bending stiffness $D_{22}$ have a larger deviation from mechanical testing, being +17% and +52%, respectively. Furthermore, the range of $D_{11}$ is relatively large, indicating the algorithm approximates the stiffness properties with considerable variations across the panel. Lastly, it can be observed that the laminate stiffness properties according to mechanical testing only marginally fall within the range of stiffness properties included in the set of reference laminates.

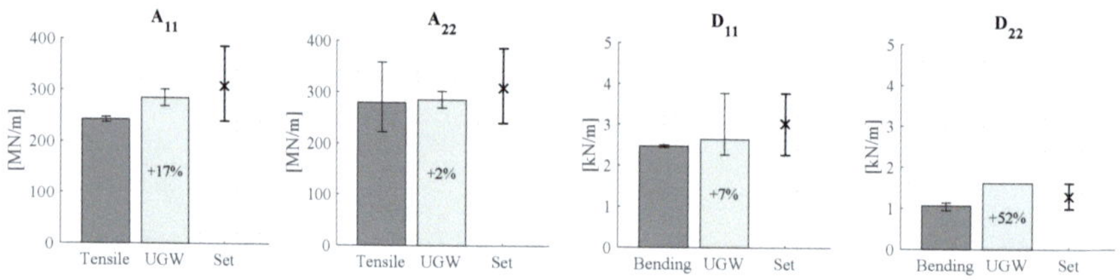


**Figure 18.** The approximated stiffness properties using the set of reference laminates based on the manufacturer's data (UGW) as well as the stiffness properties according to mechanical testing (Tensile & Bending). Also, the range of stiffness properties included in the set of reference laminates (Set) is shown.

Results Using the Mechanical Testing set of Reference Laminates

Figure 19 presents the stiffness approximation using the mechanical testing set of reference laminates. The figure demonstrates that this set of reference laminates more accurately covers the plate stiffness properties according to mechanical testing. Moreover, the approximation of $D_{22}$ shows a significant improvement. Nonetheless, notable differences between the approximated stiffness properties using UGW testing and those obtained from mechanical testing still exist. Furthermore, the results obtained using the algorithm show large variations in the stiffness approximation for $D_{11}$ and $D_{22}$.

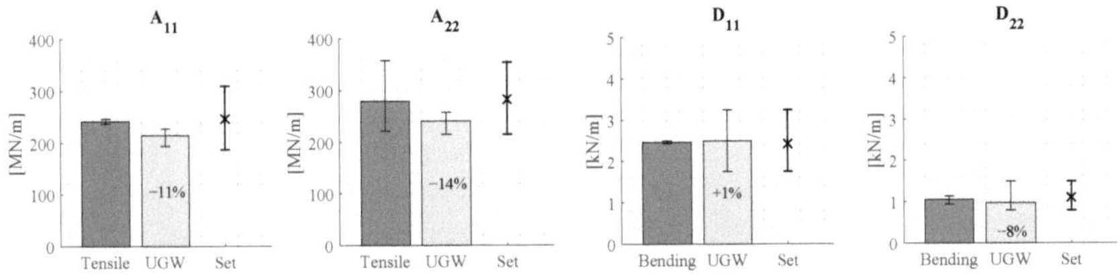


**Figure 19.** The approximated stiffness using the set of reference laminates based on the mechanical testing data (UGW) as well as the stiffness properties according to mechanical testing (Tensile and Bending). Also, the range of stiffness properties included in the set of reference laminates (Set) is shown.

These differences in stiffness assessment between the algorithm and mechanical testing, as well as the wide range of approximated stiffness properties by the algorithm, can be partly attributed to the moderate quality of the manufactured test panel. Orientation of plies may be subject to inaccuracy and additionally the plies may not be perfectly transversely isotropic. Mechanical testing revealed that stiffness variations were mainly observed in $E_{2m}$ across the panel. Furthermore, the area over which a single measurement is performed (Figure 16) is relatively large compared to the size of the panel and the coupons used for mechanical testing (Figure 5). As a result, the stiffness variations across the measured area are averaged by the measurements, and the algorithm provides an average stiffness approximation of this measured area. Therefore, an ideal comparison of the stiffness properties of a single mechanical testing coupon to those of the algorithm is not possible. To achieve this, the minimal required area for UGW testing should be reduced; this might pose, however, additional challenges in separating the two fundamental wave modes, as well as performing measurements along multiple directions.

### 4.3. Limitations and Practical Remarks

The methodology presented in this study estimates the average stiffness properties over the measurement area, which is determined by the position of the source and receivers. Consequently, if any damaged area of interest is not covered by the wave propagation path, no change in the stiffness properties will be detected either. In addition, defects that are present in the measurement area will be captured by their effect on the average stiffness properties. It is believed that this will not hamper the intended applications of the methodology for rapid inspection or scanning of designated areas of large-scale composite structures.

When applying the methodology in practice, the influence of environmental conditions should be taken into account. At first, the effect of temperature on wave propagation velocities should be considered. In the experimental evaluation and as mentioned earlier, all experiments were conducted at room temperature (20 °C). In case measurements are performed at notably different temperatures, application of temperature compensation algorithms for the extracted wave velocities may be necessary. Examples of such algorithms can be found in literature [62–66]. Secondly, the presence of background noise interfering with the input signal may influence the measurement accuracy. To mitigate the presence of background noise components, e.g., due to environmental and electromagnetic interference, a total of 30 signals were averaged in the experimental evaluation. When implementing the technology in circumstances with significant background noise, such as machinery, electrical systems, waves and wind, the number of averaged wave signals may need to be increased.

## 5. Conclusions

In this research, a new methodology for assessing the structural stiffness of FRC materials is proposed. The methodology uses an inversion algorithm that couples UGW speed to structural stiffness. The performance of the methodology is demonstrated in a numerical and experimental evaluation. For this evaluation, a glass fiber plate consisting of a symmetric and balanced cross-ply layup is manufactured. The following conclusions are drawn from the numerical evaluation:

- The stiffness approximation provides converged results when the size of the set of reference laminates is sufficiently large. A set of 2000 reference laminates is concluded sufficient for the stiffness assessment of a balanced and symmetric laminate for which the actual properties of the plies are known within a range of $\pm 20\%$. The numerical evaluation showed that each $ABD$-component can potentially be approximated within a MAPE of 10.4% compared to its actual value for 3100 test cases. The extensional stiffness can be approximated with an average MAPE of 3.6%. The bending stiffness properties have an average MAPE of 9.1%.
- To apply the technology in typical in situ environments, it is found that a system configuration which includes all ten $ABD$-components is sensitive to measurement errors in the input data. To deal with this, it is concluded that a system configuration excluding the shear stiffness components $A_{66}$ and $D_{66}$ provides better system robustness against measurement errors.

The findings of the numerical evaluation are implemented in the experimental evaluation. Measurements are performed on the test panel and compared to the stiffness approximation according to mechanical testing. The results of mechanical testing revealed that the laminate was less stiff and did not exhibit the anticipated behavior of a balanced laminate, as initially assumed. Moreover, variations in $E_{2m}$ across the panel are observed. It is concluded that the stiffness assessment using the algorithm is not in desirable agreement with mechanical testing and variations across the plate are observed. These differences are partly attributed to the moderate quality of the manufactured test panel as well as the relatively large dimensions of the measurement device compared to the test coupons. In future research, the results can be improved by optimization of the measurement device and improvement of the test panel's production quality.

**Author Contributions:** Conceptualization, M.A., A.H. and L.P.; methodology, M.A., A.H. and L.P.; software, M.A.; formal analysis, M.A.; investigation, M.A.; resources, L.P. and C.K.; writing—original draft preparation, M.A.; writing—review and editing, L.P., C.K. and J.A.A.V.; visualization, M.A.; supervision, A.H. and L.P.; project administration, L.P.; funding acquisition, L.P. All authors have read and agreed to the published version of the manuscript.

**Funding:** This research was funded by Materiel and IT Command of The Netherlands Ministry of Defence. The APC was funded by TU Delft.

**Institutional Review Board Statement:** Not applicable.

**Informed Consent Statement:** Not applicable.

**Data Availability Statement:** The data presented in this study are available upon request from the corresponding author.

**Acknowledgments:** The authors are grateful to Sajad Alimirzaei for his support regarding implementation of the semi-analytical finite element method.

**Conflicts of Interest:** The authors declare no conflicts of interest. The funders had no role in the design of the study; in the collection, analyses, or interpretation of data; in the writing of the manuscript; or in the decision to publish the results.

## References

1. Rohith, K.; Shreyas, S.; Vishnu Appaiah, K.B.; Sheshank, R.V.; Ganesha, B.B.; Vinod, B. Recent Material Advancement for Marine Application. In Proceedings of the 9th International Conference of Materials Processing and Characterization (ICMPC), Hyderabad, India, 8–10 March 2019; pp. 4854–4859.
2. Fiore, V.; Di Bella, G.; Valenza, A. Glass-basalt/epoxy hybrid composites for marine applications. *Mater. Des.* **2011**, *32*, 2091–2099. [CrossRef]
3. Mouritz, A.P.; Gellert, E.; Burchill, P.; Challis, K. Review of advanced composite structures for naval ships and submarines. *Compos. Struct.* **2001**, *53*, 21–41. [CrossRef]

4. Daniel, I.M.; Ishai, O. *Engineering Mechanics of Composite Materials*, 2nd ed.; Oxford University Press: Oxford, UK, 2006.
5. Ibrahim, M.E. Nondestructive testing and structural health monitoring of marine composite structures. In *Marine Applications of Advanced Fibre-Reinforced Composites*; Maritime Division, Defence Science and Technology Organisation: Melbourne, VIC, Australia, 2017; pp. 147–183.
6. Martens, A.; Kersemans, M.; Daemen, J.; Verboven, E.; van Paepegem, W.; Degrieck, J.; Delrue, S.; van den Abeele, K. Numerical study of the Time-of-Flight Pulsed Ultrasonic Polar Scan for the determination of the full elasticity tensor of orthotropic plates. *Compos. Struct.* **2017**, *180*, 29–40. [CrossRef]
7. Ghobadi, A. Common Type of Damages in Composites and Their Inspections. *World J. Mech.* **2017**, *7*, 24–33. [CrossRef]
8. Irfan, M.S.; Khan, T.; Hussain, T.; Liao, K.; Umer, R. Carbon coated piezoresistive fiber sensors: From process monitoring to structural health monitoring of composites—A review. *Compos. Part A Appl. Sci. Manuf.* **2021**, *141*, 106236. [CrossRef]
9. Talreja, R. Manufacturing defects in composites and their effects on performance. In *Polymer Composites in the Aerospace Industry*; Irving, P.E., Soutis, C., Eds.; Woodhead Publishing: Cambridge, UK, 2014; Chapter 4.
10. Hsiao, H.M.; Daniel, I.M. Effect of Fiber Waviness on Stiffness and Strength Reduction of Unidirectional Composites Under Compressive Loading. *Compos. Sci. Technol.* **1996**, *59*, 581–593. [CrossRef]
11. Bowles, K.J.; Frimpong, S. Void Effects on the Interlaminar Shear Strength of Unidirectional Graphite-Fiber-Reinforced Composites. *Compos. Mater.* **1992**, *26*, 1487–1509. [CrossRef]
12. Huang, H.; Talreja, R. Effects of Void Geometry on Elastic Properties of Unidirectional Fiber Reinforced Composites. *Compos. Sci. Technol.* **2005**, *65*, 1964–1981. [CrossRef]
13. Giurtiutiu, V. Structural Health Monitoring (SHM) of Aerospace Composites. In *Polymer Composites in the Aerospace Industry*; Irving, P.E., Soutis, C., Eds.; Woodhead Publishing: Cambridge, UK, 2014; Chapter 17, pp. 491–508
14. Montalvão, D.; Maia, N.M.M.; Ribeiro, A.M.R. A Review On Vibration-Based Structural Health Monitoring with Special Emphasis on Composite Materials. *Shock Vib. Dig.* **2006**, *38*, 295–324. [CrossRef]
15. Hsu, D.K.J.; Barnard, D.J.; Peters, J.J.; Dayal, V. Physical basis of tap test as a quantitative imaging tool for composite structures on aircraft. *AIP Conf. Proc.* **2000**, *509*, 1857–1864.
16. Tan, K.T.; Watanabe, N.; Iwahori, Y. X-ray radiography and micro-computed tomography examination of damage characteristics in stitched composites subjected to impact loading. *Compos. Part B* **2011**, *42*, 874–884. [CrossRef]
17. Rique, A.M.; Machado, A.C.; Oliveira, D.F.; Lopes, R.T.; Lima, I. X-ray imaging inspection of fiberglass reinforced by epoxy composite. *Nucl. Instrum. Methods Phys. Res. B* **2015**, *349*, 181–191. [CrossRef]
18. Agrai, A.N.; Girugiutiu, V. Electromechanical Impedance Modeling. In *Encyclopedia of Structural Health Monitoring*; Boller, C., Chang, F.K., Fujino, Y., Eds.; John Wiley & Sons: Hoboken, NJ, USA, 2009; Chapter 5.
19. Deng, Y.; Liu, X. Electromagnetic Imaging Methods for Nondestructive Evaluation Applications. *Sensors* **2011**, *11*, 11774–11808. [CrossRef] [PubMed]
20. Upda, S.S.; Moore, P.O. *Nondestructive Testing Handbook—Electromagnetic Testing*, 3rd ed.; American Society for Nondestructive Testing: Columbus, OH, USA, 2002.
21. Hung, Y.Y.; Yang, L.X.; Huang, Y.H. Non-destructive evaluation (NDE) of composites: Digital shearography. In *Non-Destructive Evaluation (NDE) of Polymer Matrix Composites: Techniques and Applications*; Karbhari, V.M., Ed.; Woodhead Publishing: Cambridge, UK, 2013; Chapter 5.
22. Fan, W.; Qiao, P. Vibration-based Damage Identification Methods: A Review and Comparative Study. *Struct. Health Monit.* **2011**, *10*, 83–111. [CrossRef]
23. Aggelis, D.G.; Barkoula, N.M.; Matikas, T.E.; Paipetis, A.S. Acoustic structural health monitoring of composite materials: Damage identification and evaluation in cross ply laminates using acoustic emission and ultrasonics. *Compos. Sci. Technol.* **2012**, *72*, 1127–1133. [CrossRef]
24. Scholey, J.J.; Wilcox, P.D.; Wisnom, M.R.; Friswell, M.I. Quantitative experimental measurements of matrix cracking and delamination using acoustic emission. *Compos. Part A* **2010**, *41*, 613–623. [CrossRef]
25. Giurgiutiu, V.; Santoni-Bottai, G. Structural Health Monitoring of Composite Structures with Piezoelectric-Wafer Active Sensors. *AIAA J.* **2011**, *49*, 565–581. [CrossRef]
26. Ruzek, R.; Lohonka, R.; Jironc, J. Ultrasonic C-Scan and shearography NDI techniques evaluation of impact defects identification. *NDT&E Int.* **2006**, *39*, 132–142.
27. Benammar, A.; Drai, R.; Guessoum, A. Detection of delamination defects in CFRP materials using ultrasonic signal processing. *Ultrasonics* **2008**, *48*, 731–738. [CrossRef]
28. Kersemans, M.; Lammers, N.; Martens, A.; van den Abeele, K. Identification of the Elastic Properties of Isotropic and Orthotropic Thin-Plate Materials with the Pulsed Ultrasonic Polar Scan. *Exp. Mech.* **2014**, *54*, 1121–1132. [CrossRef]
29. Alnuaimi, H.; Amjad, U.; Russo, P.; Lopresto, V.; Kundu, T. Monitoring damage in composite plate from crack initiation to macro-crack propagation combining linear and nonlinear ultrasonic techniques. *Struct. Health Monit.* **2021**, *20*, 139–150. [CrossRef]
30. Yang, H.; Yang, L.; Yang, Z.; Shan, Y.; Gu, H.; Ma, J.; Zeng, X.; Tian, T.; Ma, S.; Wu, Z. Ultrasonic detection methods for mechanical characterization and damage diagnosis of advanced composite materials: A review. *Compos. Struct.* **2023**, *324*, 117554. [CrossRef]
31. Su, Z.; Lin, Y. *Identification of Damage Using Lamb Waves*; Springer Science + Business Media: New York, NY, USA, 2009; Volume 48.
32. Pahlavan, L. Wave Propagation in Thin-Walled Composite Structures: Application to Structural Health Monitoring. Ph.D. Thesis, Delft University of Technology, Delft, The Netherlands, 2012.

33. Lee, B.C.; Staszewski, W.J. Modelling of Lamb Waves for Damage Detection in Metallic Structures: Part II. Wave Interactions with Damage. *Smart Mater. Struct.* **2003**, *12*, 815–824. [CrossRef]
34. Monkhouse, R.S.C.; Wilcox, P.W.; Lowe, M.J.S.; Dalton, R.P.; Cawley, P. The rapid monitoring of structures using interdigital Lamb wave transducers. *Smart Mater. Struct.* **2000**, *9*, 304–309. [CrossRef]
35. Rose, J.L. A Baseline and Vision of Ultrasonic Guided Wave Inspection Potential. *J. Press. Vessel Technol. Trans. ASME* **2002**, *124*, 273–282. [CrossRef]
36. Rose, J.L.; Pilarski, A.; Ditri, J.J. An Approach to Guided Wave Mode Selection for Inspection of Laminated Plate. *J. Reinf. Plast. Compos.* **1993**, *12*, 536–544. [CrossRef]
37. Pahlavan, L.; Blacquière, G. Fatigue crack sizing in steel bridge decks using ultrasonic guided waves. *NDT&E Int.* **2016**, *77*, 49–61.
38. Yilmaz, C.; Topal, S.; Ali, H.Q.; Tabrizi, I.E.; Al-Nadhari, A.; Suleman, A.; Yildiz, M. Non-destructive determination of the stiffness matrix of a laminated composite structure with lamb wave. *Compos. Struct.* **2020**, *237*, 111956. [CrossRef]
39. Santoni, A.; Schoenwald, S.; van Damme, B.; Fausti, P. Determination of the elastic and stiffness characteristics of cross-laminated timber plates from flexural wave velocity measurements. *J. Sound Vib.* **2017**, *400*, 387–401. [CrossRef]
40. Castellano, A.; Fraddosio, A.; Marzano, S.; Piccioni, M.D. Ultrasonic immersion tests for mechanical characterization of multilayered anisotropic materials. In Proceedings of the 2014 IEEE Workshop on Environmental, Energy, and Structural Monitoring Systems, Naples, Italy, 17–18 September 2014.
41. Castaings, M.; Hosten, B. Ultrasonic guided waves for health monitoring of high-pressure composite tanks. *NDT&E Int.* **2008**, *41*, 648–655.
42. Cawley, P. The rapid non-destructive inspection of large composite structures. *Composites* **1994**, *25*, 351–357. [CrossRef]
43. Toyama, N.; Noda, J.; Okabe, T. Quantitative damage detection in cross-ply laminates using Lamb wave method. *Compos. Sci. Technol.* **2003**, *63*, 1473–1479. [CrossRef]
44. Doebling, S.W.; Farrar, C.R.; Prime, M.B. A summary review of vibration-based damage identification methods. *Shock Vib. Dig.* **1998**, *30*, 91–105. [CrossRef]
45. Su, Z.; Ye, L.; Lu, Y. Guided Lamb waves for identification of damage in composite structures: A review. *J. Sound Vib.* **2006**, *295*, 753–780. [CrossRef]
46. Lemistre, M. Structural health monitoring system based on diffracted Lamb wave analysis by multiresolution processing. *Smart Mater. Struct.* **2001**, *10*, 504–511. [CrossRef]
47. Raghavan, A. Guided-wave structural health monitoring. Ph.D. Thesis, The University of Michigan, Ann Arbor, MI, USA, 2007.
48. Nayfeh, A.H. *Wave Propagation in Layered Anisotropic Media*, 1st ed.; Springer Science + Business Media: New York, NY, USA, 1995.
49. Gopalakrishnan, S.; Chakraborty, A.; Roy Mahapatra, D. *Spectral Finite Element Method*, 1st ed.; Springer: London, UK, 2008.
50. Barazanchy, B.; Giurgiutiu, V. A comparative convergence and accuracy study of composite guided-ultrasonic wave solution methods: Comparing the unified analytic method, SAFE method and DISPERSE. *Proc. Inst. Mech. Eng. Part C J. Mech. Eng. Sci.* **2017**, *231*, 2961–2973. [CrossRef]
51. Bartoli, I.; Marzani, A.; Lanza di Scalea, F.; Viola, E. Modeling wave propagation in damped waveguides of arbitrary cross-section. *J. Sound Vib.* **2006**, *295*, 685–707. [CrossRef]
52. Sale, M.; Rizzo, P.; Marzani, A. Semi-analytical formulation for the guided waves-based reconstruction of elastic moduli. *Mech. Syst. Signal Process.* **2011**, *25*, 2241–2256. [CrossRef]
53. Marzani, A.; de Marchi, L. Characterization of the elastic moduli in composite plates via dispersive guided waves data and genetic algorithms. *J. Intell. Mater. Syst. Struct.* **2013**, *24*, 2135–2147. [CrossRef]
54. Cui, R.; Lanza di Scalea, F. On the identification of the elastic properties of composites by ultrasonic guided waves and optimization algorithm. *Compos. Struct.* **2019**, *223*, 110969. [CrossRef]
55. Pahlavan, L.; Blacquière, G. Influence of asphalt on fatigue crack monitoring in steel bridge decks using guided waves. *Constr. Build. Mater.* **2016**, *120*, 593–604. [CrossRef]
56. Kamal, A.; Giurgiutiu, V. Stiffness Transfer Matrix Method (STMM) for Stable Dispersion Curves Solution in Anisotropic Composites. In Proceedings of the SPIE 9064, Health Monitoring of Structural and Biological Systems 2014 International Conference, San Diego, CA, USA, 9–13 March 2014; Volume 9064, p. 906410.
57. Baddour, N. Hamilton's Principle for the Derivation of Equations of Motion. In *Leading-Edge Applied Mathematical Modeling Research*; Alvarez, M., Ed.; Nova Science Publisher: New York, NY, USA, 2007; pp. 155–182
58. Vargalui, A.; Martinez, M.; Zarouchas, D.S.; Pant, S. Temperature effects on an acoustic emission based SHM system—Applied to composite materials. In Proceedings of the 26th International Conference on Adaptive Structures and Technologies, Kobe, Japan, 14–16 October 2015.
59. *D3039/D3039M*; Standard Test Method for Tensile Properties of Polymer Matrix Composite Materials. ASTM: West Conshohockenm, PA, USA, 2020.
60. *D6272*; Standard Test Method for Flexural Properties of Unreinforced and Reinforced Plastics and Electrical Insulating Materials by Four-Point Bending. ASTM: West Conshohockenm, PA, USA, 2020
61. Kassapoglou, C. *Design and Analysis of Composite Structures: With Applications to Aerospace Structures*, 1st ed.; John Wiley & Sons Ltd: Chichester, UK, 2010.
62. Yue, N.; Aliabadi, M.H. A scalable data-driven approach to temperature baseline reconstruction for guided wave structural health monitoring of anisotropic carbon-fibre-reinforced polymer structures. *Struct. Health Monit.* **2019**, *19*, 1487–1506. [CrossRef]

63. Lu, Y.; Michaels, J.E. A methodology for structural health monitoring with diffuse ultrasonic waves in the presence of temperature variations. *Ulstrasonics* **2005**, *43*, 717–731. [CrossRef] [PubMed]
64. Dworakowski, Z.; Ambrozinski, L.; Stepinski, T. Multi-stage temperature compensation method for Lamb wave measurements. *J. Sound Vib.* **2016**, *382*, 328–339. [CrossRef]
65. Croxford, A.J.; Moll, J.; Wilcox, P.D.; Michaels, J.E. Efficient temperature compensation strategies for guided wave structural health monitoring. *Ultrasonics* **2010**, *50*, 517–528. [CrossRef] [PubMed]
66. Giannakeas, I.N.; Khodaei, Z.S.; Aliabadi, M.H. An up-scaling temperature compensation framework for guided wave-based structural health monitoring in large composite structures. *Struct. Health Monit.* **2022**, *22*, 777–798. [CrossRef]

*sensors*

MDPI

*Article*

# Rail Flaw Detection via Kolmogorov Entropy of Chaotic Oscillator Based on Ultrasonic Guided Waves

**Ziyan Zeng [1], Jing Wu [2,*], Mingfang Zheng [3] and Hongwei Ma [2,3,4]**

1 School of Mechanics and Construction Engineering, Jinan University, Guangzhou 510632, China
2 School of Mechanical Engineering, Dongguan University of Technology, Dongguan 523808, China
3 School of Environment and Civil Engineering, Dongguan University of Technology, Dongguan 523808, China
4 Guangdong Provincial Key Laboratory of Intelligent Disaster Prevention and Emergency Technologies for Urban Lifeline Engineering, Dongguan 523808, China
* Correspondence: jingjinglulie@163.com

**Abstract:** Ultrasonic guided wave (UGW) inspection is an emerging non-destructive testing(NDT) technique for rail flaw detection, where weak UGW signals under strong noise backgrounds are difficult to detect. In this study, a UGW signal identification model based on a chaotic oscillator is established. The approach integrates the UGW response into the critical state of the Duffing system to serve as a disturbance control variable. By evaluating the system's motion state before and after introducing the UGW response, identification of UGW signals can be realized. Thus, the parameters defining the critical state of Duffing oscillators are determined by $K_e$. Moreover, an electromagnetic transducer was specifically devised to enable unidirectional excitation for UGWs targeted at both the rail base and rail head. Experimental studies showed that the proposed methodology effectively detected and located a 0.46 mm notch at the rail base and a 1.78 mm notch at the rail head. Furthermore, $K_e$ was directly proportional to the notch size, which could be used as a quantitative index to characterize the rail flaw.

**Keywords:** ultrasonic guided waves; chaotic oscillator; Kolmogorov entropy; rail; electromagnetic acoustic transducer

## 1. Introduction

Rail flaw detection is crucial due to the increased likelihood of rail damage from external loads as the service life extends. An accident can lead to significant loss of life and property. Currently, ultrasonic detection is the most widely employed technology for rail flaw detection. The fundamental operation of ultrasonic detection involves using a transducer to excite ultrasonic pulses that characterize and locate internal rail flaws through measured quantities such as the amplitude of the echo signal and time. However, a limitation of ultrasonic testing is that it requires point-to-point scanning of the rail, leading to low detection efficiency and a blind spot at the rail base.

UGW inspection is an emerging NDT technology [1]. Due to the dispersive, multimodal, and attenuating characteristics of UGWs in long range detection, actual sampled guided wave signals often appear as weak signals against a background of strong noise. Scholars have extensively researched signal processing methods for UGWs, including time–frequency analysis [2], such as short-time Fourier transform [3], 2D Fourier transform [4], wavelet transform [5,6], Hilbert–Huang transform [7], empirical modal decomposition [8], Wigner–Ville distribution [9], and artificial neural networks [10], as well as dispersion compensation methods [11], time inversion focusing methods [12], etc. Most of the above methods use noise suppression techniques to reduce the noise of the target signal and the noise signal superimposed on the overlapped signal, which can reduce the sensitivity of damage detection. Furthermore, some state-of-the-art fault detection methods [13–15] have been proposed using training data collected with ambient noise in industrial processes.

With the development of nonlinear science, some scholars have begun to study weak signal detection methods based on nonlinear systems, and one of the most representative methods is chaos detection based on chaos theory. Chaos theory discusses the unity of complexity, randomness, and certainty that prevails in nature. Lorenz identified the following three characteristics of chaos [16]: (1) an appearance of randomness, with the actual behavior determined by precise laws; (2) a sensitive dependence on initial conditions; and (3) a sensitive dependence on the intrinsic variability in initial conditions.

Typical models of chaotic dynamics include the Duffing equation [17], the Van-der-pol system [18], logistic mapping [19], and the Loren attractor [20], among which the Duffing equation is a typical model of chaotic dynamics that has garnered significant attention in the field of signal detection, due to its inclusion of a periodic excitation term. Jalilvand [21] examined the impact of frequency, phase, and noise on a weak signal in a Duffing oscillator. Nohara [22] researched the response of the Duffing system when subjected to square wave excitation, assessing its potential for square wave detection. Srinivasan [23] delved into the dynamics of the Duffing equation under sawtooth wave excitation. As research progressed, scholars started incorporating chaotic determination indices into UGW signal detection, Cheng [24] utilized the Poincaré map as a chaos indicator to identify pipe damage using the Duffing oscillator. Acknowledging the subjective nature of qualitative chaos indices in determining system motion states, some scientists began exploring quantitative chaos indexes. Zhang [25] examined the effectiveness of an enhanced Duffing system for UGW detection in pipelines by altering the nonlinear term of the Duffing equation. Hu [26,27] detected weak second harmonic signals in plates due to micro-cracks by assessing the maximum Lyapunov exponent of the Duffing equation. Wu [28] carried out simulation and experimental studies on the UGW detection of pipeline defects using the maximum Lyapunov exponent and the Lyapunov fractional dimensions as phase determination indexes, respectively. Ng [29] identified hole defects in rails using the maximum Lyapunov exponent of the Duffing equation. Additionally, Cheng [30] developed a pipeline damage detection method based on the double Duffing equation for detecting weak defect echo signals caused by pipeline defects.

Scholars have conducted extensive research on utilizing chaotic oscillators for weak signal detection. However, challenges persist in applying these methods to detect UGW signals:

- The difficulty in the quantitative determination of the system parameters of a chaotic oscillator when it is in the critical state between chaotic and periodic states;
- The commonly employed quantitative measures of chaos, such as maximum Lyapunov exponents and Lyapunov dimensions, necessitate constant re-orthogonalization during calculations, leading to computational inefficiency;
- Currently, only qualitative assessment and localization of damage can be achieved, as it is difficult to quantitatively characterize defects using chaotic oscillators.

Otherwise, piezoelectric acoustic transducers are commonly used in UGW excitation and reception technology for current rail detection. Although the efficiency of the transducer is high, it is strongly influenced by the coupling conditions, which limits its engineering applicability.Thus, it is important to develop a non-contact UGW transducer.

Given the challenges that current studies have struggled to address, this paper aimed to achieve the following research objectives:

1. To develop a quantitative method to determine the system parameters of a chaotic oscillator in the critical state;
2. To develop a computationally efficient quantitative characterization of chaos;
3. To develop a method for quantitative characterization of rail flaws;
4. To design a non-contact UGW transducer.

The outline of this paper is as follows: In Section 2, a UGW signal identification model based on the chaotic oscillator is introduced. The model incorporates a method for calculating Kolmogorov entropy($K_e$) through orthogonal triangular decomposition. Within this framework, $K_e$ is employed as a quantitative index that characterizes the motion state of the chaotic oscillator system. In Section 3, an electromagnetic transducer is designed which can achieve unidirectional excitation for UGWs at the rail base and rail head, and experimental verification confirmed that the EMAT successfully amplified forward mode signals and suppressed reverse mode signals. In section 4, experiments demonstrated that the conventional wavelet transform method is incapable of detecting weak UGW signals reflected by small-size defects. Furthermore, this study's proposal to use the Kolmogorov entropy of the Duffing oscillator for identifying rail damages was experimentally validated, highlighting its effectiveness in damage identification. Section 5 provides a summary of this study.

## 2. Detection UGW Signal Using a Chaotic Oscillator

### *2.1. Duffing Oscillator System*

Since the Duffing equation contains a cosine element and the UGW signal is excited in the form of a trigonometric function, the dynamic system expressed by the Duffing equation is used as a UGW signal detection system.The standard Duffing equation can be expressed mathematically through the state space as follows:

$$\begin{cases} \dfrac{d\psi_1}{dt} = \psi_2 \\ \dfrac{d\psi_2}{dt} = \psi_1 - \psi_1^3 - \delta\psi_2 + \gamma\cos\Omega\psi_3 + s(t) \\ \dfrac{d\psi_3}{dt} = 1 \end{cases} \tag{1}$$

where $\delta$ refers to the damping ratio; $\gamma$ refers to the amplitude of driving force of the Duffing oscillator; $\Omega$ refers to the angular frequency; $\psi_1$, $\psi_2$, and $\psi_3$ correspond to displacement, velocity, and time in state space; and s(t) is the UGW signal to be detected. In numerical calculations, s(t) can be expressed as

$$s(t) = \frac{A}{2}(1 - \cos\frac{\omega t}{10})\sin\omega t + \sigma n(t) \tag{2}$$

In Equation (2), the first term on the right-hand side illustrates the UGW signal modulated by the Hanning window, where A is the amplitude of the guided wave signal, and $\omega$ is its angular frequency. The second term represents Gaussian noise, $\sigma$ refers to the magnitude of the noise, and n(t) is white noise with standard normal distribution.

While $\Omega = \omega$, inputting the UGW signal into the Duffing system is equivalent to increasing the amplitude of the driving force. Since the Duffing oscillator is in a critical state between chaotic and periodic motion, increasing the magnitude of the driving force will cause the system to undergo a transition from a chaotic to periodic state.

Upon inputting both UGWs and noise signals (as shown in Figure 1) into the Duffing oscillator system, the motion state of the system is scrutinized using the phase portrait, as illustrated in Figure 2. While $\delta = 0.5, \gamma = 0.73493, \Omega = \omega = 1$, the Duffing system is in a critical state. Inputting UGW signals of very small magnitude at this point, the motion state of the system will change, while noise is only a perturbation of the system state.

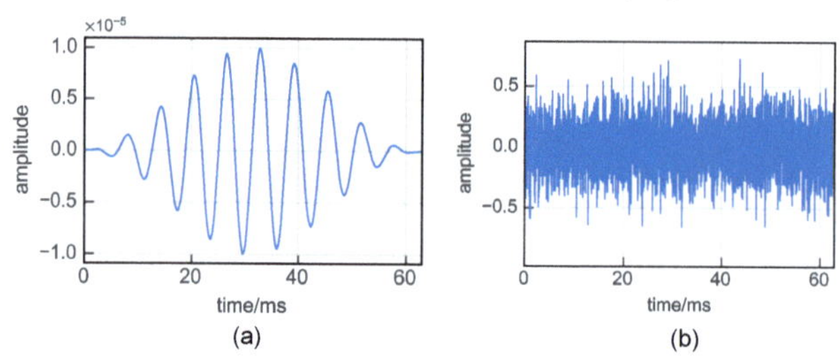


**Figure 1.** UGW signal and noise signal: (**a**) Guided wave signal where $A = 0.00001$, (**b**) Gauss white noise where $\sigma = 0.2$.

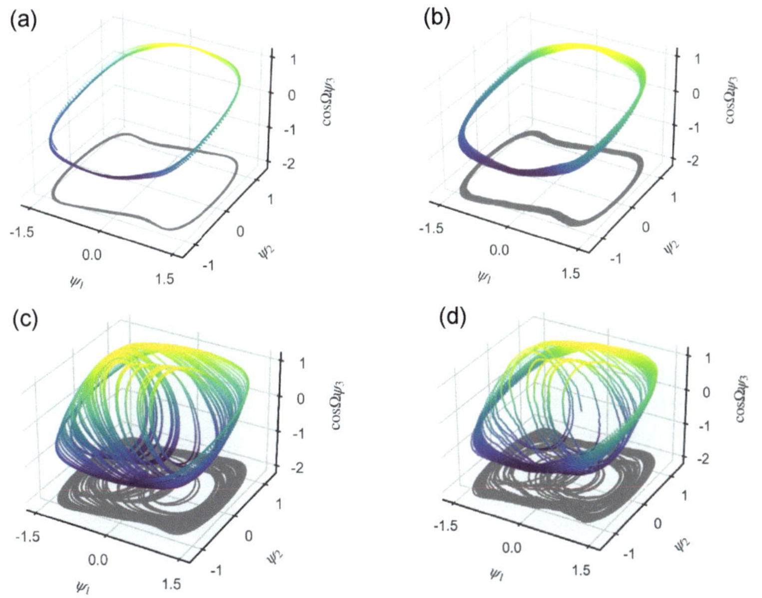


**Figure 2.** Phase portrait under different input signals: (**a**) when $A = 0, \sigma = 0$, periodic state; (**b**) when $A = 0, \sigma = 0.2$, periodic state; (**c**) when $A = 0.00001, \sigma = 0$, chaotic state; (**d**) when $A = 0.00001, \sigma = 0.2$, chaotic state.

*2.2. Kolmogorov Entropy as a Quantitative Index of Chaos*

Determining the state of motion of the Duffing system from the phase portrait is somewhat subjective, so it is preferable to use a quantitative index to describe its state. Kolmogorov entropy (later referred to as $K_e$) is an extended concept of Shannon entropy, which is a measure that describes the degree of chaos in a dynamical system and represents the average information growth of the dynamical system, as shown in Table 1.

**Table 1.** A criterion for determining the motion state of a dynamical system by means of $K_e$.

| The Value of K | Motion State |
|---|---|
| $K_e = 0$ | periodic state |
| $0 < K_e < +\infty$ | chaotic state |
| $K_e = +\infty$ | complete random state |

The more common calculation of $K_e$ is currently approximated by the generalized second-order Renyi entropy via a reconstruction-based vector space solution method, due to the fact that $K_e$ is numerically equal to the first-order Renyi entropy [31]. Practical engineering often requires a detection method that does not require a benchmark, and the above solution method is an estimation of $K_e$ [32]. In this paper, K entropy is calculated from the perspective that $K_e$ is defined as the average growth rate of the amount of information.

Consider the one-dimensional discrete mapping:

$$\varphi_{n+1} = F(\varphi_n) \tag{3}$$

Assuming that the interval of variation of the variable $\varphi$ is divided into n equal sub-intervals and that $\varphi$ has equal probability in each sub-interval, if $\varphi$ is known to be in a certain interval, the amount of information obtained is

$$H(\varphi) = -\sum_{i=1}^{n} \frac{1}{n} \ln \frac{1}{n} = \ln n \tag{4}$$

The mapping enlarges the interval of variation of the variables by a factor of $F'(\varphi_n)$ after each iteration, so that each sub-interval becomes $F'(\varphi_n)/n$ after the iteration, and the change in the amount of information in the above mapping system after one iteration is

$$\Delta H(\varphi) = \sum_{i=1}^{n/F'(\varphi_n)} \frac{|F'(\varphi_n)|}{n} \ln \frac{|F'(\varphi_n)|}{n} + \ln n = \ln|F'(\varphi_n)| \tag{5}$$

The average per-iteration information growth over the entire process as the number of iterations tends to infinity is given by

$$K_e = \Delta \bar{H}(\varphi) = \lim_{\lambda \to +\infty} \frac{1}{\lambda} \sum_{i=0}^{\lambda} \left| F'_{(i)}(\varphi_n) \right| \tag{6}$$

For an m-dimensional multidimensional system, the Kolmogorov entropy is

$$K_e^{(m)} = \int \rho(\varphi) \sum_{i=1}^{m} \Delta \bar{H}_i \mathrm{d}\varphi \tag{7}$$

where $\rho(\varphi)$ refers to the density of states function in phase space, and for an dynamic system with no concrete physical meaning, the density of states is assumed to be invariant, so that we can obtain

$$K_e^{(m)} = \sum_{i=1}^{m} \Delta \bar{H}_i \int \rho(\varphi) \mathrm{d}\varphi = \sum_{i=1}^{m} \Delta \bar{H}_i \tag{8}$$

Thus, the key point in calculating $K_e$ is to solve Equation (6). The following focuses on the $K_e$ calculation of the Duffing oscillator. The differential equation represented by the Duffing oscillator can be expressed as

$$\dot{\boldsymbol{\Psi}} = \boldsymbol{J}\boldsymbol{\Psi} \tag{9}$$

where $\boldsymbol{J}$ refers to the Jacobi matrix. Performing an orthogonal triangular decomposition of $\boldsymbol{\Psi}$, i.e., decompose $\boldsymbol{\Psi}$ into a product of an orthogonal matrix and a positive upper triangular matrix; that is,

$$\boldsymbol{\Psi} = \boldsymbol{QR} \tag{10}$$

where $\boldsymbol{Q}$ is a orthogonal matrix, $\boldsymbol{R}$ is a positive upper triangular matrix, so we obtain

$$\dot{\boldsymbol{Q}}\boldsymbol{R} + \boldsymbol{Q}\dot{\boldsymbol{R}} = \boldsymbol{JQR} \tag{11}$$

Multiplying the left by the transpose matrix of $\boldsymbol{Q}$ and the right by the inverse matrix of $\boldsymbol{R}$, we obtain

$$\boldsymbol{Q}^{\mathrm{T}}\dot{\boldsymbol{Q}} + \dot{\boldsymbol{R}}\boldsymbol{R}^{-1} = \boldsymbol{Q}^{\mathrm{T}}\boldsymbol{JQ} \tag{12}$$

Introducing the intermediate variable $\theta$, the orthogonal matrix $\boldsymbol{Q}$ is denoted as

$$\boldsymbol{Q} = \begin{pmatrix} \cos\theta & \sin\theta \\ -\sin\theta & \cos\theta \end{pmatrix} \tag{13}$$

the positive upper triangular matrix $\boldsymbol{R}$ is denoted as

$$\boldsymbol{R} = \begin{pmatrix} e^{\eta_1} & r_{12} \\ 0 & e^{\eta_2} \end{pmatrix} \tag{14}$$

The Jacobi matrix for the Duffing equation is

$$\boldsymbol{J} = \begin{pmatrix} J_{11} & J_{12} \\ J_{21} & J_{22} \end{pmatrix} = \begin{pmatrix} 0 & 1 \\ 1 - 3\varphi_1^2 & -\delta \end{pmatrix} \tag{15}$$

Combining the above equations, we obtain

$$\begin{cases} \dfrac{d\eta_1}{dt} = J_{11}\cos^2\theta - J_{21}\sin\theta\cos\theta - J_{12}\sin\theta\cos\theta + J_{22}\sin^2\theta \\ \dfrac{d\eta_2}{dt} = J_{11}\sin^2\theta + J_{21}\sin\theta\cos\theta + J_{12}\sin\theta\cos\theta + J_{22}\cos^2\theta \\ \dfrac{d\theta}{dt} = -J_{11}\sin\theta\cos\theta - J_{21}\cos^2\theta + J_{12}\sin^2\theta + J_{22}\sin\theta\cos\theta \end{cases} \tag{16}$$

The intermediate variables $\xi_1, \xi_2$ are introduced and satisfy the following equation:

$$\begin{cases} \dfrac{d\xi_1}{dt} = \dfrac{d\eta_1}{dt} + \dfrac{d\eta_2}{dt} \\ \dfrac{d\xi_2}{dt} = \dfrac{d\eta_1}{dt} - \dfrac{d\eta_2}{dt} \end{cases} \tag{17}$$

so we can obtain

$$\begin{cases} \dfrac{d\xi_1}{dt} = J_{11} + J_{22} \\ \dfrac{d\xi_2}{dt} = (J_{11} - J_{22})\cos 2\theta - (J_{12} + J_{21})\sin 2\theta \\ \dfrac{d\theta}{dt} = \dfrac{1}{2}(J_{22} - J_{11})\sin 2\theta - J_{21}\cos^2\theta + J_{12}\sin^2\theta \end{cases} \tag{18}$$

Substituting the Jacobi matrix of the Duffing system into the above equation, we can obtain

$$\begin{cases} \dfrac{d\xi_1}{dt} = -\delta \\ \dfrac{d\xi_2}{dt} = (3\psi_1^2 - 2)\sin 2\theta + \delta\cos 2\theta \\ \dfrac{d\theta}{dt} = -\dfrac{1}{2}\delta\sin 2\theta + (3\psi_1^2 - 2)\cos^2\theta + 1 \\ \dfrac{d\eta_1}{dt} = \dfrac{1}{2}(\dfrac{d\xi_1}{dt} + \dfrac{d\xi_2}{dt}) \\ \dfrac{d\eta_2}{dt} = \dfrac{1}{2}(\dfrac{d\xi_1}{dt} - \dfrac{d\xi_2}{dt}) \end{cases} \tag{19}$$

By solving the differential equations Equation (19), the $K_e$ can be calculated as:

$$K_e = \max(\eta_1, 0) + \max(\eta_2, 0) \tag{20}$$

To localize the damage, a time-shift window is employed, as illustrated in Figure 3. Only signals falling within this time-shifted window are fed into the critical state Duffing oscillator system after undergoing periodic continuation. Therefore, the presence of a signal within the window with $K_e > 0$ signifies a damage defect, while the absence indicates no defect.

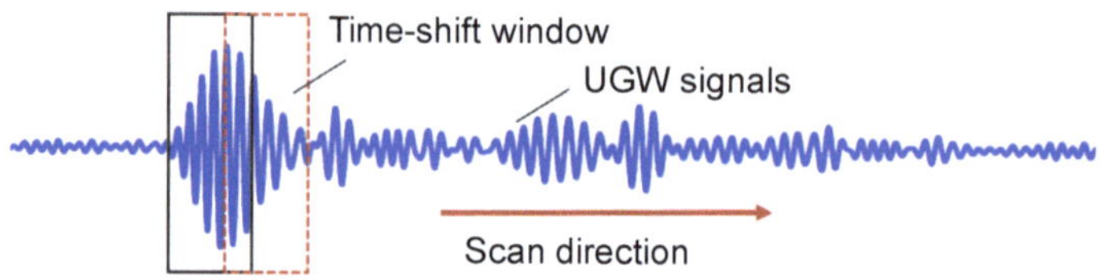


**Figure 3.** Time-shift window to localize the defect.

*2.3. Determination of the Parameters of the Duffing Equation for the Critical State*

To use the Duffing oscillator as a detection system for UGW signals, one of the keys is to set the system as a critical state between chaotic and periodic motions, and it is important to determine the parameters of the damping ratio and driving force amplitude at the critical state. A common method of determining critical state parameters is through bifurcation diagrams. The basic idea is to determine a specific damping ratio, solve the Poincare map of the Duffing oscillators for different amplitudes of driving force , and then the set of projections on the Poincare map is the bifurcation diagram. A bifurcation diagram cannot quantify the degree of chaos and is still essentially a qualitative rather than quantitative assessment method.

It has been shown in Section 2.2 that $K_e$ can be used as a quantitative characterization index of the chaotic motion state, and therefore the critical parameter can be determined by the relationship of $K_e$ to the change in the motion state of the system with the driving force amplitude $\gamma$ at different damping ratios $\delta$. In practical engineering terms, it is ideal to be able to find a quantitative indicator that is strictly positively or negatively correlated with the degree of damage.

As is shown in Figure 4, in some damping ratios, there is a large range in which $K_e$ is strictly positively correlated with the amplitude of the driving force $\gamma$, and the noise has almost no effect on the $K_e$ of the system in this range. The UGW signals will be normalized during the actual detection, and the amplitude of the guided wave at the damage is generally smaller than its peak amplitude, so the range of $\gamma$ from 0.4 to 0.9, and the range of $K_e$ strictly increasing with the amplitude of the driving force, generally covers the amplitude of the damaged UGW signals. Considering the diversity and uncertainty of an actual detection environment, we still choose the critical point of the periodic and chaotic states as the driving force amplitude of the detection system, and the critical point is taken as $\delta = 0.8, \gamma = 1.3040$. It is worth noting that, in an actual detection, due to the influence of the environment and other signal aberrations and other complex factors, it is impossible to achieve the ideal state of the simulation, and a certain threshold range should be maintained when taking the driving force amplitude. The driving force amplitude is taken in the left field of the critical point, although to some extent it will not be able to achieve an ideal state in the simulation. Although it will reduce the detection sensitivity to some extent, it can improve the stability of the Duffing detection system.

Thus, a rail flaw detection model based on the Duffing oscillator is established. By evaluating the system's motion state before and after introducing the UGW response, the identification of ultrasonic guided wave signals can be achieved. $K_e$ is a measure that describes the degree of chaos in a dynamical system. A method for calculating $K_e$ based on orthogonal triangular decomposition is proposed, making it a quantitative characterization factor of the motion state of the chaotic oscillator system. The entire flow of the above method is shown in the Figure 5.

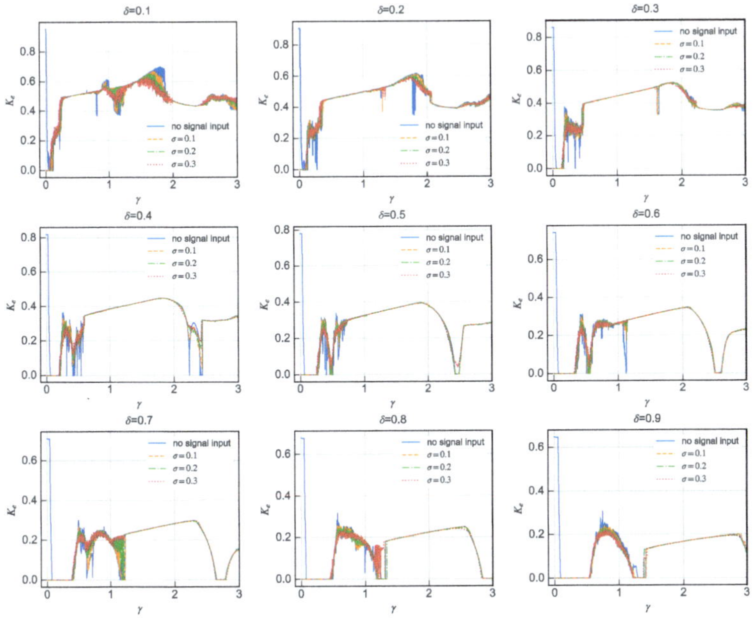


**Figure 4.** Relating $\gamma$ and $K_e$ for different damping ratios $\delta$ when $\Omega = 1$.

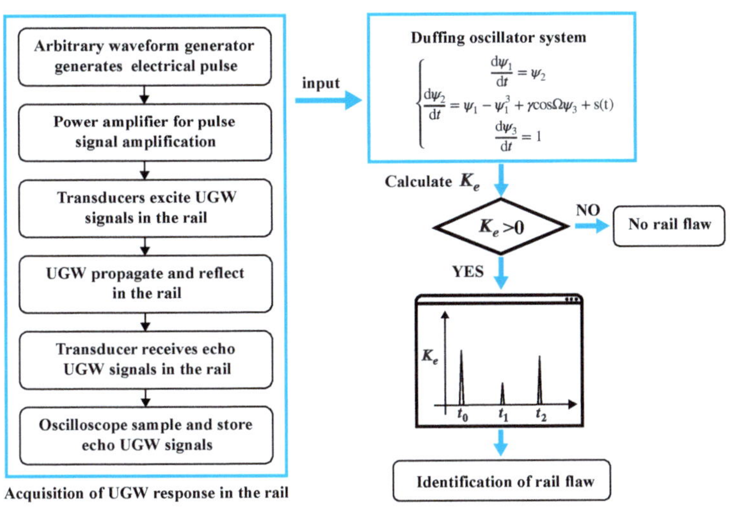


**Figure 5.** Flow chart of the rail flaw detection based on the Duffing oscillator system.

## 3. Design of an Electromagnetic Acoustic Transducer for UGWs

The ultrasonic acoustic transducers in NDT based on UGWs mainly include piezoelectric, electromagnetic, air-coupled, and laser acoustic transducers. The piezoelectric acoustic transducer is currently the most widely used type , but the disadvantage is that it is greatly affected by the coupling conditions, thus limiting its applicability to engineering sites. Therefore, designing a non-contact type transducer is of great importance for the realization of rail flaw detection.

Electromagnetic acoustic transducer(EMAT) has the advantage of good design-ability and lower material costs compared to the lower conversion efficiency of piezoelectric transducers.The principle of the excitation of the Lorentz force-based EMAT is shown in Figure 6.

Due to the complexity of the rail cross-section, it is difficult to excite a single guided wave mode in the rail, and when the guided wave encounters the boundary, the reflection will undergo a complicated mode conversion. If the guided wave propagates from both sides in the rail, the transducer will receive the reflected wave from both sides, which will greatly increase the difficulty of the subsequent signal identification and feature extraction. Therefore, in this study, a unidirectional excitation EMAT is designed to achieve enhancement of the forward guided waves modal and suppression of the reverse modal. The basic components of the EMAT are shown in Figure 7. Amplification in the forward direction and suppression in the backward direction of guided-wave signals is achieved by the arrangement of two meander coils spaced apart from each other and fed with electrical pulses with a time delay, and the arrangement of the two coils is shown in Figure 8. As is shown in Figure 9, the distance between the adjacent wires of the coil is $\lambda/2$ (where $\lambda$ is the wavelength of the guided waves), so that the guided waves generated by each wire gain each other. The distance between coil 1 and coil 2 is $\lambda/4$, and a reverse current is applied with a time delay of $T/4$ (where $T$ is the cycle of the guided waves) to achieve amplification of the forward UGW signals and suppression of the reverse signals.

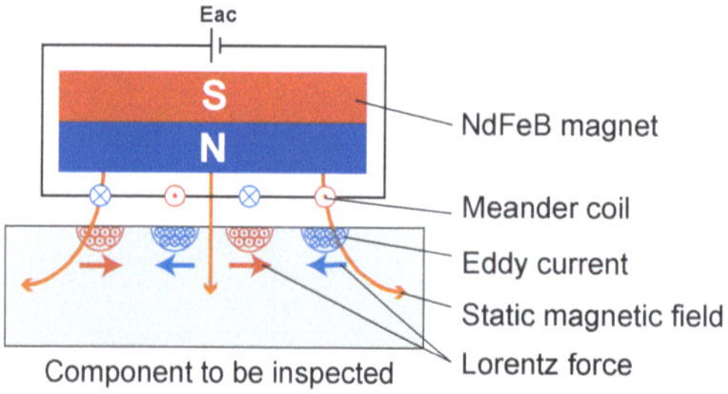


**Figure 6.** Schematic diagram of UGWs excited by EMAT.

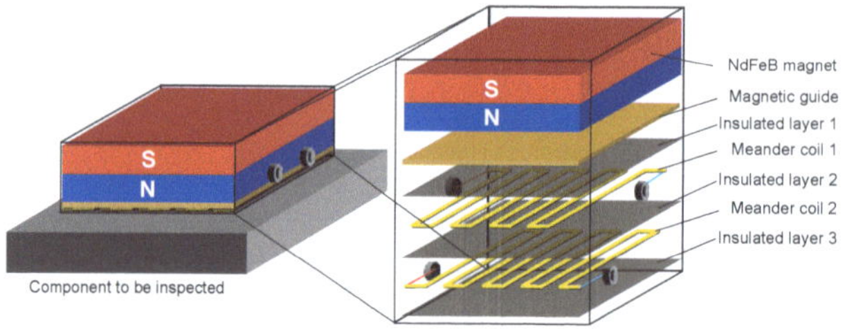


**Figure 7.** Schematic diagram of EMAT.

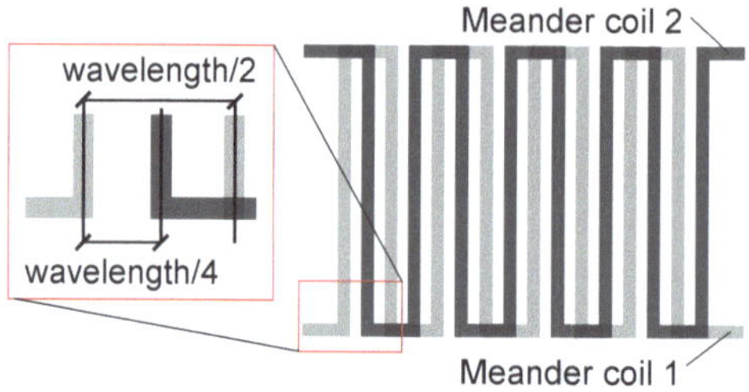


**Figure 8.** Arrangement of two meander coils.

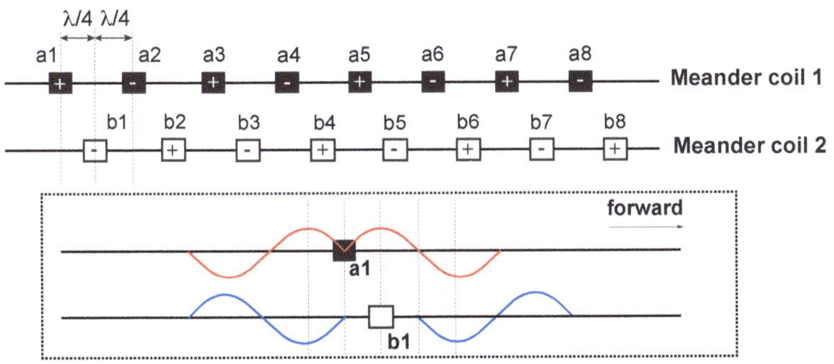


**Figure 9.** Schematic diagram of the principle of unidirectional guided wave excitation by EMAT.

The effectiveness of the EMAT in exciting UGWs was verified through an experiment at the head and base of the rail. The experimental scheme is shown in Figure 10. The effect of the unidirectional excitation of the double meander coils was verified by comparing the guided wave signals received at equidistant positions on both sides of the EMAT, and the experimental results are shown in Figure 11. The experimental results showed that the EMAT designed in this study could achieve the amplification of the forward guided wave signals and the suppression of the backward guided wave signals at the rail head and at the rail base.

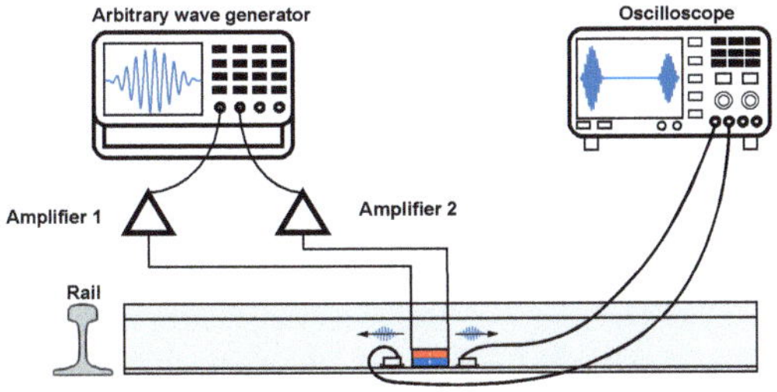


**Figure 10.** Experiment to verify the effect of the EMAT on the excitation of UGWs in the rail.

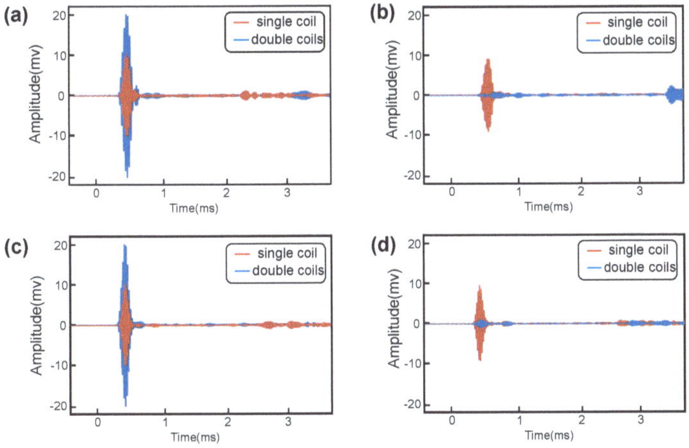


**Figure 11.** Experimental results: (**a**) Forward UGW signals at the rail base; (**b**) reverse UGW signals at the rail base; (**c**) forward UGW signals at the rail head; (**d**) reverse UGW signals at the rail head.

Thus, an EMAT was developed to enable unidirectional excitation of UGW signals at both the rail base and rail head. The strategic placement of two meander coils, separated at a specific distance, allows the amplification of guided-wave signals in the forward direction, while simultaneously suppressing them in the backward direction. This effect is achieved through feeding the coils with electrical pulses that are intentionally time-delayed.

## 4. Experiments to Detect Rail Defects

NDT technology based on UGWs includes four steps: excitation, propagation, reception, and signal processing of UGWs. In order to verify the effectiveness of the NDT method utilizing UGWs based on a chaotic oscillator, an experimental study on rail flaw detection was conducted.

### *4.1. Experimental Design*

Based on the above research, experimental schemes for damage detection of rail head and rail bottom were established. First, an arbitrary waveform generator produced a sinusoidal signal modulated by the Hanning window, which was passed through a power amplifier to obtain amplification. The amplified signal was then passed through the EMAT to excite UGWs in the rail. Next, the guided ultrasonic waves were reflected when they encountered defects during propagation in the rail. The reflected guided wave signals were then sampled as electrical signals by an oscilloscope through a receiving acoustic transducer. The sampled guided wave signals were then input into the detection system based on the Duffing oscillator described in Section 2, which ultimately enabled the identification of rail damage.

In the UGW detection experiment, it was important to reduce crosstalk between the external environmental noise and signals in each channel. To achieve this, it is recommended to use BNC radio frequency cables with shielding layers and to keep the cable length as short as possible to minimize signal distortion. Additionally, it is crucial not to share the ground of each unit and to avoid sharing the shield of the signal cable with the ground of other electrical equipment. Furthermore, it is advisable to keep power and signal cables of the equipment as far apart as possible. In cases where separation is not feasible, one should avoid cable crossings, refrain from laying cables in parallel, and if crossing is necessary, do so vertically. Furthermore, it is important to avoid setting the amplifier power too high, as excessive power can result in the generation of odd numbers of high harmonic currents, due to the non-linearity of the electronic circuit.

The experimental scheme is shown in Figure 12. For rail base detection, the EMAT was attached to the top of the rail base on both sides. For rail head detection, the EMAT was attached to the bottom of the rail head on both sides.

In the experiment, artificial notches were set at the rail base and rail head, as described in Figure 13. There were 7 different notches for the rail foot and 9 different notches for the rail head. The specific notch sizes for different cases at the rail base can be seen in Table 2, and the specific notch sizes at the rail head can be seen in Table 3.

**Table 2.** Various cases of defects at the rail base.

| Case | Case 1 | Case 2 | Case 3 | Case 4 | Case 5 | Case 6 | Case 7 |
|---|---|---|---|---|---|---|---|
| Notch width/mm | 0 | 0.46 | 0.88 | 1.07 | 1.48 | 2.07 | 2.72 |

**Table 3.** Various cases of defects at the rail head.

| Case | Case 8 | Case 9 | Case 10 | Case 11 | Case 12 | Case 13 | Case 14 | Case 15 | Case 16 |
|---|---|---|---|---|---|---|---|---|---|
| Notch width/mm | 0 | 1.78 | 2.20 | 2.75 | 3.10 | 3.76 | 4.70 | 5.08 | 5.65 |

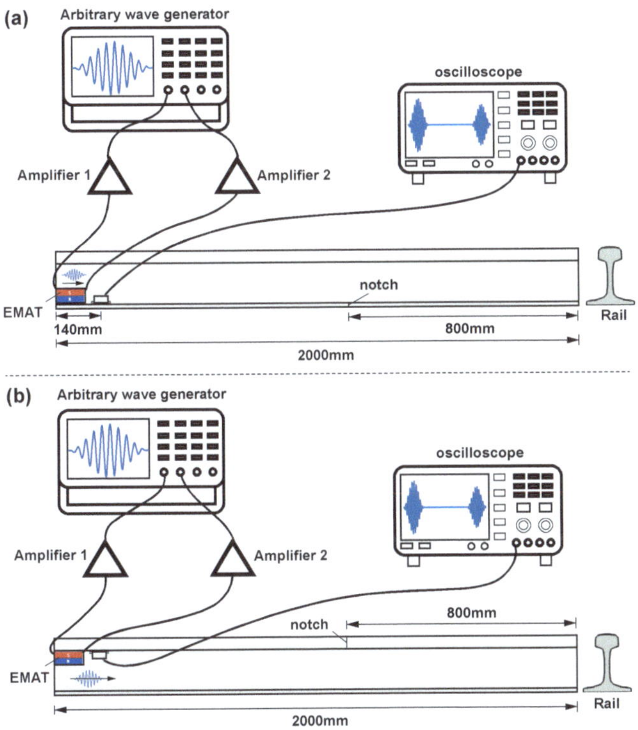


**Figure 12.** Experimental scheme for rail flaw detecting : (**a**) rail base; (**b**) rail head.

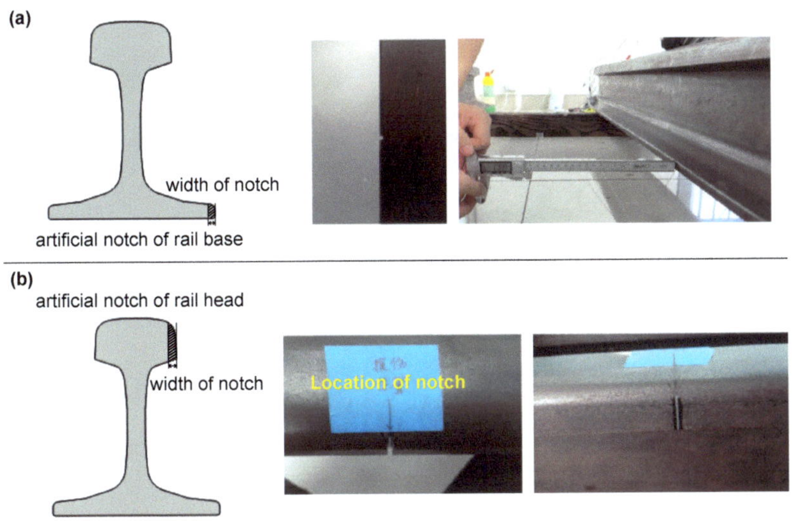


**Figure 13.** Artificial notches in the experiment: (**a**) rail base; (**b**) rail head.

## 4.2. Experimental Results

In the flaw detection experiments conducted at the rail base and rail head, the time domain signals of the guided waves were as depicted in Figures 14a and 15a. Initially, it was apparent that the time domain signal alone had limited efficacy in detecting damage, with only a 2.07 mm notch discernible at the rail base and a 3.76 mm notch at the rail head due to smaller defects being overshadowed by noise. Given this challenge, traditional signal processing approaches resort to signal cancellation techniques. Among these techniques, time–frequency analysis stands out as a commonly utilized method in current research practices. To address this limitation, the present study employed time–frequency analyses using a wavelet transform on the UGW signals. The outcomes of this analysis are displayed in Figures 14b and 15b. The application of the wavelet transform yielded a notable noise reduction effect, thereby enhancing the damage detection threshold to some extent. Nevertheless, the method still faced challenges in detecting minute damages. Specifically, the wavelet transform-based time–frequency analysis method successfully identified a 1.48 mm notch at the rail base and a 3.10 mm notch at the rail head.

Rail damage was detected using the chaotic oscillator detection system as discussed in Section 2. The incoming UGW signals were processed by the critical Duffing oscillator system, with the parameter $K_e$ serving as an indicator of the system's motion state. To localize the damage, a time-shift window was employed. Only signals falling within this time-shifted window were fed into the critical state Duffing oscillator system after undergoing periodic continuation. Therefore, the presence of a signal within the window with $K_e > 0$ signified a damage defect, while the absence indicated no defect. The results are depicted in Figures 16 and 17, revealing the system's ability to detect a 0.46 mm notch at the rail base and a 1.78 mm notch at the rail head. Upon comparing the wavelet transform method with the proposed method, it is evident that the method introduced in this paper demonstrated greater sensitivity to rail damage and enhanced the threshold for detecting rail flaws.

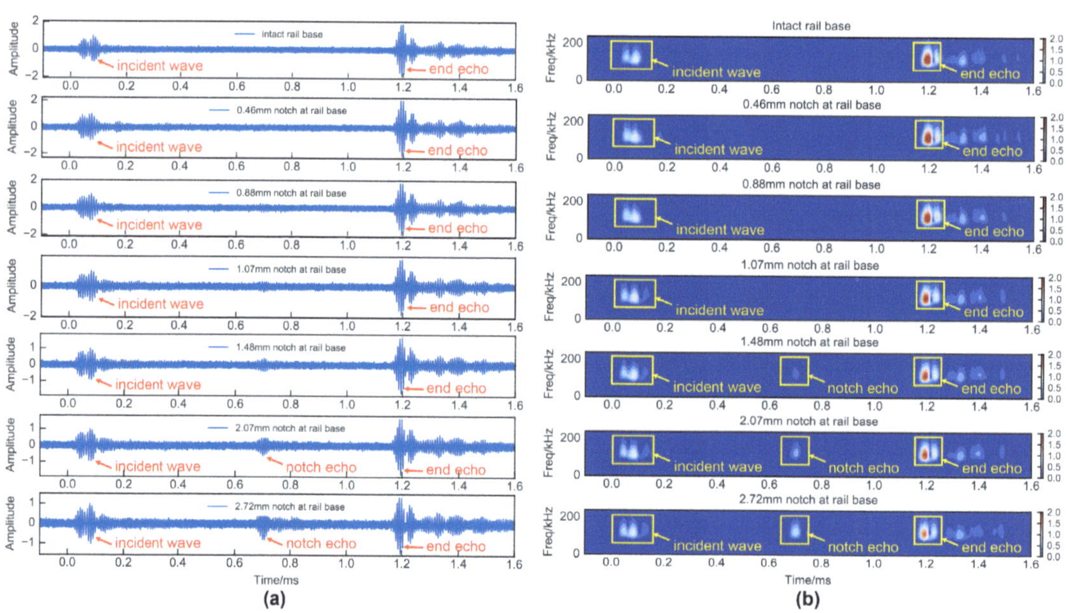


**Figure 14.** Experimental results of rail base detection: (**a**) time domain signal; (**b**) wavelet transform.

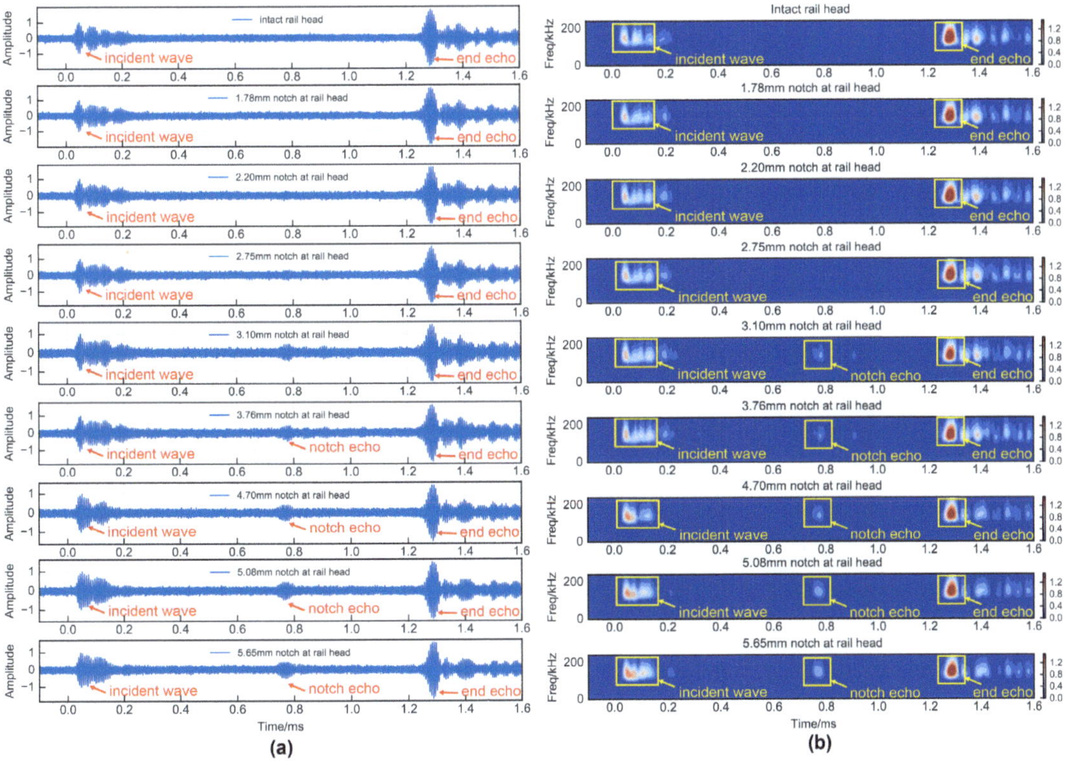


**Figure 15.** Experimental results of rail head detection: (**a**) time domain signal; (**b**) wavelet transform.

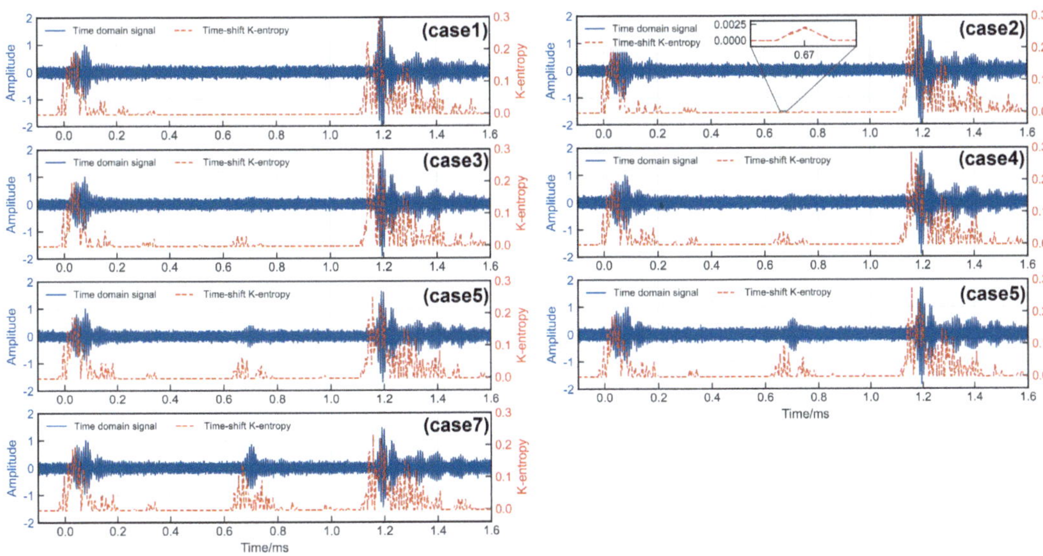


**Figure 16.** Experimental results of rail base detection via $K_e$ of Duffing oscillator.

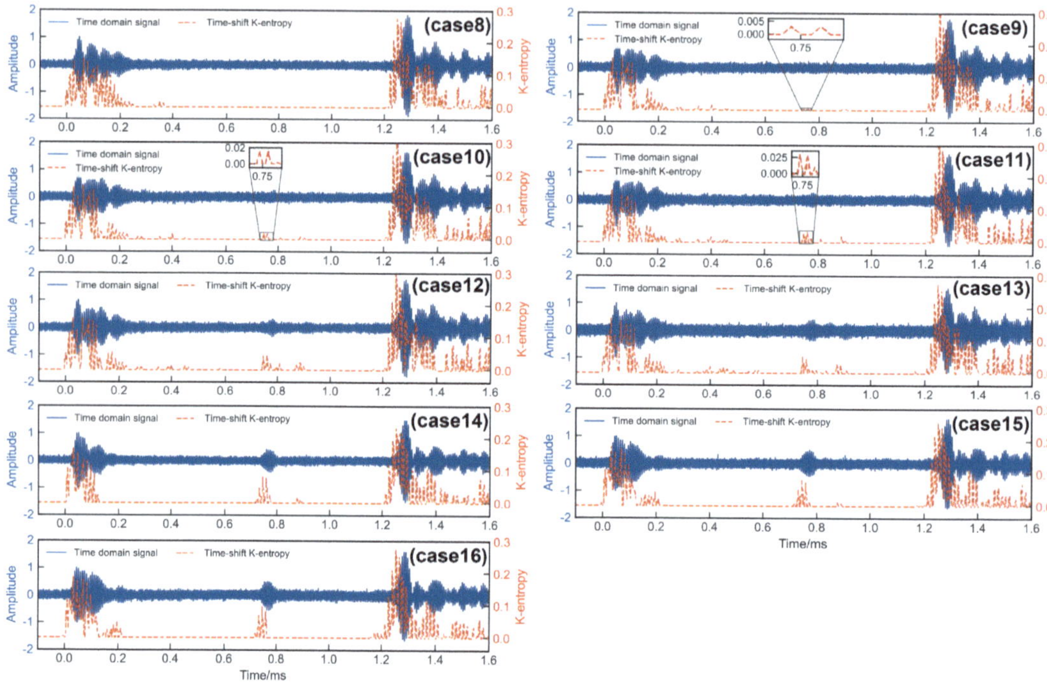


**Figure 17.** Experimental results of rail head detection via $K_e$ of Duffing oscillator.

The results of identifying and localizing rail damage using the $K_e$ of the Duffing oscillator system are presented in Tables 4 and 5.

The position error($p_e$) for the rail notch could be calculated using Equation (21) as

$$p_e = \left| \frac{(L - l_0)(t_2 - t_1)}{l_1(t_1 - t_0)} - 1 \right| \quad (21)$$

where $L$ is the axial length of the rail, $l_0$ is the axial distance between the center of the receiver transducer and the excited end face of the rail, $l_1$ is the actual axial distance between the center of the receiver transducer and the rail notch, $t_0$ represents the time of the incident wave with maximum amplitude, $t_1$ corresponds to the time of the end echo with maximum amplitude, and $t_2$ is the time corresponding to the maximum amplitude of notch echo.

**Table 4.** Results of locating the notch at the rail base.

| Case Number | Width of notch/mm | $t_0$ (ms) | $t_1$ (ms) | $t_2$ (ms) | $K_e$ | Positioning Error (%) |
|---|---|---|---|---|---|---|
| 1 | 0 | 0.038732 | 1.158122 | / | 0 | / |
| 2 | 0.46 | 0.038732 | 1.158122 | 0.670047 | 0.001992 | 1.04 |
| 3 | 0.88 | 0.038732 | 1.142207 | 0.670047 | 0.032303 | 0.39 |
| 4 | 1.07 | 0.038732 | 1.158122 | 0.670047 | 0.041133 | 1.04 |
| 5 | 1.48 | 0.038732 | 1.158122 | 0.670047 | 0.064721 | 1.04 |
| 6 | 2.07 | 0.038732 | 1.158122 | 0.670047 | 0.093227 | 1.04 |
| 7 | 2.72 | 0.038732 | 1.158122 | 0.670047 | 0.139945 | 1.04 |

**Table 5.** Results of locating the notch at the rail head.

| Case Number | Width of notch/mm | $t_0$ (ms) | $t_1$ (ms) | $t_2$ (ms) | $K_e$ | Positioning Error (%) |
|---|---|---|---|---|---|---|
| 8 | 0 | 0.027364 | 1.250584 | / | 0 | / |
| 9 | 1.78 | 0.027364 | 1.250584 | 0.745835 | 0.003014 | 3.07 |
| 10 | 2.20 | 0.027364 | 1.250584 | 0.745835 | 0.016560 | 3.07 |
| 11 | 2.75 | 0.027364 | 1.250584 | 0.745835 | 0.029862 | 3.07 |
| 12 | 3.10 | 0.027364 | 1.250584 | 0.754930 | 0.042997 | 4.37 |
| 13 | 3.76 | 0.027364 | 1.250584 | 0.745835 | 0.051369 | 3.07 |
| 14 | 4.70 | 0.031911 | 1.255131 | 0.745835 | 0.079399 | 2.41 |
| 15 | 5.08 | 0.031911 | 1.255131 | 0.670047 | 0.082339 | 2.41 |
| 16 | 5.65 | 0.031911 | 1.255131 | 0.670047 | 0.094784 | 2.41 |

Moreover, the width of notches at the rail base and rail head were directly proportional to the $K_e$, as shown in Figure 18. Thus, Kolmogorov entropy could be used as a quantitative index to characterize the rail defects. The experimental results demonstrated the effectiveness of the detection system based on a Duffing oscillator for rail flaw detection. Specifically, the Duffing oscillator system was capable of detecting a 0.46 mm notch at the rail base and a 1.78 mm notch at the rail head.

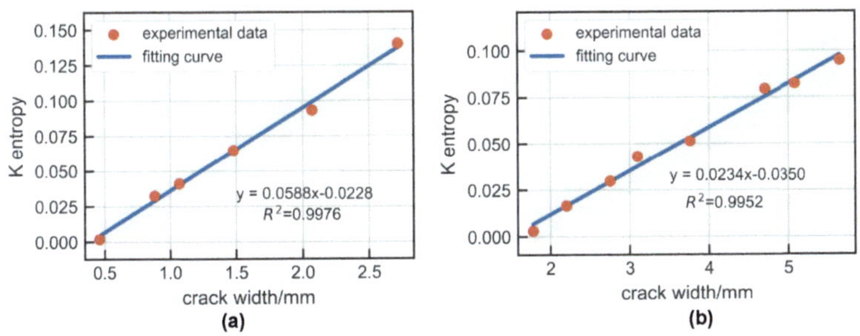


**Figure 18.** The relationship between notch width and $K_e$: (**a**) rail base; (**b**) rail head.

## 5. Conclusions and Discussion

This study addressed the challenge of identifying weak UGW signals in a strong noise background in rail flaw detection by proposing a damage identification method based on chaotic oscillators. Initially, a mathematical model for detecting UGW signals using the Duffing oscillator was introduced. The motion state of the Duffing system was characterized by the Kolmogorov entropy, and a formula for calculating this entropy was established. Subsequently, an electromagnetic UGW transducer was developed to amplify forward UGW modes and suppress unidirectional UGW modes. The efficacy of the proposed model and transducer in rail damage detection was then validated through experimental testing. The main conclusions of this study are analyzed as follows:

1. A UGW signal identification model based on the chaotic oscillator was established. The approach integrates the UGW response into the critical state of the Duffing system to serve as a disturbance control variable. This incorporation leads to alterations in the system's motion state through the exploitation of the parameter disturbance sensitivity characteristic of chaotic systems and the traversal of chaotic motion. By evaluating the system's motion state both pre- and post-introduction of the UGW response, the identification of ultrasonic guided wave signals can be realized. This methodology encapsulates the fundamental concept of employing chaotic systems for discerning faint guided wave signals in NDT applications centered on UGWs;
2. A method for calculating Kolmogorov entropy based on orthogonal triangular decomposition was proposed. Kolmogorov entropy is a measure that describes the degree

of chaos in a dynamical system and represents the average information growth of the system. $K_e$ can be used as a quantitative characterization factor of the motion state of the chaotic oscillator system. When $K_e = 0$, the system is in a state of periodic motion, and when $0 < K_e < +\infty$, the system is in a chaotic state. This method eliminates the need for reconstructing the phase space, thereby improving the efficiency of calculating Kolmogorov entropy.

3. An electromagnetic transducer was designed that can achieve unidirectional excitation for UGWs at the rail base and rail head. Amplification in the forward direction and suppression in the backward direction of guided-wave signals was achieved though the arrangement of two meander coils spaced apart from each other and fed with electrical pulses with a time delay.The distance between adjacent wires of the coil was $\lambda/2$, so that the UGW generated by each wire gained each other. The distance between coil 1 and coil 2 was $\lambda/4$, and a reverse current was applied with a time delay of $T/4$ to achieve amplification of the forward guided wave signal and suppression of the reverse signal. Experimental verification confirmed the effectiveness of the EMAT in producing the desired effects mentioned above;
4. The experimental results indicated the challenge in effectively identifying the weak UGW echoes caused by small sized damage using time-domain signals. Although the traditional signal processing method based on wavelet transform showed improved denoising capabilities, it continued to struggle in effectively distinguishing the weak UGW signals.
5. The width of notches at both the rail base and rail head were directly proportional to the $K_e$, hence Kolmogorov entropy can serve as a quantitative characterization index of rail damage. The experimental results demonstrated the effectiveness of the detection system based on a chaotic oscillator in detecting weak UGW signals. Specifically, the Duffing oscillator system was capable of detecting a 0.46 mm notch at the rail base and a 1.78 mm notch at the rail head.

In summary, this study proposed a method for detecting rail flaws using the Kolmogorov entropy of a chaotic oscillator based on UGWs. This method aims to accurately locate and quantitatively characterize defects at the rail base and rail head to enhance the sensitivity of rail flaw detection. However, in engineering applications, the method described above may encounter limitations, particularly when dealing with large rail damage. In such cases, the guided wave within the damaged area may undergo mode conversion. The presence of multiple damages on the rail further complicates the situation, making it challenging to differentiate between the modal conversion signal of the initial damage and the reflection signal produced by subsequent damages. Hence, the study of the specific interaction between rail flaws and UGWs remains a key research direction for the future application of the method proposed in this paper.

**Author Contributions:** Conceptualization, Z.Z. and J.W.; methodology, Z.Z.; software, Z.Z.; validation, Z.Z. and J.W.; writing—original draft preparation, Z.Z.; writing—review and editing, Z.Z., J.W., and M.Z.; visualization, Z.Z.; supervision, J.W., M.Z., and H.M.; project administration, J.W. and H.M.; funding acquisition, J.W. and H.M. All authors have read and agreed to the published version of the manuscript.

**Funding:** This research was funded by the National Natural Science Foundation of China (Grant No. 12202103), and the Guangdong Basic and Applied Basic Research Foundation (Grant No. 2021B1515120090 and Grant No. 2022A1515011324).

**Institutional Review Board Statement:** Not applicable.

**Informed Consent Statement:** Not applicable.

**Data Availability Statement:** The data generated during this study are currently private due to pending patent applications.

**Acknowledgments:** The authors would like to acknowledge the Institute of Science and Technology Innovation, Dongguan University of Technology for equipment support.

**Conflicts of Interest:** The authors declare no conflicts of interest.

## Abbreviations

The following abbreviations are used in this manuscript:

| | |
|---|---|
| UGW | Ultrasonic guided wave |
| NDT | Non-destructive testing |
| EMAT | Electromagnetic acoustic transducer |

## References


1. Rose, J.L. *Ultrasonic Guided Waves in Solid Media;* Cambridge University Press: Cambridge, UK, 2014. [CrossRef]
2. Dai, D.; He, Q. Structure damage localization with ultrasonic guided waves based on a time–frequency method. *Signal Process.* **2014**, *96*, 21–28. [CrossRef]
3. Luo, Z.; Zeng, L.; Lin, J. A novel time–frequency transform for broadband Lamb waves dispersion characteristics analysis. *Struct. Health Monit.* **2021**, *20*, 3056–3074. [CrossRef]
4. Li, G.; Zhang, J.; Cheng, J.; Wang, K.; Yang, D.; Yuan, Y. Multi-Order Mode Excitation and Separation of Ultrasonic Guided Waves in Rod Structures Using 2D-FFT. *Sensors* **2023**, *23*, 8483. [CrossRef] [PubMed]
5. Xu, Z.D.; Zhu, C.; Shao, L.W. Damage identification of pipeline based on ultrasonic guided wave and wavelet denoising. *J. Pipeline Syst. Eng. Pract.* **2021**, *12*, 04021051. [CrossRef]
6. Liu, Z.; Xu, K.; Li, D.; Ta, D.; Wang, W. Automatic mode extraction of ultrasonic guided waves using synchrosqueezed wavelet transform. *Ultrasonics* **2019**, *99*, 105948. [CrossRef] [PubMed]
7. Shi, Z.M.; Liu, L.; Peng, M.; Liu, C.; Tao, F.; Liu, C. Non-destructive testing of full-length bonded rock bolts based on HHT signal analysis. *J. Appl. Geophys.* **2018**, *151*, 47–65. [CrossRef]
8. Dushyanth, N.; Suma, M.; Latte, M.V. Detection and localization of damage using empirical mode decomposition and multilevel support vector machine. *Appl. Phys. A* **2016**, *122*, 1–9. [CrossRef]
9. Rizvi, S.H.M.; Abbas, M. An advanced Wigner–Ville time–frequency analysis of Lamb wave signals based upon an autoregressive model for efficient damage inspection. *Meas. Sci. Technol.* **2021**, *32*, 095601. [CrossRef]
10. Mardanshahi, A.; Nasir, V.; Kazemirad, S.; Shokrieh, M. Detection and classification of matrix cracking in laminated composites using guided wave propagation and artificial neural networks. *Compos. Struct.* **2020**, *246*, 112403. [CrossRef]
11. Liu, L.; Yuan, F. A linear mapping technique for dispersion removal of Lamb waves. *Struct. Health Monit.* **2010**, *9*, 75–86. [CrossRef]
12. Kannusamy, M.; Kapuria, S.; Sasmal, S. An efficient Lamb wave-based virtual refined time-reversal method for damage localization in plates using broadband measurements. *Ultrasonics* **2022**, *124*, 106767. [CrossRef] [PubMed]
13. Yu, W.; Zhao, C. Robust monitoring and fault isolation of nonlinear industrial processes using denoising autoencoder and elastic net. *IEEE Trans. Control. Syst. Technol.* **2019**, *28*, 1083–1091. [CrossRef]
14. Yu, W.; Zhao, C.; Huang, B.; Wu, M. A robust dissimilarity distribution analytics with Laplace distribution for incipient fault detection. *IEEE Trans. Ind. Electron.* **2023**, *70*, 12752–12761. [CrossRef]
15. Yu, W.; Zhao, C.; Huang, B. MoniNet with concurrent analytics of temporal and spatial information for fault detection in industrial processes. *IEEE Trans. Cybern.* **2021**, *52*, 8340–8351. [CrossRef] [PubMed]
16. Lorenz, E.N. Deterministic nonperiodic flow. *J. Atmos. Sci.* **1963**, *20*, 130–141.<0130:DNF>2.0.CO;2. [CrossRef]
17. Salas, A.H. Exact solution to duffing equation and the pendulum equation. *Appl. Math. Sci.* **2014**, *8*, 8781–8789. [CrossRef]
18. Cveticanin, L. On the Van der Pol oscillator: An overview. *Appl. Mech. Mater.* **2013**, *430*, 3–13. [CrossRef]
19. Kumara, S.; Ranjan, P.; Surana, A.; Narayanan, V. Decision making in logistics: A chaos theory based analysis. *CIRP Ann.* **2003**, *52*, 381–384. [CrossRef]
20. Yang, S.K.; Chen, C.L.; Yau, H.T. Control of chaos in Lorenz system. *Chaos Solitons Fractals* **2002**, *13*, 767–780. [CrossRef]
21. Jalilvand, A.; Fotoohabadi, H. The application of Duffing oscillator in weak signal detection. *ECTI Trans. Electr. Eng. Electron. Commun.* **2011**, *9*, 1–6. [CrossRef]
22. NOHARA, B.T.; ARIMOTO, A. Periodic Solutions of the Duffing Equation with the Square Wave External Force. *Theor. Appl. Mech. Japan* **2012**, *60*, 359–379. [CrossRef]
23. Srinivasan, K.; Thamilmaran, K.; Venkatesan, A. Effect of nonsinusoidal periodic forces in Duffing oscillator: Numerical and analog simulation studies. *Chaos Solitons Fractals* **2009**, *40*, 319–330. [CrossRef]
24. Cheng, M.; Zhang, W.; Zhang, C.; Ma, H. Application of chaotic information entropy for ultrasonic guided wave detection in pipe. *J. Civ. Struct. Health Monit.* **2024**, *14*, 29–39. [CrossRef]
25. Zhang, W.; Hao, H.; Wu, J.; Li, J.; Ma, H.; Li, C. Detection of minor damage in structures with guided wave signals and nonlinear oscillator. *Measurement* **2018**, *122*, 532–544. [CrossRef]
26. Liu, X.; Bo, L.; Liu, Y.; Zhao, Y.; Zhang, J.; Hu, N.; Fu, S.; Deng, M. Detection of micro-cracks using nonlinear lamb waves based on the Duffing-Holmes system. *J. Sound Vib.* **2017**, *405*, 175–186. [CrossRef]
27. Liu, X.; Bo, L.; Liu, Y.; Zhao, Y.; Zhang, J.; Deng, M.; Hu, N. Location identification of closed crack based on Duffing oscillator transient transition. *Mech. Syst. Signal Process.* **2018**, *100*, 384–397. [CrossRef]

28. Wu, J.; Wang, Y.; Zhang, W.; Nie, Z.; Lin, R.; Ma, H. Defect detection of pipes using Lyapunov dimension of Duffing oscillator based on ultrasonic guided waves. *Mech. Syst. Signal Process.* **2017**, *82*, 130–147. [CrossRef]
29. Ng, K.; Ghafoor, I.; Tse, P. A novel laser-based duffing oscillator system to identify weak ultrasonic guided wave signals related to rail defects. *Opt. Lasers Eng.* **2022**, *157*, 107111. [CrossRef]
30. Cheng, M.; Zhang, W.; Wu, J.; Ma, H. Analysis and application of weak guided wave signal detection based on double Duffing oscillators. *Mech. Syst. Signal Process.* **2023**, *191*, 110196. [CrossRef]
31. Karmakar, C.; Udhayakumar, R.; Palaniswami, M. Entropy profiling: A reduced—Parametric measure of kolmogorov—sinai entropy from short-term HRV signal. *Entropy* **2020**, *22*, 1396. [CrossRef]
32. Kiss, G.; Bakucz, P. Using Kolmogorov Entropy to Verify the Description Completeness of Traffic Dynamics of Highly Autonomous Driving. *Appl. Sci.* **2024**, *14*, 2261. [CrossRef]

sensors

MDPI

Article

# A Dual-Mode Surface Acoustic Wave Delay Line for the Detection of Ice on 64°-Rotated Y-Cut Lithium Niobate

**Philipp Schulmeyer [1], Manfred Weihnacht [2] and Hagen Schmidt [1,*]**

[1] Leibniz Institute for Solid State and Materials Research, 01069 Dresden, Germany; p.schulmeyer@ifw-dresden.de
[2] innoXacs, 01744 Dippoldiswalde, Germany; innoxacs@online.de
* Correspondence: h.schmidt@ifw-dresden.de

**Abstract:** Ice accumulation on infrastructure poses severe safety risks and economic losses, necessitating effective detection and monitoring solutions. This study introduces a novel approach employing surface acoustic wave (SAW) sensors, known for their small size, wireless operation, energy self-sufficiency, and retrofit capability. Utilizing a SAW dual-mode delay line device on a 64°-rotated Y-cut lithium niobate substrate, we demonstrate a solution for combined ice detection and temperature measurement. In addition to the shear-horizontal polarized leaky SAW, our findings reveal an electrically excitable Rayleigh-type wave in the X+90° direction on the same cut. Experimental results in a temperature chamber confirm capability for reliable differentiation between liquid water and ice loading and simultaneous temperature measurements. This research presents a promising advancement in addressing safety concerns and economic losses associated with ice accretion.

**Keywords:** surface acoustic wave (SAW); SH-LSAW; Rayleigh wave; ice; lithium niobate; piezoelectric; sensor; finite-element method (FEM); non-destructive testing (NDT)

## 1. Introduction

The accumulation of ice on infrastructure and machines compromises their performance and reliability, leading to safety concerns and economic losses. This applies to various industrial fields concerned with road, rail track and power line surveillance, wind power and aviation as well as with environmental and condition monitoring and instruments providing machine vision for unmanned vehicles. To address these issues, the development of robust and effective ice detection systems is crucial. Such systems not only allow the detection of iced surfaces but also serve as an essential prerequisite for de-icing procedures and provide indicators for the malfunction of anti- and de-icing systems. The purpose of this paper is to show a new non-destructive surface guided wave test approach in combined ice detection and temperature measurement by theoretical and experimental investigations using a surface acoustic wave (SAW) dual-mode delay line. Its significance is underlined by the technology's benefits of small sensor designs, passive and wireless operation, retrofit options and capabilities for mass production [1]. The findings also strongly relate to investigations of liquid characteristics like required for lab-on-a-chip, biomedical or chemical process applications [2].

The concept of a SAW-based sensor for ice detection was first introduced by [3] using a delay line device deploying a shear-horizontal polarized leaky surface acoustic wave (SH-LSAW) on 36°-rotated Y-cut $LiTaO_3$ (36LT) with propagation the along X direction. It was followed by the first devices deploying a Love wave in a $SiO_2$ waveguide layer on ST-X+90° quartz [4] and 31°-rotated Y-cut quartz [5], respectively. Today's aspirations in further developing SAW ice sensors still cover Love wave [6] as well as acoustic plate mode devices [7,8]. Moreover, dual-mode delay line devices are proposed by [9], consisting of a ZnO/quartz structure making use of the difference of Rayleigh-type wave's dominating sagittal polarization strongly coupling with liquids for actuation (mixing) and the Love wave's horizontal polarization weakly coupling with liquids for sensing purposes. Another dual-mode approach consisting of two separate acoustic propagation paths perpendicular to each other is proposed deploying Rayleigh-type waves and SH-SAW on 36LT for the determination of the acoustic properties of liquids [10].

In contrast, the work presented here demonstrates how Rayleigh-type wave and SH-LSAW utilized together in a dual-mode delay line (DM-DL) device on 64°-rotated Y-cut lithium niobate (64° YX $LiNbO_3$, 64LN) with an electromechanical coupling coefficient approximately twice as large compared to 36LT can be used for ice detection. While recent investigations deploying SH-modes on 64LN deal with protein detection [11], hydrogen sulfide [12] or hydrogen sensing [13] as well as with microfluidic actuators [14] the choice of this substrate here has a different motivation. Due to its strong electromechanical coupling, 64LN supports efficient excitation of different modes providing capabilities for a wireless sensor device able to reliably differentiate between liquid water and ice loading and simultaneously measuring temperature.

## 2. Modal Analysis

64°-rotated Y-cut $LiNbO_3$ is well known for its shear-horizontal polarized leaky SAW (SH-LSAW) with propagation along the crystallographic *X*-axis. A modal analysis is performed to find electrically excitable acoustic surface modes exhibiting a dominant polarization within the sagittal plane spanned by the propagation direction and the surface-normal direction. These can be either a Rayleigh-type wave (RW) with pure sagittal polarization or a generalized SAW (GSAW) with a significant sagittal polarization accompanied by a less dominant horizontal component.

Finite-element (FEM) software COMSOL Multiphysics 6.2 is used to perform a modal analysis to search for configurations supporting different modes [15] and, more specifically, where both SH-LSAW and sagittal-polarized wave can be electromechanically excited on the same cut and which directions are beneficial for the excitation of sagittal-polarized wave and of SH-LSAW, respectively. The 3D model (Figure 1) consists of a lithium niobate block of 1 μm in length with 20 mesh elements, exactly one element of 0.3 μm in width and 4 μm in height with 80 elements. The 80 elements have an exponential grow, rate where the lowest element is 5 fold the size of the element at the top. This is set to ensure a finer mesh in the surface region. Surface domains facing length and width directions have periodic boundary conditions. The material constants for lithium niobate are chosen according to [16]; see Table A1. Additionally, a perfectly matched layer (PML) domain of 1 μm height and ten elements in global z direction is used at the bottom to suppress ground reflection of bulk components at the lower boundary with fixed constraint. Crystallographic *X*-axis of lithium niobate corresponds to global x-coordinate and the 64°-rotated Y-cut corresponds to the x-y plane of the model.

An eigenfrequency study with a parametric angular sweep about the global *z*-axis reveals for a rotation of 0°, the existence of an SH eigenmode (Figure 1b) with a complex eigenfrequency of $f = Re(f) + j{\cdot}Im(f) = (4695.96 + 4.5361j)$ MHz, where the very small imaginary component indicates the related acoustic damping caused by leakage into bulk acoustic waves (BAW). This eigenmode corresponds to the well-known shear-horizontal polarized leaky wave (SH-LSAW) propagating along the *X*-axis of 64LN. Considering as wavelength $\lambda$ the spatial periodicity of 1 μm along the length coordinate as defined by the periodic boundary condition and the real part of complex eigenfrequency $f$, Equation (1) yields the real part of the complex phase velocity $v$ of the corresponding SH-LSAW mode for free (electrically open) surface, which results in 4695.96 m/s (Figure 2a). This value differs by only 0.084% compared to [14].

$$v = \frac{\lambda}{Re(f)} \tag{1}$$

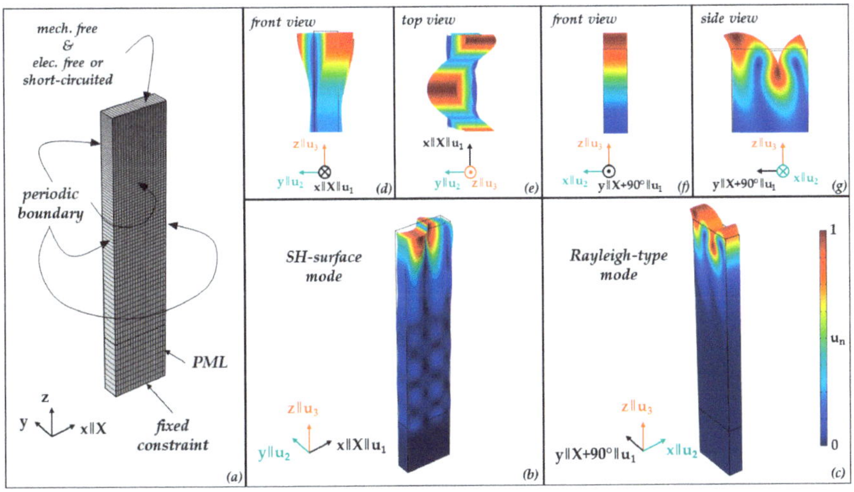


**Figure 1.** 3D FEM eigenfrequency study: (**a**) model of 64LN block with mesh and boundary domain definitions, (**b**) shape of leaky SH-surface mode and (**c**) shape of Rayleigh-type mode. (**d**) Front and (**e**) top view of (**b**); (**f**) front and (**g**) side view of (**c**); color bar corresponds to normalized displacement magnitude $u_n = \frac{u_{res}}{u_{res,max}}$ with $u_{res} = \sqrt{u_1^2 + u_2^2 + u_3^2}$ for all color plots.

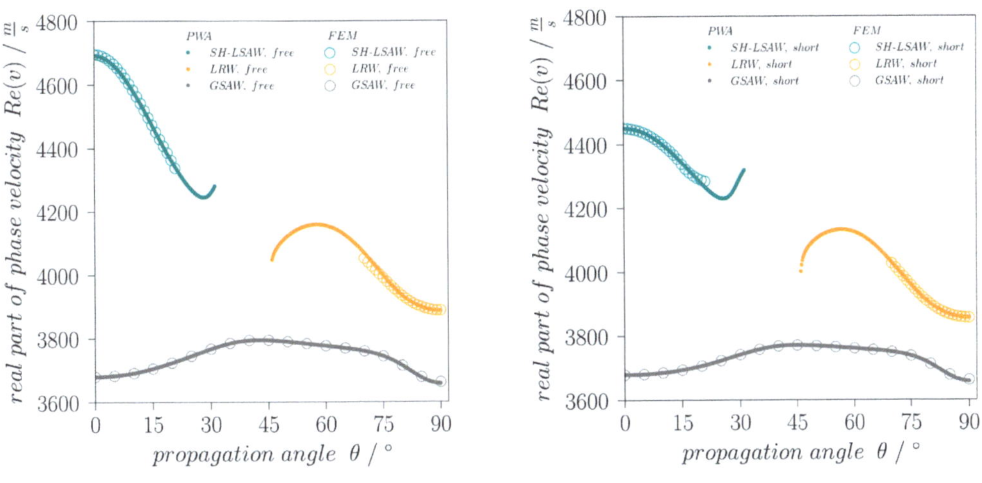


**Figure 2.** Angular dispersion of phase velocities of SH-LSAW, leaky Rayleigh-type (LRW) wave and generalized SAW (GSAW) branches with (**a**) free and (**b**) short-circuited surface.

In addition to the eigenfrequency study, a partial-wave analysis (PWA; see [17] for details) is performed with a custom made software package based on [18] proving the suitability of the FEM model for further investigation.

In case of the finite-element analysis at a rotation angle of 90°, an eigenmode with $f$ = (3887.71 + 0$j$) MHz indicates a corresponding SAW mainly polarized within the sagittal plane, i.e., a Rayleigh-type wave (Figure 1c). As this mode is non-leaky only for this special direction but leaky for all other angles, it will be referred to in the following as leaky Rayleigh-type wave (LRW) branch. Additionally, there exists a third, generalized surface wave mode (GSAW) branch with a polarization significantly changing its character over propagation direction and with a phase velocity always lower than SH-LSAW and LRW (Figure 2). To further investigate the possibility to excite the identified SAW modes electrically, the electromechanical coupling coefficient $K^2$ that is formally defined in terms of the relative phase velocity difference for electrically open and short-circuited surface by Equation (2) [19]

$$K^2 = \frac{2\left(v_{free} - v_{short}\right)}{v_{free}} \tag{2}$$

is calculated for the modes as a function of the propagation angle. For this, eigenfrequency analysis and PWA are performed for the whole angular range under the same conditions like before but with an electrically short-circuited substrate surface. $K^2$ for the leaky SH-SAW at 0° has been calculated to 10.46% which is very close to the value of 11% given in [19]. The leaky Rayleigh-type wave has a $K^2$ of 1.6% at 90° propagation angle whereas the coupling of the GSAW mode for the angles 0° and 90° is zero (Figure 3a). This means it is practically non-piezoelectric for these directions and can therefore not be excited by electrical fields but by scattering effects [14] making it useless for devices. An interesting effect is also seen in case of the SH-LSAW for angles with high attenuation, where $K^2$ even reaches negative values when calculated formally using Equation (2) [20]. Note for comparison that commonly used STX-quartz has a $K^2$ of only 0.12% [19]. Another important feature for the technical deployment of waves is their attenuation, what is especially important for all leaky modes. Figure 3b shows the attenuation coefficient $\alpha_{dB/\lambda}$ calculated from complex phase velocity components using Equation (3).

$$\alpha_{dB/\lambda} = 10 \cdot lg(e) \cdot 4\pi \frac{Im(v)}{Re(v)} . \tag{3}$$

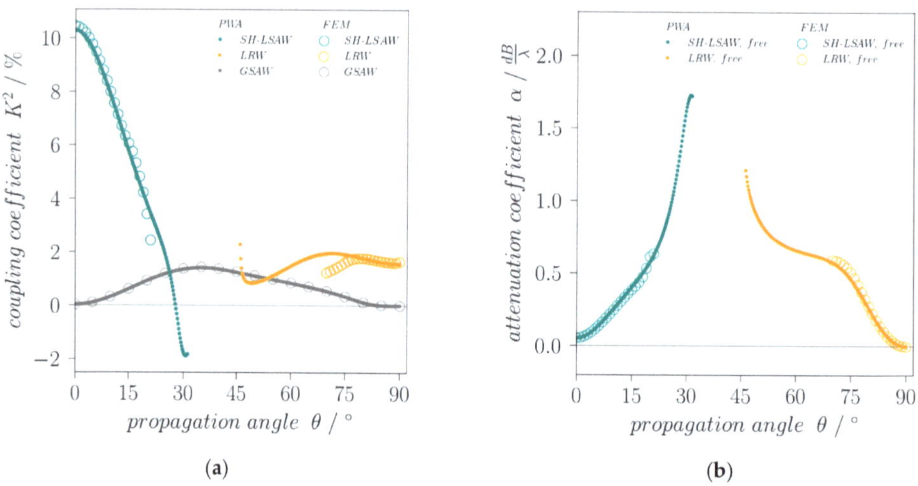


**Figure 3.** Angular dispersion of (**a**) electromechanical coupling coefficient $K^2$ for SH-LSAW, leaky Rayleigh-type wave (LRW) and generalized SAW (GSAW) branches and of (**b**) attenuation coefficient $\alpha$ for SH-LSAW and LRW for free surface.

For the SH-LSAW propagation along $X$-axis the attenuation has an acceptable minimum of 0.05 $dB/\lambda$ whereas for the leaky Rayleigh-type wave (LRW) it is zero for 90° propagation angle $\theta$.

### 3. Device Characterization

Based on simulation results, dual-mode delay line devices are prepared on 64°-rotated Y-cut lithium niobate with Euler angles (0°, −26°, 0°) for the SH-SAW and (0°, −26°, 90°)

for the attenuation-free Rayleigh wave (RW) as an exceptional case of the LRW branch. The single-side polished piezoelectric substrate has a thickness of 500 μm with the electrode metallization with an overall thickness of 300 nm (295 nm Al on 5 nm Ti) on top. The interdigital transducer (IDT) structures are patterned by mask-less photolithography (MLA 100, Heidelberg Instruments, Heidelberg, Germany) using conventional lift-off technique after e-beam evaporation. All four solid finger IDTs (4 ports, 4P) are of identical design, consisting of 33 finger pairs with a finger width and gap of $\lambda/4$ = 37.5 μm ($\lambda_{SAW}$ = 150 μm) and an aperture of 2 mm. Both delay lines have a length of 50$\lambda$ face-to-face. To reduce coherent reflections at the device edges these are cut at a 30-degree angle to the transducer orientation and covered with highly viscous photoresist. Generally, the radiation of bulk acoustic waves (BAW) into the depth of the substrate as a result of parasitic SH-BAW excitation in the case of Rayleigh wave IDTs as well as the energy leakage of the SH-LSAW can lead to undesired acoustic interference at the surface due to back-reflection from the substrate rear side. To avoid this, a waffle-weave pattern [14] has been cut into the backside of the substrate to diffuse BAW scattering and thereby minimizing negative effects on the sensor signal.

All IDTs are electrically pair-wise characterized in respect to their orientation by S-parameter measurements using a vector network analyzer (VNA, E5070B, Keysight Technologies, Santa Rosa, CA, USA). The SAW dual-mode delay line device (DM-DL) is mounted on a custom-made sample holder and electrically connected via gold-plated spring-loaded pins soldered to a printed circuit board (PCB) with conductor-backed coplanar waveguides (CBCPW) of 50 Ω characteristic impedance and SMA connectors. A short-open-load-through (SOLT) calibration is performed at the SMA connector level and the electrical delay caused by the CBCPWs is compensated. For both delay lines, all scattering parameters are measured at the corresponding ports 1, 2 (SH-LSAW) and 3, 4 (RW), i.e., reflection in terms of $S_{11}$, $S_{22}$, $S_{33}$, $S_{44}$ and transmission $S_{21}$, $S_{12}$, $S_{43}$, $S_{34}$, with all determined at room temperature for unloaded substrate surface. Due to the DM-DL structure, the relations $S_{xx} = S_{yy}$ and $S_{yx} = S_{xy}$ are valid for the corresponding ports and are experimentally confirmed. The $|S_{11,SH\text{-}LSAW}|$ and $|S_{33,RW}|$ minimum frequencies are 29.72 MHz and 25.69 MHz, respectively. While the SH-LSAW IDT reflection coefficient shows a minimum of 0.31, the $|S_{33,RW}|$ minimum is 0.72 (Figure 4). This is because all IDTs are of identical design and no specific impedance matching was introduced for the Rayleigh wave direction.

Mechanical treatment is applied to the DM-DL devices in order to reduce influences of parasitic BAW contributions as well as distortions caused by edge-reflected surface wave components. Comprising a waffle-weave pattern cut into the chip backside and applying photoresist at the edge-facing end of all IDTs this treatment proves to be effective as the curves of reflection coefficients $|S_{xx}|$ and transmission coefficients $|S_{yx}|$ are significantly less distorted (Figures 4 and 5). Moreover, the destructive acoustic interference in the SH-LSAW transmission curve at approximately 30 MHz is also prevented (Figure 5a,b). Further improvement of all S-parameter curves is achieved by signal processing that includes Fourier transformation, application of appropriate gating in the time domain to suppress non-mode-specific contributions and back-transformation into the frequency domain.

The wavefield of an DM-DL device is experimentally investigated by Laser-Doppler-Vibrometry (LDV, UHF-120, Polytec, Waldbronn, Germany) to validate results of the mode simulations. Figure 6a shows measured displacement amplitudes in surface-normal direction (i.e., parallel to global $z$-axis) for a Rayleigh mode pattern along $y\|X$+90° direction at 25.69 MHz with both IDTs activated to check their correct operation. The maximum displacement of the resultant standing wave pattern reaches $u_{3,max}$ = 120 pm. Figure 6b shows displacement amplitudes for the SH-LSAW at 29.72 MHz under the same operation conditions. As expected from modal analysis, the surface-normal displacements ($u_{3,max}$ = 37 pm) are much smaller here compared to the Rayleigh mode. Nevertheless, the characteristic standing wave pattern can be clearly identified within the IDT aperture.

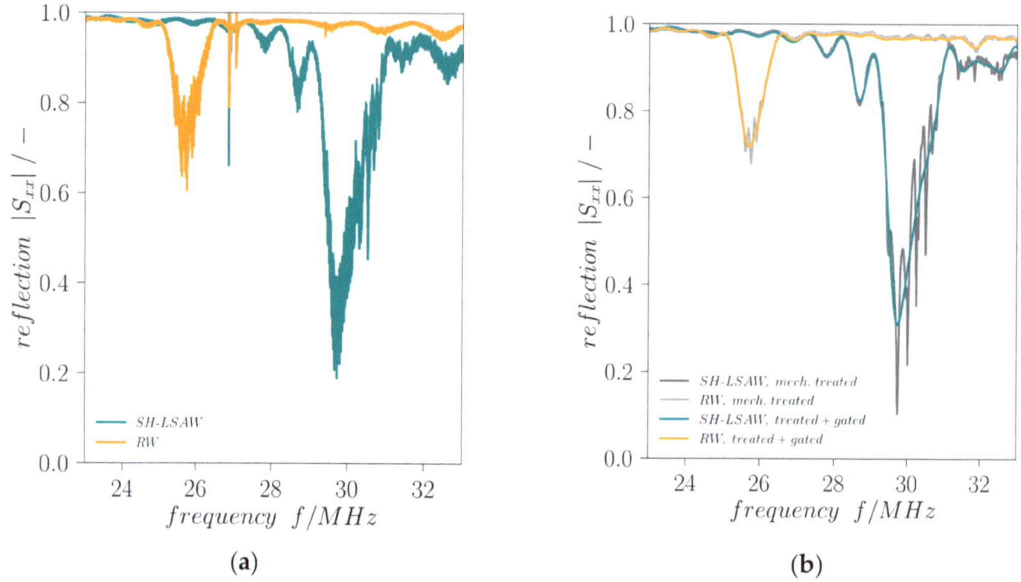


**Figure 4.** Scattering parameter $|S_{xx}|$ vs. frequency (**a**) without and (**b**) with mechanical treatment and gating.

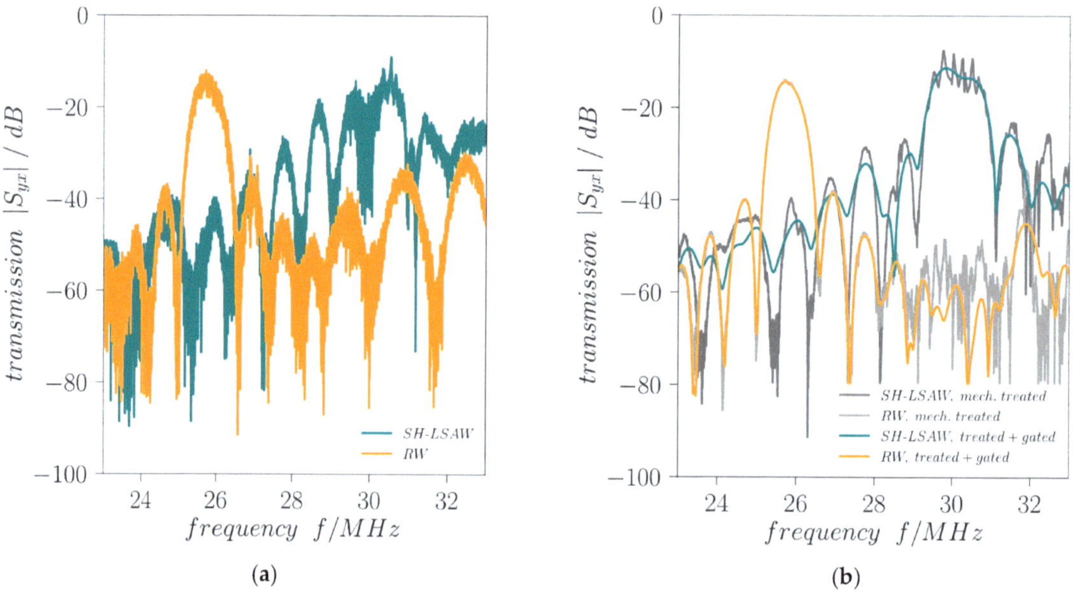


**Figure 5.** Scattering parameter $|S_{yx}|$ vs. frequency (**a**) without and (**b**) with mechanical treatment and gating.

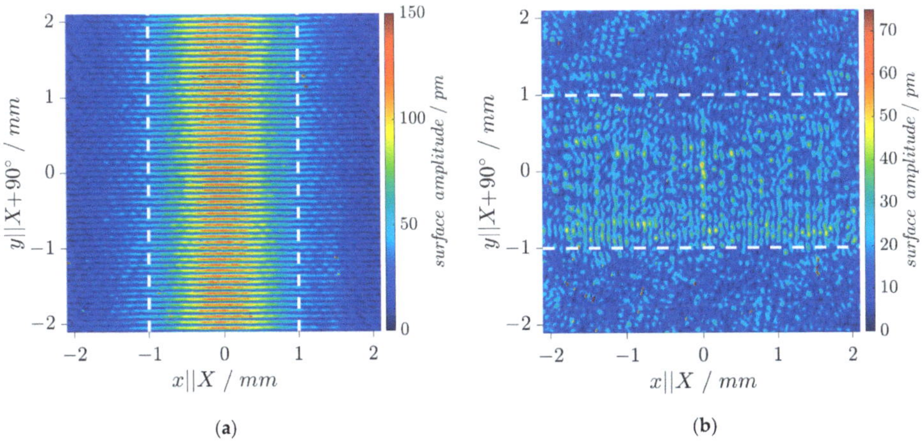


**Figure 6.** LDV wavefield measurements: Surface displacement amplitudes in surface-normal direction for (**a**) Rayleigh-type wave at 25.69 MHz and (**b**) SH-LSAW at 29.72 MHz (device under test without mechanical treatment; IDT aperture shown by white dashed lines).

## 4. Sensing Experiments

The experimental test setup to characterize the dual-mode delay line behavior with liquid water and ice loading consists of a temperature chamber (VT 4002, Vötsch, Balingen, Germany) and VNA. Temperature behavior of the DM-DL device within the range of −30 °C to 50 °C is measured in the state without surface loading, resulting in temperature coefficients of frequency (TCF) for both modes. A type-K thermocouple (TC) inserted underneath the custom-made sample holder with its tip directly touching the bottom-side of the piezoelectric chip measures the substrate temperature. Substrate and aluminum sample holder are thermally decoupled by a PEEK plate (Figure 7).

The frequency shift of the minimum of reflection $|S_{xx}|$ provides information about the actual device temperature which can be deployed as additional indicator for the presence of icing conditions. Figure 8 shows the according frequency shifts of time-gated reflection curves for SH-LSAW and Rayleigh-type waves (path directions X and X+90°, respectively). There is a linear temperature behavior obtained for both modes in the investigated range of −30 °C to 50 °C yielding temperature coefficients (TCF) calculated from the shift of minimum $|S_{xx}|$ of $TCF_X = -61.62$ ppm/°C and $TCF_{X+90°} = -82.11$ ppm/°C w.r.t 20 °C as the reference.

To investigate the dual-mode device response on ice loading the transmission behavior of both propagation paths is measured. Figure 9 shows the temperature of the DM-DL device measured underneath the piezoelectric substrate during the experiment (grey curve) as well as the regarding time-gated transmission coefficients $|S_{yx}(f_{op})|$ at operating frequency $f_{op}$ defined by the center frequency between the zeros next to the transmission main lobe. While the device is still in dry conditions, without any surface load the temperature chamber is cooled down to −30 °C until the piezoelectric substrate temperature stabilizes as can be concluded from stable S-parameters. The cooling takes approximately 40 min. After 120 min, the chamber door is opened to apply a drop of tap water (non-deionized, room temperature) with a volume of 40 μL by pipette (Eppendorf Research, Hamburg, Germany) onto the sensitive surface of the sensor device including both wave propagation paths (Figure 7b,c). Considered the device geometry, a 40 μL drop ensures complete wetting of the whole sensitive area. The chamber door is closed immediately after the deposition. Due to the difference between water and the DM-DL substrate temperature, a rise of 7.0 °C quickly reverting to approximately −30 °C is measured, and is indicated in Figure 9 by a small peak. During the droplet application, comparatively warm and humid air streams into the chamber, leading to a deposition of a thin and weakly adherent ice layer on all parts with a sub-freezing temperature including the SAW device. Here, a small decrease in transmission for the Rayleigh wave can be seen as the thin ice layer covers the full device including IDTs and attenuates the sagittal polarized Rayleigh wave there. The shear-horizontal polarized wave is not influenced due to the very low adhesion of the ice to the sensitive surface. As the air inside the chamber cools down again quickly and becomes cooler than the device setup the thin ice layer vanishes due to sublimation. The applied liquid water load freezes in an approximate time frame of less than 5 min, resulting in a cone-shaped ice tip (Figure 7d). Underlaying this procedure is the assumption that the surface area covered by ice is equal to that formerly covered by the liquid water. Moreover, thermal expansion of the water as well as changes in the substrate surface energy during the freezing process are neglected.

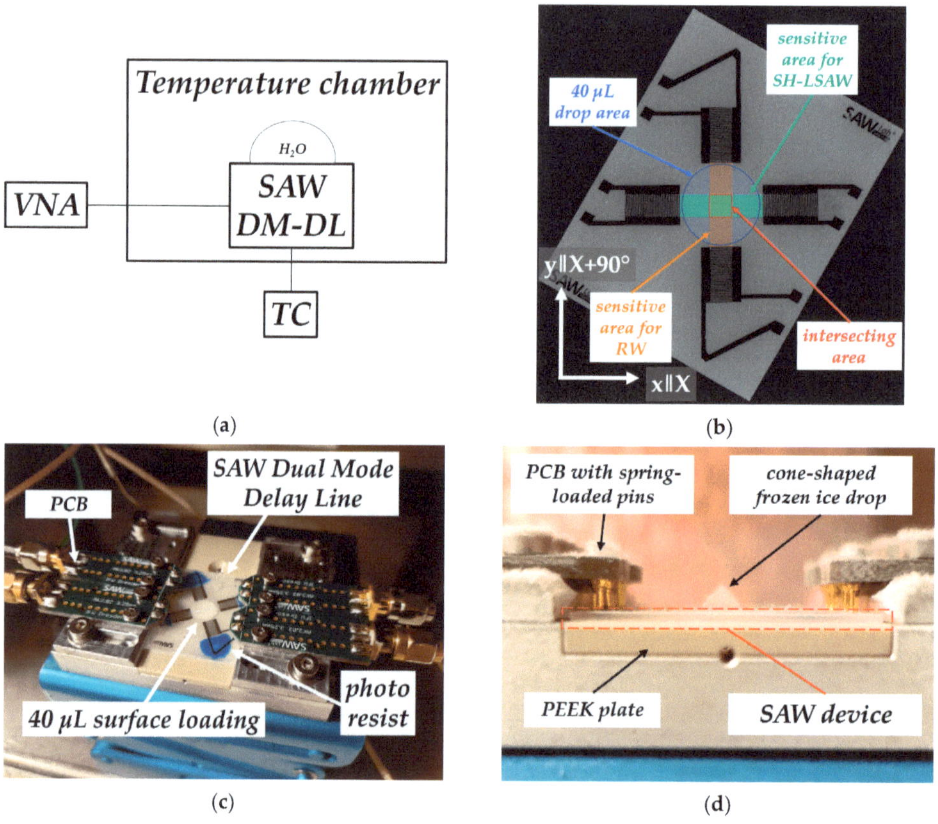


**Figure 7.** (**a**) Scheme of experimental setup; (**b**) dual-mode delay line device with acoustic area and loaded area; (**c**) device setup inside temperature chamber; (**d**) frozen drop on device.

A waiting time of 20 min after water drop application is held to ensure stable S-parameter until the heating to room temperature is initialized. The increase in air temperature inside the chamber leads to dew formation on all surfaces which include the DM-DL device transducers, leading to additional attenuation while the water droplet is still mostly frozen. As soon as the device temperature reaches 15 °C, the dew vaporized again clearing the IDTs from any liquid. After the heating period to room temperature (23 °C) of approximately 30 min, the device sensitive area is loaded with melted water only. A final S-parameter measurement is performed after completely removing the liquid water to ensure same transmission values before and after the experiment.

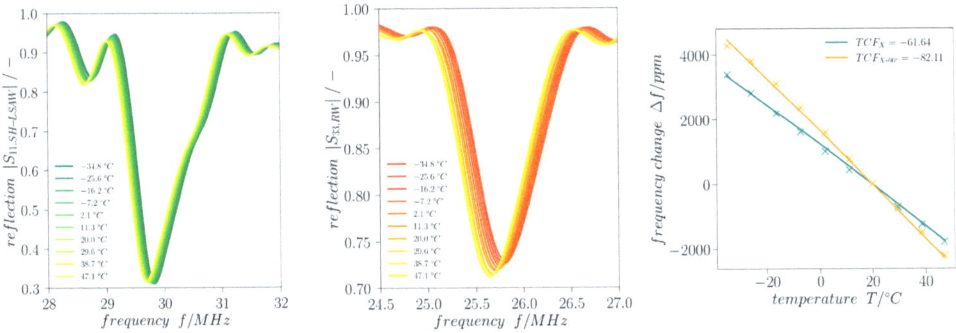


(a) (b) (c)

**Figure 8.** Temperature dependence of time-gated (**a**) $|S_{11,SH\text{-}LSAW}|$ and (**b**) $|S_{33,RW}|$ within the range −30 °C to 50 °C. (**c**) Derived temperature coefficients of frequency (TCF) for X and X+90° direction.

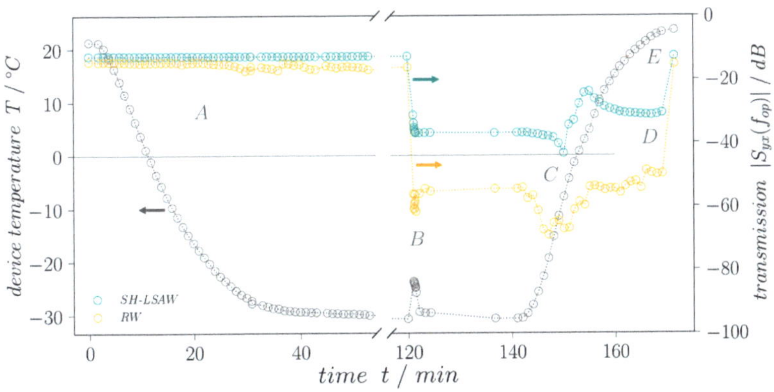


**Figure 9.** DM-DL device temperature (grey) and time-gated transmission behavior over time for SH-LSAW (green) and RW (yellow) propagation paths. (**A**): cooling in dry conditions, (**B**): water drop application and freezing, (**C**): heating with dew formation and melting, (**D**): water drop completely liquid, and (**E**): dry surface. Dotted lines are for eye guidance only.

Figure 10 illustrates the measured transmission curves at the beginning ($t$ = 0) of the experiment without surface load (dry), at approximately 137 min with ice loading and at end just before drying the surface with liquid water loading. Both acoustic modes propagating in perpendicular directions on the DM-DL device experience a different degree of attenuation in response to surface loading by liquid water which leads to a decrease in the corresponding transmission curves. Moreover, for both modes freezing of the liquid leads to a further decrease in the transmission. Important to mention, an ideal SAW attenuation is indicated by a nearly constant decrease in transmission over the whole frequency range whereas here also (changing) dielectric properties of the water drop loading will influence the characteristics [21]. As expected, the shear-horizontal polarized wave shows a lower attenuation compared to the Rayleigh-type wave. This is due to the neglectable out-of-plane displacement of an ideal SH-LSAW at the substrate surface, where the Rayleigh wave has its maximum surface-normal particle displacement, leading to strong acoustic leakage by radiation of longitudinal BAW into the surface load for the latter case. Moreover, an SH-LSAW shows only weak shear coupling to liquid loads like water with a low dynamic viscosity of approximately 1 mPas at room temperature [22]. Nevertheless, in the device under test the SH-LSAW experiences acoustic leakage as the real wavefield shows also minor particle displacement in surface-normal direction as revealed by LDV measurement

(Figure 6b). This non-ideal behavior leads to a mechanical coupling to the liquid water load, which results in the decrease in $|S_{21,SH\text{-}LSAW}|$. With increasing viscosity during the freezing process also the applied shear forces lead to shear stress at the interface between load and substrate surface what results in a further decrease in the transmission coefficient. The sagittal polarized Rayleigh-type wave with its major particle displacement in surface-normal direction witnesses a high attenuation $|S_{43,RW}|$ under liquid water loading right away which even increases during the freezing process. A quantitative view on the load-dependent increase in attenuation for both modes provides Figure 11. At the DM-DL operating frequencies defined by the center frequency between the zeros next to the transmission main lobe a dramatic increase in almost 36 dB in Rayleigh-mode attenuation is present for water loading, further increasing by 5.5 dB when the water freezes. In contrast, SH-LSAW experiences generally a much lower increase in attenuation due to surface loading with differences of ~18 dB for liquid water and ~23 dB when it freezes.

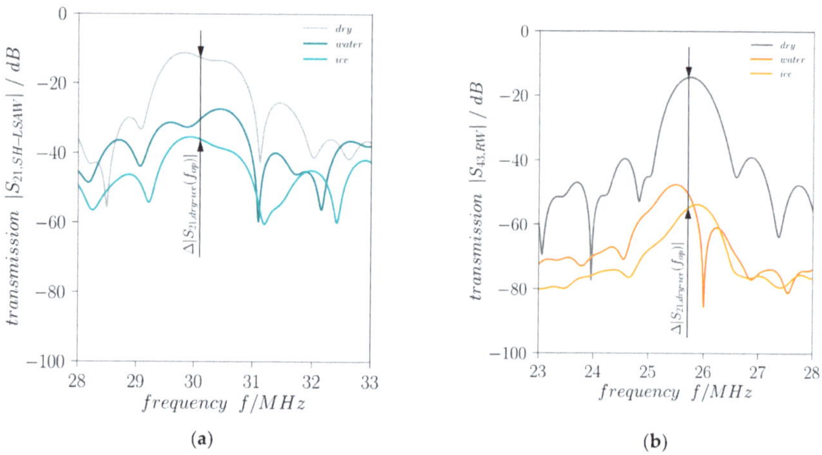


**Figure 10.** Effect of surface loading on DM-DL transmission characteristics (**a**) $|S_{21,SH\text{-}LSAW}|$ and (**b**) $|S_{43,RW}|$ (dry: no surface load, water: 40 µL water with complete wetting (both at 23 °C); ice: 40 µL ice (−30 °C); all curves are time-gated).

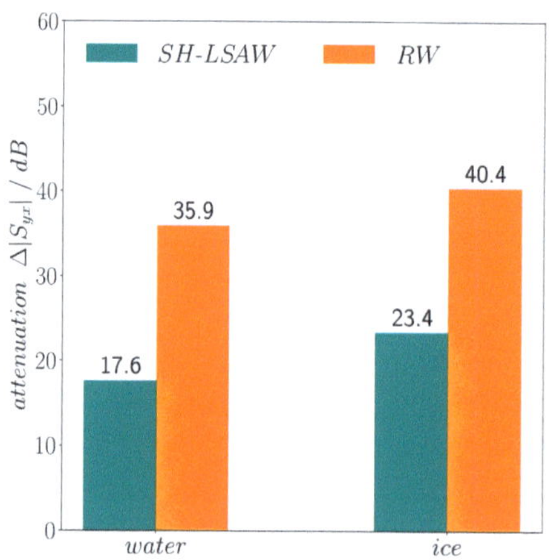


**Figure 11.** Increase in SAW attenuation $\Delta|S_{21,mode}|$ dependent on mode and surface loading for the DM-DL operating frequencies indicated in Figure 10 w.r.t. unloaded surface as reference.

## 5. Conclusions

The discussed results show that it is possible to electrically excite Rayleigh-type waves on 64°-rotated Y-cut lithium niobate, additionally to the commonly used shear-horizontal polarized wave, at a propagation angle of 90° relative to the X direction. These findings allow the fabrication of a 4-port dual-mode delay line device. A waffle-weave patterned cut on the device backside to minimize leaking bulk wave interferences as well as the application of photoresist as an acoustic absorber removed noise from the measured S-parameters and eased the evaluations. The 4-port arrangement allowed the determination of reflection coefficients as well as transmission coefficients. While the IDT reflection coefficient $|S_{xx}|$ remain unaffected by surface loading of the sensitive area within the propagation path), it delivers important information on the real device temperature. Additionally, the different change in transmission characteristics $|S_{yx}|$ for both acoustic modes in response on surface loading allow the precise differentiation of liquid water and ice load. In combining both effects, the presented DM-DL device demonstrates its potential for a sensor to simultaneously determine surface load condition and temperature for reliable ice detection on industrial surfaces.

**Author Contributions:** Conceptualization, P.S. and H.S.; methodology, P.S. and H.S.; investigation, all authors; writing—original draft preparation, all authors; writing—review and editing, all authors. All authors have read and agreed to the published version of the manuscript.

**Funding:** Supported by funds from Deutsche Bundesstiftung Umwelt (DBU grant 38120/01-24/0).

**Institutional Review Board Statement:** Not applicable.

**Informed Consent Statement:** Not applicable.

**Data Availability Statement:** Data are contained within the article.

**Acknowledgments:** P.S. and H.S. thank Deutsche Bundesstiftung Umwelt (DBU) for financial support, A. Büst and D. Bieberstein for device manufacturing, SAWLab and MST Lab at IFW Dresden as well as cp.max Rotortechnik for technical advice.

**Conflicts of Interest:** The authors declare no conflict of interest. Author M.W. was with company innoXacs. The remaining authors declare that the research was conducted in the absence of any commercial or financial relationships that could be construed as a potential conflict of interest. The funders had no role in the design of the study; in the collection, analyses, or interpretation of data; in the writing of the manuscript; or in the decision to publish the results.

## Appendix A

**Table A1.** Material constants for $LiNbO_3$ at room temperature (T = 26 °C) from [16].

| Elastic Constants in $10^{10}$ N/m$^2$ | |
|---|---|
| $c^E_{11}$ | 19.839 ± 0.089 |
| $c^E_{12}$ | 5.472 ± 0.097 |
| $c^E_{13}$ | 6.513 ± 0.193 |
| $c^E_{14}$ | 0.788 ± 0.004 |
| $c^E_{33}$ | 22.790 ± 0.324 |
| $c^E_{44}$ | 5.965 ± 0.008 |
| **Piezoelectric constants in C/m$^2$** | |
| $e_{15}$ | 3.69 ± 0.06 |
| $e_{22}$ | 2.42 ± 0.04 |
| $e_{31}$ | 0.30 ± 0.08 |
| $e_{33}$ | 1.77 ± 0.12 |
| **Dielectric constants in $\varepsilon_0$** | |
| $\varepsilon^S_{11}$ | 45.6 ± 1.5 |
| $\varepsilon^S_{33}$ | 26.3 ± 1.6 |
| **Mass density in kg/m$^3$** | |
| ρ | 4628 |

## References


1. Bruckner, G.; Bardong, J.; Binder, A.; Nicolay, P. SAW Delay Lines as Wireless Sensors for Industrial Applications. In Proceedings of the 8th Conference on Smart Structures and Materials, SMART 2017, Madrid, Spain, 5–8 June 2017.
2. Huang, Y.; Das, P.K.; Bhethanabotla, V.R. Surface acoustic waves in biosensing applications. *Sens. Actuators Rep.* **2021**, *3*, 100041. [CrossRef]
3. Varadan, V.V.; Varadan, V.K.; Bao, X.-Q. IDT sensors for detection of ice on rotorcraft. *Proc. SPIE* **1998**, *3328*, 49–58. [CrossRef]
4. Gangadaran, S.; Varadan, V.V.; Varadan, V.K.; Jose, K.A.; Atashbar, M.Z. Love-wave-based ice sensor. *Proc. Smart Struct. Mater.* **1999**, *3673*, 354279. [CrossRef]
5. Vellekoop, M.J.; Jakoby, B.H.; Bastemeijer, J. A Love-wave ice detector. *Proc. IEEE Ultrason. Symp.* **1999**, *1*, 453–456. [CrossRef]
6. Yin, Y.; Wang, W.; Jia, Y.; Liang, Y. Development of a novel wireless and passive Love wave based ice sensor. In Proceedings of the IEEE International Ultrasonics Symposium (IUS), Glasgow, UK, 6–9 October 2019; pp. 2553–2556.
7. Anisimkin, V.; Kolesov, V.; Kuznetsova, A.; Shamsutdinova, E.; Kuznetsova, I. An Analysis of the Water-to-Ice Phase Transition Using Acoustic Plate Waves. *Sensors* **2021**, *21*, 919. [CrossRef] [PubMed]
8. Karimzadeh, A.; Weissker, U.; del Moral, J.; Winkler, A.; Borrás, A.; González-Elipe, A.R.; Jacob, S. Harnessing a Vibroacoustic Mode for Enabling Smart Functions on Surface Acoustic Wave Devices—Application to Icing Monitoring and Deicing. *Adv. Mater. Technol.* **2024**, 2301749. [CrossRef]
9. Nicolay, P.; Sarry, F.; Le Brizoual, L.; Bouvot, L.; Elmazria, O.; Beyssen, D. P4K-4 Novel Layered SAW Structure for Droplet Multidirectional Actuating and Sensing. In Proceedings of the IEEE Ultrasonics Symposium, New York, NY, USA, 28–31 October 2007.
10. Nomura, T.; Uchiyama, U.; Saitoh, A.; Furukawa, S. Surface acoustic wave liquid sensor using dual mode delay line. In Proceedings of the Eleventh IEEE International Symposium on Applications of Ferroelectrics, Montreux, Switzerland, 24–27 August 1998; pp. 289–292.
11. Jeng, M.-J.; Li, Y.-C.; Sharma, M.; Chen, C.-W.; Tsai, C.-L.; Lin, Y.-H.; Huang, S.-F.; Chang, L.-B.; Lai, C.-S. A Surface Acoustic Wave Sensor with a Microfluidic Channel for Detecting C-Reactive Protein. *Chemosensors* **2021**, *9*, 106. [CrossRef]
12. Asad, M.; Sheikhi, M.H. Surface acoustic wave based $H_2S$ gas sensors incorporating sensitive layers of single wall carbon nanotubes decorated with Cu nanoparticles. *Sens. Actuators B Chem.* **2014**, *198*, 134–141. [CrossRef]
13. Atashbar, M.Z.; Sadek, A.Z.; Wlodarski, W.; Sriram, S.; Bhaskaran, M.; Cheng, C.J.; Kaner, R.B.; Kalantar-zadeh, K. Layered SAW gas sensor based on CSA synthesized polyaniline nanofiber on AlN on 64° YX $LiNbO_3$ for $H_2$ sensing. *Sens. Actuators B Chem.* **2009**, *138*, 85–89. [CrossRef]
14. Weser, R.; Darinskii, A.N.; Schmidt, H. Polarization manipulation of surface acoustic waves by metallization patterns on a piezoelectric substrate. *Appl. Phys. Lett.* **2020**, *117*, 143502. [CrossRef]
15. Darinskii, A.N.; Weihnacht, M. Acoustic waves in bounded anisotropic media: Theorems, estimations, and computations. *IEEE Trans. Ultrason. Ferroelectr. Freq. Control* **2005**, *52*, 702–801. [CrossRef]
16. Kovacs, G.; Anhorn, M.; Engan, H.E.; Visintini, G.; Ruppel, C.C.W. Improved material constants for $LiNbO_3$ and $LiTaO_3$. In Proceedings of the IEEE Symposium on Ultrasonics, Honolulu, HI, USA, 4–7 December 1990; pp. 435–438.
17. Farnell, G.W. Types and properties of surface waves. In *Acoustic Surface Waves*; Topics in Applied Physics; Springer: Berlin/Heidelberg, Germany; New York, NY, USA, 1978; Volume 24.
18. Wobst, R. The generalized eigenvalue problem and acoustic surface wave computations. *Computing* **1987**, *39*, 57–69. [CrossRef]
19. Morgan, D.; Paige, E.G.S. *Surface Acoustic Wave Filters*, 2nd ed.; Academic Press: Cambridge, MA, USA, 2007.
20. Mozhaev, V.G.; Weihnacht, M. Incredible negative values of effective electromechanical coupling coefficient for surface acoustic waves in piezoelectrics. *Ultrasonics* **2000**, *37*, 687–691. [CrossRef] [PubMed]
21. Leidl, A.; Oberlack, I.; Schaber, U.; Mader, B.; Drost, S. Surface acoustic wave devices and applications in liquid sensing. *Smart Mater. Struct.* **1997**, *6*, 680–688. [CrossRef]
22. Haynes, W.M.; Lide, D.R.; Bruno, T.J. *CRC Handbook of Chemistry and Physics*, 95th ed.; CRC Press: Boca Raton, FL, USA, 2014.

sensors

MDPI

*Article*

# Optimal Design of Sparse Matrix Phased Array Using Simulated Annealing for Volumetric Ultrasonic Imaging with Total Focusing Method

**Dmitry Olegovich Dolmatov * and Vadim Yurevich Zhvyrblya**

School of Non-Destructive Testing, National Research Tomsk Polytechnic University, 30 Lenin Avenue, 634050 Tomsk, Russia
* Correspondence: dolmatovdo@tpu.ru

**Abstract:** The total focusing method (TFM) is often considered to be the 'gold standard' for ultrasonic imaging in the field of nondestructive testing. The use of matrix phased arrays as probes allows for high-resolution volumetric TFM imaging. Conventional TFM imaging involves the use of full matrix capture (FMC) for ultrasonic signals acquisition, but in the case of a matrix phased array, this approach is associated with a huge volume of data to be acquired and processed. This severely limits the frame rate of volumetric imaging with 2D probes and necessitates the use of high-end equipment. Thus, the aim of this research was to develop a novel design method for determining the optimal sparse 2D probe configuration for specific conditions of ultrasonic imaging. The developed approach is based on simulated annealing and involves implementing the solution of the sparse matrix phased array layout optimization problem. In order to implement simulated annealing for the aforementioned task, its parameters were set, the acceptance function was introduced, and the approaches were proposed to compute beam directivity diagrams of sparse matrix phased arrays in TFM imaging. Experimental studies have shown that the proposed approach provides high-quality volumetric imaging with a decrease in data volume of up to 84% compared to that obtained using the FMC data acquisition method.

**Keywords:** ultrasonic nondestructive testing; ultrasonic imaging; total focusing method; matrix phased arrays; sparse phased arrays; optimization task; sparse array optimization; stochastic optimization methods; simulated annealing

## 1. Introduction

Phased arrays are increasingly being used in ultrasonic testing because of their flexibility in solving a variety of materials inspection tasks. This type of transducer consists of a number of individual piezoelectronic elements; therefore, different approaches to signal acquisition can be used. One of the approaches is FMC [1], which involves the sequential transmission of ultrasonic waves by each element of the phased array and the reception of the reflected signals by all elements of the probe. The resulting set of signals can then be post-processed to restore the image of the internal structure of the test object. A combination of FMC and post-processing using algorithms based on the synthetic aperture focusing technique was first proposed by Holmes et al. [2] and was referred to as TFM. The combined use of TFM and FMC is often referred to as TFM/FMC, and this approach is widely regarded as the benchmark for ultrasonic imaging in nondestructive testing [3].

Real-time imaging with TFM/FMC can be challenging when large arrays are used and when inspections are to be carried out at high scanning speeds (e.g., railway inspection [4]). This is due to the time required to acquire the signals using the FMC and the time required to restore the image by post-processing the acquired signals. Increasing the speed of TFM imaging is therefore a hot research topic in non-destructive testing. For this purpose, computationally efficient post-processing algorithms can be applied [5,6], and computations within post-processing algorithms could be transferred to field-programmable gate arrays [7,8] or graphics processing units [9,10]. Another approach to increase the speed of TFM imaging is to use sparse arrays [11] when a limited number of elements are used to transmit ultrasonic waves and receive the reflected signals. The speed of imaging performed using sparse arrays is higher compared to the FMC method, since a limited volume of signals to be post-processed reduces the ultrasonic data acquisition time and image restoration time. In addition, a limited number of active elements can reduce the requirements for the electronic units used, thus reducing their cost. Finally, the configuration of the sparse phased array can be varied with respect to the ultrasonic imaging conditions, proving the versatility of this approach.

The key factor that determines the efficiency of ultrasonic imaging using sparse arrays is the determination of their layout (configuration) to obtain a high-quality result. Therefore, several approaches have been proposed to determine the sparse configuration in the case of two-dimensional imaging using TFM and linear phased arrays. In reference [12], Bannouf et al. implemented an iterative procedure to determine the linear sparse array layout. At each iteration, a new element was introduced into the configuration to maximally reduce the side lobe level (SLL) and the main lobe width (MLW) of the point spread function of the sparse array. The iteration process was completed when the point spread function of the sparse phased array reached the specified parameters. Another approach to determine the optimal sparse array configuration is the application of stochastic optimization methods. Hu et al. [13,14] proposed a genetic algorithm for this purpose. The authors introduced the beam directivity function of the sparse array and optimized the parameters of the beam directivity diagram using the genetic algorithm. The genetic algorithm was also used to determine the layout of sparse arrays in reference [15] by Bazulin et al. In this case, the optimization is performed based on the difference between the image obtained using the considered configuration of the sparse phased array and the restored results when FMC is applied. Zhang et al. [16] considered particle swarm optimization to determine the optimal sparse array configuration. Optimization of the beam directivity function parameters was also applied in this research.

Volumetric ultrasonic imaging with matrix phased arrays is another example of an area where real-time acquisition of the results is challenging [6]. Since the pitch of uniform phased arrays should not exceed half a wavelength [11], 2D probes must contain a large number of elements in order to have a sufficient aperture size. This also increases the demands placed on the electronics units used. Similar to the 2D ultrasonic TFM imaging with linear phased arrays, the application of sparse probes can be considered for the volumetric case. Although sparse matrix phased array configuration determination is a hot topic in medical imaging [17–21], it has not been considered in the field of nondestructive testing and TFM imaging. Determination of the sparse array configuration can be regarded as the optimization problem, which can be solved using stochastic optimization methods. On the basis of the research conducted in the medical [19,22,23] and nondestructive testing [13–15] fields, two stochastic optimization algorithms have found widespread application: simulated annealing and the genetic algorithm. The performance of these stochastic optimization methods largely depends on their implementation and the problem to be solved, yet simulated annealing is the more highly recommended of these two methods [24].

Thus, the aim of this research was to develop a method for determining the optimal sparse matrix phased array configuration in volumetric TFM imaging. The aforementioned task within the method to be developed was treated as an optimization problem, which was solved using simulated annealing. The method developed based on this approach provides high-quality ultrasonic imaging under various inspection conditions. An effective implementation of simulated annealing for the solution of the sparse matrix phased layout optimization problem requires the introduction of the acceptance function. Within the aforementioned problem, this function should apply the parameters of the acoustic field produced by sparse 2D probes in TFM imaging. This requires the implementation of these approaches to calculate the above parameters. Furthermore, the parameters of simulated annealing must be determined to obtain a fast and global solution of the optimization problem. All the considered issues have not been studied in sufficient detail in the scientific literature in relation to volumetric TFM imaging with matrix phased arrays. This necessitated the research reported in this paper.

## 2. Theory

### *2.1. Beam Directivity Function of Matrix Phased Arrays*

The optimum sparse array configuration should be determined with regard to ultrasonic imaging conditions. Consider the data acquisition mode of the matrix phased array layout, which involves sequential emission of ultrasonic waves by each active element of the configuration and recording of reflected acoustic signals by all active elements. This can be either a sparse 2D probe or a full matrix phased array where all the elements of the transducer are active.

Consider an element of the matrix phased array with a center located at a point with coordinates ($x$, $y$, $z$ = 0), and consider a point in space with coordinates ($x_p$, $y_p$, $z_p$). Let $\overline{R}$ be the vector from the center of the element to the point in the medium with coordinates ($x_p$, $y_p$, $z_p$), let $\overline{r}$ be the vector from the point in space to the origin of the coordinate axis, and let $\overline{s}$ be the vector from the origin of the coordinate system to the center of the element of the probe (Figure 1).

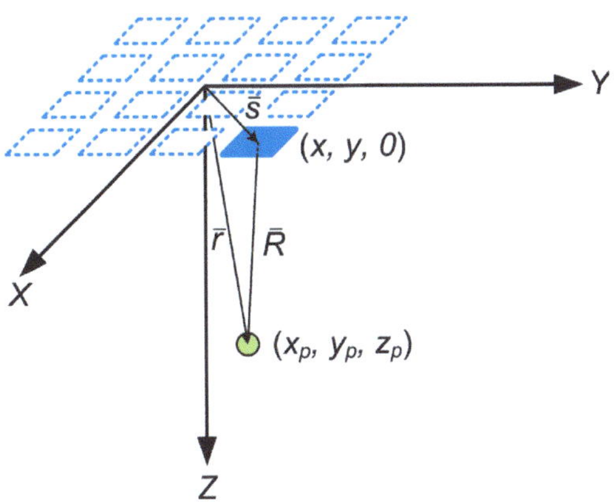


**Figure 1.** Vectors $\overline{R}$, $\overline{s}$, and $\overline{r}$ for the considered element of the matrix phased array.

The law of cosine can be written as follows:

$$||\overline{R}||^2 = ||\overline{r}||^2 + ||\overline{s}||^2 - 2 \cdot ||\overline{r}|| \cdot ||\overline{s}|| \cdot \cos(\alpha) \tag{1}$$

where $\alpha$ is the angle between vectors $\overline{r}$ and $\overline{s}$.

In the far field distance, $||\overline{R}||$ is sufficiently large. Thus, $||\overline{R}||$ can be approximated according to reference [25] to the first order as follows:

$$||\overline{R}|| \approx ||\overline{r}|| - ||\overline{s}|| \cdot \cos(\alpha), \tag{2}$$

The cosine of angle $\alpha$ can also be expressed as follows:

$$\cos(\alpha) = \frac{x_p \cdot x + y_p \cdot y + z_p \cdot 0}{||\overline{r}|| \cdot ||\overline{s}||}, \tag{3}$$

Consider the non-conventional spherical coordinate system reported in reference [26]. In this coordinate system, each point is described by two angular coordinates and by the distance between the origin of the coordinate system and the considered point. In this case, angle $\theta_x$ describes the angle between the projection of the distance on the XOZ plane and the OZ axis, while angle $\theta_y$ describes the angle between the projection of the distance on the YOZ plane and the OZ axis (Figure 2).

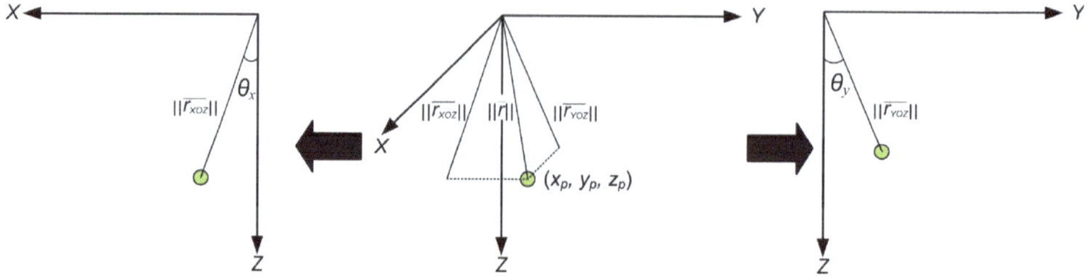


**Figure 2.** Angles $\theta_x$ and $\theta_y$ in the considered coordinate system.

Thus, the coordinates $x_p$ and $y_p$ in the considered non-conventional coordinate system could be expressed as follows [26]:

$$x_p = \frac{||\bar{r}|| \cdot \tan(\theta_x)}{\sqrt{1 + \tan^2(\theta_x) + \tan^2(\theta_y)}}, \tag{4}$$

$$y_p = \frac{||\bar{r}|| \cdot \tan(\theta_y)}{\sqrt{1 + \tan^2(\theta_x) + \tan^2(\theta_y)}}, \tag{5}$$

Inserting Equations (4) and (5) into Equation (3) gives:

$$\cos(\alpha) = \frac{\tan(\theta_x) \cdot x + \tan(\theta_y) \cdot y}{\sqrt{1 + \tan^2(\theta_x) + \tan^2(\theta_y)} \cdot ||\bar{s}||}, \tag{6}$$

In turn, inserting Equation (6) into Equation (2) gives the following dependence between $||\bar{R}||$ and $||\bar{r}||$:

$$||\bar{R}|| = ||\bar{r}|| - \frac{\tan(\theta_x) \cdot x + \tan(\theta_y) \cdot y}{\sqrt{1 + \tan^2(\theta_x) + \tan^2(\theta_y)}}, \tag{7}$$

A pressure field from the point source at a distance $||\bar{R}||$ can be expressed as follows:

$$p(||\bar{R}||, t) = \frac{P_0}{||\bar{R}||} \cdot \exp\left[j\left(wt - k||\bar{R}||\right)\right], \tag{8}$$

The total pressure field at a given point generated by a matrix phased array element can be obtained by integrating the pressure fields generated by point sources along the area of the element. Assuming that the phased array element is a square with side size $a$, the following integral can be written:

$$P(||\bar{R}||, t) = \int_{x-a/2}^{x+a/2} \int_{y-a/2}^{y+y/2} p(||\bar{R}||, t)\, dx\, dy, \tag{9}$$

Carrying out the integration, considering that $a << R$, $a << r$, and taking into account relation 7 yields the following:

$$\begin{gathered} P(\|\bar{r}\|,\theta_x,\theta_y,x,y,t) = -\frac{f_3^2\cdot\exp(-ikr)\cdot\exp(i\omega t)}{f_1\cdot f_2\cdot k^2\cdot\|\bar{r}\|} \\ \cdot\left(\exp\left(\frac{ikf_1(x-a/2)}{f_3}\right)-\exp\left(\frac{ikf_1(x+a/2)}{f_3}\right)\right) \\ \cdot\left(\exp\left(\frac{ikf_2(y-a/2)}{f_3}\right)-\exp\left(\frac{ikf_2(y+a/2)}{f_3}\right)\right), \end{gathered} \tag{10}$$

where:

$$\begin{gathered} f_1 = \tan(\theta_x) \\ f_2 = \tan(\theta_y) \\ f_3 = \sqrt{1+\tan^2(\theta_x)+\tan^2(\theta_y)}. \end{gathered} \tag{11}$$

The synthesized sound pressure can be determined using the following equation:

$$P_{array}(\|\bar{r}\|,\theta_x,\theta_y,t) = \sum_{i=1}^{N} g_i\cdot P_i(\|\bar{r}\|,\theta_x,\theta_y,x_i,y_i,t). \tag{12}$$

where $N$ is the number of elements in the matrix phased array and $g$ is a binary coefficient. The coefficient is equal to 1 if the element is active in the considered configuration of the sparse array, and it is equal to 0 if the element is excluded from the processes of transmission of ultrasonic waves and reception of echo signals.

In this case, the directivity function can be written as follows:

$$H(\theta_x,\theta_y) = \left|\frac{P(\|\bar{r}\|,\theta_x,\theta_y,t)}{p(\|\bar{r}\|,\theta_{xm},\theta_{ym},t)}\right|, \tag{13}$$

where $\theta_{xm}$ and $\theta_{ym}$ are steering angles. For the TFM, these angles can be assumed as 0. Equation (14) can be used to obtain the beam directivity diagram of the considered matrix phased array configuration. This diagram can be used to evaluate the imaging performance of the considered 2D probe layout. An example of the beam directivity diagram obtained using Equation (13) is shown in Figure 3.

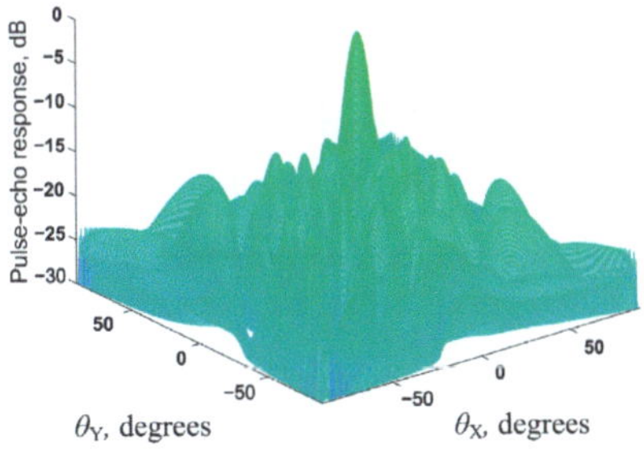


**Figure 3.** Example of beam directivity diagram.

## 2.2. Simulated Annealing Algorithm

Simulated annealing is one of the best-known stochastic optimization methods for tasks where the objective function is not explicitly defined. This algorithm can be used to determine the global optima and exclude the selection of local optima as the solution for the optimization task [27]. The basic block diagram of the simulated annealing algorithm is shown in Figure 4. The efficiency of solving optimization problems using simulated annealing depends heavily on the appropriate choice of the parameters of this algorithm. Temperature is one of the most important parameters in simulated annealing used to find the optimal solution to the optimization problems. At the first iterations of the algorithm, the temperature parameter (T) is typically set high. This avoids the selection of local optima as the final solution. As the algorithm runs, this parameter is reduced according to the cooling schedule. The execution of the algorithm is stopped when the temperature parameter reaches the defined limit. The decision to accept the trial state as the current solution is made using the values of the acceptance function.

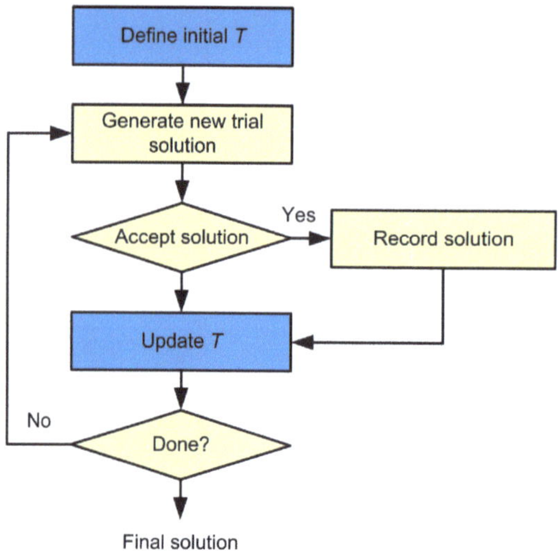


**Figure 4.** Block diagram of the simulated annealing algorithm.

## 3. Determination of the Optimal Sparse Matrix Phased Array Configuration Using the Simulated Annealing Algorithm

### *3.1. Simulated Annealing Algorithm Implementation*

The simulated annealing algorithm was implemented to determine the optimal sparse matrix phased array configurations. At the first stage, a sparse matrix phased array configuration was randomly determined and considered as the current solution. An iterative procedure was then implemented. At each iteration, a new configuration of the sparse matrix phased array was determined, which was considered as a trial solution at this iteration. Then, the beam directivity diagram of the current trial solution was determined using Equation (14), and its parameters were evaluated. The parameters evaluated included the main lobe width (MLW) and the side lobe level (SLL, Figure 5). In this case, the main lobe width was evaluated using the full width at half maximum parameter. The parameters obtained from the beam directivity diagram of the trial configuration of the sparse matrix phased array were then used to evaluate the acceptance function for this configuration and to make a decision regarding its acceptance as the current solution.

The algorithm used for the determination of the sparse matrix phased array configuration was implemented using the MATLAB R2020b software. The algorithm requires test conditions as initial parameters. The test conditions to be defined are as follows:

1. The number of elements in the matrix phased array;
2. The center frequency of the probe elements;
3. The pitch of the array and the dimensions of each element;
4. The velocity of the ultrasonic waves in the test object;
5. The number of active elements in the sparse matrix phased array configuration.

The appropriate choice of simulated annealing parameters is another important factor to determine the efficiency of the optimization solution. During algorithm execution, the temperature parameter was changed from 0.5 to 0.00025. For each constant value of the temperature parameter, five iterations were performed. The temperature parameter changed according to the following law:

$$T_{k+1} = 0.98 \cdot T_k \tag{14}$$

The execution of the algorithm was completed when the temperature parameter reached the lower limit.

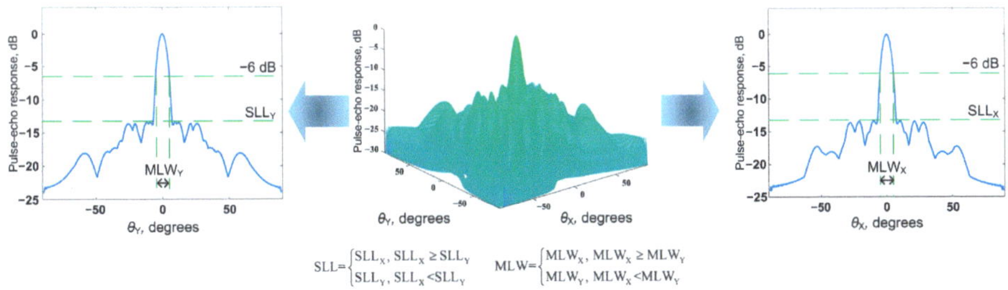


**Figure 5.** Determination of the beam directivity diagram parameters.

The acceptance function is used to make a decision regarding the acceptance of the sparse matrix phased array configuration as the current solution at each iteration of the algorithm execution. This function should be defined with respect to the objective of the optimization task; in this case, the task is to obtain the sparse matrix phased array configuration with a beam directivity diagram that has low SLL and MLW. The acceptance function implemented in the developed algorithm takes the following form:

$$P = \begin{cases} 1;\ \Delta MLW < 0,\ \Delta SLL < 0 \\ \exp(-\Delta MLW/T_k);\ \Delta MLW > 0,\ \Delta SLL < 0 \\ \exp(-\Delta SLL/T_k);\ \Delta MLW < 0,\ \Delta SLL > 0 \\ \exp(-\Delta MLW/T_k) \cdot \exp(-\Delta SLL/T_k);\ \Delta MLW > 0,\ \Delta SLL > 0 \end{cases} \tag{15}$$

where $\Delta MLW$ and $\Delta SLL$ are the differences in MLW and SLL, respectively, between the beam directivity diagram of the trial configuration of the sparse matrix phased array considered at a current iteration of the algorithm execution and the beam directivity diagram of the current solution.

### *3.2. Results of Simulated Annealing Algorithm Execution*

The initial data used to determine the optimal sparse matrix phased array configuration by the developed algorithm are presented in Table 1. These data include the parameters of the probe and the speed of the longitudinal ultrasonic waves in the medium (aluminum).

**Table 1.** Matrix phased array parameters for which the optimization task was solved.

| Parameter | Value |
|---|---|
| Number of elements | $8 \times 8$ |
| Center frequency, MHz | 5 |
| Pitch, mm | 1 |
| Element dimensions, mm | $0.8 \times 0.8$ mm |
| Speed of ultrasonic waves in the media | 6350 m/s |

In total, three sparse phased array configurations with 49, 36, and 25 elements were determined using the developed algorithm that employed simulated annealing. These configurations are shown in Figure 6. The beam directivity diagrams of the obtained configurations are shown in Figure 7.

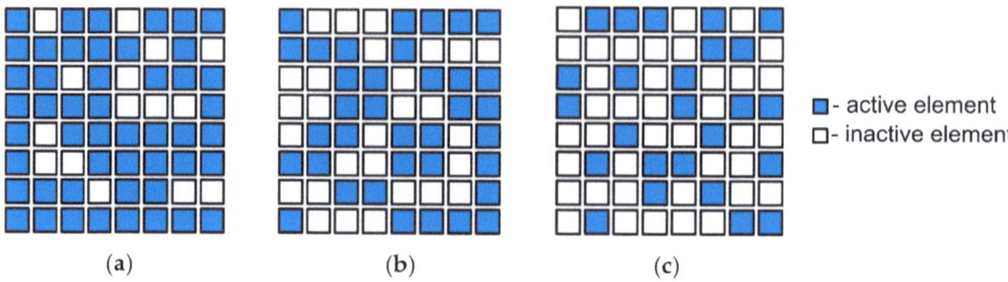


**Figure 6.** Sparse matrix phased configurations obtained using the developed algorithm: (**a**) sparse matrix phased array configuration with 49 elements; (**b**) sparse matrix phased array configuration with 36 elements; (**c**) sparse matrix phased array configuration with 25 elements.

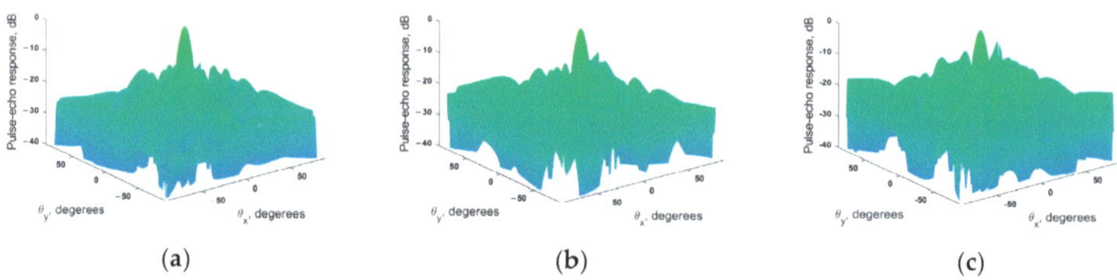


**Figure 7.** Beam directivity diagrams of sparse matrix phased array configurations obtained using the developed algorithm: (**a**) sparse matrix phased array configuration with 49 elements; (**b**) sparse matrix phased array configuration with 36 elements; (**c**) sparse matrix phased array configuration with 25 elements.

Table 2 shows the parameters of the beam directivity diagrams (MLW and SLL) of the obtained sparse matrix phased array configuration and the full array.

**Table 2.** Parameters of beam directivity diagrams of all the considered configurations of the matrix phased arrays.

| Number of Elements in Configuration | MLW, Degrees | SLL, dB |
|---|---|---|
| 64 | 10.88 | −13.23 |
| 49 | 10.96 | −12.53 |
| 36 | 11.13 | −13.29 |
| 25 | 11.15 | −11.72 |

Thus, the directivity diagram parameters of the sparse phased array configurations were close to these of the full matrix phased array. The difference in MLW between all the sparse phased array configurations and the full array did not exceed 0.27 degrees. At the same time, the difference in SLL between the sparse probes and the full array did not exceed 1.51 dB. The obtained differences in the beam directivity diagram parameters of the sparse and full arrays should lead to the restoration of flaw images with a comparable quality.

## 4. Experiments

The imaging performance of the obtained sparse matrix phased array configurations was verified via in situ experiments in conditions corresponding to the input parameters used during implementation of the simulated annealing algorithm. The experiments employed the matrix phased array Doppler 5M8 × 8BP1.0 (Figure 8). The probe consisted of 64 elements, forming a matrix of 8 × 8 elements. The dimensions of each element were 0.8 × 0.8 mm, and its center frequency was 5 MHz.

**Figure 8.** Matrix phased array Doppler 5M8 × 8BP1.0.

The Optus electronic unit (I-Deal technologies GmbH, Saarbrücken, Germany, Figure 9) was used for signal acquisition using various configurations of matrix phased arrays. This electronic unit has 128 multiplexed channels, allowing the Doppler 5M8 × 8BP1.0 to be used in FMC and sparse phased array modes of signal acquisition.

**Figure 9.** Optus electronic unit.

Two sections of a 17 mm thick aluminum test block were used for experimental verification. Both sections contained flat-bottom holes that were 2 mm in diameter and 10 mm deep. The first section contained one flaw, and the second section contained two flaws located at a distance of 4 mm from each other. The location of the flaws in the test sections is shown in Figure 10.

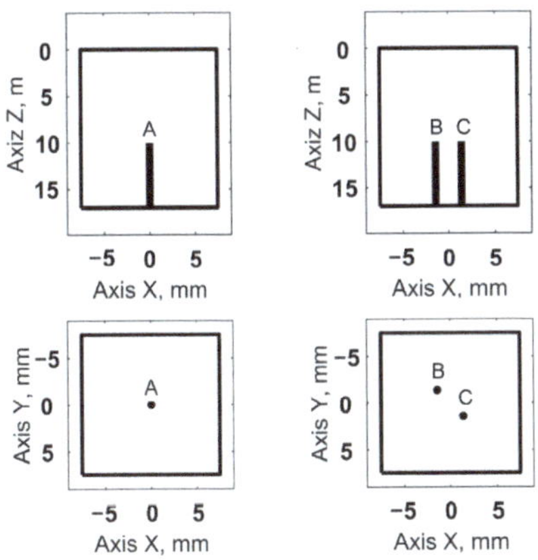


(a) (b)

**Figure 10.** Location of the flaws in the test sections: (**a**) test section 1; (**b**) test section 2.

The experiments considered full array and sparse matrix phased array configurations that were obtained using the developed algorithm. The arrangements of the active elements in the sparse matrix phased array configurations are shown in Figure 6. For all the configurations, the signal acquisition procedure was the same and was based on FMC, taking into account that not all the elements of the probe may be active in the considered transducer configuration. In all the experiments, the probe was initially placed in the center of the test section. The data acquisition was performed in several stages according to the number of active elements in the configuration. At each stage, one active element of the configuration transmitted ultrasonic waves into the specimen, while all the active elements received the reflected signals. This resulted in a set of ultrasonic data that was post-processed. For this purpose, the post-processing algorithm discussed in reference [2] was adapted for volumetric imaging and sparse matrix phased array application.

For all the configurations, the post-processing results were volumetric images of the flaws in the test sections. To evaluate the performances of the different sparse matrix phased array configurations, metrics characterizing the resolution and signal-to-noise ratio were determined. The resolution of the images was evaluated using the array performance indicator (API). This value was calculated for each of the flaws using the following equation [6]:

$$API = \frac{N_{-6\mathrm{dB}} D_x D_y D_z}{\lambda^3} \tag{16}$$

where $D_x$, $D_y$, and $D_z$ are the dimensions of the voxels along the $X$, $Y$, and $Z$ axes, respectively; $N_{-6\mathrm{dB}}$ is the number of voxels with an amplitude exceeding the −6 dB threshold, where 0 dB corresponds to the maximum obtained amplitude of the flaw; and $\lambda$ is the wavelength of the longitudinal ultrasonic waves in the specimen.

The signal-to-noise ratio can also be evaluated for each of the flaws using the following equation [28]:

$$SNR = 20\log_{10}\left(\frac{A_s}{A_n}\right) \tag{17}$$

where $A_s$ is the maximum amplitude of the signal from the flaw and $A_n$ is the maximum amplitude of the noise at the depth of the flaw location. In the experiments, the value of $A_n$ was obtained by scanning the flawless zone.

The above metrics were used to evaluate the quality of the images obtained using different configurations of matrix phased arrays. According to reference [29], a high quality NDT ultrasonic image should meet the following requirements:

- SNR is higher than 20 dB;
- the level of artifacts relative to the amplitude of the flaw in the image is less than 20 dB;
- the image of the entire flaw boundary is restored.

## 5. Results and Discussion

As mentioned above, the results of ultrasonic imaging performed using different configurations of matrix phased arrays are the volumetric images of the test sections of the aluminum specimens used in the experiments. Examples of such images for both test sections are shown in Figure 11.

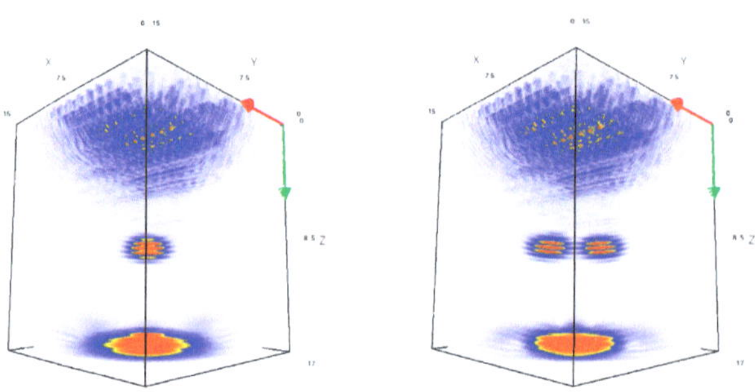

(a) (b)

**Figure 11.** Examples of volumetric results: (**a**) test section 1; (**b**) test section 2.

Figures 12 and 13 show the imaging results for test section 1. Figure 12 shows the projections of the volumetric images obtained using the configurations of sparse arrays on the XY plane. The profiles of these projections on the X and Y axes are shown in Figure 13. Similar results for test section 2 with two closely spaced flat-bottom holes are shown in Figures 14 and 15.

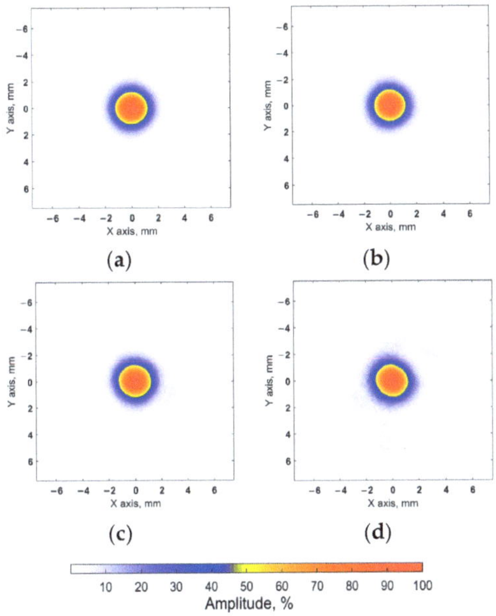


**Figure 12.** Projections of volumetric images of test section 1 in the XY plane: (**a**) full array; (**b**) 49-element sparse array; (**c**) 36-element sparse array; (**d**) 25-element sparse array.

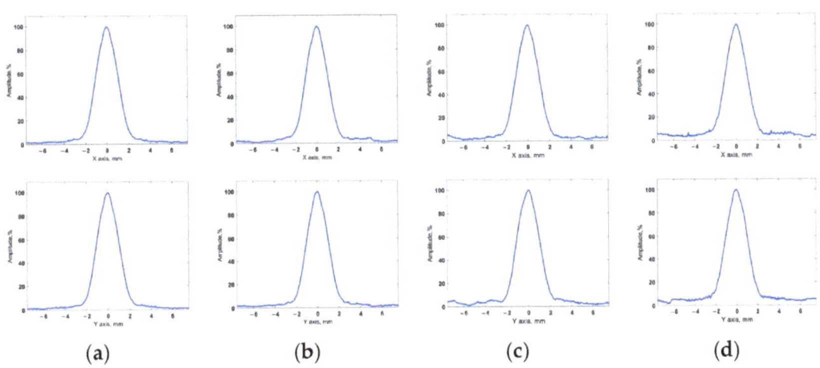


**Figure 13.** Profiles of obtained volumetric images of test section 1: (**a**) full array; (**b**) 49-element sparse array; (**c**) 36-element sparse array; (**d**) 25-element sparse array.

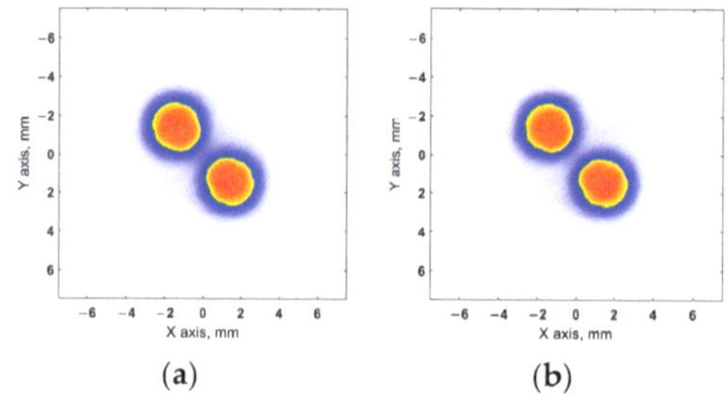


**Figure 14.** *Cont.*

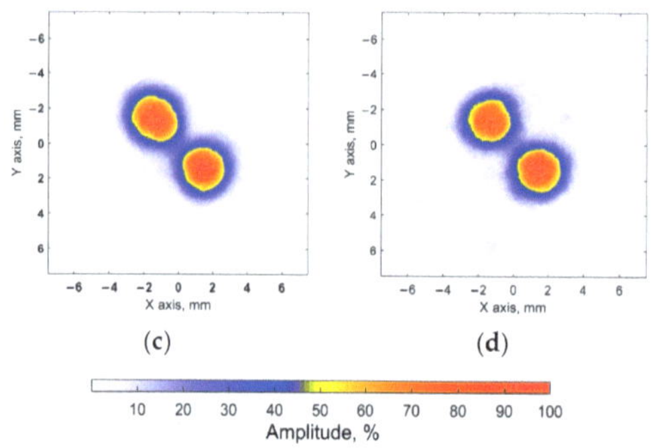


**Figure 14.** Projections of volumetric images of test section 2 in the XY plane: (**a**) full array; (**b**) 49-element sparse array; (**c**) 36-element sparse array; (**d**) 25-element sparse array.

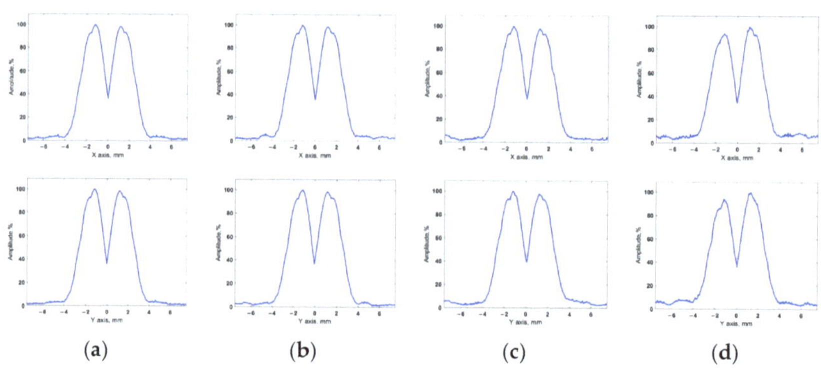


**Figure 15.** Profiles of obtained volumetric images of test section 2: (**a**) full array; (**b**) 49-element sparse array; (**c**) 36-element sparse array; (**d**) 25-element sparse array.

For the obtained results, the API and SNR for each of the flaws were evaluated using Equations (16) and (17). The resulting values are presented in Table 3, where the flaws are indicated, similar to the flaws in Figure 10. Furthermore, the obtained API and SNR for the flaws in the images that were restored using the sparse array data were compared with the corresponding values for the flaws in the images obtained using the full matrix phased array.

**Table 3.** API and SNR values for the obtained results.

| Configuration | Defect | API | | SNR | |
|---|---|---|---|---|---|
| | | Value | Difference, % | Value, dB | Difference, dB |
| Full array | A | 0.623 | 0 | 36.21 | 0 |
| | B | 0.649 | 0 | 33.89 | 0 |
| | C | 0.668 | 0 | 33.18 | 0 |
| Sparse 49-element | A | 0.614 | 1.44 | 33.49 | −2.72 |
| | B | 0.640 | 1.39 | 32.71 | −1.18 |
| | C | 0.670 | 0.30 | 30.68 | −2.5 |
| Sparse 36-element | A | 0.620 | 0.48 | 32.61 | −3.6 |
| | B | 0.661 | 1.85 | 29.67 | −4.22 |
| | C | 0.650 | 2.69 | 29.48 | −3.7 |
| Sparse 25-element | A | 0.629 | 0.96 | 27.98 | −8.23 |
| | B | 0.634 | 2.31 | 26.35 | −7.54 |
| | C | 0.645 | 3.44 | 26.83 | −6.35 |

Thus, the images that were restored using the sparse matrix phased array configurations had a resolution that was close to that of the images obtained using the full matrix array. The difference in the API between the considered cases did not exceed 3.44%. Furthermore, all the sparse matrix phased array configurations used were able to resolve closely spaced flaws in test section 2 using the −6dB drop method. The SNR of all the results obtained with sparse probes was greater than 20 dB, and these images did not contain artifacts with amplitudes exceeding −20 dB relative to the maximum amplitude of the flaw. The obtained results included high-quality images of both test sections for all the applied sparse arrays configurations. At the same time, the application of sparse arrays reduced the amount of ultrasonic data compared to the full array. In our experiments, the use of 49-, 36-, and 25-element sparse arrays reduced the set of signals 1.7-, 3.1-, and 6.6-fold, respectively. The obtained results demonstrate a high imaging performance of the resulting sparse matrix phased array configurations and the efficiency of the proposed approach for determining these layouts with regard to the conditions of ultrasonic testing.

## 6. Conclusions

TFM/FMC provides high imaging performance, but it is time consuming. This paper proposed a method to design a sparse matrix phased array using the simulated annealing algorithm for TFM imaging. Within the framework of the considered approach, the parameters of the beam directivity diagram were optimized in order to determine the appropriate matrix phased array layout. The beam directivity diagrams were evaluated in the modified spherical coordinate system using far-field approximation.

The considered approach was applied for proper conditions of TFM imaging. As a result, sparse matrix phased array configurations with 49, 36, and 25 elements were determined. The parameters of the beam directivity diagrams of these sparse phased arrays were close to the corresponding parameters of the full array. The difference in the main lobe width for the above configuration did not exceed 0.27 degrees compared to the same parameters for the 64-element full matrix phased array when all the probe elements were active. A similar comparison performed for the side lobe level showed that the maximum difference between the obtained sparse arrays and the full array was 1.51 dB.

The performance of the obtained sparse matrix phased array configurations was evaluated via in situ experiments. The experimental conditions corresponded to the ultrasonic imaging parameters considered for the determination of the sparse matrix phased array configuration using the developed method. As a result, high-resolution images were obtained for both test sections with flat-bottom holes using the obtained sparse matrix phased array configurations. The difference in the API between the results obtained using sparse matrix phased arrays and the full array did not exceed 3.44% for all the cases considered. In addition, in all the images obtained for test section 2, the closely spaced defects were resolved using the −6dB drop method. This is another indicator of the achieved resolution of the restored images. In addition, the SNR for the flaws in the obtained images was greater than 20 dB, with no artefacts of amplitudes exceeding −20 dB found in the results. The evaluated quality metrics of the volumetric images demonstrate the high quality of the results obtained using all the determined sparse matrix phased array configurations and the efficiency of the proposed sparse matrix phased array design method. At the same time, the use of sparse matrix phased arrays allows for a significant reduction in the volume of data to be processed compared to the full array. The number of signals was reduced 1.7-fold (49 active elements), 3.1-fold (36 active elements), and 6.6-fold (25 elements) for the sparse matrix phased arrays. In practice, this should significantly reduce the ultrasonic imaging time.

The results obtained demonstrate the efficiency of the proposed approach and serve as a basis for further research and development. The next stage of research should be focused on the effects of the sparse phased array on the imaging time during post-processing with parallelization of calculations using GPU or FPGA. Furthermore, the adaptation of the developed design method to ultrasonic imaging of complex-shaped objects can also be considered.

**Author Contributions:** Conceptualization, D.O.D.; methodology, D.O.D.; validation, V.Y.Z. and D.O.D.; experimental, V.Y.Z.; writing—original draft preparation, D.O.D.; writing—review and editing, D.O.D.; supervision, D.O.D.; project administration, D.O.D.; funding acquisition, D.O.D. All authors have read and agreed to the published version of the manuscript.

**Funding:** This work was supported by the Russian Science Foundation, project no. 22-79-00244 (https://rscf.ru/en/project/22-79-00244/, accessed on 9 February 2024).

**Data Availability Statement:** The data that support the findings of this study are available from the corresponding author upon reasonable request.

**Acknowledgments:** This work was supported by the Russian Science Foundation, project no. 22-79-00244 (https://rscf.ru/en/project/22-79-00244/, accessed on 9 February 2024).

**Conflicts of Interest:** The authors declare no conflicts of interest.

## References

1. Long, R.; Russell, J.; Cawley, P. Ultrasonic phased array inspection using full matrix capture. *Insight* **2012**, *54*, 380–385. [CrossRef]
2. Holmes, C.; Drinkwater, B.W.; Wilcox, P.D. Post-processing of the full matrix of ultrasonic transmit–receive array data for non-destructive evaluation. *NDT E Int.* **2005**, *38*, 701–711. [CrossRef]
3. Cruza, J.F.; Camacho, J. Total Focusing Method with virtual sources in the presence of unknown geometry interfaces. *IEEE Trans. Ultrason. Ferroelectr. Freq. Control* **2016**, *63*, 1581–1592. [CrossRef] [PubMed]
4. Clark, R. Rail flaw detection: Overview and needs for future developments. *NDT E Int.* **2004**, *37*, 111–118. [CrossRef]
5. Hunter, A.J.; Drinkwater, B.W.; Wilcox, P.D. The wavenumber algorithm for full-matrix imaging using an ultrasonic array. *IEEE Trans. Ultrason. Ferroelectr. Freq. Control* **2008**, *55*, 2450–2462. [CrossRef]
6. Marmonier, M.; Robert, S.; Laurent, J.; Prada, C. Real-time 3D imaging with Fourier-domain algorithms and matrix arrays applied to non-destructive testing. *Ultrasonics* **2022**, *124*, 106708. [CrossRef]
7. Njiki, M.; Elouardi, A.; Bouaziz, S.; Casula, O.; Roy, O. A multi-FPGA architecture-based real-time TFM ultrasound imaging. *J. Real-Time Image Process.* **2019**, *16*, 505–521. [CrossRef]
8. Wang, C.; Mao, J.; Leng, T.; Zhuang, Z.Y.; Wang, X.M. Efficient acceleration for Total Focusing Method based on advanced parallel computing in FPGA. *Int. J. Acoust. Vib.* **2017**, *22*, 536–540. [CrossRef]
9. Rougeron, G.; Lambert, J.; Iakovleva, E.; Lacassagne, L.; Dominguez, N. Implementation of a GPU accelerated total focusing reconstruction method within CIVA software. *AIP Conf. Proc.* **2014**, *1581*, 1983–1990.
10. Lambert, J.; Pédron, A.; Gens, G.; Bimbard, F.; Lacassagne, L.; Iakovleva, E. Analysis of multicore CPU and GPU toward parallelization of total focusing method ultrasound reconstruction. In Proceedings of the DASIP 2012-Conference on Design and Architectures for Signal and Image Processing, Karlsruhe, Germany, 23–25 October 2012.
11. Moreau, L.; Drinkwater, B.W.; Wilcox, P.D. Ultrasonic imaging algorithms with limited transmission cycles for rapid nondestructive evaluation. *IEEE Trans. Ultrason. Ferroelectr. Freq. Control* **2009**, *56*, 1932–1944. [CrossRef] [PubMed]
12. Bannouf, S.; Robert, S.; Casula, O.; Prada, C. Data set reduction for ultrasonic TFM imaging using the effective aperture approach and virtual sources. *J. Phys. Conf. Ser.* **2013**, *457*, 012007. [CrossRef]

13. Hu, H.; Du, J.; Ye, C.; Li, X. Ultrasonic phased array sparse-TFM imaging based on sparse array optimization and new edge-directed interpolation. *Sensors* **2018**, *18*, 1830. [CrossRef]
14. Hu, H.; Du, J.; Xu, N.; Jeong, H.; Wang, X.H. Ultrasonic sparse-TFM imaging for a two-layer medium using genetic algorithm optimization and effective aperture correction. *NDT E Int.* **2017**, *90*, 24–32. [CrossRef]
15. Bazulin, E.G.; Medvedev, L.V. Increasing rate of recording of echo signals with ultrasonic antenna array using optimum sparsing of switching matrix with a genetic algorithm. *Russ. J. Nondestruct. Test.* **2021**, *57*, 945–952. [CrossRef]
16. Zhang, H.; Bai, B.; Zheng, J.; Zhou, Y. Optimal design of sparse array for ultrasonic total focusing method by binary particle swarm optimization. *IEEE Access* **2020**, *8*, 111945–111953. [CrossRef]
17. Jensen, J.A.; Schou, M.; Jorgensen, L.T.; Tomov, B.G.; Stuart, M.B.; Traberg, M.S.; Taghavi, I.; Oygaard, S.H.; Ommen, M.L.; Steenberg, K.; et al. Anatomic and Functional Imaging Using Row–Column Arrays. *IEEE Trans. Ultrason. Ferroelectr. Freq. Control* **2022**, *69*, 2722–2738. [CrossRef] [PubMed]
18. Sobhani, M.R.; Ghavami, M.; Ilkhechi, A.K.; Brown, J.; Zemp, R. Ultrafast orthogonal row–column electronic scanning (uFORCES) with bias-switchable top-orthogonal-to-bottom electrode 2-D arrays. *IEEE Trans. Ultrason. Ferroelectr. Freq. Control.* **2022**, *69*, 2823–2836. [CrossRef] [PubMed]
19. Roux, E.; Ramalli, A.; Tortoli, P.; Cachard, C.; Robini, M.C.; Liebgott, H. 2-D ultrasound sparse arrays multidepth radiation optimization using simulated annealing and spiral-array inspired energy functions. *IEEE Trans. Ultrason. Ferroelectr. Freq. Control* **2016**, *63*, 2138–2149. [CrossRef] [PubMed]
20. de Souza, J.C.E.; Parrilla Romero, M.; Higuti, R.T.; Martínez-Graullera, O. Design of ultrasonic synthetic aperture imaging systems based on a non-grid 2D sparse array. *Sensors* **2021**, *21*, 8001. [CrossRef] [PubMed]
21. Masoumi, M.H.; Kaddoura, T.; Zemp, R. Costas Sparse 2D Arrays for High-resolution Ultrasound Imaging. *IEEE Trans. Ultrason. Ferroelectr. Freq. Control* **2023**, *70*, 460–472. [CrossRef] [PubMed]
22. Diarra, B.; Liebgott, H.; Robini, M.; Tortoli, P.; Cachard, C. Optimized 2D array design for Ultrasound imaging. In Proceedings of the 20th European Signal Processing Conference (EUSIPCO), Bucharest, Romania, 27–31 August 2012.
23. Trucco, A. Thinning and weighting of large planar arrays by simulated annealing. *IEEE Trans. Ultrason. Ferroelectr. Freq. Control* **1999**, *46*, 347–355. [CrossRef]
24. Skiena, S.S. *The Algorithm Design Manual*, 2nd ed.; Springer: London, UK, 2012.
25. Schmerr, L.W. *Fundamentals of Ultrasonic Phased Arrays. Solid Mechanics and Its Applications*; Springer: Berlin/Heidelberg, Germany, 2015; Volume 215.
26. Kazys, R.; Jakevicius, L.; Mazeika, L. Beamforming by means of 2D phased ultrasonic arrays. *Ultragarsas/Ultrasound* **1998**, *29*, 12–15.
27. Glover, F.; Kochenberger, G. *Handbook of Metaheuristics*; Vol. 57 of International Series in Operations Research and Management Science; Kluwer Academic Publishers: Dordrecht, The Netherlands, 2003.
28. Zhang, J.; Drinkwater, B.W.; Wilcox, P.D.; Hunter, A.J. Defect detection using ultrasonic arrays: The multi-mode total focusing method. *NDT E Int.* **2010**, *43*, 123–133. [CrossRef]
29. Bazulin, E.G. On the possibility of using the maximum entropy method in ultrasonic nondestructive testing for scatterer visualization from a set of echo signals. *Acoust. Phys.* **2013**, *59*, 210–227. [CrossRef]

*sensors*

MDPI

*Article*

# Frequency-Resolved High-Frequency Broadband Measurement of Acoustic Longitudinal Waves by Laser-Based Excitation and Detection

**Felix Brand and Klaus Stefan Drese ***

Institute of Sensor and Actuator Technology, Coburg University of Applied Sciences and Arts, Am Hofbräuhaus 1b, 96450 Coburg, Germany; felix.brand@hs-coburg.de
* Correspondence: klaus.drese@hs-coburg.de; Tel.: +49-9561-317-522

**Abstract:** Optoacoustics is a metrology widely used for material characterisation. In this study, a measurement setup for the selective determination of the frequency-resolved phase velocities and attenuations of longitudinal waves over a wide frequency range (3–55 MHz) is presented. The ultrasonic waves in this setup were excited by a pulsed laser within an absorption layer in the thermoelastic regime and directed through a layer of water onto a sample. The acoustic waves were detected using a self-built adaptive interferometer with a photorefractive crystal. The instrument transmits compression waves only, is low-contact, non-destructive, and has a sample-independent excitation. The limitations of the approach were studied both by simulation and experiments to determine how the frequency range and precision can be improved. It was shown that measurements are possible for all investigated materials (silicon, silicone, aluminium, and water) and that the relative error for the phase velocity is less than 0.2%.

**Keywords:** laser ultrasound; photoacoustics; frequency resolved phase velocity; attenuation; dispersion; laser ultrasonic spectroscopy; ndt; longitudinal waves; compression waves; acoustics

## 1. Introduction

In industry and academic research, optoacoustics has already been established in many applications [1]. The advantages of this method is the fact that the measurements are non-destructive and non-contact [1–3]. It should also be emphasised that photoacoustic measurements are very broadband due to laser excitation, which means that frequency-dependent analyses can be carried out in a range from kHz to GHz [4]. This allows macrostructures and thin films, as well as microstructures, to be probed [1]. Photoacoustics is, therefore, a promising technique for the study and analysis of complex media properties [3].

Bulk acoustic waves are typically non-dispersive, and measurements are performed with a single frequency or in a narrow frequency band. The recent literature has shown that the dispersion of bulk acoustic waves can be utilised to measure material properties such as material inhomogeneity. Karabutov et al. [5] presented a method that allows conclusions about the porosity of isotropic metal matrix composites from the dispersion of the phase velocity and the frequency-resolved attenuation of longitudinal waves. Podymova et al. [6] have demonstrated that the effect of porosity on the dispersion of longitudinal waves is also present in aluminium alloy matrix composites. Other researchers have characterised 3D-printed photopolymers by measuring frequency-resolved acoustic properties [7,8]. The dispersion of longitudinal waves inside of Polyvinylchloride (PVC) has been shown by Demcenko et al. [9].

In addition, during the development and testing of acoustic metamaterials, a frequency-resolved broadband characterisation of materials is necessary. Examples of metamaterials with dispersive longitudinal waves can be found in the literature [10–12]. Thus, broadband, frequency-resolved measurements of sound velocity and attenuation open up new possibilities for materials analysis.

One method of broadband excitation of acoustic waves is excitation via a nanosecond pulsed laser. This excitation method enables non-destructive and non-contact measurements with a broad frequency spectrum and high amplitude signals [13,14]. This excitation technique for short ultrasonic pulses is used for the excitation of longitudinal [15–17], shear [18–20], Rayleigh [21–25], and lamb waves [26–29]. All studies require high-quality experimental data and a deep understanding of the limitations of the experiment and the analysis.

In this paper, we present an optical method for precise, low-contact, non-destructive, sample-independent, broadband, and frequency-resolved measurement of the dispersion and attenuation of compression waves. The components of the measurement setup are introduced, and their purpose and impact on the measurement are discussed. Simulations and experiments were used to identify factors influencing the generation of sound by a pulsed laser. Further experiments were performed to analyse the accuracy of the method.

In the following section, we first introduce the idea of the measurement method and explain the components, starting with detection and going backwards to excitation. Special attention is paid to the measuring cell, its components, and its influence on the measurement. The influence on the measurement is first demonstrated using simulations and, later, using experimental measurements. The article concludes with dispersion measurements of various materials.

## 2. Materials and Methods

### 2.1. Concept of the Measurement Setup

The acoustic measurement cell is the centrepiece of the measurement setup. It consists of an absorption layer on top of a glass substrate, the specimen, and water in contact with the absorption layer and the specimen. There are several objectives that are addressed with the measuring cell presented below (see Figure 1):

- The measurement of the frequency-resolved phase velocity and attenuation of acoustic compression waves.
- The excitation and detection of an acoustic compression pulse independent of the sample material and over a wide frequency range.
- Non-destructive and ideally contactless measurement.
- The suppression of other acoustic excitations, e.g., shear waves.

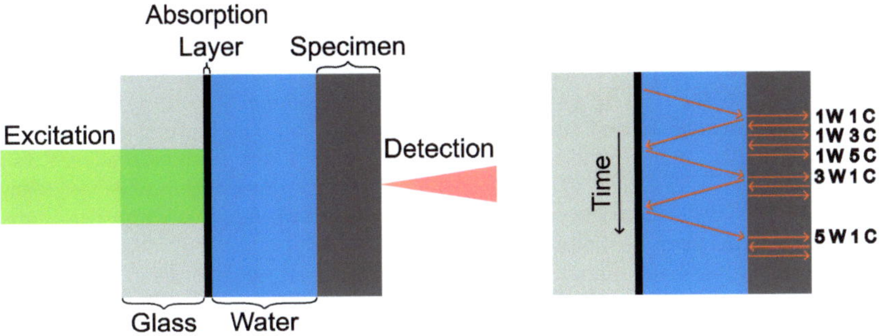


**Figure 1.** Scheme of acoustic measurement cell. (**Left**) Components of measurement cell. (**Right**) Scheme of possible acoustic paths. C: Path of compression wave through specimen. W: Path of compression wave through water. Time axis is not scaled.

The ultrasonic pulse is generated at the absorption layer via an ns pulsed laser. From this layer, the pulse travels through the water and the specimen and will be detected with an adaptive interferometer with a photorefractive crystal. The complete optical setup is visualised in Figure 2.

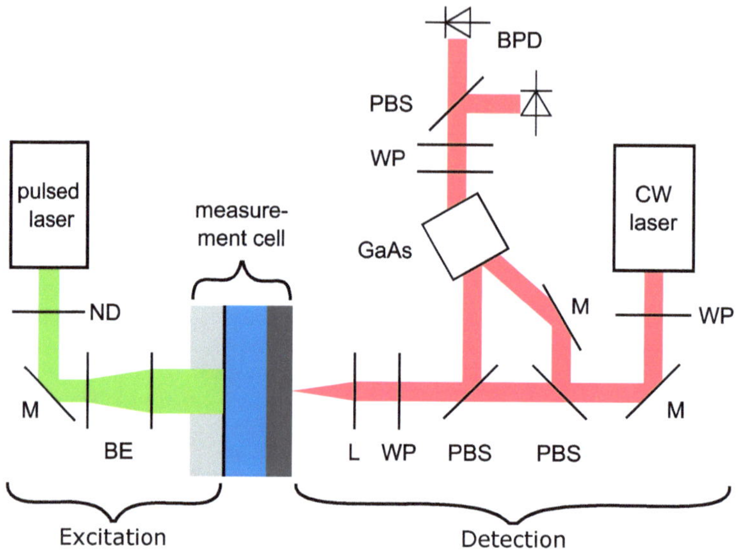


**Figure 2.** Complete measurement setup. Green: excitation, red: detection; ND: neutral density filter, BE: beam expander, M: mirror, WP: waveplate, PBS: polarizing beam splitter, L: lens, BPD: balanced photodetector.

In the following section, we discuss the individual components, their purpose, and their influence on the measurement, starting at detection and going backwards to excitation.

### 2.2. Detection

The ultrasonic waves were detected using a self-built adaptive interferometer with a photorefractive crystal. These interferometers, used for the detection of ultrasonic waves or vibrations, are well described in the literature [30–32]. In our setup, we used an undoped GaAs crystal ($5 \times 5 \times 10\,\text{mm}^3$) in a diffusion-dominated regime with an anisotropic diffraction setup. The optical setup for the detection is shown in Figure 2.

This kind of interferometer has several advantages. The wavefront of the diffracted beam is identical to the wavefront of the nondiffracted beam of the sample beam, allowing both beams to interfere with exactly the same wavefronts [30,33]. A possible source of wavefront distortion can be, for example, the surface roughness of the sample. The dynamic hologram inside the crystal acts as a high-pass filter with a cut-off frequency $f_c$ with a time constant $\tau$ of the crystal ($f_c = 1/(2\pi\tau)$). The cut-off frequency depends on the setup and is in the range of 3.5 and 10 kHz for GaAs crystals [30]. The high-pass suppresses low-frequency disturbances, such as air fluctuations, thermal deformation, acoustic noise, and other sources of vibration [30,33]. In addition the interferometer has a wide frequency response [33], with a flat transfer function above its cut-off frequency and below the bandwidth of the balanced photodetector [32,34].

Detection is the main component, which defines the noise level of the setup. The noise level is mainly determined by the laser power of the interferometer, the number of averages, the detector, and the electronic and acoustic background noise.

In our setup, we used a 1064 nm laser with a maximum CW power of 10 W (Azurlight Fiber Laser, Azurlight Systems, Bordeaux, France), operating at 3 W, and an InGaAs photodetector (Femto HBPR-200M-30K-IN, FEMTO Messtechnik GmbH, Berlin, Germany), with a bandwidth of DC up to 200 MHz.

### 2.3. Measurement Cell

This section describes the individual components of the measuring cell (see Figure 1) and their influence on the measurement.

### 2.3.1. Specimen

The measuring cell was designed in such a way that the measurement is as sample-independent as possible.

It is only necessary for the evaluation (see Section 2.4.1) that two reflections of the same pulse (e.g., 1W 1C (the pulse travels one time through the water and the compression wave travels one time through the sample) and 1W 3C, as compared in Figure 1) can be measured. This sets the requirements for the sample in terms of acoustic attenuation and thickness.

### 2.3.2. Fluid

The sound pulse is transmitted through the liquid to the sample. Only longitudinal waves can propagate in the liquid, which means that primarily longitudinal waves can be observed in the sample.

A further task of the fluid is to delay the reflection of the wave at the absorption layer (3W 1C, as compared in Figure 1) so that at least a number of $n_S$ reflections can be measured within the sample before the first interfering reflection occurs. This delay can be adjusted by the fluid thickness $x_F$ and its velocity $c_F$.

$$x_F > c_F \cdot \left( n_S \cdot \frac{x_S}{c_S} + \frac{\tau_{Pulse}}{2} \right), \quad (1)$$

where $x_S$ and $c_S$ are the thickness and acoustic velocity of the sample, and $\tau_{Pulse}$ is the acoustic pulse duration. The first term of the equation describes that $n_S$ reflections can be measured before the wave propagating in the fluid is detected for the first time. The second term ensures that the two waves do not overlap due to the pulse duration.

Considering the acoustic attenuation in the fluid, the aim should be to minimise the fluid thickness while still fulfilling the condition above. The literature typically describes the frequency-dependent attenuation $\alpha$ by a power law [35,36]. For water, a quadratic dependence on frequency is commonly assumed [35,37–42].

$$\alpha_{Water} = k \cdot f^2 \quad (2)$$

The literature often provides values for $k$ in the range of $0.2\,\frac{\text{dB}}{\text{m}\,\text{MHz}^2}$, measured for frequencies up to 15 MHz [37,39–41]. Other sources give slightly higher attenuations of $0.4\,\frac{\text{dB}}{\text{m}\,\text{MHz}^2}$ to $1\,\frac{\text{dB}}{\text{m}\,\text{MHz}^2}$ [43] or $<1\,\frac{\text{dB}}{\text{m}\,\text{MHz}^2}$ [44] up to very high values of $44\,\frac{\text{dB}}{\text{m}\,\text{MHz}^2}$ [36]. Sometimes, a linear dependence is also assumed with values from $0.2\,\frac{\text{dB}}{\text{m}\,\text{MHz}}$ [45] to $30\,\frac{\text{dB}}{\text{m}\,\text{MHz}}$ [46].

The reflection and transmission factor and, therefore, the acoustic impedance difference at the fluid–sample interface influence the signal strength of the different reflections inside the sample. On the one hand, a low reflection coefficient is good for the transmission of the acoustic wave from the fluid into the sample, and on the other hand, a high reflection coefficient is good for additional reflections within the sample. Neglecting the transmission between the sample and the surrounding air, the optimal reflection coefficient $R_{opt}$ for the $n_S$ reflection inside the sample is

$$R_{opt} = \frac{n_S - 1}{n_S} \quad (3)$$

This fact must be considered when selecting the liquid. The choice of liquid can be influenced by other factors, such as potentially hazardous substances, solvent vapours, and the material compatibility of the sample with the liquid. Equally important is the cleaning effort required to remove the liquid from the cell and the sample.

We used deionised water and hydraulic oil in our setup. Unless stated otherwise, the fluid layer was water and 9 mm thick.

### 2.3.3. Absorption Layer

The absorption layer is the main part that influences the generation of acoustic waves. Excitation via an absorption layer has two advantages. Firstly, the material of the absorption layer can be freely chosen. Secondly, the excitation of the acoustic wave is independent of the sample.

The absorption layer should not be damaged; therefore, the excitation had to be selected to be in the thermoelastic regime (see Section 2.3.5). The damage threshold of the absorption layer defines the maximum pulse energy of the laser (Equation (7)).

The acoustic impedance of the absorption layer $Z_{AL}$ should match the impedance of the fluid $Z_F$ to enable good transmission of the acoustic pulse into the fluid.

$$Z_{AL} \approx Z_F \tag{4}$$

In Section 3.1, we simulated the influence of different material properties on the elongation and frequency spectra of the excited acoustic pulse (see Formulas (19)–(21)). The elongation of the pulse increases with the thermal expansion coefficient while decreasing with the density, specific heat capacity, Young's modulus, and optical absorption coefficient. The material parameters of polymer-based materials, with the exception of specific heat capacity and absorption coefficient, result in higher acoustic amplitudes compared to metal-based materials. Polymer-based materials also fulfil the impedance condition mentioned above better than metallic materials.

To achieve measurements with a high bandwidth, the decrease in the frequency spectrum should be small. According to the simulations (see Formula (20)), a high acoustic velocity and high optical absorption are advantageous. The desired slow decrease in the frequency spectrum is, therefore, inverse to the elongation in terms of the density and optical absorption coefficient. The frequency spectra (see Appendix A) show that the increasing power compensates for the stronger drop at high frequencies. Polymer-like materials are, therefore, preferable.

Another aspect to consider is the thickness of the absorber. If the absorber is slightly transparent, an acoustic pulse will be generated at the glass–absorption layer interface and the absorption layer–fluid interface. The influence of reflections within the absorption layer can be minimised by a layer that is as thin as possible.

We used two kinds of absorption layers. The aim of the first coating was to obtain a thin layer and a small slope in the frequency spectrum. A sputtering process was chosen for this purpose. Based on the absorption coefficient at the excitation wavelength, the coefficient of thermal expansion, the specific heat capacity, and the density, chromium was chosen for the absorption layer. The thickness was 125 nm.

The aim of the second layer was to excite waves with high amplitudes. Polydimethylsiloxane (PDMS) was chosen because of its very high coefficient of thermal expansion, low density, and low elastic modulus compared to metals. Carbon was added to the PDMS to increase optical absorption. The spin-coated layer was approximately 50 µm thick.

#### 2.3.4. Glass

A glass substrate has several functions. It acts as a support for the absorption layer and also as a constraining layer (see Section 2.3.5). A constraining layer increases the amplitude of the generated acoustic pulse. The glass needs to be transparent for the wavelength of the pulsed laser. The glass substrate must have a minimum thickness so that the reflection at the rear end of the substrate is sufficiently delayed until the wave reaches the detector position.

As discussed earlier, we want a minimum of $n_S$ pulses inside the sample before the first unwanted signal occurs. This results in a minimum glass thickness $x_G$ of

$$x_G > (n_S - 1)x_S \cdot \frac{c_G}{c_S} + c_S \cdot \tau_{Pulse} \tag{5}$$

where $c_G$ and $c_S$ are the longitudinal velocity inside the glass and sample, respectively; $x_S$ is the thickness of the sample; and $\tau_{Pulse}$ is the acoustic pulse duration. The signal strength is also positively influenced by a high reflection factor between the glass and the absorption layer. A high impedance difference should be aimed for.

$$Z_G \gg Z_{AL} \qquad \text{or} \qquad Z_G \ll Z_{AL} \tag{6}$$

where $Z_G$ and $Z_{AL}$ are the acoustic impedance of the glass and the absorption layer.

2.3.5. Excitation

For laser excitation, there are several points to consider: the excitation regime, the excitation spot size, the laser wavelength, and the pulse duration.

There are two regimes for laser excitation of ultrasonic waves. The first is thermoelastic excitation, which is truly non-destructive because the absorbed laser pulse is below the damage threshold of the material [15]. The second regime is ablative excitation, which is destructive because the laser pulse energy is above the damage threshold and vaporises the material [15].

Aussel et al. [47] show that material evaporation due to excessively absorbed power densities $AW$ occurs under the following condition:

$$AW \geq \frac{\pi}{4}\sqrt{\frac{\lambda_{th}\rho c_p}{\tau_L}(T_v - T_i)} \tag{7}$$

where $\lambda_{th}$ is the thermal conductivity, $\rho$ is the density, $c_p$ is the specific heat capacity, $\tau_L$ is the laser pulse duration, $T_v$ is the vaporisation temperature, and $T_i$ is the initial temperature.

For a point laser source, the excited waves under a propagation angle $\theta$ can be described by analytical formulas [13].

For the thermoelastic excitation of a point source [13]:

$$\begin{aligned} u_{c,thermoelastic} &\sim \frac{\sin 2\theta \cdot \left(k^2 - \sin^2\theta\right)^{0.5}}{\left(k^2 - 2\sin^2\theta\right)^2 + 4\sin^2\theta\left(1 - \sin^2\theta\right)^{0.5} \cdot \left(k^2 - \sin^2\theta\right)^{0.5}} \\ u_{s,thermoelastic} &\sim \frac{k \sin 4\theta}{k\left(1 - 2\sin^2\theta\right)^2 + 4\sin^2\theta\left(1 - \sin^2\theta\right)^{0.5} \cdot \left(1 - k^2\sin^2\theta\right)^{0.5}} \end{aligned} \tag{8}$$

And, for the ablative excitation of a point source [13]:

$$\begin{aligned} u_{c,ablative} &\sim \frac{2k^2 \cos\theta\left(k^2 - 2\sin^2\theta\right)}{\left(k^2 - 2\sin^2\theta\right)^2 + 4\sin^2\theta\left(1 - \sin^2\theta\right)^{0.5} \cdot \left(k^2 - \sin^2\theta\right)^{0.5}} \\ u_{s,ablative} &\sim \frac{\sin 2\theta\left(1 - k^2\sin^2\theta\right)^{0.5}}{k\left(1 - 2\sin^2\theta\right)^2 + 4\sin^2\theta\left(1 - \sin^2\theta\right)^{0.5} \cdot \left(1 - k^2\sin^2\theta\right)^{0.5}} \end{aligned} \tag{9}$$

where the wavevector $k = c_c/c_s$. The index $c$ represents the compression wave, and the index $s$ represents the shear wave. The analytical formulas are visualised in Figure 3 for aluminium.

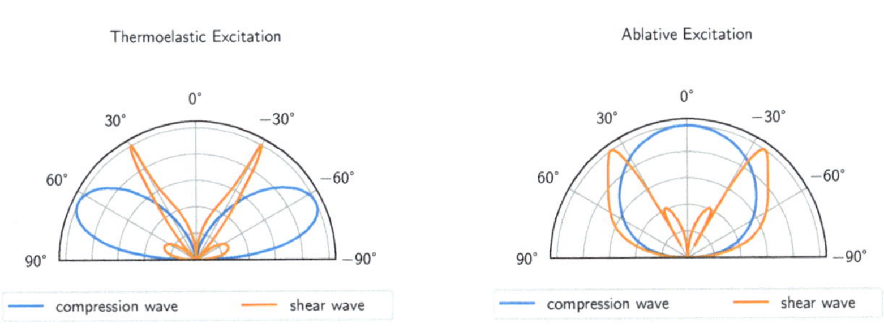


**Figure 3.** Directivity of compression and shear waves for point sources in case of aluminium.

The measuring cell presented in this paper is based on compression waves that propagate in a normal direction. Therefore, a point source in the thermoelastic excitation regime is not suitable due to the directivity of the compression waves (see Figure 3 (left)). Ablative excitation has a suitable radiation characteristic but cannot be used because of the minor surface damage it causes, which would damage the thin absorption layer.

Since the use of point sources is not advantageous in our case, extended sources were investigated. The excited waves using extended sources can be calculated using FEM. In our case, we used COMSOL 6.1. The propagated waves after 75 ns are visualised in Figure 4. The extended source results in a plane compression wave (PC) as well as additional waves (head (H), compression (C), shear (S), and Rayleigh (R) waves) at the edge of the excitation [47]. These waves do not contribute to the sensing concept but can interfere with the plane compression waves.

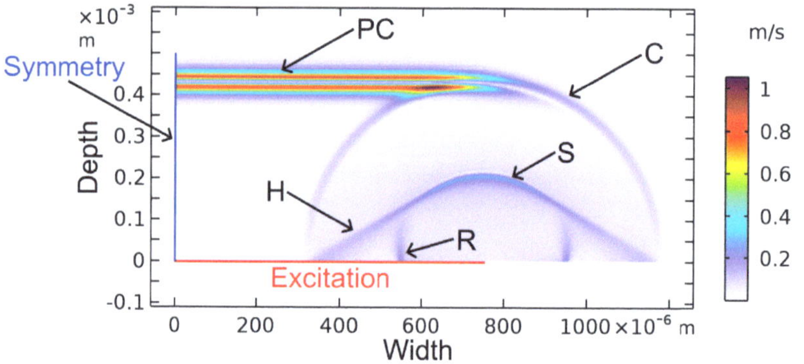


**Figure 4.** Simulation of excited waves in aluminium via heat impulse after 75 ns. C: compression wave; S: shear wave; R: Rayleigh wave; H: head wave; PC: plane compression wave. Colour scale corresponds to particle velocity.

A plane compression wave, whose plane wave front propagates parallel to the excitation, is the desired wave for this application. This plane compression wave is not obtained in point excitation.

In addition, the strength of the excitation amplitude can be increased by using a transparent constraining layer, resulting in a buried ultrasonic source. This enables the generation of normal stresses, resulting in an enhanced generation of compression waves [48]. Hutchins [49] used light oil, silicone resin, water, and acetone as a constraining layer and increased the compression wave amplitude by 21 to 25 dB compared to the unmodified surface. The use of a glass slide, cemented to the surface, increased the amplitude by 30 dB. The use of a buried source is, therefore, advantageous and was used in the experiments. In our case, the glass substrate and the water acted as a constraining layer. Deionised water was used to minimise the interaction with the specimen and reduce the cleaning effort compared to other fluids like oil.

If the measurement is not completely in the near field of the acoustic waves, there are two negative aspects to consider. In the far field, sound energy decreases with distance $z$ by $\propto 1/z^2$ [50], while it remains constant in the near field. In addition, the ultrasonic wave experiences a phase shift of $\pi/2$ in the transition between near and far field [51,52].

If the waves were measured in the transition area or in the far field, both effects had to be taken into account when calculating the attenuation or phase velocity. The near-field length $L$ can be used as an approximation for the transition between both fields [53,54].

$$L = \frac{D^2}{4 \cdot \lambda} \tag{10}$$

where $D$ represents the excitation diameter, and $\lambda$ represents the wavelength of the acoustic wave. High-frequency components are normally in the near field and therefore not affected.

The transition to the far field for low-frequency components resulted in a lower cut-off frequency, at which errors occurred.

The laser wavelength was of secondary importance. The laser light should have low absorption in the glass and high absorption in the absorption layer.

Pulse duration influenced the amplitude and frequency spectra of the excited waves. Longer pulses resulted in a higher energy input into the absorption layer and increased the amplitude of the acoustic wave while decreasing the frequency bandwidth (Formulas (20) and (21) and Figure A6). For measurements in the mid to upper double-digit MHz range, a single digit pulse duration of a few nanoseconds was appropriate. Short pulses with high power may damage the sample.

In our experiments, we used a 4 ns pulsed laser (Brilliant, Quantel-Laser, Les Ulis Cedex, France) at 532 nm and a pulse energy of about 18 mJ (after some neutral density filter). The beam diameter was chosen to be 18 mm to ensure diffraction-free propagation of the acoustic beam. A second pulsed laser (Wedge, Bright Solutions, Cura Carpignano, Italy) with a 1 ns pulse duration, at 532 nm, and a 1 mJ pulse energy was used to investigate the influence of pulse duration. Due to the low pulse energy, a beam cross-section of 2 mm was selected for this laser.

### 2.4. Data Analysis Methods

As mentioned in the introduction, the target quantity is a frequency-resolved measurement of phase velocity and attenuation over a wide frequency range. This chapter presents a method for calculating phase velocity and attenuation, determining an SNR level, and a method for correcting diffraction.

#### 2.4.1. Determination of Phase Velocity

One method for calculating a frequency-resolved phase velocity $c_{ph}$ is a phase spectrum analysis (PSA). This method uses two signals of the same acoustic wave $s_1$ and $s_2$ with a certain path difference $\Delta x$ and compares the phase change between both signals. This is achieved with the following formula [55,56]:

$$c_{ph}(f) = \frac{2\pi f \Delta x}{\Delta \phi}. \tag{11}$$

The phase difference $\Delta\phi$ can be calculated from the Fourier-transformed signals of both windowed signals [56]:

$$\Delta\phi = \arctan \frac{\Im(\mathcal{F}(s_1) \cdot \mathcal{F}(s_2)^*)}{\Re(\mathcal{F}(s_1) \cdot \mathcal{F}(s_2)^*)} \tag{12}$$

where $\mathcal{F}$ is the operator for the fast Fourier transformation, $*$ is the operator for the complex conjugated value, and $\Im$ and $\Re$ are operators for the imaginary and real parts, respectively.

#### 2.4.2. Determination of Noise Level

To decide up to which frequency there was still sufficient signal strength, a frequency-dependent noise level was calculated by measuring noise $S_{noise}$ without any acoustic signal. The noise spectrum is Gaussian-like and is determined by the measurement setup and the digital filter used. The noise spectrum $N(f)$ is calculated by

$$N(f) = 20\,\mathrm{dB}\log_{10}|\mathcal{F}(S_{noise})| \tag{13}$$

several times, and a quadratic formula was fitted to the averaged noise level. Three times the standard deviation of the fit residuals on top of the quadratic fit indicates the border between noise and signal (Figure 5).

The signal $S_{measurement}$ can be approximated by a linear decay in the logarithmic frequency spectrum [57]. Low frequencies are neglected in the linear fit due to diffraction effects (see Section 2.4.4). The intersection between the linear approximation of the signal and the approximation of the noise level $N_{noiselevel}$ defined the bandwidth of the signal. Every frequency above was interpreted as noise and was neglected.

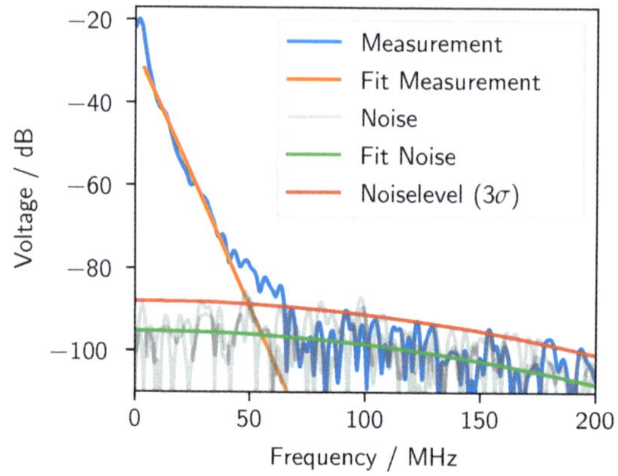


**Figure 5.** Determination of noise level from numerous noise measurements (grey signals) and the quadratic fit. The intersection of the noise level and the linear fit of the measurement data defined the maximum frequency of the signal.

The energy spectra shown in Section 3 are plotted as $SNR$.

$$SNR = S_{measurement} - N_{noiselevel} \tag{14}$$

### 2.4.3. Determination of Attenuation

In addition to the phase velocity, it was also possible to calculate the frequency-resolved attenuation $\alpha$ of the material within the path $\Delta x$.

$$\alpha = \frac{20\,\mathrm{dB}\log\frac{|\mathcal{F}(s_1)|}{|\mathcal{F}(s_2)|}}{\Delta x} \tag{15}$$

For the attenuation, the bandwidth was defined by the $SNR$ of both signals $s_1$ and $s_2$.

Within the setup described in Section 2.1, the analysed ultrasound waves were reflected at the sample–air interface and the sample–water interface. These reflections and the associated energy transfer to the water and air had to be taken into account when calculating the attenuation. The energy remaining in the sample depends on the reflection coefficient $R$ [58]

$$R = \left(\frac{Z_2 - Z_1}{Z_1 + Z_2}\right)^2 \tag{16}$$

where the acoustic impedance $Z = \rho \cdot c_{ph}$. Each reflection results in an offset of the attenuation spectra and needs to be corrected. This results in the following formula

$$\alpha = \frac{20\,\mathrm{dB}\log\frac{|\mathcal{F}(s_1)|}{|\mathcal{F}(s_2)|} + \sum_i 10\,\mathrm{dB}\log(R_i)}{\Delta x} \tag{17}$$

where $i$ represents the reflections that occur between the signals $s_1$ and $s_2$ of the acoustic pulse.

### 2.4.4. Diffraction Correction

For small beam diameters, long propagation paths, or low frequencies, the acoustic wave can transition from the near to the far field. This transition occurs at the approximated near field length $L$ (see Equation (10)).

A phase difference of $\pi/2$ of the acoustic wave between the near and far field distorts the determination of the phase velocity. The calculation of the attenuation in the far field is also subject to errors, as the energy decreases with distance. A correction of the phase and amplitude can be performed by calculating the diffracted wave $A(x,y,z)$ [52]

$$A(x,y,z) = \frac{i}{\lambda z} \exp\left(-\frac{i\pi}{\lambda z}\left(x^2+y^2\right)\right) \cdot \mathcal{F}\left(A_0(x,y) \exp\left(-\frac{i\pi}{\lambda z}\left(x^2+y^2\right)\right)\right) \quad (18)$$

where $z$ is the propagation direction, $A_0(x,y)$ is the acoustic source distribution, and $i$ is the imaginary unit. To the best of our knowledge, there is no better correction for the approach taken.

## 3. Results and Discussion

With the use of simulation, we investigated the influence of the material properties of the absorption layer and the laser parameters on the excitation of the acoustic waves. These results were verified by experiments. The phase velocities of various materials were determined from further measurements.

### 3.1. Simulation of Material and Laser Influence on the Excitation

Simulations were carried out with the aim of characterising the influence of the material parameters of the excited object to maximise the strength of the planar compression wave. The second criterion was a high-frequency bandwidth of the acoustic pulse to maximise the frequency range from which later information can be deduced. The simulations were 2D FEM simulations (similar to Figure 4), employing both the structural mechanics and heat transfer modules from COMSOL software. (V 6.1) Both were coupled using the formula for thermal expansion.

The optical excitation was simulated by a time-dependent heat flux along the excitation line of 750 µm and an exponential decay of the heat flux in the depth direction with a damping factor $\alpha_{op}$. The thermal pulse had a constant intensity, a duration $\tau_L$ of 4 ns, and was completely absorbed. Different from what is usually found in the literature, we calculated the acoustic pulse in a solid absorption layer instead of a weakly absorbing fluid [59,60].

The simulations used aluminium as the material, with the parameters given in Table 1. The simulation area was larger than the penetration depth of the optical pulse.

**Table 1.** Used material properties for the FEM simulation.

| | |
|---|---|
| Elastic modulus $E$ | 70 GPa |
| Density $\rho$ | 2700 $\frac{\text{kg}}{\text{m}^3}$ |
| Poisson ratio $\nu$ | 0.33 |
| Coefficient of thermal expansion $\alpha_{th}$ | $23 \times 10^{-6}$ $\frac{1}{\text{K}}$ |
| Thermal conductivity $\lambda_{th}$ | 238 $\frac{\text{W}}{\text{m K}}$ |
| Specific heat capacity $c_p$ | 900 $\frac{\text{J}}{\text{kg K}}$ |

In the simulations, the material and optical parameters were varied to determine the influence on the excited acoustic wave in the time and frequency domain. The optical parameters were laser pulse length $\tau_L$ and optical material attenuation $\alpha_{op}$. The laser power was kept constant as the pulse length was varied. The material parameters (thermal conductivity $\lambda_{th}$, coefficient of thermal expansion $\alpha_{th}$, specific heat capacity $c_p$, density $\rho$, and Young's modulus $E$) were varied to cover the entire range of common materials. Appendix A contains the simulated signals and shows the dependence of these parameters on the sum of the squared elongations $s_{ac}^2$, which is proportional to the energy of the excited planar compression wave.

$$s_{ac}^2 \propto \frac{\alpha_{th}^2 \cdot \tau_L^{1.5}}{c_p^2 \cdot \alpha_{op}^{1.5} \cdot \rho^{1.25} \cdot E^{0.75}} \quad (19)$$

For short optical pulses, the linearly scaled frequency spectrum of the elongations can be approximated using a Gaussian function [57]. The standard deviation $\sigma$ and the amplitude $A$ of the Gaussian function $A \exp\left(-f^2/\sigma^2\right)$ of the autocorrelated elongations were determined. This led to the following expressions for the plane compression wave:

$$\sigma \propto \sqrt{\frac{\alpha_{op}}{\tau_L} \cdot \sqrt{\frac{E}{\rho}}} \tag{20}$$

$$A \propto \frac{\alpha_{th}^2 \cdot \tau_L^2}{c_p^2 \cdot \alpha_{op}^2 \cdot \rho \cdot E} \tag{21}$$

The formula for the squared elongations in the time domain is identical to the integral of the Gaussian function. The aim was to obtain a high elongation and a high $\sigma$ for high-quality broadband measurements. A higher bandwidth can be achieved by a higher acoustic velocity ($c \propto \sqrt{\frac{E}{\rho}}$), stronger optical absorption, or shorter pulse durations. With the exception of the thermal expansion coefficients and the specific heat capacity, the energy cannot be optimised independently of the bandwidth. In fact, an optimisation of $s_{ac}^2$ can simultaneously cause a worsening of $\sigma$. However, a smaller $\sigma$ can often be neglected due to the higher elongations (see spectra in Appendix A).

### 3.2. *Characterisation of Acoustic Measurement Cell*

Measurements were carried out to characterise the effect of the laser pulse width, the fluid (width and type) and the excitation layer.

#### 3.2.1. Excitation Layer

As mentioned in Section 2.3.3, measurements were carried out with two different excitation layers (125 nm chromium, 50 µm PDMS). The results are shown in Figure 6.

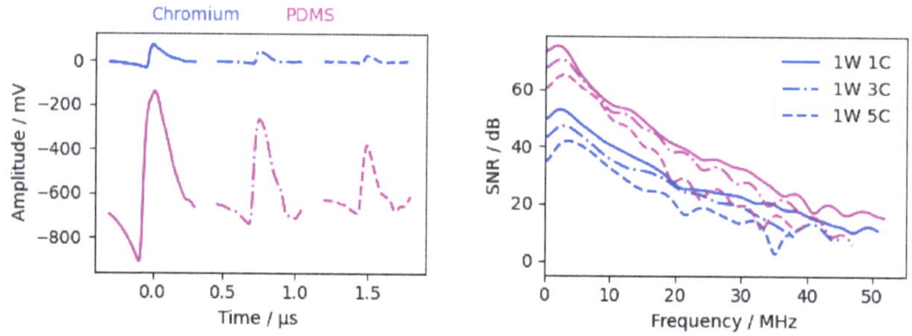


**Figure 6.** Measurements with different excitation layers. (**Left**) time signal from interferometer. PDMS signals have an offset of −610 mV. (**Right**) calculated SNR according Section 2.4.2.

The material parameters of the PDMS layer were not known, so a comparison with Formulas (19)–(21) cannot be made. Nevertheless, it is clear that the waves excited by the PDMS layer have a higher amplitude due to the different material parameters, probably mainly due to the higher coefficient of thermal expansion and the lower specific heat capacity, Young's modulus, and density. The amplitude of the PDMS layer is about 7 to 9 times higher than the chromium layer. As long as the SNR level is high enough, the calculated phase velocity is independent of the excitation layer.

The decrease in signal strength with higher frequencies is higher for the PDMS layer than for the chromium layer. This is predicted by the simulations and Formula (20). The lower amplitude combined with the slower decrease in signal strength resulted in a similar maximum frequency for PDMS and chromium. The waviness in the PDMS spectrum is caused by an additional wave within the signal peak. The additional wave can be seen clearer at the second peak in the time domain. This wave is even more visible in the measurement with a 1 ns pulsed laser (Figure 7, red circle). The reason for this additional wave is the thicker PMDS layer in which the wave is reflected.

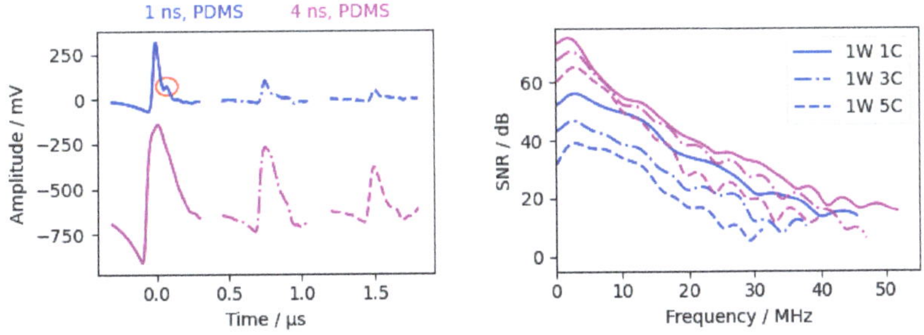


**Figure 7.** Measurements with different excitation layers. (**Left**) time signal from interferometer. The 1 ns signal was amplified by 3, the 4 ns signals have an offset of −610 mV, and the red circle shows the reflection within the PDMS layer. (**Right**) calculated SNR according Section 2.4.2.

Normally, a superposition of two waves is not desired for the phase spectrum analysis. It is, therefore, preferable to use the chromium layer as long as the amplitudes are sufficiently high.

### 3.2.2. Laser Pulse Width

In addition to the 4 ns pulse laser, we performed measurements with a 1 ns pulse laser (1 mJ pulse energy). Due to the lower pulse energy, the beam diameter had to be reduced to 2 mm to observe an acoustic wave (the 4 ns laser had a beam diameter of 18 mm). The measurements are shown in Figure 7. For better visibility, the measurement signal from the 1 ns laser is amplified by a factor of 3.

Due to the shorter 1 ns pulse, the additional reflection within the PDMS layer was clearly visible in the 1W1C signal in the time domain (Figure 7, red circle). The time delay between the two peaks is 0.076 µs, corresponding to a PDMS layer thickness of 42 µm. The subsequent waves decay rapidly and are barely observed. The decay between the different waves is greatly enhanced due to the small beam diameter and, therefore, the high diffraction of the acoustic pulse. The lower amplitude and slower decay in the frequency domain are analogous to the simulations (see Figure A6). The frequency spectrum of the 1W1C wave had a similar bandwidth between the different pulse lasers. This suggests that the acoustic bandwidth can be further increased by using a higher pulse energy and a 1 ns pulsed laser.

This suggests that the bandwidth could be further increased with a 1 ns laser pulse if the laser has a higher pulse energy.

### 3.2.3. Water Layer Thickness

Figure 8 shows three measurements with different water layer thicknesses (with chromium as the excitation layer). The measurements showed an unexpected behaviour, as the amplitudes are significantly higher with a 9.5 mm water layer than with a 7 mm water layer, although the attenuation in the 9.5 mm measurement should be higher. In the calculated attenuation spectrum (Figure 9), the attenuation between 1W1C and 3W1C increased with increasing water thickness. Therefore, the unexpected behaviour is probably due to different levels of excitation due to laser power fluctuations. In summary, the water layer should be as thin as possible but thick enough to clearly separate the waves within the sample and the waves within the water.

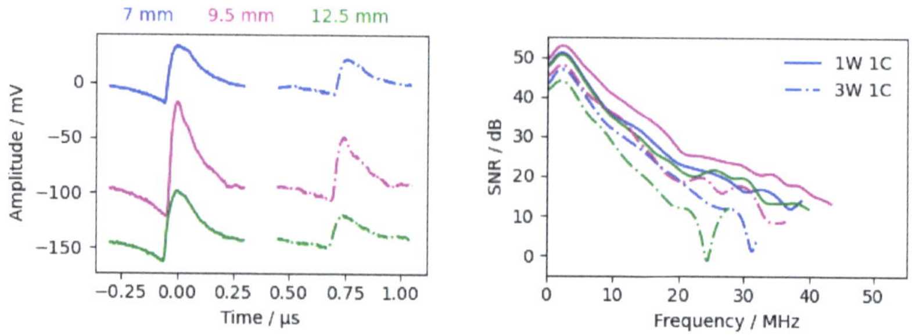


**Figure 8.** Measurements with different water layer thicknesses with chromium as excitation layer. (**Left**) time signal from the interferometer. The 9.5 mm measurement has an offset of −90 mV, and the 12.5 mm measurement has an offset of −140 mV. (**Right**) calculated SNR according Section 2.4.2.

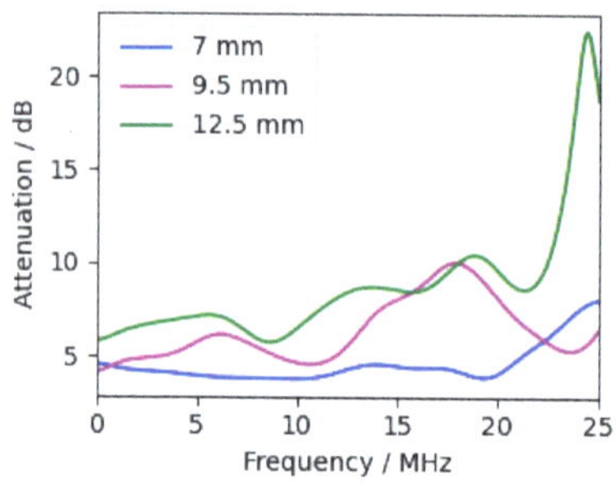


**Figure 9.** Frequency-resolved attenuation of measurements with different water layer thicknesses (Equation (15)). Compared peaks are 1W1C and 3W1C. Excitation layer was chromium.

### 3.2.4. Type of Fluid

The last variable investigated was the fluid between the excitation layer and the sample. The water was replaced by a mineral hydraulic oil (HLP 46, viscosity 46 $\frac{\text{mm}^2}{\text{s}}$), and the results are shown in Figure 10. The oil significantly attenuated higher frequencies, whereas lower frequencies (<3.5 MHz) were amplified by up to 3 dB. This suggests that oil may be a better choice for low-frequency measurements but is not suitable for frequencies above 5 MHz. This is consistent with the literature, in which it has been reported that water typically has an attenuation of ~0.2 $\frac{\text{dB}}{\text{MHz}^2\,\text{m}}$ [37,39,40] and mineral oil has an attenuation of 24 $\frac{\text{dB}}{\text{MHz}^2\,\text{m}}$ [61].

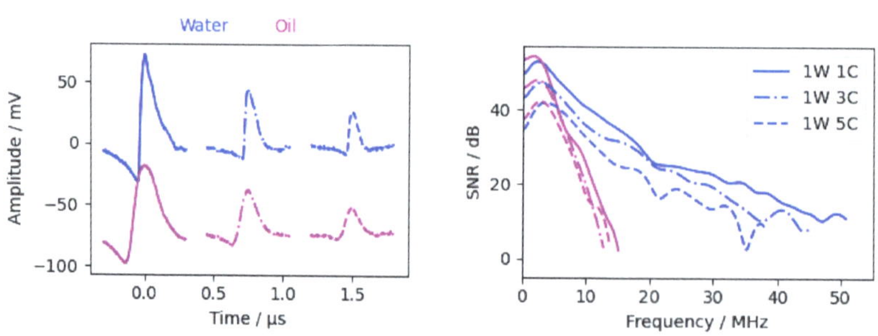


**Figure 10.** Measurements with oil and water between excitation layer and specimen. Chromium was used as excitation layer. (**Left**) time signal from the interferometer. Oil signals have an offset of −70 mV. (**Right**) calculated SNR according to Section 2.4.2.

### 3.3. Frequency Resolved Measurements in Different Materials

In this section, we show that frequency-resolved phase velocity measurements were possible in virtually any material. For this reason, we have chosen a high-velocity material (silicon single crystal), a metal (aluminium), a soft low-velocity material (silicone), and a liquid (water).

#### 3.3.1. Silicon

The sample used was a silicon window with a 25.4 mm diameter and a 5 mm thickness. The acoustic pulse travelled along the 5 mm axis, which corresponds with the [111]-orientation of the silicon crystal. The measurement is visualised in Figure 11. The PDMS absorption layer was used.

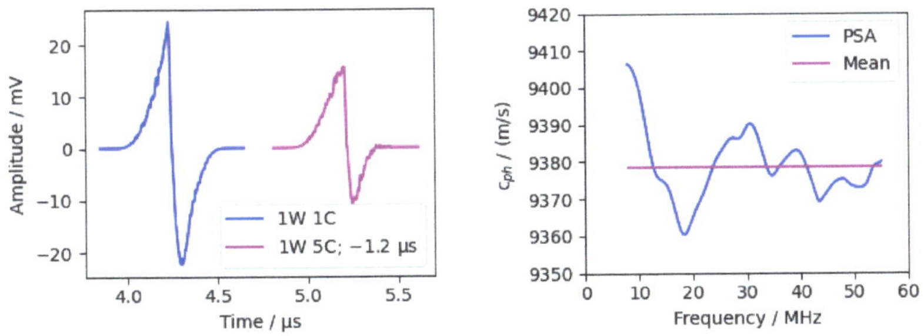


**Figure 11.** Measurement of 5 mm thick silicon sample. (**Left**) time domain with shifted second peak. (**Right**) phase velocity calculated via phase spectrum analysis (PSA).

The calculated phase velocity was approximated constant over the frequency with a mean value of 9378.4 $\frac{\text{m}}{\text{s}}$ and a standard deviation of 9.2 $\frac{\text{m}}{\text{s}}$.

The anisotropic phase velocity of silicon crystal in the [111]-direction in material libraries vary from 9362 $\frac{\text{m}}{\text{s}}$ [62] to 9372 $\frac{\text{m}}{\text{s}}$ [63].

Considering the inaccuracy of the thickness measurement (using a micrometre screw), the measurements were in good agreement with the literature values. The percentage deviations were 0.17 % [62] and 0.07% [63].

#### 3.3.2. Aluminium

The aluminium sample was 12 mm thick in the measurement direction and wide enough to suppress any reflections from the sample boundaries. The absorption layer used was chromium. The results are shown in Figure 12.

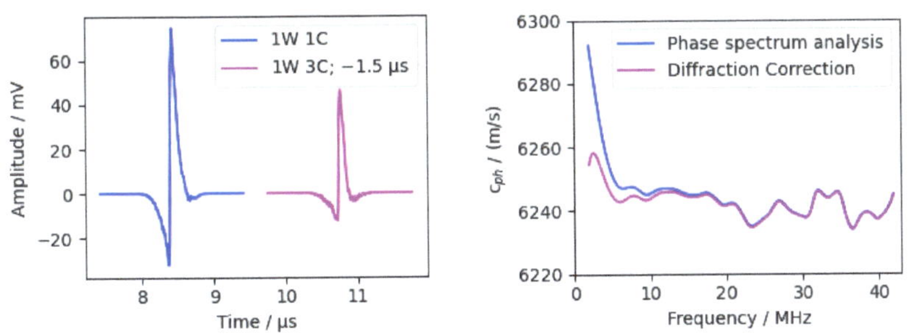


**Figure 12.** Measurement of 12 mm thick aluminium sample. (**Left**) time domain with shifted second peak. (**Right**) phase velocity calculated via phase spectrum analysis (PSA).

The phase velocity showed a constant value for frequencies above 9 MHz with an increasing velocity towards lower frequencies. This was due to the transition between the near and far field and the resulting phase change of $\pi/2$. The diffraction correction (Section 2.4.4) significantly reduces the increase at low frequencies and, thus, improves the result. No comparison was made between the measured sound speed and the literature data, as the literature data differ greatly due to the different alloys used (from 6200 $\frac{m}{s}$ to 6400 $\frac{m}{s}$ [35]).

#### 3.3.3. Silicone

The silicone sample was 2.2 mm thick and translucent. The silicone used was SF00 from "Silikonfabrik". The absorption layer used was PDMS.

As can be seen in Figure 13, measurements in the soft silicone layer were possible up to a frequency of 15 MHz. The measurement was made more challenging by the fact that the acoustic impedance of water and silicone is very similar; therefore, only a small proportion of the sound pulse was reflected at the interface (R∼2%). This resulted in a low SNR and a low maximum frequency. With a velocity of ∼1090 $\frac{m}{s}$, the measurements agree quite well with other measurements of similar materials in the literature [64–66].

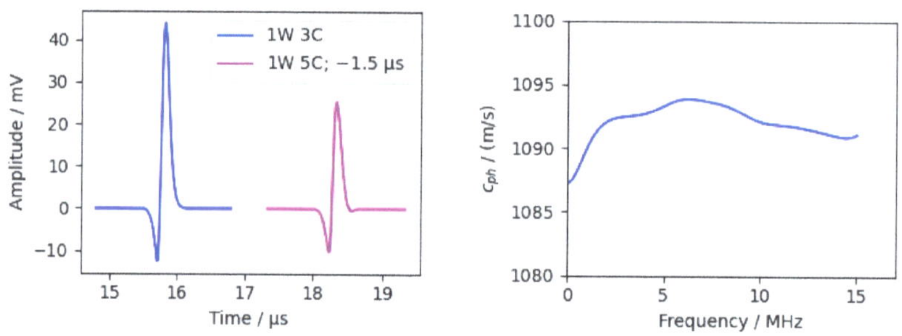


**Figure 13.** Measurement of 2.2 mm thick silicone sample. (**Left**) time domain with shifted second peak. (**Right**) phase velocity calculated via phase spectrum analysis (PSA).

#### 3.3.4. Water

In addition, the fluid between the excitation layer and the sample can be analysed. The measurement of the silicon sample was used for this, and the reflections in the water were analysed. The results are shown in the following figures (Figure 14).

The mean velocity was 1483.8 $\frac{m}{s}$ with a standard deviation of 1 $\frac{m}{s}$. The literature value for water at 20 °C is 1482.3 $\frac{m}{s}$ [67]. Considering the accuracy of the thickness measurement and the temperature uncertainty, the measurement is in good agreement.

In contrast to the previous measurements, the attenuation is also shown, as there are literature values for water with which our measuring cell can be compared. The "Reflection correction" takes into account the reflection coefficients (see Section 2.4.3). As mentioned in Section 2.3.2, the attenuation in water is usually described by a quadratic dependence on the frequency. The attenuation coefficients vary in a wide range between 0.2 $\frac{dB}{m\,MHz^2}$ [37,39–41] and 1 $\frac{dB}{m\,MHz^2}$ [43,44] and up to 44 $\frac{dB}{m\,MHz^2}$ [36].

The measurement presented here had an attenuation coefficient of 0.36 $\frac{dB}{m\,MHz^2}$, which is between the lower and slightly higher literature values.

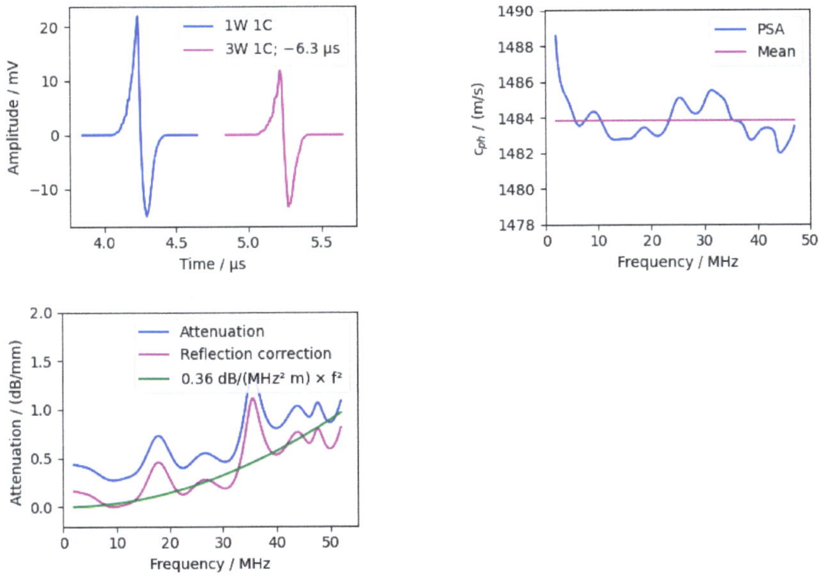


**Figure 14.** Measurement of 5.4 mm thick water layer between PDMS absorption layer and silicon window. (**Left**) time domain with shifted second peak. (**Right**) phase velocity calculated via phase spectrum analysis (PSA). (**Bottom**) attenuation of water with quadratic fit.

## 4. Conclusions

In this work, we presented a method to measure the longitudinal phase velocity and attenuation contactless over a high-frequency range (∼3 to 55 MHz, depending on material). We have investigated all components of the acoustic measuring cell and identified their influencing factors on the measurement. Simulations were used to analyse the excitation of compression waves. An unfocused beam with a large beam diameter is required to excite the intended planar compression waves. For the excitation of pulses with high energies, an absorption layer with a low elastic modulus, low density, low specific heat capacity, and high thermal expansion should be chosen. However, these material parameters also cause the energy of the excited acoustic waves to decrease more sharply at higher frequencies (Section 3.1). We were able to show this behaviour in a simulation and an experiment.

Therefore, polymers with additives to increase optical absorption would be suitable to excite high-energy waves. The greater decrease in energy at higher frequencies for polymers is partially offset by the increased energy. However, the experiments have shown that polymers need to be applied as a thin layer while still having a high optical absorption to suppress additional reflections of the acoustic pulse at the polymer–water and polymer–glass interfaces. These additional pulses interfere with the determination of the phase velocity and attenuation. The thin sputtered chromium layer (125 nm) did not show the additional pulse, but the amplitude decreased by a factor of 7 to 9 in comparison with the PDMS layer. Future experiments should, therefore, consider a new polymer (-like) material for the absorption layer, which can be applied thinly and has a high optical absorption.

The water layer fulfilled the intended purpose: transmitting only the compression wave, enabling the excitation of broadband acoustic pulses independent of the sample material, and separating the waves in the time domain. The interferometer with a photorefractive crystal also showed that it can measure the acoustic pulse independent of the sample material up to high frequencies, even with weakly reflective samples.

The accuracy of the setup was demonstrated by measuring silicon and water, the sound velocities of which have been extensively analysed in the literature. The relative error was below 0.2%. The accuracy of the attenuation measurement cannot be estimated since there are no good reference materials with consistent attenuation values found in the literature. Nevertheless, the attenuation measurement for water is between the attenuation values specified in the literature.

In summary, we have developed a measurement technique that can measure the frequency-resolved phase velocity and attenuation of a sample contactlessly (only in contact with water), non-destructively, independently of the sample, and over a wide frequency range.

**Author Contributions:** Conceptualization, F.B.; methodology, F.B.; software, F.B.; validation, F.B.; writing—original draft preparation, F.B.; writing—review and editing, F.B. and K.S.D.; visualization, F.B.; supervision, K.S.D.; project administration, K.S.D.; funding acquisition, K.S.D. All authors have read and agreed to the published version of the manuscript.

**Funding:** This research was funded by the funding program "Photonics Research Germany" of the German Federal Ministry of Education and Research (grant number: 13N15899).

**Institutional Review Board Statement:** Not applicable.

**Informed Consent Statement:** Not applicable.

**Data Availability Statement:** The data presented in this study are available on request from the corresponding author.

**Acknowledgments:** The authors would like to thank Sandra Ebert for making the absorption layers.

**Conflicts of Interest:** The authors declare no conflicts of interest.

## Appendix A. Simulation of Laser-Based Excitation of Ultrasonic Waves

This appendix details the simulations in Section 3.1 to characterise the influence of the absorption layer.

Simulations were performed in which different material and laser properties were varied in order to observe the effect on the amplitude and the frequency spectrum of the excited acoustic compression pulse. The material parameters were selected to cover the entire range of common materials.

The time signals of the elongation on the symmetry axis in a depth of 0.25 mm is plotted. The elongations of the time signals are scaled to the elongation of the original material properties (see Table 1). The frequency spectra of the pulses are equally normalised to the original material parameters.

The influence of the thermal material properties (thermal conductivity, coefficient of thermal expansion, and specific heat capacity) can be seen in Figures A1–A3. The thermal conductivity has no influence on the elongation nor the frequency spectrum. This is due to the fact, that the heat diffusion is rather slow compared to the penetration depth, the duration of the time of the heat pulse and the acoustic wave propagation. This results in an almost stable temperature profile.

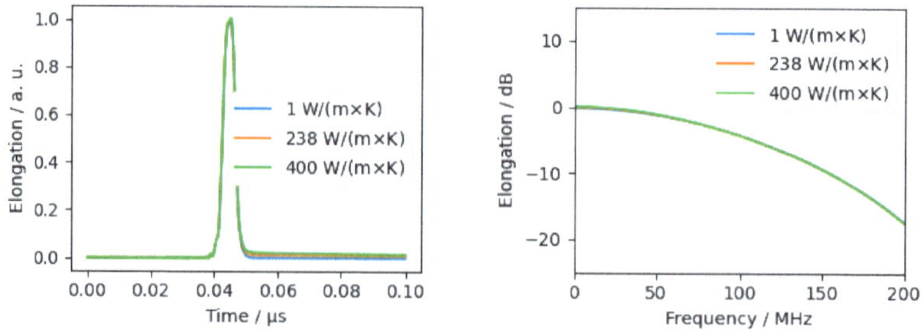


**Figure A1.** FEM-Simulation with different thermal conductivity.

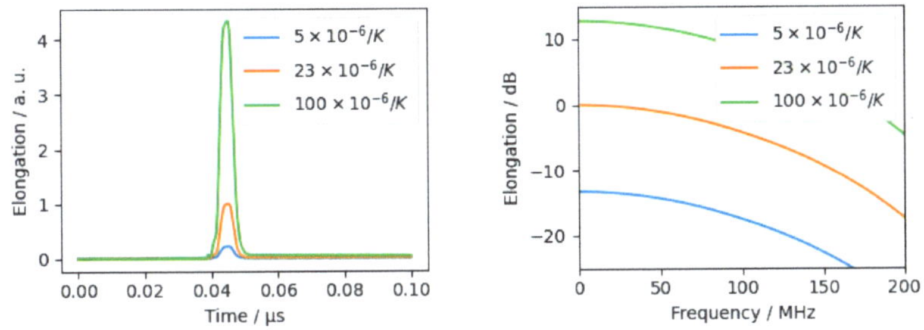


**Figure A2.** FEM-Simulation with different coefficient of thermal expansion.

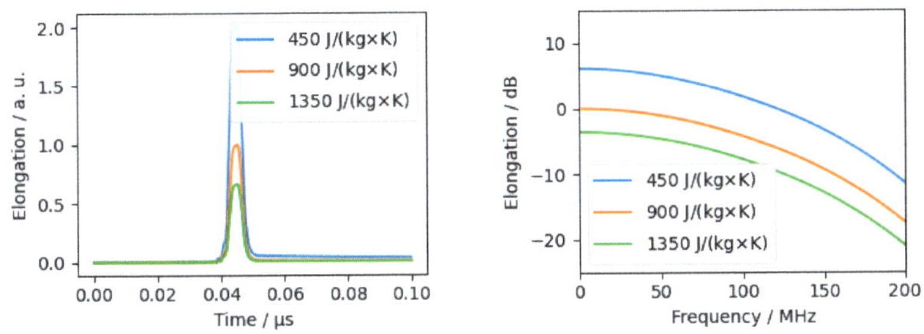


**Figure A3.** FEM-Simulation with a different specific heat capacity.

The thermal expansion and the specific heat capacity had a strong effect on the excited elongation $s_{ac}^2 \propto \alpha_{th}^2 / c_p^2$. Both were quite intuitive, as the amplitude of the acoustic pulse depends on the temperature increase during thermal excitation and the associated thermal expansion. As the pulse shape does not change, the frequency spectrum is the same with the exception of an offset.

The mechanical properties (density and Young's modulus) affected the speed of sound in the material. It can be seen that the mechanical parameters influenced the converted sound amplitude and the frequency spectra in both the time and frequency domain.

The excitation parameters (laser, respectively, heat pulse duration and the optical absorption coefficient) are shown in Figures A6 and A7. The pulse duration had a significant effect on the frequency bandwidth as well as the elongation. It should be noted that when varying the pulse duration, the introduced thermal power was kept constant in the simulation and therefore the pulse energy increases at longer pulse durations.

For the variation of the attenuation of the optical material, the inserted thermal power was kept constant for all absorption coefficients.

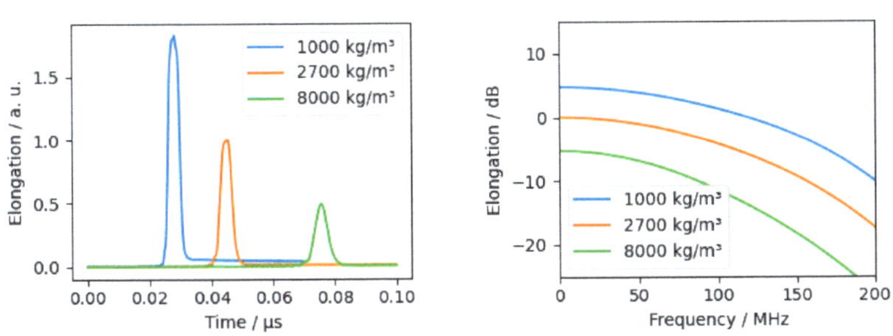


**Figure A4.** FEM-Simulation with different densities.

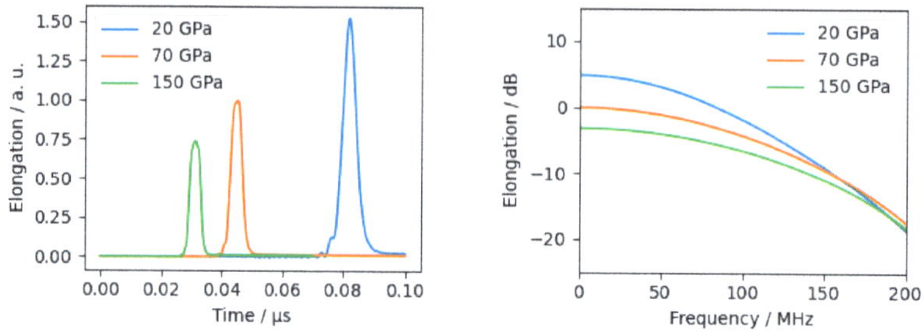


**Figure A5.** FEM-Simulation with different Young's moduli.

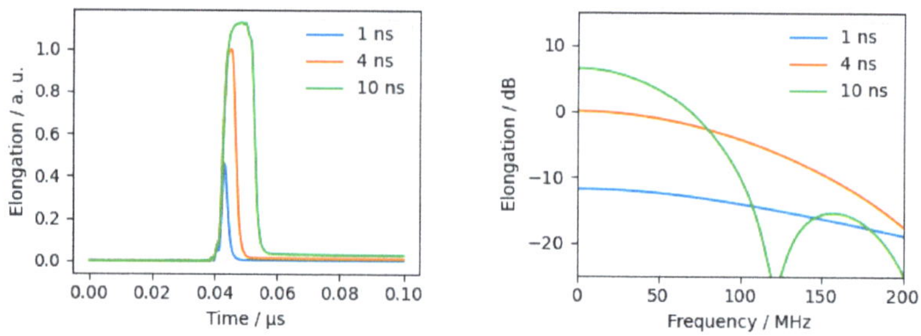


**Figure A6.** FEM-Simulation with different heat pulse length.

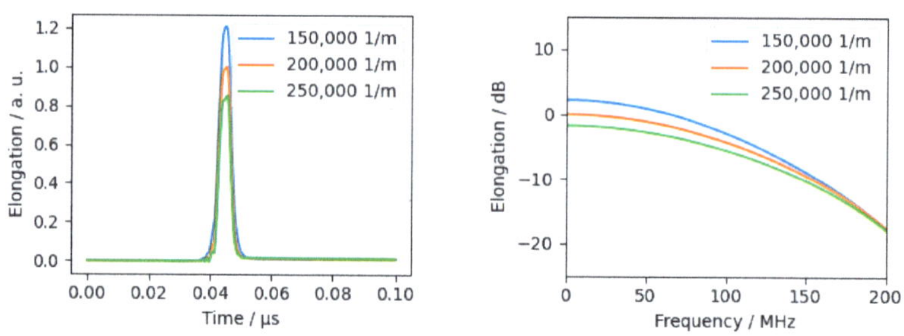


**Figure A7.** FEM-Simulation with different material absorption coefficient.

On the basis of the time signals and spectra shown in Figures A1–A7 it was possible to derive various dependencies between the sum of the squared elongation $s_{ac}^2$, the standard deviation $\sigma$ and the amplitude $A$ of the Gaussian fit in the linear scaled spectrum. The dependencies are visualised in the following figures and summarised in Formulas (19)–(21) in Section 3.1.

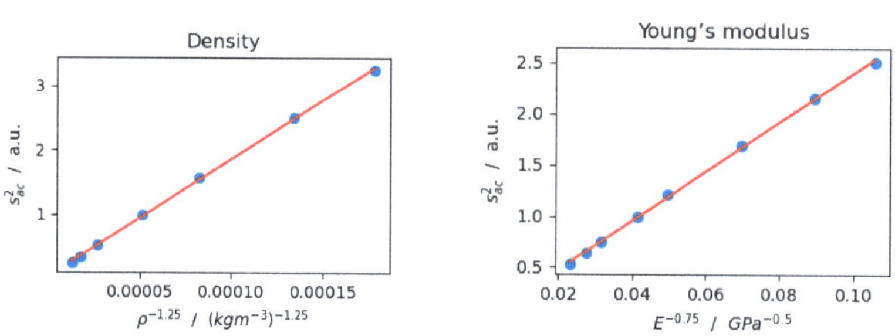


**Figure A8.** *Cont.*

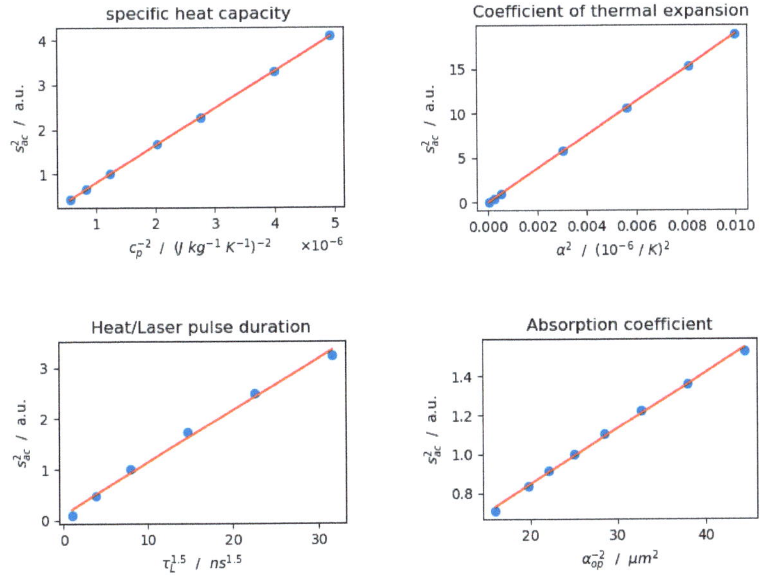


**Figure A8.** Dependencies of material and excitation parameters on the elongation $s_{ac}^2$.

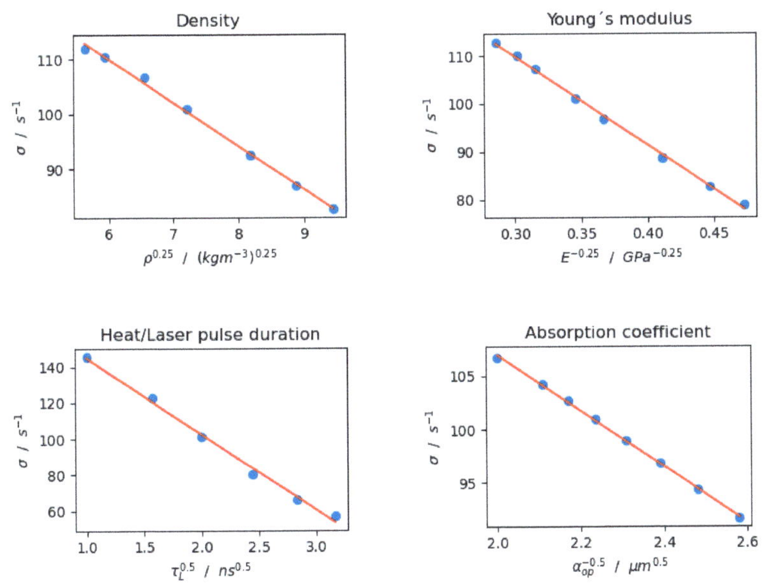


**Figure A9.** Dependencies of the material and excitation parameters on the $\sigma$ parameter.

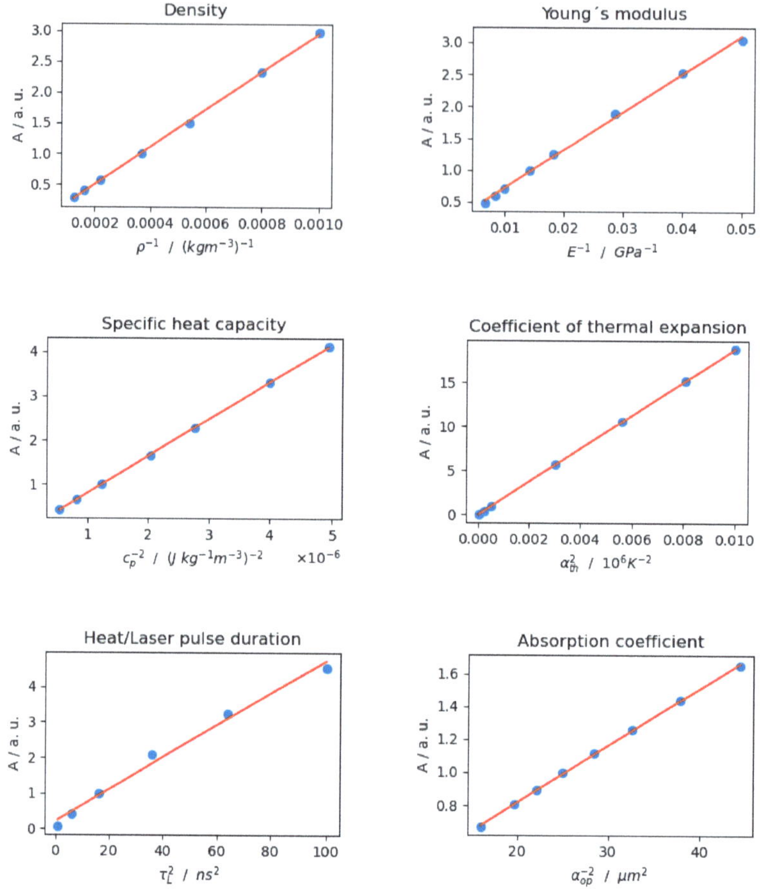


**Figure A10.** Dependencies of material and excitation parameters on the $A$ parameter.

## References


1. Sharpe, W.N. (Ed.) *Springer Handbook of Experimental Solid Mechanics: With 50 Tables*; Springer Handbooks; Springer: New York, NY, USA, 2008.
2. Barbosa, R.C.S.; Mendes, P.M. A Comprehensive Review on Photoacoustic-Based Devices for Biomedical Applications. *Sensors* **2022**, *22*, 9541. [CrossRef] [PubMed]
3. Isaiev, M.; Mussabek, G.; Lishchuk, P.; Dubyk, K.; Zhylkybayeva, N.; Yar-Mukhamedova, G.; Lacroix, D.; Lysenko, V. Application of the Photoacoustic Approach in the Characterization of Nanostructured Materials. *Nanomaterials* **2022**, *12*, 708. [CrossRef] [PubMed]
4. Wissmeyer, G.; Pleitez, M.A.; Rosenthal, A.; Ntziachristos, V. Looking at sound: Optoacoustics with all-optical ultrasound detection. *Light Sci. Appl.* **2018**, *7*, 53. [CrossRef] [PubMed]
5. Karabutov, A.A.; Podymova, N.B. Influence of the porosity on the dispersion of the phase velocity of longitudinal acoustic waves in isotropic metal matrix composites. *Acoust. Phys.* **2017**, *63*, 288–296. [CrossRef]
6. Podymova, N.B.; Karabutov, A.A. Combined effects of reinforcement fraction and porosity on ultrasonic velocity in SiC particulate aluminum alloy matrix composites. *Compos. Part B Eng.* **2017**, *113*, 138–143. [CrossRef]
7. Bakaric, M.; Miloro, P.; Javaherian, A.; Cox, B.T.; Treeby, B.E.; Brown, M.D. Measurement of the ultrasound attenuation and dispersion in 3D-printed photopolymer materials from 1 to 3.5 MHz. *J. Acoust. Soc. Am.* **2021**, *150*, 2798–2805. [CrossRef]
8. Jacquet, J.R.; Levassort, F.; Ossant, F.; Gregoire, J.M. 3D printed phantom for high frequency ultrasound imaging. In Proceedings of the IEEE International Ultrasonics Symposium (IUS), Taipei, Taiwan, 21–24 October 2015; pp. 1–4.
9. Demčenko, A.; Akkerman, R.; Nagy, P.B.; Loendersloot, R. Non-collinear wave mixing for non-linear ultrasonic detection of physical ageing in PVC. *NDT E Int.* **2012**, *49*, 34–39. [CrossRef]

10. Goldsberry, B.M.; Haberman, M.R. Negative stiffness honeycombs as tunable elastic metamaterials. *J. Appl. Phys.* **2018**, *123*, 091711. [CrossRef]
11. Wang, J.; Allein, F.; Boechler, N.; Friend, J.; Vazquez-Mena, O. Design and Fabrication of Negative-Refractive-Index Metamaterial Unit Cells for Near-Megahertz Enhanced Acoustic Transmission in Biomedical Ultrasound Applications. *Phys. Rev. Appl.* **2021**, *15*, 024025. [CrossRef]
12. Li, Z.; Yang, D.Q.; Liu, S.L.; Yu, S.Y.; Lu, M.H.; Zhu, J.; Zhang, S.T.; Zhu, M.W.; Guo, X.S.; Wu, H.D.; et al. Broadband gradient impedance matching using an acoustic metamaterial for ultrasonic transducers. *Sci. Rep.* **2017**, *7*, 42863. [CrossRef]
13. Scruby, C.B.; Drain, L.E. *Laser Ultrasonics: Techniques and Applications*; Hilger: Bristol, UK, 1990.
14. Gusev, V.E.; Karabutov, A.A. *Laser Optoacoustics*; American Institute of Physics: College Park, MD, USA, 1993.
15. Monchalin, J.P. Laser-Ultrasonics: Principles Furthermore, Industrial Applications. *E-J. Nondestruct. Test.* **2020**, *25*, 1–43.
16. Scherleitner, E.; Reitinger, B.; Hettich, M.; Berer, T.; Burgholzer, P. *Laser Ultrasonics*; Springer eBook Collection; Springer: Cham, Switzerland, 2019.
17. Treeby, B.E.; Cox, B.T.; Zhang, E.Z.; Patch, S.K.; Beard, P.C. Measurement of broadband temperature-dependent ultrasonic attenuation and dispersion using photoacoustics. *IEEE Trans. Ultrason. Ferroelectr. Freq. Control* **2009**, *56*, 1666–1676. [CrossRef]
18. Bai, X.; Ma, J.; Xu, W.; Wang, J.; Liu, S. Off-epicentral measurement of laser-ultrasonic shear-wave velocity and its application to elastic-moduli evaluation. *Ultrasonics* **2023**, *127*, 106850. [CrossRef]
19. Messaoudi, M.; Pouet, B. Recent development in all-optical ultrasonic shear wave inspection. *Nondestruct. Test. Diagn.* **2018**, *3*, 31–34.
20. Yi, D.; Pei, C.; Liu, T.; Chen, Z. Inspection of cracks with focused angle beam laser ultrasonic wave. *Appl. Acoust.* **2019**, *145*, 1–6. [CrossRef]
21. Singer, F. Laser-Ultrasonic Measurement of Elastic Properties of Anodized Aluminum Coatings. *Phys. Procedia* **2015**, *70*, 334–337. [CrossRef]
22. Singer, F.; Kufner, M. Model based laser-ultrasound determination of hardness gradients of gas-carburized steel. *NDT E Int.* **2017**, *88*, 24–32. [CrossRef]
23. Singer, F.; Kufner, M.; Drese, K. The Influence of Excitation and Detection Methods in Laser-Ultrasonics on Precise Material Characterization. In Proceedings of the Sensoren und Messsysteme—19. ITG/GMA-Fachtagung, Nürnberg, Germany, 26–27 June 2018.
24. Schneider, D. *Laser-Induced Surface Acoustic Waves for Material Testing*; Springer eBook Collection; Springer: Cham, Switzerland, 2019.
25. Veres, I.A.; Berer, T.; Burgholzer, P. Numerical modeling of thermoelastic generation of ultrasound by laser irradiation in the coupled thermoelasticity. *Ultrasonics* **2013**, *53*, 141–149. [CrossRef]
26. Dewhurst, R.J.; Edwards, C.; McKie, A.D.W.; Palmer, S.B. Estimation of the thickness of thin metal sheet using laser generated ultrasound. *Appl. Phys. Lett.* **1987**, *51*, 1066–1068. [CrossRef]
27. Burrows, S.E.; Dutton, B.; Dixon, S. Laser generation of Lamb waves for defect detection: Experimental methods and finite element modeling. *IEEE Trans. Ultrason. Ferroelectr. Freq. Control* **2012**, *59*, 82–89. [CrossRef]
28. Hosoya, N.; Umino, R.; Kanda, A.; Kajiwara, I.; Yoshinaga, A. Lamb wave generation using nanosecond laser ablation to detect damage. *J. Vib. Control* **2018**, *24*, 5842–5853. [CrossRef]
29. Gao, W.; Glorieux, C.; Thoen, J. Laser ultrasonic study of Lamb waves: Determination of the thickness and velocities of a thin plate. *Int. J. Eng. Sci.* **2003**, *41*, 219–228. [CrossRef]
30. Kamshilin, A.A.; Romashko, R.V.; Kulchin, Y.N. Adaptive interferometry with photorefractive crystals. *J. Appl. Phys.* **2009**, *105*, 031101. [CrossRef]
31. Blouin, A.; Monchalin, J.P. Detection of ultrasonic motion of a scattering surface by two-wave mixing in a photorefractive GaAs crystal. *Appl. Phys. Lett.* **1994**, *65*, 932. [CrossRef]
32. Campagne, B.; Blouin, A.; Pujol, L.; Monchalin, J.P. Compact and fast response ultrasonic detection device based on two-wave mixing in a gallium arsenide photorefractive crystal. *Rev. Sci. Instrum.* **2001**, *72*, 2478–2482. [CrossRef]
33. Dewhurst, R.J.; Shan, Q. Optical remote measurement of ultrasound. *Meas. Sci. Technol.* **1999**, *10*, R139–R168. [CrossRef]
34. Monchalin, J.P. Laser-Ultrasonics: From the Laboratory to Industry. *AIP Conf. Proc.* **2004**, *700*, 3–31.
35. Ono, K. A Comprehensive Report on Ultrasonic Attenuation of Engineering Materials, Including Metals, Ceramics, Polymers, Fiber-Reinforced Composites, Wood, and Rocks. *Appl. Sci.* **2020**, *10*, 2230. [CrossRef]
36. Martinez, R.; Leija, L.; Vera, A. Ultrasonic attenuation in pure water: Comparison between through-transmission and pulse-echo techniques. In Proceedings of the 2010 Pan American Health Care Exchanges IEEE, Lima, Peru, 15–19 March 2010; pp. 81–84.
37. Liu, Y.; Bond, L.J.; Hu, H. Ultrasonic-Attenuation-Based Technique for Ice Characterization Pertinent to Aircraft Icing Phenomena. *AIAA J.* **2017**, *55*, 1602–1609. [CrossRef]
38. Krautkrämer, J.; Krautkrämer, H. *Ultrasonic Testing of Materials*, 4th ed.; Springer: Berlin/Heidelberg, Germany, 1990.
39. Zell, K.; Sperl, J.I.; Vogel, M.W.; Niessner, R.; Haisch, C. Acoustical properties of selected tissue phantom materials for ultrasound imaging. *Phys. Med. Biol.* **2007**, *52*, N475–N484. [CrossRef]
40. Lai, T.Y.; Bruce, M.; Averkiou, M.A. Modeling of the Acoustic Field Produced by Diagnostic Ultrasound Arrays in Plane and Diverging Wave Modes. *IEEE Trans. Ultrason. Ferroelectr. Freq. Control* **2019**, *66*, 1158–1169. [CrossRef]
41. Laugier, P.; Haïat, G. *Bone Quantitative Ultrasound*; Springer: Dordrecht, The Netherlands, 2011.

42. Ye, S.G.; Harasiewicz, K.A.; Pavlin, C.J.; Foster, F.S. Ultrasound characterization of normal ocular tissue in the frequency range from 50 MHz to 100 MHz. *IEEE Trans. Ultrason. Ferroelectr. Freq. Control* **1995**, *42*, 8–14. [CrossRef]
43. Richardson, E.G. *Ultrasonic Physics*, 2nd ed.; Elsevier Science: Burlington, ON, Canada, 1962.
44. Tang, X.M.; Toksöz, M.N.; Cheng, C.H. Elastic wave radiation and diffraction of a piston source. *J. Acoust. Soc. Am.* **1990**, *87*, 1894–1902. [CrossRef]
45. Schlegel, W.; Karger, C.; Jäkel, O.; Bachert, P. (Eds.) *Medizinische Physik: Grundlagen-Bildgebung-Therapie-Technik*; Lehrbuch, Springer Spektrum: Berlin, Germany, 2018.
46. Humphrey, V.F. Ultrasound and matter–physical interactions. *Prog. Biophys. Mol. Biol.* **2007**, *93*, 195–211. [CrossRef]
47. Aussel, J.D.; Le Brun, A.; Baboux, J.C. Generating acoustic waves by laser: Theoretical and experimental study of the emission source. *Ultrasonics* **1988**, *26*, 245–255. [CrossRef]
48. Davies, S.J.; Edwards, C.; Taylor, G.S.; Palmer, S.B. Laser-generated ultrasound: Its properties, mechanisms and multifarious applications. *J. Phys. D Appl. Phys.* **1993**, *26*, 329–348. [CrossRef]
49. Hutchins, D.; Dewhurst, R.; Palmer, S. Laser generated ultrasound at modified metal surfaces. *Ultrasonics* **1981**, *19*, 103–108. [CrossRef]
50. Schlengermann, U. Über die Verwendung der Begriffe Nahfeld und Fernfeld in der Ultraschall-Werkstoffprüfung. *Mater. Test.* **1973**, *15*, 161–166. [CrossRef]
51. Lee, T.; Cheong, Y.; Baac, H.W.; Guo, L.J. Origin of Gouy Phase Shift Identified by Laser-Generated Focused Ultrasound. *ACS Photonics* **2020**, *7*, 3236–3245. [CrossRef]
52. Singer, F.; Kufner, M.; Drese, K. Measurement of Stress in Mechanical Components with a Novel Laser-Ultrasound Measurement System. In Proceedings of the 12th European Conference on Non-Destructive Testing (ECNDT 2018), Gothenburg, Sweden, 11–15 June 2018; Volume 23.
53. Schiebold, K. Ultraschallprüfsysteme. In *Zerstörungsfreie Werkstoffprüfung—Ultraschallprüfung*; Schiebold, K., Ed.; Springer: Berlin/Heidelberg, Germany, 2015; pp. 35–69.
54. Kossoff, G. Analysis of focusing action of spherically curved transducers. *Ultrasound Med. Biol.* **1979**, *5*, 359–365. [CrossRef] [PubMed]
55. Sachse, W.; Pao, Y.H. On the determination of phase and group velocities of dispersive waves in solids. *J. Appl. Phys.* **1978**, *49*, 4320–4327. [CrossRef]
56. Singer, F. Mess- und Analysemethoden in der Laserakustik bei breitbandiger Laseranregung. *Tm—Tech. Mess.* **2015**, *82*, 45–51. [CrossRef]
57. Lee, T.; Baac, H.W.; Li, Q.; Guo, L.J. Efficient Photoacoustic Conversion in Optical Nanomaterials and Composites. *Adv. Opt. Mater.* **2018**, *6*, 1800491. [CrossRef]
58. Rose, J.L. *Ultrasonic Guided Waves in Solid Media*; Cambridge University Press: New York, NY, USA, 2014.
59. Ostrovskaya, G.V. Efficiency of optical-to-acoustic energy conversion upon the interaction of a pulsed laser radiation with a liquid: I. Calculation of the efficiency upon acoustooptic interaction. *Tech. Phys.* **2002**, *47*, 1299–1305. [CrossRef]
60. Li, Q.; Zhao, X.; Fang, S.; Liang, F.; Chen, Z. Theoretical analysis for conversion efficiency of laser generation of acoustic waves in liquid. In Proceedings of the 2016 International Conference on Numerical Simulation of Optoelectronic Devices (NUSOD), Sydney, Australia, 11–15 July 2016; pp. 169–170.
61. Liang, S.; Lashkari, B.; Choi, S.S.S.; Ntziachristos, V.; Mandelis, A. The application of frequency-domain photoacoustics to temperature-dependent measurements of the Grüneisen parameter in lipids. *Photoacoustics* **2018**, *11*, 56–64. [CrossRef] [PubMed]
62. Ioffe Physico-Technical Institute. *Silicon: Mechanical Properties, Elastic Constants, Lattice Vibrations*. Available online: https://www.ioffe.ru/SVA/NSM/Semicond/Si/mechanic.html (accessed on 17 November 2023).
63. COMSOL Multiphysics. *Material Library: Si-Silicon (Single-Crystal, Anisotropic)*; COSMOL: Stockholm, Sweden, 2023.
64. Folds, D.L. Speed of sound and transmission loss in silicone rubbers at ultrasonic frequencies. *J. Acoust. Soc. Am.* **1974**, *56*, 1295–1296. [CrossRef]
65. Cafarelli, A.; Miloro, P.; Verbeni, A.; Carbone, M.; Menciassi, A. Speed of sound in rubber-based materials for ultrasonic phantoms. *J. Ultrasound* **2016**, *19*, 251–256. [CrossRef] [PubMed]
66. Xu, G.; Ni, Z.; Chen, X.; Tu, J.; Guo, X.; Bruus, H.; Zhang, D. Acoustic Characterization of Polydimethylsiloxane for Microscale Acoustofluidics. *Phys. Rev. Appl.* **2020**, *13*, 054069. [CrossRef]
67. Lubbers, J.; Graaff, R. A simple and accurate formula for the sound velocity in water. *Ultrasound Med. Biol.* **1998**, *24*, 1065–1068. [CrossRef]

*sensors*

MDPI

*Article*

# Quantitative Detection of Pipeline Cracks Based on Ultrasonic Guided Waves and Convolutional Neural Network

**Yuchi Shen [1], Jing Wu [2,*], Junfeng Chen [3], Weiwei Zhang [2], Xiaolin Yang [1] and Hongwei Ma [2,4]**

1. Department of Civil Engineering, Qinghai University, Xining 810016, China; sycamore007s@outlook.com (Y.S.)
2. Department of Mechanical Engineering, Dongguan University of Technology, Dongguan 523808, China
3. School of Mechanics and Construction Engineering, Jinan University, Guangzhou 510632, China
4. Guangdong Provincial Key Laboratory of Intelligent Disaster Prevention and Emergency Technologies for Urban Lifeline Engineering, Dongguan 523808, China

* Correspondence: jingjinglulie@163.com

**Abstract:** In this study, a quantitative detection method of pipeline cracks based on a one-dimensional convolutional neural network (1D-CNN) was developed using the time-domain signal of ultrasonic guided waves and the crack size of the pipeline as the input and output, respectively. Pipeline ultrasonic guided wave detection signals under different crack defect conditions were obtained via numerical simulations and experiments, and these signals were input as features into a multi-layer perceptron and one-dimensional convolutional neural network (1D-CNN) for training. The results revealed that the 1D-CNN performed better in the quantitative analysis of pipeline crack defects, with an error of less than 2% in the simulated and experimental data, and it could effectively evaluate the size of crack defects from the echo signals under different frequency excitations. Thus, by combining the ultrasonic guided wave detection technology and CNN, a quantitative analysis of pipeline crack defects can be effectively realized.

**Keywords:** convolutional neural network; ultrasonic guided wave; pipeline; crack defects

## 1. Introduction

Owing to its evident advantages, including minimal environmental impacts, large carrying capacities, high efficiencies, and low costs, pipeline transportation is expected to play an increasingly important role in future transportation engineering [1]. However, in pipeline engineering, the integrity and safety of pipeline structures form the premise of smooth transportation. In their absence, a loss of transported materials caused by damaged pipeline structures is expected to not only affect production and daily life, causing huge economic losses, but also exert a huge impact on the environment, even leading to major safety risks such as collapse and explosion [2,3]. Therefore, the nondestructive testing of pipeline structures is of considerable significance. In this regard, ultrasonic guided wave technology is widely used in long-distance pipeline inspection owing to its advantages, such as easy excitation, a long propagation distance, wide coverage, high detection accuracy, and low detection cost [4]. However, owing to the complex characteristics of ultrasonic guided waves, such as multimode and dispersion, and the influence of noise, quantitative evaluations of defect echoes in guided wave signals represent a significant challenge [5,6].

Interestingly, the majority of previous research on ultrasonic guided waves focuses on the localization and qualitative analysis of defects, relying on the extraction of a series of characteristic parameters from the original signals [7,8]. By contrast, quantitative research on defects directly based on the original guided waves is rather limited. In particular, only a few quantitative studies have been conducted on defects by directly using the original guided wave signals. Zhan et al. [7] used a deep learning approach to classify and detect pipe welds in noisy environments. Li et al. [8] used the CNN-LSTM hybrid model to classify pipeline defects. This research has proposed a solution for classifying pipeline defects, but it has not thoroughly explored the quantification of pipeline damage. Quantifying pipeline damage often requires a large amount of data to perform a regression analysis across the full range of damage. Determining its impact on simulation and experimental data regarding working condition diversity is a considerable challenge. Davies et al. [9] analyzed the relationship between defect sizes of cracks and small apertures and the defect echo amplitude when evaluating synthetic path image focusing and source imaging methods. Zheng [10] used a matching pursuit algorithm to quantitatively analyze the axial defect size of a pipeline. Both studies employed quantitative methods based on theoretical calculations, which may be subject to errors after environmental changes. Neural networks trained on real-time data can learn to adapt to specific environmental situations with greater compatibility. Li [11] utilized a 2D blind convolution method to estimate the dimensions of axial defects using multiple sets of data obtained from an axial sensor array. Li [12] proposed a quantitative reconstruction method for ultrasonic guided wave defects based on deep learning. This was achieved by combining the theoretical method of shear wave quantitative reconstruction of plate thinning defects, wave number space domain transform, and a convolutional neural network (CNN) using local fusion. Acciani et al. [13] used wavelet transforms and neural networks to quantitatively evaluate pipe surface damage. Preprocessing ultrasonic guided wave signals is necessary for these methods, which may exclude significant information from the original signals. A CNN network that extracts information directly from the ultrasound-guided wave response signal is highly effective in avoiding this issue. Huang [14] proposed a damage detection method based on a CNN-LSTM network for laser ultrasonic guided wave scanning detection. Miorelli et al. [15] proposed an automatic method for localizing and quantifying structural health monitoring defects based on guided wave imaging by combining convolutional neural networks. Yin et al. [16] automated the detection of pipeline defects using closed-circuit television (CCTV) and deep learning. Both studies relied on structural damage imaging and used training samples directly derived from 2D images. This method is computationally demanding and relies on image-based recognition, which can make predicting results challenging due to the effects of image imaging. The use of CNNs for quantitative defect identification can effectively prevent the masking of critical information and reduce the technical difficulty of engineering applications.

Neural network-based algorithms have made it more convenient, efficient, and accurate to quantitatively identify structural damage using ultrasonic guided wave technology. With this background, this paper proposes a method for the end-to-end quantitative characterization of pipeline defects by directly inputting the original ultrasonic guided waves signal into a network model without any preprocessing. The goal of the method is to predict the angles of radial defects on the pipe by regressing the angles of the defects directly to the prediction. To achieve this task, ceramic piezoelectric sheets are used to line a section of the pipe symmetrically at equal intervals for the excitation and reception of ultrasonic guided waves. The received signals are then used as inputs for a neural network for feature learning. The size of the penetrating crack in the pipeline is quantified using the defect damage angle, and the amount of crack damage in the ultrasonic guided wave signal is determined using a neural network method. By arranging the ceramic piezoelectric sheet symmetrically and in the same direction on the pipe, it is possible to excite ultrasonic guided waves of L mode. This mode is more sensitive to circumferential cracks in the pipe, making it better suited for the quantitative characterization of such cracks.

## 2. Defect Quantification Method Based on Ultrasonic Guided Waves and CNN

An Artificial Neural Network (ANN) represents a type of simulation and approximation of a biological neural network. It is an adaptive nonlinear dynamic network system composed of a large number of neurons connected through mutual connections, and it is primarily composed of input, hidden, and output layers [17]. The most basic unit of an ANN is a neuron, which can be expressed as Equation (1).

$$y = f(\omega_1 x_1 + \omega_2 x_2 + \omega_3 x_3 \ldots\ldots \omega_n x_n + b) \tag{1}$$

where $x_i$ denotes the input n features in a neuron, $\omega_i$ denotes the weight value of the input feature $x_i$ connected to the neuron, $b$ denotes the internal bias of the neuron, $y$ denotes the output value of the neuron, and $f(\ldots\ldots)$ denotes the activation function. The more common activation functions include the rectified linear unit (ReLU) [18], Sigmoid, Tanh(x), and radial basis functions [19]. According to the findings of Acciani [11], the ReLU function is directly selected as the activation function for this study.

A CNN is a typical deep learning method developed in recent years, with wide applications in fields such as pattern recognition and medical engineering [20]. The basic structure of a CNN consists of input, convolutional, pooling, fully connected, and output layers [21]. Because each neuron of the output feature surface in the convolutional layer is locally connected to its input, and the input value of the neuron is obtained based on the weighted sum of the corresponding connection weight and the local input plus the bias value, this process is equivalent to the convolution process, from which its name is derived [22]. CNNs are primarily used for the feature recognition of two-dimensional images, whereas 1D-CNNs have only one dimension; therefore, they are widely used for feature recognition and time-series extraction. Although a 1D-CNN only has a single dimension, it demonstrates the same advantages as a CNN [14]. Specifically, CNNs not only present the advantages of traditional neural networks, such as a strong self-learning ability and good adaptability, but also the advantages of weight sharing and easy model training [23]. This study employs a CNN model, illustrated in Figure 1, which comprises an input layer, two convolutional and pooling layers, and two fully connected and output layers.

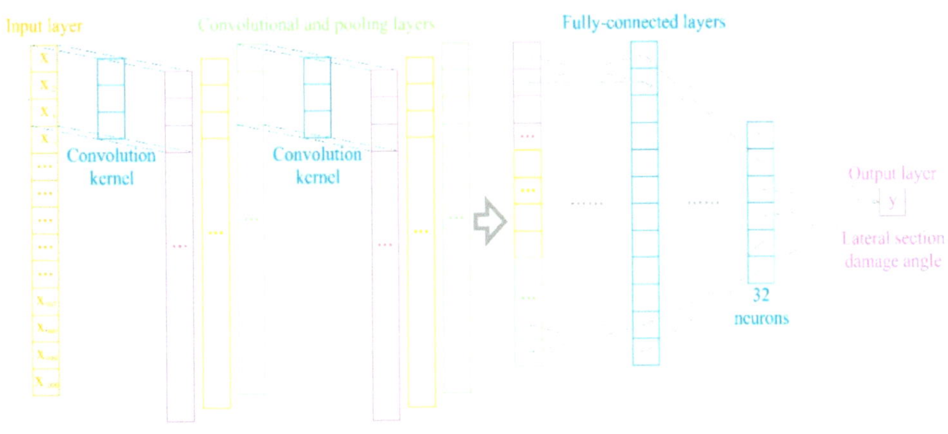


**Figure 1.** Schematic diagram of a CNN.

In order to compare the prediction performance of different types of neural networks, this study architects a multilayer perceptron (MLP) neural network model, as shown in Figure 2. It consists of an input layer of 1000 sets of guided wave signals, a first hidden layer of 128 neurons, a second hidden layer of 64 neurons, and an output layer.

The neural networks were trained using computers equipped with R9-5900HS processors and RTX-3050ti graphics cards to prevent external factors from affecting the detection results. The neural networks were trained using several program libraries, including Pytorch, NumPy, Pandas, and Scikit-learn. Additionally, the Adaptive Moment Estimation optimization algorithm was utilized.

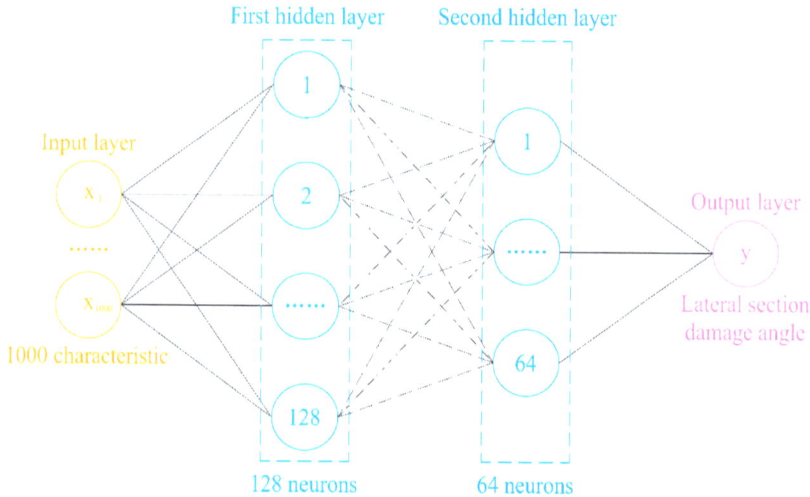


**Figure 2.** Schematic diagram of an MLP.

The convolution layer performs a valid convolution operation using a 3 × 1 kernel, and the convolution operation is like a mathematical operation on two functions, which produces the mapping relationship of the third function, and its mathematical expression is Equation (2). $f$ and $g$ represent two different mapping functions. The data pass through two convolutional layers and two fully connected layers before outputting the final prediction.

$$(f \cdot g)[n] = \sum_{m=-\infty}^{\infty} f[n-m]g[m] \quad (2)$$

The time-domain signals of ultrasonic guided waves were used as direct inputs in our quantitative characterizations of pipeline crack defects. The input signal range included the excitation wave, defect echo, and first-end face echo. In the numerical simulation stage, the displacement signal within this period was directly specified in the analysis step, and 1000 groups of amplitude signals were collected as data. During the acquisition of experimental data, the oscilloscope was set to a sampling rate of 5 M/S, resulting in a total of 10 K sampled points. To decrease the calculation time of the neural network model, each experimental signal was divided into seven data groups, each with a length of 1000. The division started from the 800th point, which was where the excitation of ultrasonic guided wave signals appeared, and ended at the 8000th point, which was the end of the first end-face echo, as shown in Figure 3. This method can increase the diversity of the data as much as possible on the basis of extracting real guided wave information.

The construction of the 1D-CNN in this paper required the following hyperparameters to be determined: the number of nodes in the hidden layer, the batch size for testing, the random seeds, the learning rate, and the number of training generations. For the selection of the number of hidden nodes, empirical Formulas (3)–(5) were used to calculate and set the number of hidden nodes with the best results. $n$ is the number of nodes in the input layer; $l$ is the number of nodes in the output layer; $m$ is the number of hidden nodes; and $a$ is a constant between 1 and 10. The remaining hyperparameters were determined through empirical formulas, which was carried out near the empirical values, and the optimal training results were selected as the parameters.

$$m = \sqrt{n+l} + a \quad (3)$$

$$m = log_2 n \quad (4)$$

$$m = \sqrt{nl} \tag{5}$$

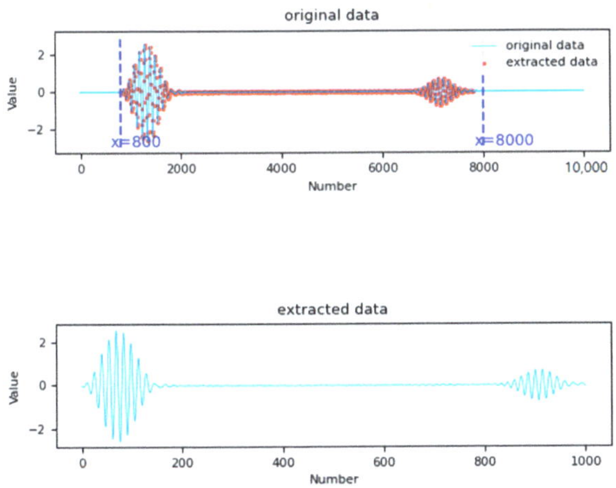


**Figure 3.** Splitting of experimental data.

Determining the hyperparameters is crucial for evaluating network performance. Excellent hyperparameters correspond to excellent performance indicators and determine the success of the network model.

To set the hyperparameters of the neural network model, certain tricks for modulating these hyperparameters were obtained from the literature [24]. The neural network model architecture adopted a structure of 2–3 hidden layers, and ReLU was used as the activation function. For the simulation and experimental data, a multi-layer perceptron (MLP) and CNN were used for training and comparison, respectively. The root mean square error (RMSE), mean absolute percentage error (MAPE), and coefficient of determination (R-square) were used to comprehensively evaluate the regression performance of the neural networks. Note that the RMSE and MAPE reflect the error between the predicted and actual values of the network. The RMSE increases with larger errors, while the MAPE decreases. The R-square value indicates the quality of the model fit; the closer it is to one, the better the fit. Based on these parameters, the advantages and disadvantages of the two different neural networks in the quantitative prediction of pipeline defects were comprehensively compared. The expressions for the above parameters are shown in (6), (7), and (8). Here, $\hat{y}$ represents the true value, $y$ represents the predicted value, and $n$ represents the number of test samples.

$$RMSE = \sqrt{\frac{1}{n}\sum_{i=1}^{n}(y_i - \hat{y}_i)^2} \tag{6}$$

$$MAPE = \frac{100\%}{n}\sum_{i=1}^{n}\left|\frac{\hat{y}_i - y_i}{y_i}\right| \tag{7}$$

$$R^2 = 1 - \frac{\sum_i(\hat{y}_i - y_i)^2}{\sum_i(\overline{y}_i - y_i)^2} \tag{8}$$

## 3. Numerical Simulation

### 3.1. Pipeline Modeling

The JMatPro (Version 7.0) and Abaqus (Version 6.16-1) software were used for the numerical simulations. The pipe used in the experiment was a 304 seamless steel pipe with a length of $l = 3$ m, an outer diameter of $d = 60$ mm, and a pipe wall thickness of $t = 2$ mm.

To ensure consistency between the simulation and experiment, the JMatPro software was used to predict the performance parameters of the pipeline material to be selected. The density of the material used in the experiment was obtained as $\rho = 7.855\ \mathrm{g/cm^3}$, Young's modulus was obtained as $E = 211.3$ GPa, Poisson's ratio was obtained as $\nu = 0.2969$, and the data corresponding to the material performance parameters were input into the Abaqus software. Simultaneously, a pipe model with a length of 3 m, an outer diameter of 60 mm, and a wall thickness of 2 mm was established using the Abaqus software, and the dynamic display analysis method was used to perform the calculation.

To conveniently select the central frequency of the excitation signal, the dispersion curve of the simulated pipeline was drawn using the Disperse (Version 2.0.16c) software, as depicted in Figure 4. According to the dispersion curve, the group velocity and phase velocity of the guided wave signal tended to be stable in the L (0, 2) mode near the central frequency of 70 kHz, and the wave velocity was approximately 5.5 km/s; hence, an excitation signal with a central frequency of 70 kHz was used to generate a single-mode guided wave.

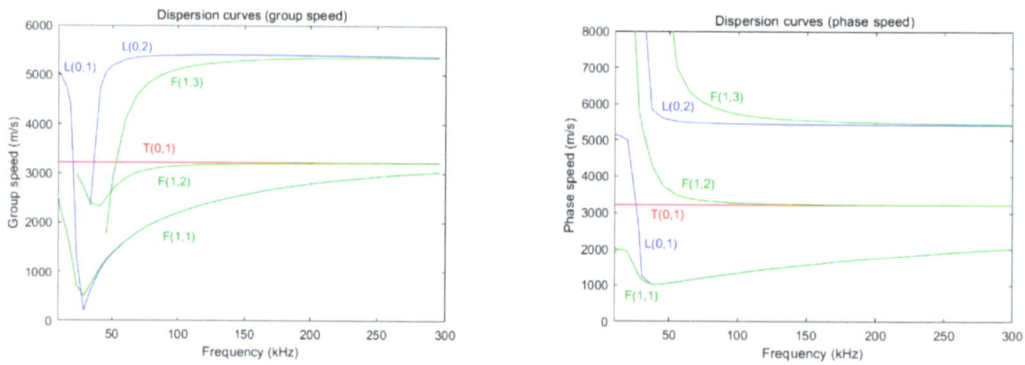


(a) (b)

**Figure 4.** Dispersion curves for (**a**) group speed and (**b**) phase speed.

In this study, a hexahedral structured mesh was used for the non-defective part of the pipeline, and a hexahedral swept mesh was used for the defective part. Generally, to control the propagation error of a waveform, the presence of multiple elements in one wavelength is essential, and the grid size of the axial elements should satisfy the conditions in Equation (9).

$$l_e < \frac{min(C_p)}{10 \times f} \tag{9}$$

where:

$l_e$: the axial grid cell length;

$C_p$: the phase velocity of the guided wave;

$f$: the frequency of the guided wave.

Based on the calculations, the size of the largest grid axial element was found to be approximately 7 mm. To ensure that the number of grids in the global mesh generation was an integer, the global mesh size was finally determined to be 5 mm. Considering that the gradient of the lateral section damage angle growth in the numerical simulation was one, and the increment of its reaction in the mesh size was approximately 0.5 mm, the mesh size was locally encrypted to 0.5 mm in the defective part (Figure 5).

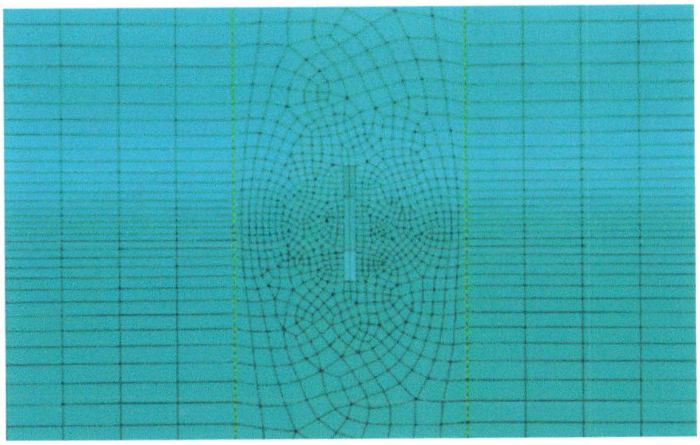

**Figure 5.** Mesh division with local refinement.

To facilitate grid division when setting defects, the lateral section damage angle was adopted as the defect measurement index (Figure 6). For this, a pipeline model with a lateral section damage angle of 0–180° was established with a gradient of one. In total, we had 181 models. Notably, this type of model only considers penetrating cracks in the pipeline. Considering the pipeline loss along the crack-depth direction, the number of models can be increased in multiple ways. Meanwhile, considering the fact that the crack depth of non-penetrating cracks cannot be accurately measured and controlled in the experimental stage, the quantification of non-penetrating cracks was not considered in this study.

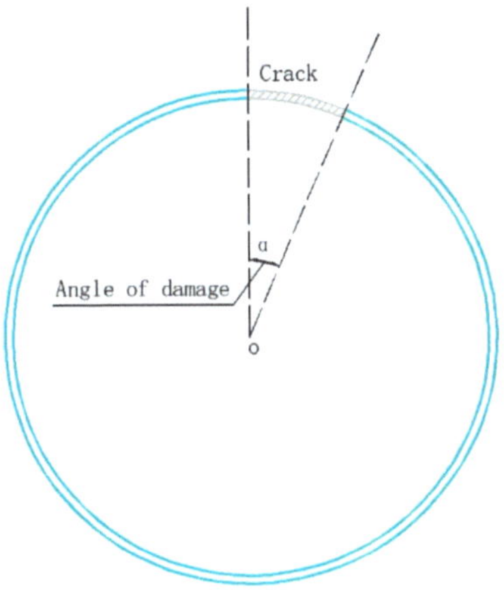


**Figure 6.** Lateral section damage angle.

To ensure that the complete waveform image, including the excitation wave, defect echo, and first-end face echo, could be observed in the analysis time, the analysis step time was required to meet the conditions in Equation (10).

$$T_s > \frac{2L}{min(C_g)} \tag{10}$$

where:

$T_s$—the analysis step time;
$L$—the length of the pipe;
$C_g$—the group velocity of the guided wave.

### 3.2. Excitation Signal

The excitation signal was a 10-cycle sinusoidal signal modulated by a cosine function with a central frequency of 70 kHz, as depicted in Figure 7. This signal energy appeared to be more concentrated near the central frequency, which is conducive to signal identification. The signal function expression is given in Equation (11) [25].

$$f(t) = 0.5\left[1 - cos\left(\frac{2\pi f_c t}{n}\right)\right] sin(2\pi f_c t) \tag{11}$$

where:

$f_c$—the center frequency of the excitation signal;
$n$—the number of cycles.

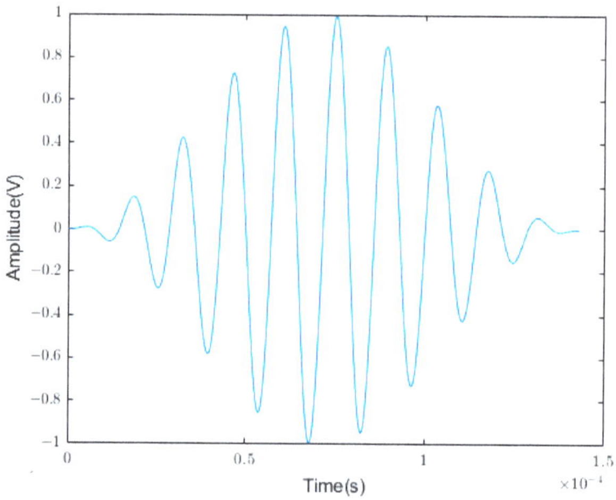


**Figure 7.** Time-domain signal.

### 3.3. Defect Quantification with ANN

The numerical simulation results indicated the presence of 181 groups of guided wave signals. The data were divided into training, testing, and validation sets in a ratio of 6:2:2. As stated, a 1D-CNN was used to learn these data and compare the performance with the traditional MLP network. To reduce the training time of the network, the entire guided wave signal was arranged in 1000 groups of characteristic values to be input into the network according to the time sequence, and the damage angle of the crack section corresponding to the guided wave signal was considered as an output value. The RMSE, MAPE, and R-square values were used to evaluate the regression performance of the neural networks.

As stated, a 1D-CNN was used to learn the simulation data, and the ReLU function was used as the activation function. The loss curves and training results are indicated in Figure 8. The results for the regression performance indicators of the model are as follows: RMSE = 0.948, MAPE = 0.015, and R-square = 0.999. The regression performance indexes of the model revealed that the regression performance of the 1D-CNN model was better, the error between the prediction results and the real section damage angle was approximately 0.9°, and the error was less than 0.5%. The loss curve and prediction results indicate the presence of some overfitting in the model. The results for the regression performance indicators of the MLP model are as follows: RMSE = 1.17, MAPE = 0.014, and R-square = 0.999. The regression performance indexes of the model indicated that the regression performance of the MLP model was excellent, the error between the prediction results and real section damage angle was approximately 1.2°, and the error was less than

0.8%, and its parameter characterization shows that its model prediction performance is not as good as that of the 1D-CNN model.

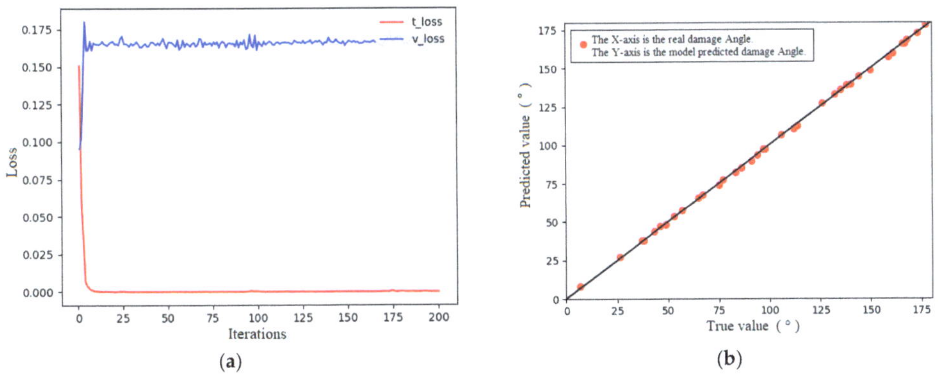


**Figure 8.** Loss curve and prediction results of the 1D-CNN. (**a**) Loss curves. (**b**) Predicted results.

To clarify whether the model could effectively predict the section damage angle corresponding to the defect echo signal when the guided wave signal at the defect echo was input into the neural network as a feature in the simulation stage, 120 groups of data at the defect echo were extracted from the original simulation data as feature inputs (Figure 9) to the neural network. Following this, the regression effect was compared with that of the entire guided wave signal as a feature input to the neural network.

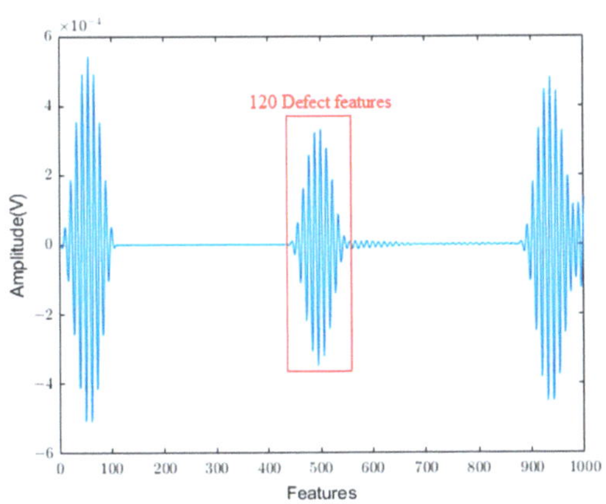


**Figure 9.** Signal at defect echo.

Table 1 summarizes the final evaluation results for each parameter of the neural network models. In this table, MLP denotes the performance index of the MLP when the full-section guided wave signal is used as the feature input, while 120-MLP represents the performance index of the MLP when the guided wave signal at the defect echo is used as the feature input. Next, 1D-CNN denotes the performance index of the 1D-CNN when the entire guided wave signal is used as the feature input, while 120-1D-CNN denotes the performance index of the 1D-CNN when the guided wave signal at the defect echo is used as the feature input. The numerical results indicate that the prediction performance of the neural networks appears better when the entire guided wave signal is used as the feature input, and the size of the defect is more closely related to the entire guided wave signal. Based on this result, it can be inferred that when only the signal at the echo of the defect is used as the feature input to the network, abundant useful information may be lost, resulting in the inability of the neural network to effectively analyze defects. The 1D-CNN model was more effective in identifying the pipeline damage value from the analog signal.

**Table 1.** Performance index evaluation.

| Name | RMSE | MAPE | R-Square |
|---|---|---|---|
| MLP | 1.17 | 0.014 | 0.999 |
| 120-MLP | 2.45 | 0.026 | 0.997 |
| 1D-CNN | 0.95 | 0.015 | 0.999 |
| 120-1D-CNN | 2.26 | 0.027 | 0.998 |

## 4. Experimental Verification

To verify the effectiveness of the proposed defect quantification method using ultrasonic guided waves in pipes based on neural networks, experimental studies on different crack defect sizes in pipes were conducted. In other words, the neural network method was used to directly predict the sizes of the corresponding pipeline crack defects in the guided wave signal from the entire guided wave signal.

### *4.1. Experimental Design*

The instruments and materials used in this study are listed in Table 2 below. Before the experiment, 32 piezoelectric ceramic slices were welded. To ensure that the piezoelectric ceramic slices would remain undamaged under the high welding temperature, the welding temperature was kept below 260 °C during welding. After welding, each piezoelectric ceramic slice was tested using an impedance analyzer to ensure that the frequency interval of the impedance mutation of each piezoelectric ceramic slice remained the same. Generally, after preparing a piezoelectric ceramic slice, treating the pipe to be pasted with this slice is essential. Thus, the cuts at both ends of the pipe were polished using an electric grinder, and the interior and exterior of pipe were cleaned. To place the piezoelectric ceramic slice in close contact with the pipe, it was carefully wiped with a disinfectant alcohol tablet approximately 10 cm from one end of the pipeline to be pasted with the piezoelectric ceramic slice, after which the ceramic piezoelectric slice was pasted evenly and tightly with epoxy resin adhesive. To ensure a clear signal, the excitation and receiving piezoelectric ceramic slices had to be placed at the same axial position as the pipeline and evenly distributed. Sixteen excitation and sixteen receiving piezoelectric ceramic slices were arranged at axial intervals of 5 mm, as depicted in Figure 10. A waveform generator was used to generate a signal modulated by a Hanning window at the excitation end of the pipeline, which was strengthened by a power amplifier and acted on the piezoelectric ceramic slice at the excitation end of the pipeline, to allow the ultrasonic guided wave to traverse all positions along the pipeline. Finally, a time curve of propagation of the ultrasonic guided wave in the pipeline was recorded using an oscilloscope. The actual experiment, instruments, and piping are illustrated in Figure 11.

**Table 2.** Experimental equipment.

| Name | Quantity | Type | Notes |
|---|---|---|---|
| Seamless steel pipe | 1 | Stainless steel | Length of 3 m, outer diameter of 60 mm, wall thickness of 2 mm |
| Piezoelectric ceramic slice | 32 | YF3-239-01 | Size: 15.5 mm × 3.5 mm × 1 mm |
| Arbitrary wave function generator | 1 | AFG31102 | Bandwidth: 100 Mhz Sampling rate: 1 GSa/S |
| Ultrasonic power amplifier | 1 | AG1020 | Frequency: 10 KHz–20 MHz |

**Table 2.** *Cont.*

| Name | Quantity | Type | Notes |
|---|---|---|---|
| MDO, mixed domain oscilloscope | 1 | MDO4054B-3 | Bandwidth: 500 Mhz<br>Sampling rate: 2.5 GSa/S |
| Precision impedance analyzer | 1 | 6632 | Error: ±0.08% |
| Constant temperature welding station | 1 | SS-257 | - |
| High-precision vernier caliper | 1 | DYX-DM90150 | Precision: 0.01 mm<br>Error: ±0.02 mm |
| Speed-regulating electric motorcycle | 1 | DYX-DM7765 | Speed: 6000–34,000 RPM/min |
| Epoxy resin adhesive | Noggin | 9911 | - |

**Figure 10.** Arrangement of piezoelectric ceramic slice.

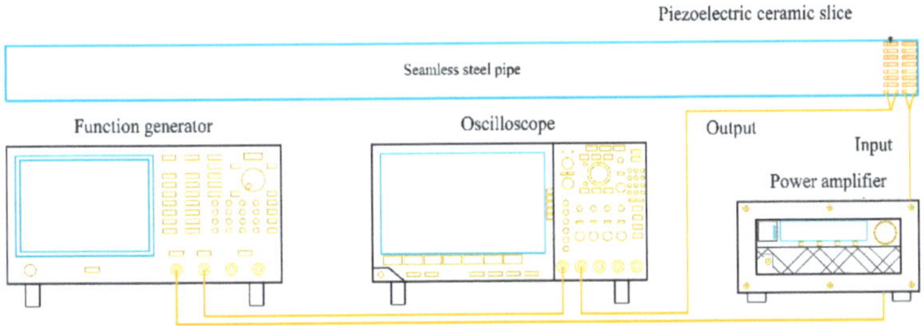


(**a**)

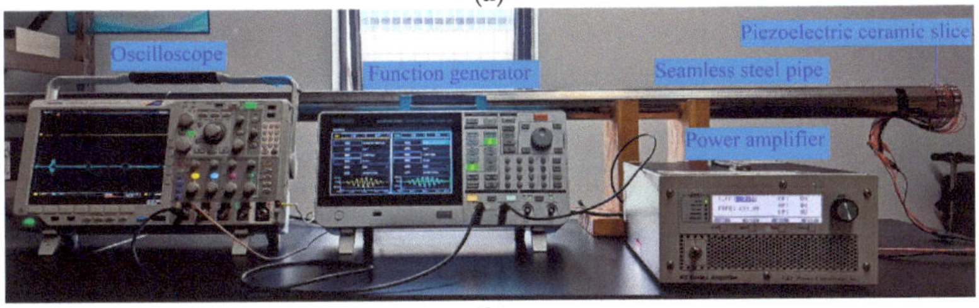


(**b**)

**Figure 11.** Schematic of the experiment setup. (**a**) Schematic. (**b**) Real object.

In the experiment, cutting was carried out first and then measurements were conducted to determine the angle of the section damage. In the absence of damage to the pipeline, the echo signal of the complete pipeline was first obtained, following which the pipeline was cut using an electric friction machine equipped with a special pipeline-cutting blade. After each cut, the crack size was measured and recorded using Vernier calipers (Figure 12). To prevent the cutting-induced temperature rise from affecting guided wave detection, the pipeline was allowed to stand for 20 min after each cutting. After the temperature of the cutting surface was lowered until it was close to the indoor temperature, guided wave detection was performed, and the relevant guided wave data were recorded using an oscilloscope.

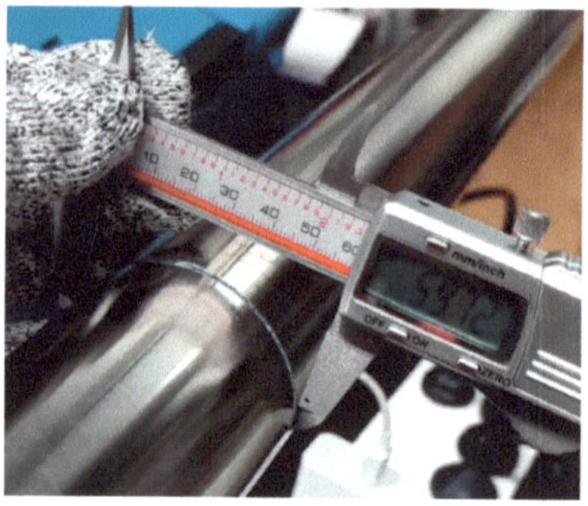

**Figure 12.** Defect measurement.

To ensure that the excitation wave, defect echo, and the first-end echo could be clearly observed on the oscilloscope, and the guided wave signal could be completely collected, the horizontal scanning time base of the oscilloscope was selected to be 200 μs, and the number of sampling points was 10,000. We selected 31 excitation signals with a central frequency within the range of 50–200 kHz and considered 5 kHz as the optimal step to conduct the frequency sweep operation on the pipeline. During the frequency sweep, we observed that when the central frequency was 80 kHz, the guided wave signal was clear, and no interference was observed from the other modal waveforms, as indicated in Figure 13. Therefore, 80 kHz was selected as the optimal central frequency for the guided wave experiment on this pipe. A total of 21 groups of guided wave data were collected using excitation signals with central frequencies ranging from 70 to 90 kHz, with steps of 1 kHz above and below 80 kHz. To match the numerical simulation, the experiment was terminated when the section damage angle reached 180°. The section damage angles recorded in the experiment are listed in Table 3.

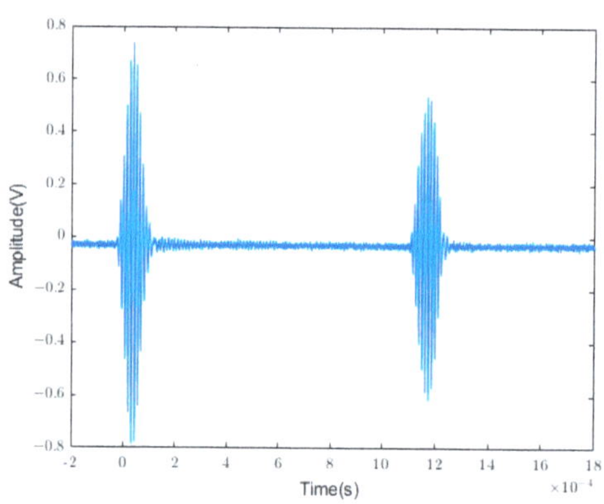


**Figure 13.** Experimental guided wave data at the optimal central frequency.

**Table 3.** Experimental data.

| Serial Number | The Length of the Damage Tangent (mm) | The Lateral Section Damage Angle (°) | Serial Number | The Length of the Damage Tangent (mm) | The Lateral Section Damage Angle (°) |
|---|---|---|---|---|---|
| 1 | 0 | 0 | 19 | 42.62 | 91 |
| 2 | 8.04 | 15 | 20 | 44.77 | 97 |
| 3 | 10.31 | 20 | 21 | 45.83 | 100 |
| 4 | 12.81 | 25 | 22 | 46.82 | 103 |
| 5 | 14.92 | 29 | 23 | 47.78 | 106 |
| 6 | 18.08 | 35 | 24 | 48.39 | 108 |
| 7 | 20.24 | 40 | 25 | 49.25 | 111 |
| 8 | 22.32 | 44 | 26 | 50.1 | 114 |
| 9 | 24.1 | 48 | 27 | 51.35 | 119 |
| 10 | 26.78 | 53 | 28 | 52.21 | 122 |
| 11 | 28.55 | 57 | 29 | 53.18 | 126 |
| 12 | 30.2 | 61 | 30 | 54.33 | 131 |
| 13 | 32.39 | 66 | 31 | 55.09 | 135 |
| 14 | 34.42 | 70 | 32 | 56.93 | 145 |
| 15 | 36.42 | 75 | 33 | 58.18 | 154 |
| 16 | 38.14 | 79 | 34 | 59.23 | 165 |
| 17 | 40.65 | 86 | 35 | 59.72 | 180 |
| 18 | 41.53 | 88 | | | |

### 4.2. Analysis of Experimental Results

Through the above experimental operation, 10,920 groups of data were collected for neural network training, and 210 groups of guided wave data were collected under optimal frequency conditions. Based on the consistency between the experiment and simulation, the guided wave data at the optimal frequency were analyzed, and the data were divided into training, testing, and verification sets at a ratio of 6:2:2. The MLP and 1D-CNN were used to learn these data, and the regression performance of the two different networks on these data was compared. The final results obtained through network learning and training are as follows.

The training results of the 1D-CNN on the experimental data are shown in Figure 14, and the results of the regression performance indicators of the obtained model are as follows: RMSE = 3.25, MAPE = 0.039, and R-square = 0.993. The regression performance indexes of the model indicate that the regression performance of the 1D-CNN model was not as good as that of the simulated data when the experimental data were used. The error between the predicted results and actual section damage angle was approximately 3.3°, and the error reached 1.8%, which was approximately 1.7 mm when converted into arc length. The results of the regression performance indicators of the MLP are as follows: RMSE = 3.99, MAPE = 0.043, and R-square = 0.990. In general, a neural network can accomplish the quantitative characterization of pipeline damage both from simulation and experimental data, and the regression performance of the 1D-CNN is better than that of the MLP in this task.

Owing to the characteristics of multimodal and frequency dispersions of ultrasonic guided waves, an evident defect echo cannot be observed in the guided wave echo signal under a nonoptimal excitation frequency, as presented in Figure 15. To evaluate whether the neural network could identify ultrasonic guided-wave signal data at other frequencies, 10,920 data groups were input into the network for learning. The corresponding results are presented in Table 4 and Figure 16.

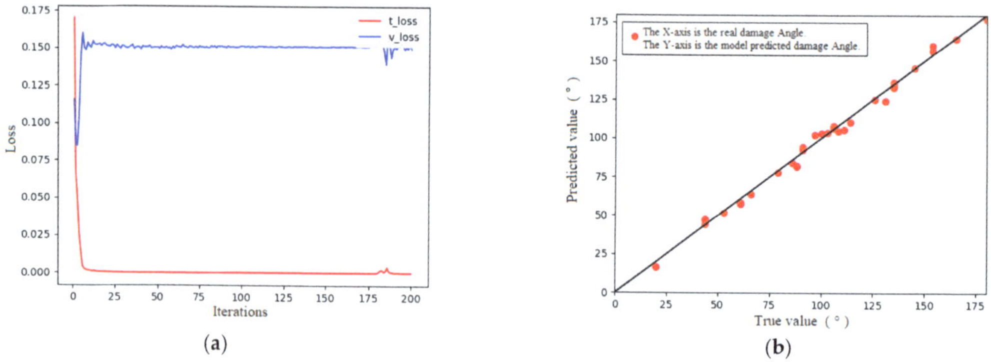


**Figure 14.** Loss curve and prediction results of the 1D-CNN. (**a**) Loss curves. (**b**) Predicted results.

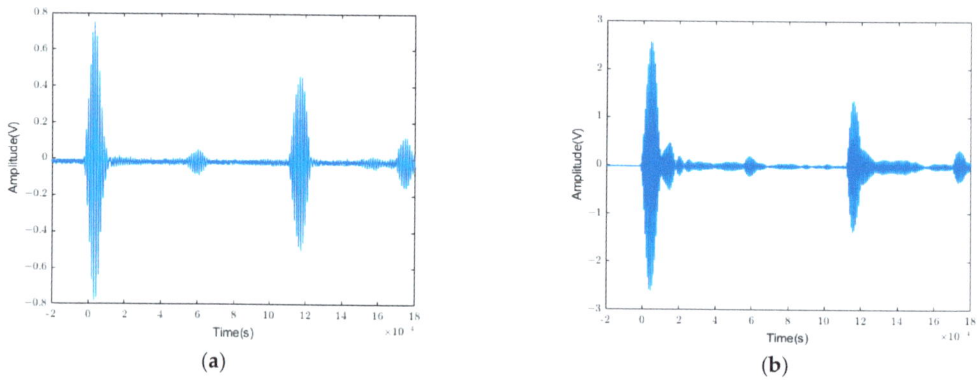


**Figure 15.** Comparison diagram of ultrasonic guided wave signals. Guided wave echo of an excitation signal with central frequencies of (**a**) 80 kHz and (**b**) 120 kHz.

**Table 4.** Performance index evaluation.

| Name | RMSE | MAPE | R-Square |
|---|---|---|---|
| MLP | 7.13 | 0.106 | 0.974 |
| 1D-CNN | 3.70 | 0.071 | 0.993 |

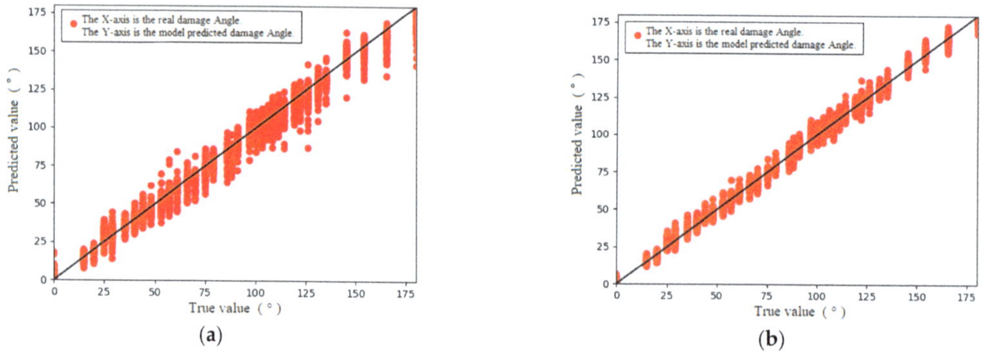


**Figure 16.** Prediction results. (**a**) MLP; (**b**) 1D-CNN.

The regression performance indexes of the model indicate that the error between the prediction result of the MLP model and the real section damage angle was approximately 7.13°, and the error reached 3.9%, which was approximately 3.7 mm when converted into arc length. The error between the prediction result of the 1D-CNN model and the real section damage angle was approximately 3.7°, and the error reached 2%, which was approximately 1.9 mm when converted into arc length. These calculation results indicate that a neural network has the ability to identify guided wave echo signal data under nonoptimal excitation frequencies, and it can extract the defect quantity from the data. The 1D-CNN is better in this respect, and its regression performance is basically the same as that of the guided wave signal under the optimal excitation frequency.

Finally, the CNN model was generated, encapsulated, and used to recognize a new dataset formed by integrating and disrupting simulated and experimental data. The recognition results are shown in Figure 17, with performance index scores of RMSE = 12.46, MAPE = 0.241, and R-Square = 0.922. As shown in the figure, the validation data are distributed on both sides of the accurate prediction value. When compared to the validation using only the experimental dataset, the prediction error increases, resulting in an error of 12.46°, up from 7.13°. Additionally, the direction of the arc length also increases by 3. The data have been transformed from the original 2000 groups of validation sets to the current 11,001 groups of validation sets. As a result, the margin of error has increased from 3.7 mm to 6.4 mm. However, this error is still relatively small considering the large amount of accumulated data. These results correspond to an arc length error of 6 mm, further demonstrating the effectiveness of the CNN.

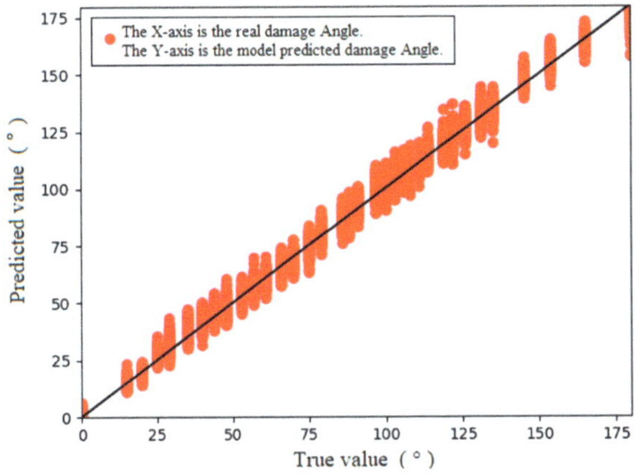


**Figure 17.** Predicted results for mixed data.

## 5. Conclusions

In this study, the quantitative characterization of pipeline crack defects was realized by combining a 1D-CNN and ultrasonic guided wave technology. Through a numerical simulation and experimental study, a 1D-CNN was used to quantitatively verify the dimension of a pipeline crack defect by constructing an ultrasonic guided wave experimental platform. Ultrasonic guided wave signals with different crack defects in the pipeline were collected and quantitatively analyzed. The primary conclusions of this analysis are as follows:

(1) A quantitative analysis of pipeline crack defects can be realized from end to end by combining a 1D-CNN with ultrasonic guided wave technology.
(2) Using the entire guided wave signal, including the incident wave, defect echo, and first-end face echo signal, the feature input can improve the accuracy of the 1D-CNN in identifying the size of the pipeline crack defect.

(3) The 1D-CNN is more suitable for identification training of the defect size of the pipeline than the MLP.
(4) The 1D-CNN can effectively identify the defect size from the echo signal under an excitation signal with different central frequencies, and it can predict the defect size with an error of less than 2%.

In summary, the combination of a 1D-CNN and ultrasonic guided waves can effectively identify pipeline crack damage. Further studies are needed to quantify other types of pipe damage. When the training data samples are insufficient, the CNN tends to produce significant deviations. It is necessary to make breakthroughs in this aspect in the future since the unexplainability of the network makes it impossible to analyze the training data. CNNs can identify various degrees of damage from the response signals of different dispersion states when the data samples are sufficient, which is a unique advantage of CNNs. However, the principle requires in-depth study.

**Author Contributions:** Conceptualization, Y.S. and J.W.; methodology Y.S. and J.W.; software, Y.S.; validation, Y.S. and J.C.; formal analysis, Y.S. and J.W.; investigation, Y.S. and J.C.; resources, J.W., W.Z., X.Y. and H.M.; data curation, Y.S.; writing—original draft preparation, Y.S.; writing—review and editing, Y.S. and J.W.; visualization, Y.S.; supervision, J.W., W.Z., X.Y. and H.M.; project administration, J.W. and H.M.; funding acquisition, J.W. All authors have read and agreed to the published version of the manuscript.

**Funding:** This research was funded by the National Natural Science Foundation of China (Grant Nos. 12202103), and the Guangdong Basic and Applied Basic Research Foundation (Grant Nos. 2021B1515120090 and Grant Nos. 2022A1515011324).

**Institutional Review Board Statement:** Not applicable.

**Informed Consent Statement:** Not applicable.

**Data Availability Statement:** The data generated during this study are currently private due to pending patent applications.

**Acknowledgments:** The authors would like to acknowledge Institute of Science & Technology Innovation Dongguan University of Technology for equipment support.

**Conflicts of Interest:** The authors declare no conflict of interest.

## References

1. Bai, B.; Li, X.; Yuan, Y. A New Cost Estimate Methodology for Onshore Pipeline Transport of $CO_2$ in China. *Energy Procedia* **2013**, *37*, 7633–7638. [CrossRef]
2. Xu, C.; Gong, P.; Xie, J.; Shi, H.; Chen, G.; Song, G.B. An acoustic emission based multilevel approach to buried gas pipeline leakage localization. *J. Loss Prev. Process Ind.* **2016**, *44*, 397–404. [CrossRef]
3. Wu, J.; Hao, H.; Li, J.; Wang, Y.; Wu, Z.; Ma, H. Defect detection in pipe structures using stochastic resonance of Duffing oscillator and ultrasonic guided waves. *Int. J. Press. Vessel. Pip.* **2020**, *187*, 104168. [CrossRef]
4. He, C.; Zheng, M.; Lv, Y.; Deng, P.; Zhao, H.; Liu, X.; Song, G.; Liu, Z.; Jiao, J.; Wu, B. Development applications and challenges in ultrasonic guided waves testing technology. *Chin. J. Sci. Instrum.* **2016**, *37*, 1713–1735.
5. Wilcox, P.; Lowe, M.; Cawley, P. The effect of dispersion on long-range inspection using ultrasonic guided waves. *NDT E Int.* **2001**, *34*, 1–9. [CrossRef]
6. Soleimanpour, R.; Ng, C.T. Scattering of the fundamental anti-symmetric Lamb wave at through-thickness notches in isotropic plates. *J. Civ. Struct. Health Monit.* **2016**, *6*, 447–459. [CrossRef]
7. Zhang, Z.; Pan, H.; Wang, X.; Lin, Z. Deep Learning Empowered Structural Health Monitoring and Damage Diagnostics for Structures with Weldment via Decoding Ultrasonic Guided Wave. *Sensors* **2022**, *22*, 5390. [CrossRef] [PubMed]
8. Shang, L.; Zhang, Z.; Tang, F.; Cao, Q.; Pan, H.; Lin, Z. CNN-LSTM Hybrid Model to Promote Signal Processing of Ultrasonic Guided Lamb Waves for Damage Detection in Metallic Pipelines. *Sensors* **2023**, *23*, 7059. [CrossRef]
9. Davies, J.; Cawley, P. The application of synthetically focused imaging techniques for high resolution guided wave pipe inspection, AIP conference proceedings. *Am. Inst. Phys.* **2007**, *894*, 681–688. [CrossRef]
10. Zheng, G.; Tang, Z.; Wang, Y.; Lv, F. Evaluation of Pipe Axial Defeat through the Use of Ultrasonic Guided Wave and Matching Pursuit. *J. Mech. Eng.* **2013**, *49*, 1–5. [CrossRef]
11. Li, J. On circumferential disposition of pipe defects by long-range ultrasonic guided waves. *J. Press. Vessel Technol.* **2005**, *127*, 530–538. [CrossRef]

12. Li, Q.; Ji, Y.; Wang, B.; Liu, D.; Qian, Z. Deep Learning-assisted Accurate Defect Reconstruction Using Ultrasonic Guided Waves. *Chin. J. Solid Mech.* **2021**, *42*, 33–44.
13. Acciani, G.; Brunetti, G.; Fornarelli, G.; Giaquinto, A. Angular and axial evaluation of superficial defects on non-accessible pipes by wavelet transform and neural network-based classification. *Ultrasonics* **2010**, *50*, 13–25. [CrossRef] [PubMed]
14. Huang, L.; Hong, X.; Yang, Z.; Liu, Y.; Zhang, B. CNN-LSTM network-based damage detection approach for copper pipeline using laser ultrasonic scanning. *Ultrasonics* **2022**, *121*, 106685. [CrossRef] [PubMed]
15. Miorelli, R.; Fisher, C.; Kulakovskyi, A.; Chapuis, B.; Mesnil, O.; Almeida, O.D. Defect sizing in guided wave imaging structural health monitoring using convolutional neural networks. *NDT E Int.* **2021**, *122*, 102480. [CrossRef]
16. Yin, X.; Chen, Y.; Bouferguene, A.; Zaman, H.; Al-Hussein, M.; Kurach, L. A deep learning-based framework for an automated defect detection system for sewer pipes. *Autom. Constr.* **2020**, *109*, 102967. [CrossRef]
17. Zhou, F.; Jin, L.; Dong, J. Review of Convolutional Neural Network. *Chin. J. Comput.* **2017**, *40*, 1229–1251.
18. Nair, V.; Hinton, G.E.; Farabet, C. Rectified Linear Units Improve Restricted Boltzmann Machines. In Proceedings of the 27th International Conference on Machine Learning (ICML-10), Haifa, Israel, 21–24 June 2010; pp. 807–814.
19. Hagan, M.T.; Demuth, H.B.; Beale, M. *Neural Network Design*; Dai, K., Translator; China Machine Press: Beijing, China, 2002.
20. Sawant, S.; Sethi, A.; Banerjee, S.; Tallur, S. Unsupervised learning framework for temperature compensated damage identification and localization in ultrasonic guided wave SHM with transfer learning. *Ultrasonics* **2023**, *130*, 106931. [CrossRef]
21. Yi, J.; Hou, J.; Huang, L.; Shi, H.; Hu, J. Partial Occlusion Face Recognition Based on CNN and HOG Feature Fusion. In Proceedings of the 2021 IEEE 4th International Conference on Electronics and Communication Engineering (ICECE), Xi'an, China, 17–19 December 2021; pp. 55–59.
22. Lecun, Y.; Bottou, L.; Bengio, Y.; Haffner, P. Gradient-based learning applied to document recognition. *Proc. IEEE* **1998**, *86*, 2278–2324. [CrossRef]
23. Bao, J.; Dong, Y.; Liu, H. Review of Development of Convolutional Neural Networks, Proceedings of the Twenty-fourth Annual Conference on New Network Technologies and Applications. *Netw. Appl. Branch China Comput. Users Assoc.* **2020**, 16–21.
24. Montavon, G.; Orr, G.; Müller, K.R. *Neural Networks: Tricks of the Trade*, 2nd ed.; Springer: London, UK, 2012; pp. 49–131.
25. Cong, M.; Wu, X.; Sun, P.; Wu, L. Simulation analysis of surface excitation reception of ultrasonic guided wave pipeline. *Nondestruct. Test.* **2014**, *38*, 24–26.

sensors

MDPI

Article

# A Grating Interferometric Acoustic Sensor Based on a Flexible Polymer Diaphragm

**Linsen Xiong [1,2] and Zhi-mei Qi [1,2,*]**

1 State Key Laboratory of Transducer Technology, Aerospace Information Research Institute, Chinese Academy of Sciences, Beijing 100190, China; xionglinsen17@mails.ucas.ac.cn

2 School of Electronic, Electrical and Communication Engineering, University of Chinese Academy of Sciences, Beijing 100049, China

* Correspondence: zhimei-qi@mail.ie.ac.cn; Tel.: +86-10-588-871-96

**Abstract:** This study presents a grating interferometric acoustic sensor based on a flexible polymer diaphragm. A flexible-diaphragm acoustic sensor based on grating interferometry (GI) is proposed through design, fabrication and experimental demonstration. A gold-coated polyethylene terephthalate diaphragm was used for the sensor prototype. The vibration of the diaphragm induces a change in GI cavity length, which is converted into an electrical signal by the photodetector. The experimental results show that the sensor prototype has a flat frequency response in the voice frequency band and the minimum detectable sound pressure can reach 164.8 $\mu$Pa/$\sqrt{\text{Hz}}$. The sensor prototype has potential applications in speech acquisition and the measurement of water content in oil. This study provides a reference for the design of optical interferometric acoustic sensor with high performance.

**Keywords:** optical acoustic sensor; grating interferometry; flexible diaphragm

## 1. Introduction

Optical interferometric acoustic sensors have been widely studied recently and have advanced applications in areas such as sound source localization [1–3], speaker recognition [4] and medical devices [5,6]. The most dominant acoustic sensors are based on the principle of capacitive detection. The performance of capacitive acoustic sensors is limited by the scaling law. In contrast, optical interferometric acoustic pressure detection methods are immune to the scaling laws as electrical methods [7]. Optical interferometric acoustic sensors are mainly based on Sagnac interferometer (SI) [8], Fabry-Perot interferometer (FPI) [9–11], laser self-mixing interferometer (LSMI) [12], and grating interferometer (GI) [13,14]. The common implementation of the SI and the FPI requires optical fibers, which makes it difficult to integrate [15]. The LSMI has a high degree of integration. However, due to the complexity of its demodulation algorithm, the LSMI is not suitable for detecting high-frequency acoustic signals and is limited to detecting infrasound [16]. Compared to other optical interferometric acoustic sensors, GI-based acoustic sensors could be designed with high integration, wide bandwidth, and high acoustic fidelity [17,18].

According to previous studies, the diaphragm materials used to construct GI acoustic sensors are monocrystalline silicon [19,20], polycrystalline silicon [21,22], and silicon nitride [23]. In order to have wide bandwidth for the acoustic sensor, the diaphragm needs to be properly tensile pre-stressed. Methods to adjust the tensile pre-stress of silicon-based diaphragms include annealing, doping, and stress compensation [24,25]. The way to apply tensile pre-stress to flexible polymer diaphragms is relatively simple. Due to the flexibility of the polymer diaphragm, the diaphragm tensile pre-stress can be applied by mechanical stretching [26]. Therefore, a GI acoustic sensor based on a flexible polymer diaphragm is proposed and experimentally demonstrated in this study. The flexible diaphragm is made of polyethylene terephthalate (PET). The PET diaphragm is coated with a metal film underneath, which, together with the fabricated metal grating, forms the GI. The vibration of the diaphragm introduces a change in the GI cavity length, which is detected by a photodetector (PD) and output as a voltage signal. The performance of the fabricated GI acoustic sensor is also theoretically simulated and experimentally tested. The subsequent section describes the principle, design, fabrication and experimental results of the proposed optical interferometric acoustic sensor.

## 2. Principle and Design

The general structure of the GI-based acoustic sensor is schematically shown in Figure 1. The sensor is composed of a diaphragm, spacer, grating with 1/2 duty ratio, laser and PD. The diaphragm and grating form the GI with a cavity length of $d_{gap}$. The initial $d_{gap}$ is determined by the thickness of the spacer. When the sensor operates, the laser is emitted by a laser diode (LD) or vertical cavity surface emitting laser (VCSEL). The laser first reaches the grating surface and is then split into two parts. One part is reflected by the grating surface. The other part enters the GI cavity and is reflected back to the grating by the diaphragm. Both parts diffract on the grating surface. The part that enters the GI cavity travels a longer optical path than the part that diffracts directly from the grating. As a result, optical interference occurs below the grating surface, and a series of diffraction spots (0th, ±1st, ±3rd, etc.) are produced. According to the scalar diffraction theory, the normalized interferometric optical intensities of the 0th and ±1st diffraction spots are as follows:

$$\frac{I_0}{I_{in}} = \frac{1}{2}\left[1 + \cos\left(\frac{4\pi d_{gap}}{\lambda}\right)\right] \quad (1)$$

$$\frac{I_{\pm 1}}{I_{in}} = \frac{2}{\pi^2}\left[1 - \cos\left(\frac{4\pi d_{gap}}{\lambda}\right)\right] \quad (2)$$

where $I_0$ and $I_{\pm 1}$ are the interferometric optical intensities of the 0th and ±1st diffraction spots, respectively, $I_{in}$ is the intensity of the laser emitted from the LD or VCSEL, and $\lambda$ is the wavelength of the laser. Since it is necessary to reflect the laser, a reflector is uniformly plated or coated on the lower surface of the diaphragm to increase the reflectivity. The reflector is usually a thin metal film.

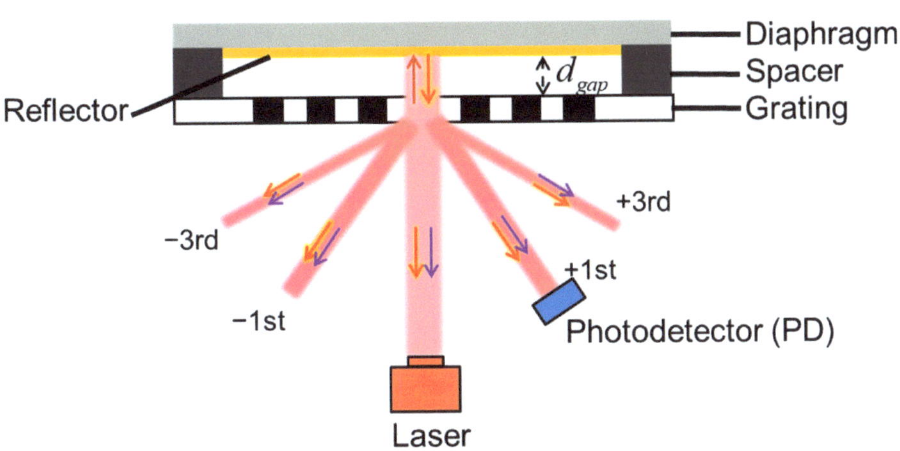


**Figure 1.** Schematic diagram of the grating interferometer (GI)-based acoustic sensor.

When acoustic pressure is applied to the GI acoustic sensor, the vibration of the diaphragm causes a change of $d_{gap}$. According to the vibration theory [27,28], the displacement amplitude $d_r$ and the first resonant frequency $f_1$ of the circular diaphragm are as follows:

$$d_r = \frac{p_{in}\rho}{\omega^2 {P_d}^2 h}\left[\frac{J_0(kr)}{J_0(kr_d)} - 1\right] \tag{3}$$

$$f_1 = \frac{2.405}{2\pi r_d}\sqrt{\frac{P_d}{\rho}} \tag{4}$$

where $r$ is the radial variable, $p_{in}$ is the amplitude of sound pressure, $J_0$ is the zero-order column Bessel function, $\omega$ is the angular frequency of vibration, $P_d$ is the magnitude of the radial tensile pre-stress, $r_d$, $h$, $\rho$ are the radius, thickness, density, of the diaphragm, respectively, the tension is $T_d = P_d h$, and $k = \sqrt{\omega^2 P_d / \rho}$ is the wave number. When $kr_d << 1$, the displacement at the center of the diaphragm $d_0$ is:

$$d_0 = \frac{p_{in}{r_d}^2}{4P_d h} \tag{5}$$

which is theoretically the change in $d_{gap}$. According to Equations (1) and (2), the change in $d_{gap}$ causes a variation of the interferometric optical intensity. The optical intensity is detected by the PD and then converted to a current signal. Therefore, the acoustic pressure can be measured by the electrical signal output from the PD. Considering the spatial location and diffraction efficiency, the $\pm$1st diffraction spot was chosen as the detection signal in the actual sensor design.

First resonance frequency and mechanical sensitivity are important indicators of an acoustic sensor. The first resonant frequency is usually designed to be higher than the operating band range to give the sensor a wide operating bandwidth. And, mechanical sensitivity is usually used to define the magnitude of the mechanical response of the diaphragm to sound pressure. According to Equation (3), the mechanical sensitivity $S_m$ is:

$$S_m = \frac{d_r}{p_{in}} \tag{6}$$

Theoretically, the GI acoustic sensor detects the displacement at the center of the diaphragm. Therefore, according to Equations (5) and (6), mechanical sensitivity is:

$$S_{m0} = \frac{{r_d}^2}{4P_d h} \tag{7}$$

In order to design a diaphragm with a higher first resonant frequency and higher mechanical sensitivity, the influence of diaphragm thickness, radius, density, and tensile stress on performance is analyzed based on Equations (4) and (7). The analysis results are shown in Figure 2. Figure 2a indicates that when the applied tension to the diaphragm is fixed, a thinner diaphragm thickness leads to a higher first resonant frequency. Moreover, the change in diaphragm thickness has no effect on the mechanical sensitivity. However, since the linear deformation range of the diaphragm is 30% of its thickness [29], a thinner diaphragm results in a smaller sound pressure measurement range. Therefore, the design of the diaphragm thickness still requires a trade-off consideration. Figure 2b shows that as the radius increases, the first resonant frequency and mechanical sensitivity of the diaphragm exhibit opposite trends. In practical applications, diaphragms are typically manufactured as standard acoustic sensors with package sizes of 1/2 inch, 1/4 inch, and so on. Therefore, this study chooses to design the diaphragm radius as 3mm for the production of common 1/2-inch acoustic sensors. Figure 2c indicates that a lower diaphragm material density leads to a higher first resonant frequency. Therefore, a higher first resonant frequency can be designed by selecting an appropriate diaphragm material. Figure 2d shows that a larger tensile stress results in a higher first resonant frequency but also leads to a lower mechanical sensitivity. Therefore, the magnitude of the tension applied to the diaphragm is determined by the design objectives of the acoustic sensor.

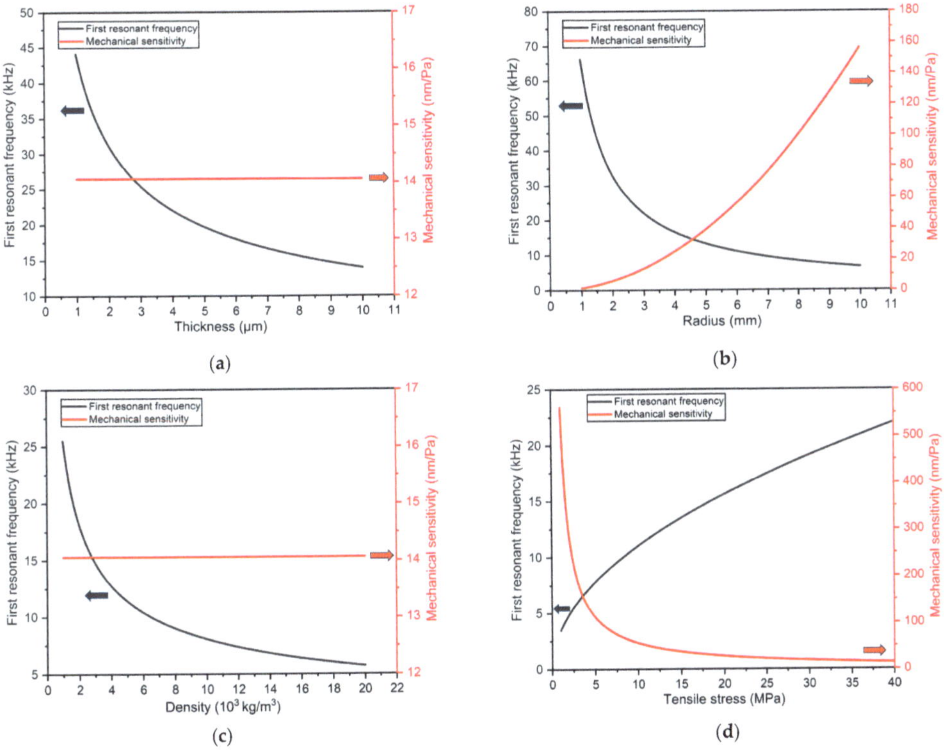


**Figure 2.** Theoretical analysis of the first resonant frequency and mechanical sensitivity for different (**a**) thickness, (**b**) radius, (**c**) density, (**d**) tensile pre-stress of diaphragm.

Considering the above design factors, PET has been chosen as the material for the diaphragm. This type of flexible polymer materials is widely used in the production of flexible devices [30]. The materials used in previous studies for the diaphragms of GI acoustic sensors are nickel, monocrystalline silicon, polycrystalline silicon, and silicon nitride. Compared to these materials, PET has a lower density, as shown in Table 1. Therefore, with the same diaphragm size and tensile pre-stress, the diaphragm made of PET has a higher first resonance frequency according to Equation (4) and Figure 2c.

**Table 1.** Properties of diaphragm materials.

| Material | PET | Nickel | Monocrystalline Silicon | Polycrystalline Silicon | Silicon Nitride |
|---|---|---|---|---|---|
| Density (kg/m$^3$) | 1340 | 8800 | 2300 | 2330 | 3000 |

In order to conduct further analysis of the mechanical performance of the designed sensor, the basic characteristics of the diaphragm structure were simulated using the finite element method. The structural diagram used in the simulation is shown in Figure 3. The modeling area is shown in the red dashed box. In order to make effective use of computational resources in simulation, a two-dimensional axisymmetric model is created, and the red dashed box in Figure 3 shows the modeling area. The parameters of the simulated structure are summarized in Table 2. As shown in Table 2, based on the theoretical analysis presented in Figure 2, the simulation parameter for the diaphragm thickness is selected as 4 μm in order to achieve a good dynamic range for the diaphragm. Then, in order to observe the response of the diaphragm in the low-frequency band, the parameter of tensile pre-stress should be set to a larger value, which is set to 40 MPa in this case. Additionally, considering the use of a 0.5 mm thick glass plate as the backplate in subsequent experiments, the backplate thickness is set to 0.5mm in the simulation.

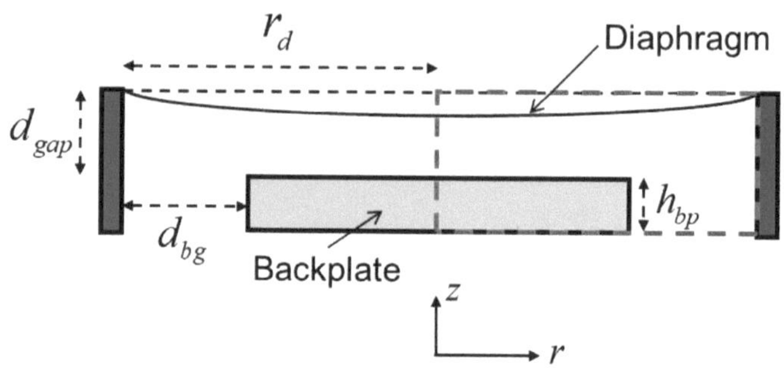


**Figure 3.** Sketch of the acoustic sensor including variables and coordinate system. The red dashed box indicates the modeled region.

**Table 2.** Parameters of the simulated model.

| Description | Parameter | Value or Range |
|---|---|---|
| (Fixed) Radius of diaphragm | $r_d$ | 3 mm |
| (Fixed) Thickness of diaphragm | $h$ | 4 μm |
| (Fixed) Tensile pre-stress of diaphragm | $P_d$ | 40 MPa |
| (Variable) Air gap | $d_{gap}$ | 20~400 μm |
| (Fixed) Thickness of backplate | $h_{bp}$ | 0.5 mm |
| (Fixed) Amplitude of sound pressure | $p_{in}$ | 1 Pa |

Figure 4 shows the simulated variation in diaphragm deflection with applied sound pressure (1 Pa) at 1 kHz. It is worth noting that the displacement in the center region of the diaphragm is larger than the displacement in the edge region. The center of the diaphragm has a maximum deflection of 14.094 nm. According to Equation (5), the theoretical value of the diaphragm center displacement is 14.062 nm, which is in good agreement with the simulated value. Additionally, the linear deformation range of the diaphragm is 30% of its thickness, which is 1.2 μm. Therefore, the maximum linear measurement range of the PET diaphragm for sound pressure is approximately 85.1 Pa.

Figure 5 shows the simulation results of modal analysis performed on the diaphragm. Figure 5a–d respectively presents the mode shapes and corresponding frequencies of mode 1, mode 2, mode 6, and mode 17. Among them, the frequency corresponding to mode 1 is usually the first resonance frequency. Therefore, the first resonance frequency obtained from the simulation is 22.043 kHz, which is consistent with the value (22 kHz) calculated by Equation (4). It is worth noting that, compared to the subsequent frequency response analysis simulations, the frequency of mode 1 obtained from modal analysis can more accurately calculate the first resonance frequency with fewer computational resources.

Figure 6 shows the results of the frequency response analysis simulation. Considering that the amplitude of the input sound pressure is 1 Pa, the corresponding displacement can be converted into mechanical sensitivity. A comparison of the average value of the mechanical sensitivity of the diaphragm with the center value is shown in Figure 6a. Figure 6a shows that the center value of the diaphragm displacement is larger than the average value in the frequency range of 20 Hz to 100 kHz. The curves in the range of ±3 dB fluctuation with the mechanical sensitivity at 1 kHz as a reference were defined as flat frequency band. In the flat frequency band, the center value (14.094 nm/Pa) of mechanical sensitivity is about twice the average value (7.031 nm/Pa). Moreover, the capacitive-based displacement detection principle usually detects the average displacement of the whole diaphragm, while the GI-based displacement detection principle detects the center displacement. Therefore, the mechanical sensitivity of the GI acoustic sensor is twice that of the capacitive acoustic sensor when the same diaphragm is used. In addition, Figure 6 shows three frequency resonance peaks, with corresponding frequencies of approximately 22 kHz, 50 kHz, and 80 kHz, respectively. Combined with the analysis of the modal shapes results, it can be deduced that the three frequency resonance peaks correspond to mode 1, mode 6, and mode 17 presented in Figure 5. Through the analysis results of Figures 5 and 6, it can be found that the acoustic sensor is not suitable for detecting wide-band sound pressure signals above the frequency of mode 1 due to the changes in different modal shapes. Therefore, in order to achieve wide-band signal detection, the sensor should operate within the flat frequency band.

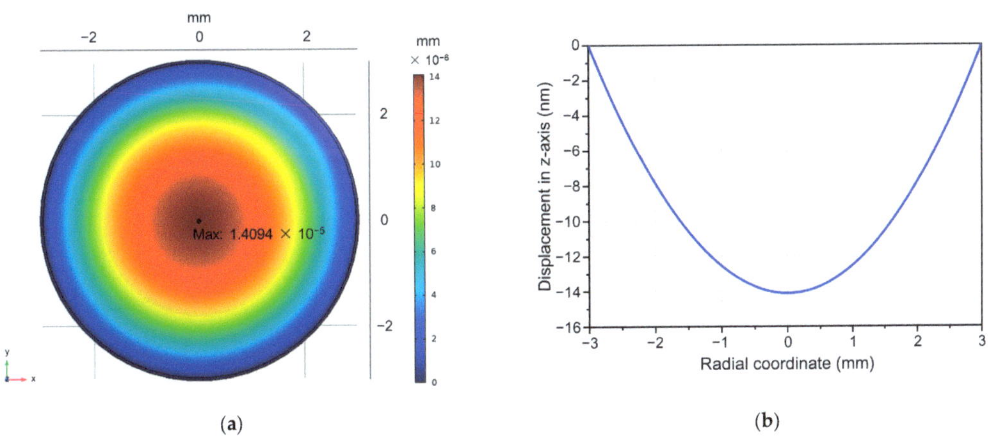


**Figure 4.** Simulated deformation of the diaphragm at a sound pressure of 1 Pa at 1 kHz. (**a**) Displacement distribution in the plane of the diaphragm. (**b**) Displacement distribution on the radial direction of the diaphragm (z-component).

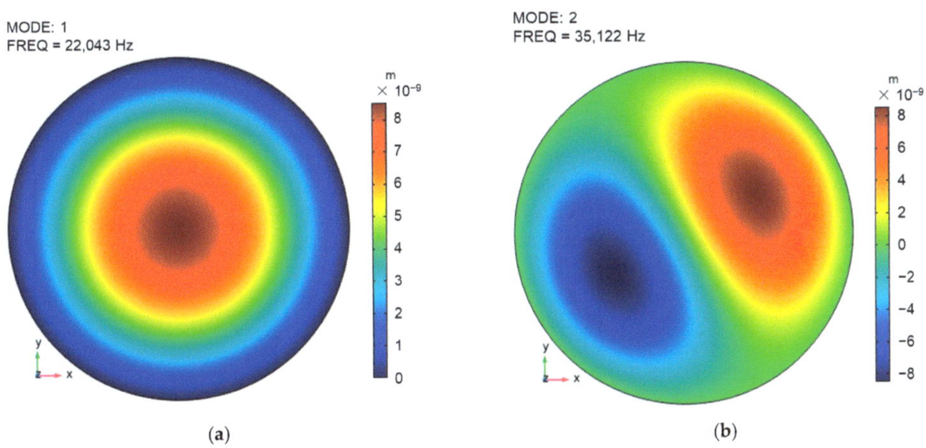


**Figure 5.** *Cont.*

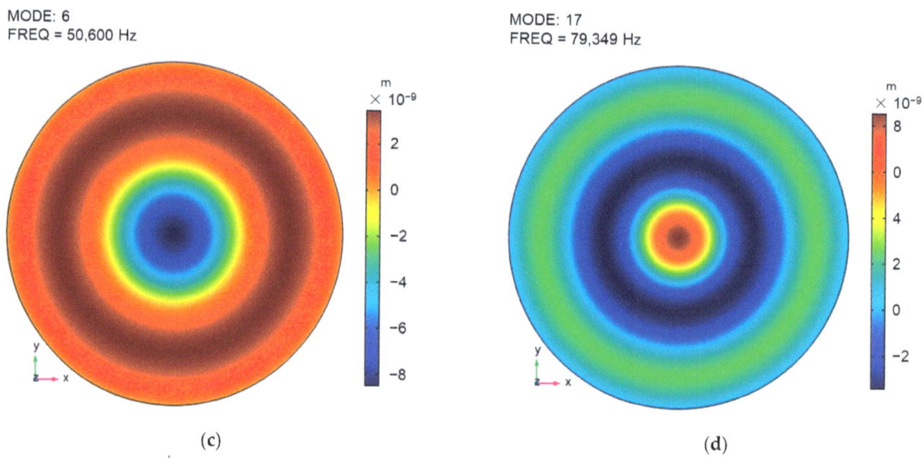


**Figure 5.** Simulated mode shapes and corresponding modal frequencies of the diaphragm: (**a**) mode 1, (**b**) mode 2, (**c**) mode 6, (**d**) mode 17.

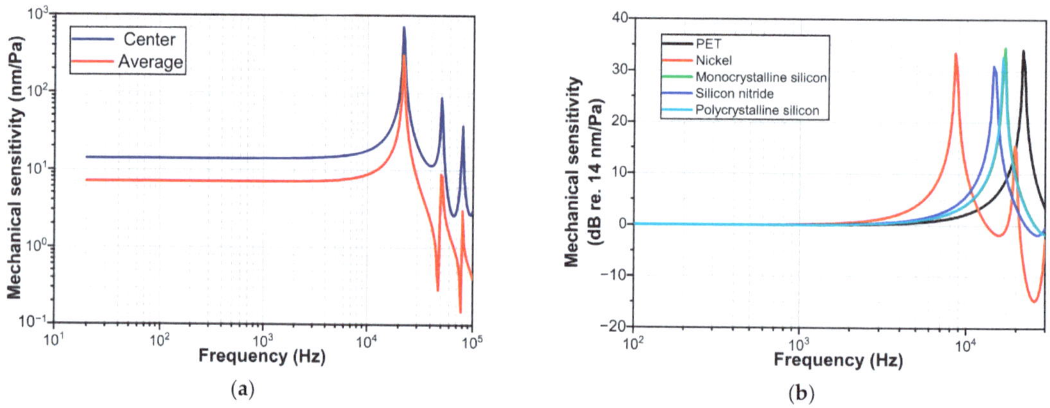


**Figure 6.** (**a**) Simulated frequency response curve of mechanical sensitivity. (**b**) Simulated frequency response curves of mechanical sensitivity of diaphragms (center point) of different materials.

Frequency response curves were also obtained for diaphragms of different materials displaced at the center point, as shown in Figure 6b. Figure 6b shows that the mechanical sensitivity at the center point of the diaphragm for all five materials in the flat frequency band is about 14 nm/Pa. Thus, the mechanical sensitivity is independent of the material density, which is consistent with Equation (4). In addition, the first resonant frequency of the PET diaphragm is the highest among the diaphragms of these five different materials. This is due to the fact that the density of PET is the lowest among the five materials.

Since thin PET diaphragms are usually transparent, it is necessary to coat a certain thickness of metal reflector on the diaphragm to increase its optical reflectivity. Considering the effect of the metal reflector on the mechanical response of the diaphragm, PET diaphragms coated with different thicknesses of metal reflector were simulated using the equivalent single layer theory [31]. In this simulation, the metal reflector used is gold, and its density is 19,300 kg/m$^3$. The simulation results are shown in Figure 7a. Figure 7a illustrates that a thicker gold coating results in a lower first resonance frequency, but does not affect the mechanical sensitivity of the flat frequency band. A 100 nm thick Au layer causes the first resonance frequency of the PET to decrease from 22 kHz to 19 kHz.

The composite density of 4 um thick PET coated with 100 nm thick Au is approximately 1778 kg/m$^3$, which is greater than the density of PET. Therefore, according to Equation (4), the first resonance frequency of this composite structure is 18.9 kHz, which is close to the simulated value. As the thickness of the gold plating increases, the composite density further increases, resulting in a decrease in the first resonance frequency. Therefore, a thinner metal reflector should be used while ensuring that the diaphragm has a certain optical reflectivity.

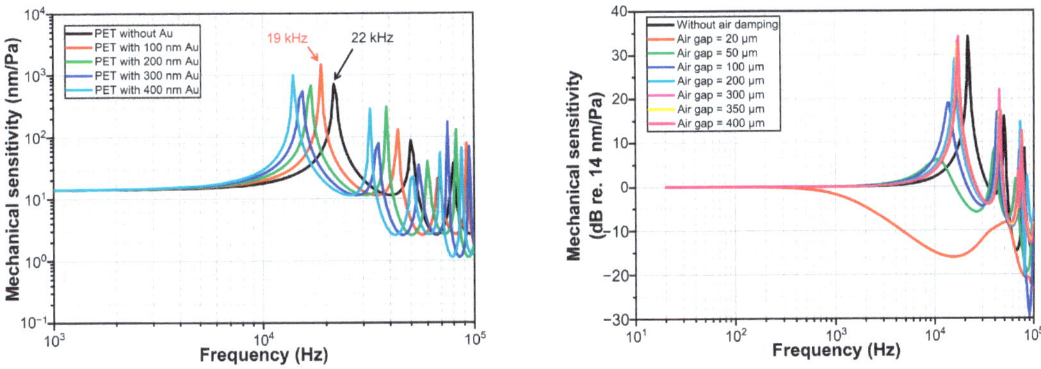


(**a**) (**b**)

**Figure 7.** Simulated frequency response of diaphragm (center point) mechanical sensitivity for different thickness of: (**a**) gold reflector, (**b**) air gap.

The diaphragm vibration discussed above does not involve the influence of the surrounding air. In fact, the vibration of the diaphragm will drive the air vibration and is also damped by the air. The effect of air damping cannot be easily ignored. In order to evaluate the effect of air damping, the frequency response of the mechanical sensitivity at the center of the diaphragm is simulated for different thicknesses of the air gap, as shown in Figure 7b. Figure 7b shows that the presence of an air gap has an effect on the first resonant frequency of the diaphragm. A small air gap (e.g., 20 μm) results in a reduction in the flat frequency band. A suitable air gap (e.g., 50 μm) can widen the flat band. A large air gap (e.g., 400 μm) reduces the effect of air damping on the resonance frequency. Therefore, in order to obtain a good frequency response, it is necessary to design an air gap.

In addition to the analysis of the diaphragm, a further analysis of the grating interferometer was also conducted. Considering the position and optical utilization, the optical intensity of the first diffraction spot was chosen as the detection signal. The optical interference curve of the first intensity analyzed based on Equation (2) is shown in Figure 8a. The cavity gap selected for analysis is 50 μm. The optical curve is sine-like in the case of the monochromatic laser beam. It can be found that the sensor's operating point depends on the wavelength when the cavity gap is fixed. The responses of the GI-based acoustic sensor operating at different operating points of the interference curve are analyzed with the signal frequency of 1 kHz. The analysis results are shown in Figure 8b. It can be observed that the responses at different operating points are different in one cycle. The responses are distorted when the acoustic sensor works at points A, B, C, and D. It should be noted that the response of the sensor operating at point B is completely distorted. Therefore, the sensor should work in the linear range centered at quadrature points (Q1, Q2 & Q3) for a high-fidelity output signal.

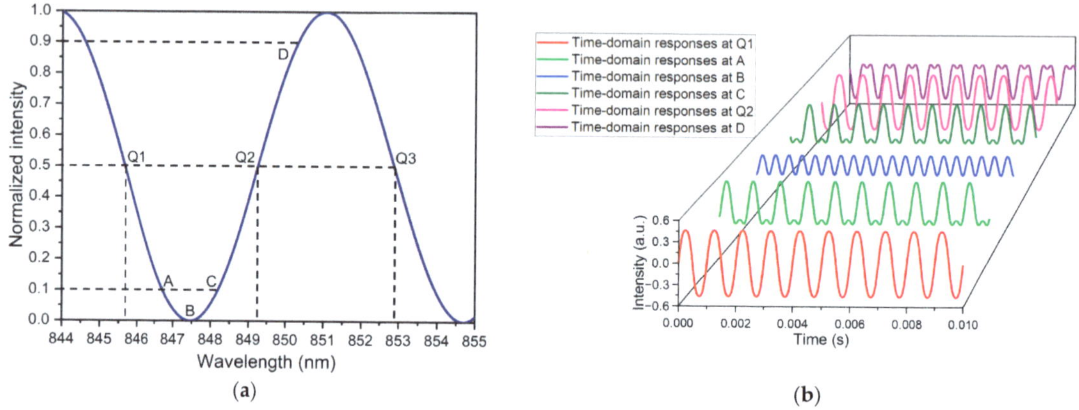


**Figure 8.** (**a**) Optical interference curve of the GI-based acoustic sensor. (**b**) The responses of GI-based acoustic sensor work at different operating points.

The cavity gap usually has a machining error during the fabrication of the sensor [14,19,32]. Thus, the initial operating point is uncertain. The operating point can be adjusted by tuning the wavelength of the LD, which can make the GI-based acoustic sensor work to achieve maximum high fidelity. According to Figure 8a, the half cycle of the optical interference curve is 3.6 nm (from 847.5 nm to 851.1 nm). Thus, the LD selected for building sensors with a cavity length of 50 μm should have a wavelength tuning range over 3.6 nm.

## 3. Fabrication and Packaging

Figure 9 shows a schematic of the fabrication process of the proposed flexible-diaphragm acoustic sensor chip. The fabrication process consists of two main steps. Firstly, the fabrication process of the chromium (Cr) grating on the backplate substrate has been detailed in Figure 10a. A 50 nm Cr film for grating fabrication was sputtered onto a high-transparency glass substrate [33]. Then, a photoresist mask with grating patterns was formed on the Cr film using lithography technology. Afterwards, the Cr film not covered by the photoresist was removed through ion beam etching (IBE). The wafer after IBE is shown in Figure 10b, where the grating pattern has been formed on the glass substrate. And, the grating period is 2.4 μm. Next, the photoresist was cleaned off to retain the Cr film with the grating pattern. Subsequently, perforations were formed on the glass substrate using laser cutting, and a silicon wafer was cut to form spacers. The glass substrate and silicon spacers were assembled together using anodic bonding, as schematically shown in Figure 10c. Finally, the wafer after anodic bonding was subjected to laser cutting to form individual "grating-spacer" structures, as illustrated in Figure 10d.

According to Figure 9, the second step involves assembling the PET diaphragm with the "grating-spacer" structure. First, a 100 nm thick gold reflector was deposited on a large-size PET membrane with a thickness of 4 μm using magnetron sputtering technology. Then, a large-size PET membrane was pre-tensioned using membrane tensioning ring [26], which is a method of applying pre-stress through mechanical stretching. The pre-tensioning force was applied to the PET membrane with a diameter of 68 mm. Next, the "grating-spacer" structure was tightly adhered to the large-size PET membrane using a UV adhesive, as shown in Figure 11a. Finally, the small-size PET diaphragm with a "grating-spacer" structure was cut out using a blade. The small-size PET diaphragm with an effective radius of 3 mm is shown in Figure 11b. In this manufacturing process, the tension of the pre-tensioned PET membrane with a diameter of 68 mm can be measured using the resonant method [34]. The experimental setup for tension measurement is shown in Figure 11c, and its principle is to derive the tensile pre-stress through the measured resonant frequency of the large-size membrane using Equation (4). The test result of the membrane with a diameter of 68 mm are shown in Figure 11d. Figure 11d shows that the resonant frequency of the membrane is 546.4 Hz, and thus indicates that the tension applied to the pre-tensioned PET membrane is 3.16 MPa. Therefore, according to Equations (4) and (7), the small-size PET diaphragm with an effective radius of 3 mm, cut from the large-size membrane, has a theoretical first resonant frequency of 6.2 kHz and a mechanical sensitivity of 178 nm/Pa. Since the maximum linear deformation of the 4 μm thick PET diaphragm is 1200 nm, the maximum measurable sound pressure of this PET diaphragm is about 6.7 Pa.

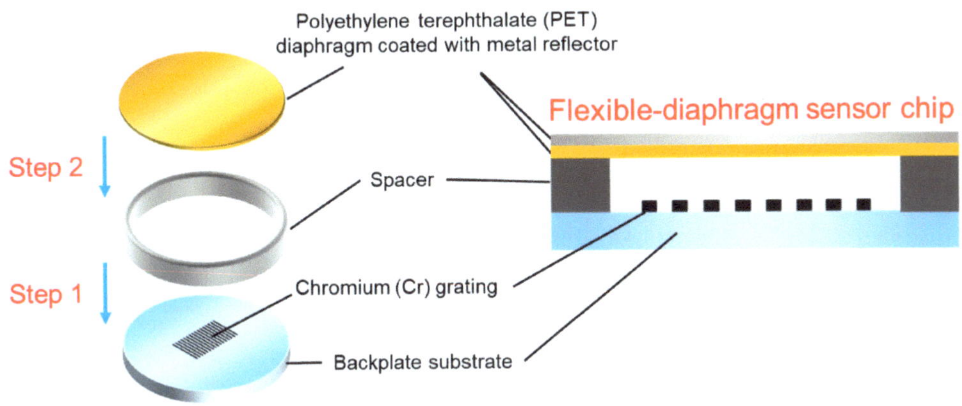


**Figure 9.** Schematic diagram of the fabrication process of the proposed flexible-diaphragm acoustic sensor chip.

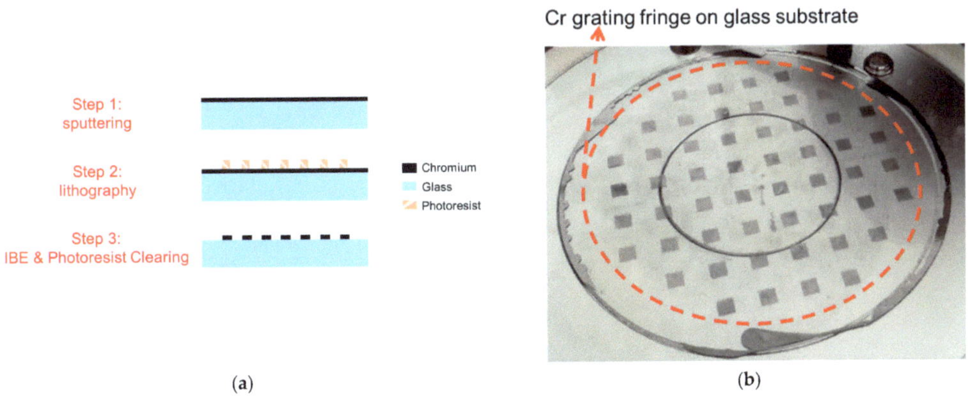


**Figure 10.** *Cont.*

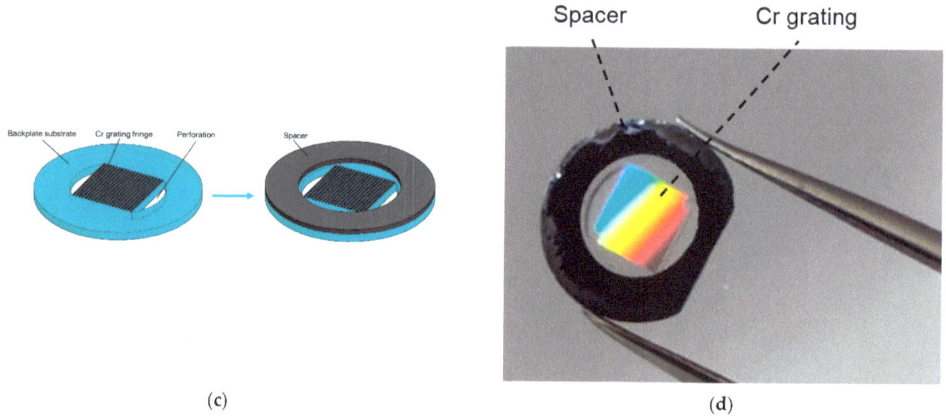


**Figure 10.** (**a**) Fabrication process of the chromium grating on the glass substrate. (**b**) Photograph of the grating on the glass substrate after ion beam etching. (**c**) Schematic diagram of the combination of grating and spacer. (**d**) Photograph of the "grating-spacer" structure.

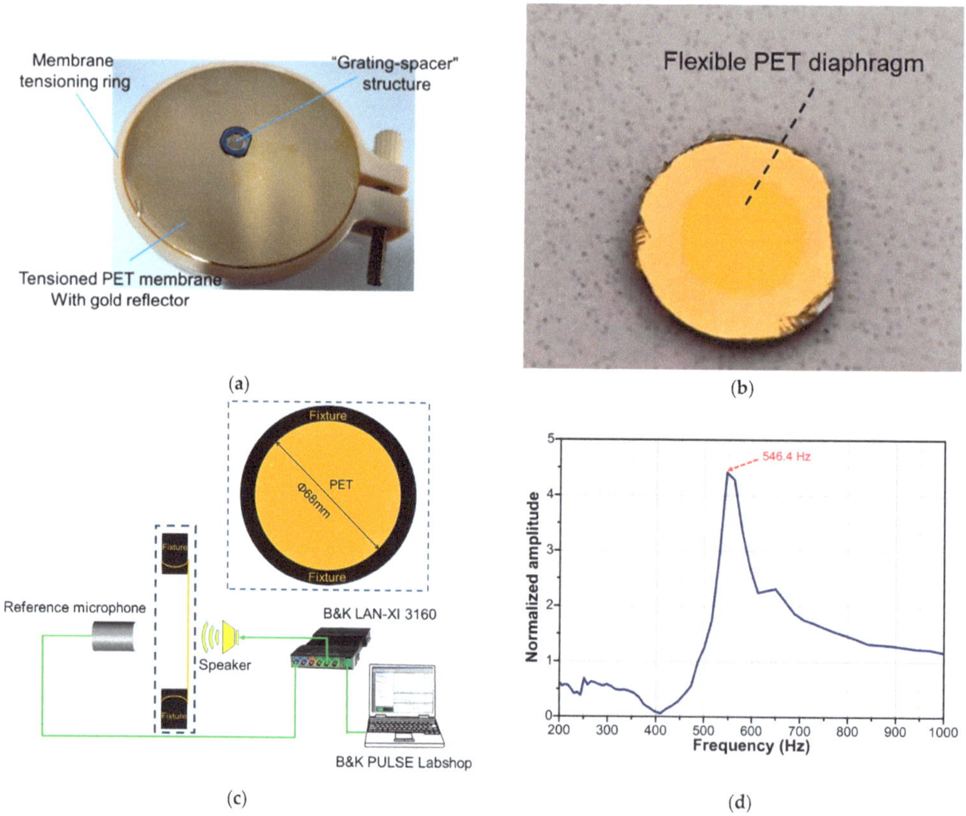


**Figure 11.** (**a**) Photograph of assembly of PET membrane with "grating-spacer" structure. (**b**) Photograph of proposed flexible-diaphragm acoustic sensor chip. (**c**) Schematic diagram of experimental setup for testing resonant frequency of tensioned large-size membrane. (**d**) Resonant frequency of the tensioned large-size membrane.

For experimental testing, the sensor chip was packaged. A schematic of the packaged acoustic sensor is shown in Figure 12a. The sensor chip was placed on top of the holder, while the diaphragm was exposed to perceive sound pressure, as shown by the red dashed box in Figure 12a. The central area of the diaphragm was aligned with the optical window of the LD. A PD was placed on one side of the LD to receive the diffracted spot. Based on the theory of diffraction, the distance (d1) between the center of the LD and the center of the PD, as well as the height (h1) from the surface of the PD to the surface of the grating was determined by the angle (θ) of the first diffraction order of the grating. Both the holder and optical components were mounted on a printed circuit board (PCB), which interfaced with a peripheral circuit. This peripheral circuit consisted of a driving electrical module connected to the LD and a signal processing module connected to the PD. The output voltage signal of the signal processing module was then used to detect the acoustic signal. In this work, the infrared LD (Model D6-6-850-50, Egismos Inc., Chinese Taipei, Taiwan region) was utilized as the laser source, as depicted in Figure 12b. The wavelength tuning range of the LD, measured through a spectrometer (Model AQ6370B, Yokogawa Inc., Tokyo, Japan), is shown in Figure 12c. Figure 12c shows that the wavelength varies by 14.2 nm as the LD voltage changes from 1.5 V to 4 V. The inset illustrates the laser spectrum measured by the spectrometer at a driving voltage of 1.5 V, indicating a wavelength of 850.6 nm. The large tuning range of wavelengths allows the sensor to operate at the quadrature working point. The photograph of the sensor chip and its packaging is presented in Figure 12d.

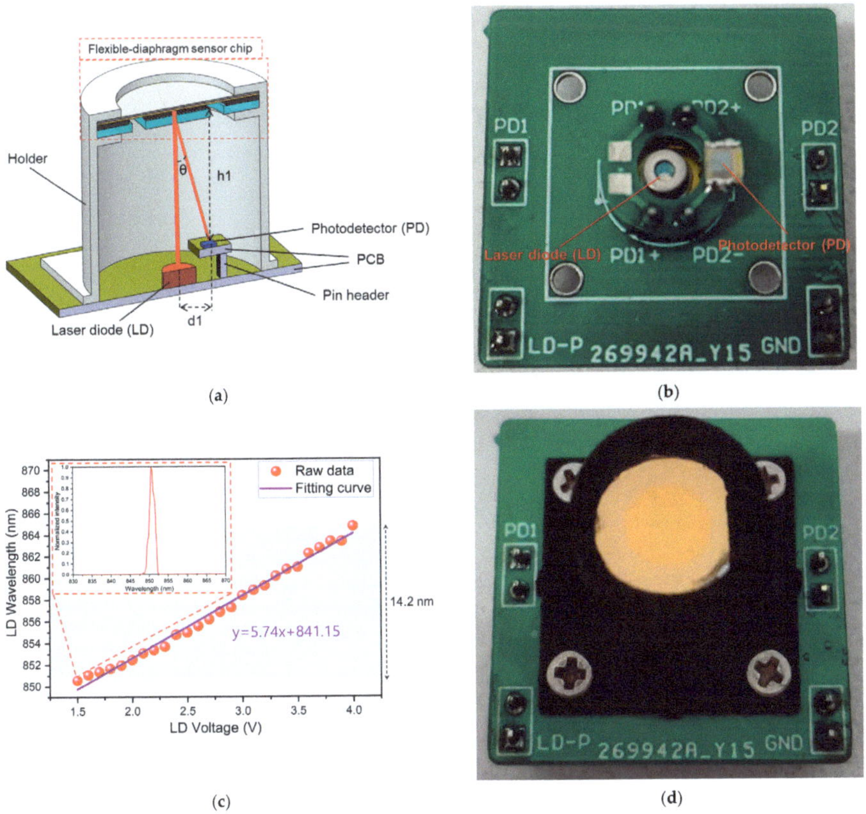


**Figure 12.** (**a**) Schematic diagram of the packaged acoustic sensor. (**b**) Photograph of the laser diode (LD) and photodetector (PD) utilized in the work. (**c**) Measured wavelength range of the LD. (**d**) Photograph of the packaged acoustic sensor.

## 4. Experiment and Discussion

To characterize the packaged acoustic sensor prototype, an acoustic test system was built, as shown in Figure 13. The test system consists of the sensor to be tested, a signal processing module (powered by a lithium battery), a reference acoustic sensor, a loudspeaker, a multifunctional module and signal analysis software. The proposed sensor and a reference acoustic sensor (B&K 4193-L-004) are placed side by side in front of a loudspeaker, as shown by the blue dashed box in Figure 13. The signal generator of the multifunctional module (B&K LAN-XI 3160) drives the loudspeaker to produce a single-frequency acoustic signal with a certain sound pressure. A grating interferometric acoustic sensor excited by the sound pressure outputs a photocurrent signal. The photocurrent signal is converted to a voltage signal by a signal processing module and then acquired by a multifunctional module. The output signal of the reference acoustic sensor is acquired simultaneously. The acquired signals are transferred to a PC with the electroacoustic equipment test system software (B&K PULSE Labshop Version 21.0.0.671) and analyzed by the software.

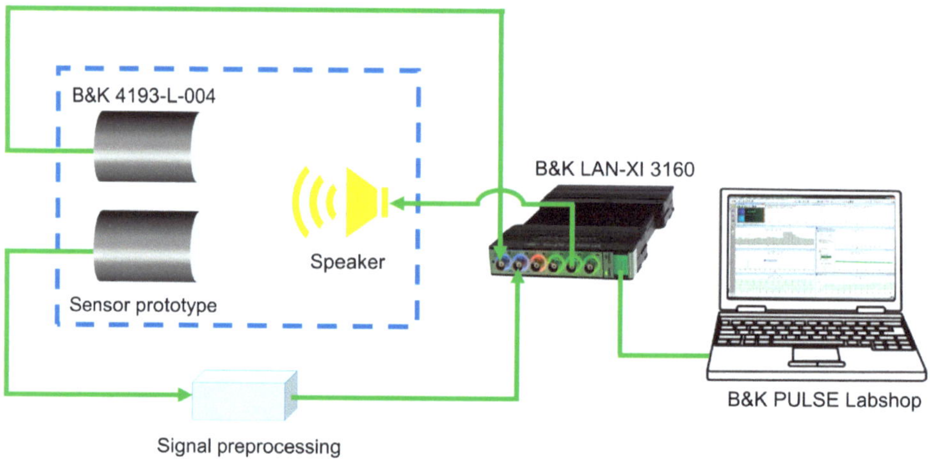


**Figure 13.** The experimental setup for acoustic characterization.

The single-frequency response of the sensor was tested based on the experimental setup as shown in Figure 14. Sinusoidal acoustic waves with frequencies of 250 Hz, 500 Hz, 1.0 kHz, and 5.0 kHz were generated, respectively. The time domain signals of the sensor are shown in Figure 14a–d, and the corresponding frequency domain spectra are shown in Figure 14e–h. It can be seen that the main peaks of the FFT spectra corresponding to the signal waveforms at each frequency are obvious, and the main peak frequencies are consistent with the corresponding test acoustic frequencies. In addition, the minimum detectable sound pressure (MDP) is a key performance of the acoustic sensor. MDP can be calculated from the FFT spectrum. In Figure 14g, the applied 1 kHz acoustic waveform is 77.9 mPa. For a frequency resolution of 0.5 Hz, the background noise is about −94.8 dB, and the measured SNR is 56.5 dB. Therefore, the calculated MDP is 164.8 $\mu Pa/\sqrt{Hz}$.

In addition to the MDP, a flat frequency response is a key performance of the acoustic sensor. The frequency response of the proposed acoustic sensor was tested from 50 Hz to 6.4 kHz. The test results are shown in Figure 15a. The response of the sensor proposed in this work is flat in the range of 50 Hz to 6.4 kHz with a fluctuation range of no more than 3 dB. This frequency response performance is a significant improvement over the previous GI-based acoustic sensor [13,35]. And the prototype in [13] is based on a monocrystalline silicon diaphragm. Due to the brittle of monocrystalline silicon diaphragm, it is impossible to apply tensile pre-stress through mechanical stretching [36]. Meanwhile, applying pre-stress on monocrystalline silicon via micro-electro-mechanical system (MEMS) technology remains a relatively complex technique [24]. Compared to monocrystalline silicon, PET possesses flexibility and stretchability, allowing for a simpler method to apply tensile pre-stress, and thus, offers a better frequency response than uncontrolled pre-stressed monocrystalline silicon diaphragm [13]. The response test points of the acoustic sensor in the voice frequency band (300 Hz to 4.0 kHz) are shown in Figure 15b. As shown in Figure 15b, the sensitivity at 1 kHz is 155.6 mV/Pa. The sensitivity does not fluctuate by more than 3.2% in the voice frequency band, which indicates that the sensor has potential applications in situations related to speech acquisition [37,38].

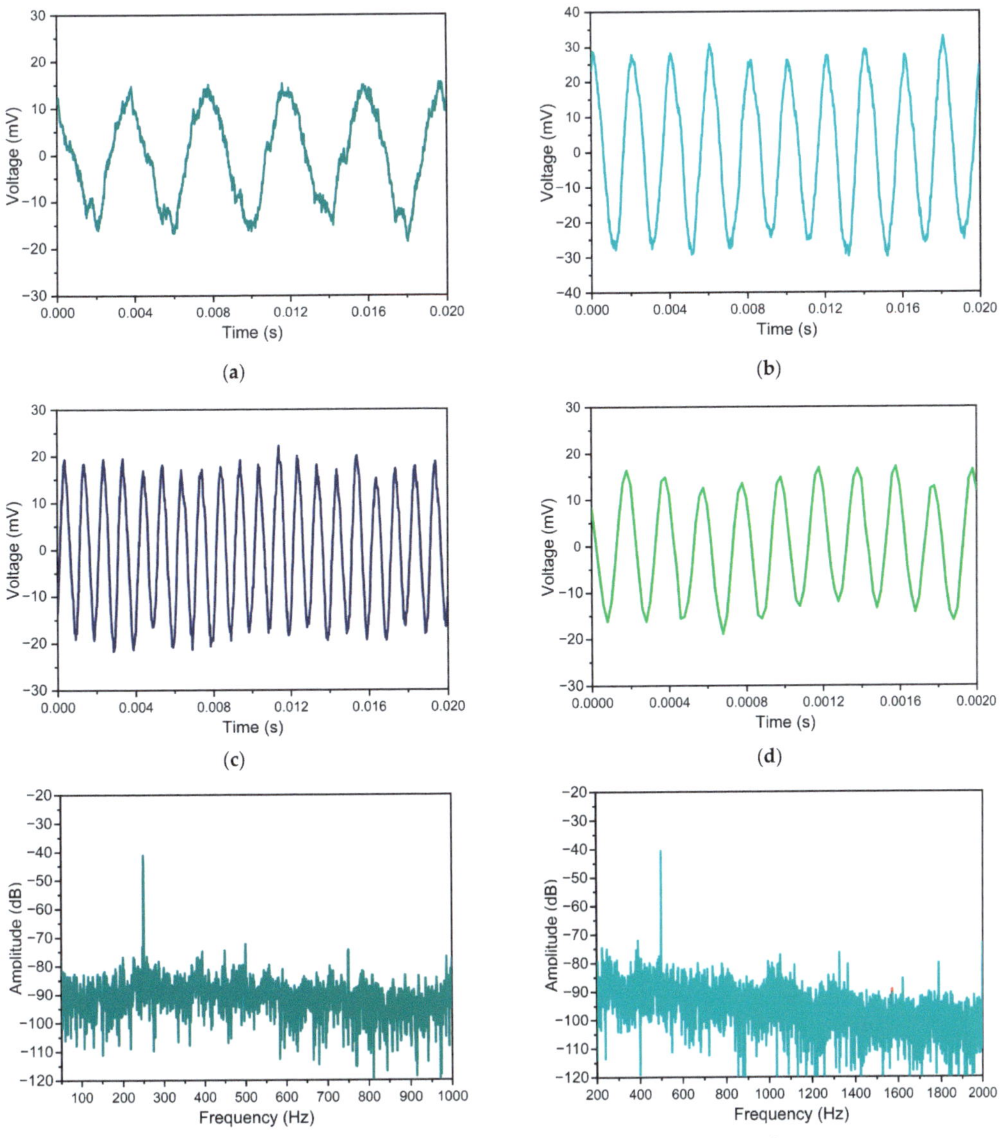


**Figure 14.** *Cont.*

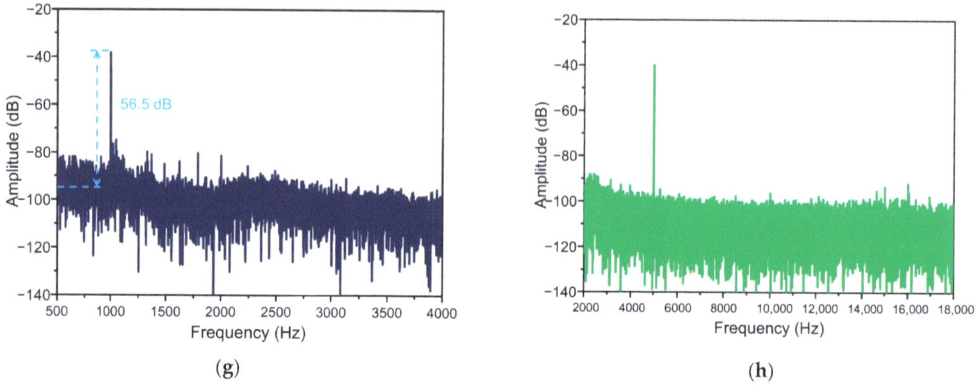


**Figure 14.** Responses of the proposed sensor to acoustic signals with different frequencies and corresponding frequency domain results obtained by Fourier transform processing. (**a**,**e**) 250 Hz; (**b**,**f**) 500 Hz; (**c**,**g**) 1000 Hz; (**d**,**h**) 5000 Hz.

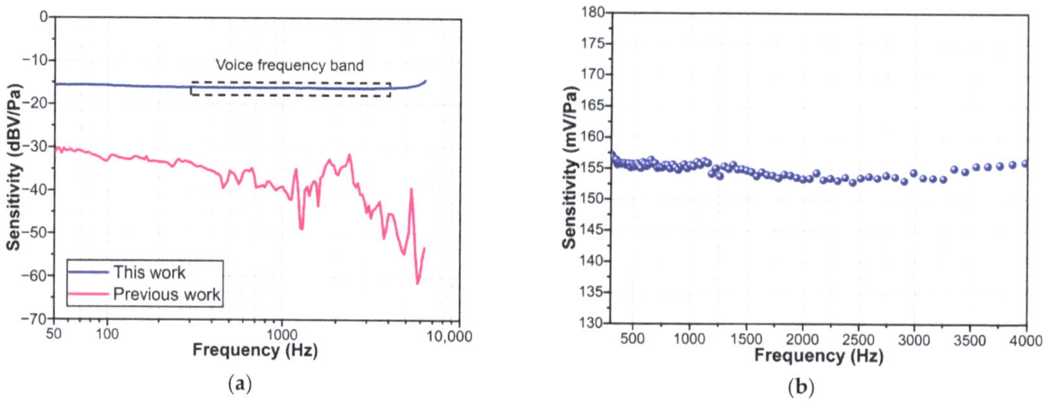


**Figure 15.** (**a**) Frequency response curve from 50 Hz to 6.4 kHz. The previous work is Ref. [13]. (**b**) Frequency response test points over voice frequency band.

In addition, the permeability of water and oil to PET is very low; thus, the acoustic sensor with PET as the diaphragm has potential applications in the measurement of water content in oil [39]. And various polymers can be further investigated for achieving high performances of flexible-diaphragm acoustic sensor. There may be other polymer materials made diaphragms that enable the sensor to have a wider frequency band or lower MDP [40,41].

## 5. Conclusions

In conclusion, a flexible PET diaphragm acoustic sensor has been proposed based on the GI. Firstly, the measurement principle of GI and the theory of membrane vibration are analyzed and explained. The size, density, and pre-stress of the diaphragm are theoretically analyzed to illustrate the impact of these design parameters on sensor performance. Meanwhile, the PET diaphragm is modeled and analyzed using finite element simulation. The simulation results provide design guidance for the metal coating and air gap of the diaphragm. Furthermore, by analyzing the response of GI at different operating points, a wavelength-tuning method is employed to control the operating point. Then, the flexible diaphragm sensor chip is fabricated by MEMS process and machining. The sensor chip was packaged in a prototype for experimental testing. The experimental results show that the packaged sensor prototype can reach a minimum detectable sound pressure of 164.8 $\mu Pa/\sqrt{Hz}$ and a sensitivity of 155.6 mV/Pa at 1 kHz. Moreover, the sensor prototype has a flat frequency response with a fluctuation of no more than 3.2% in the voice frequency band, which has potential application in speech acquisition. And, the low water and oil permeability of the PET diaphragm suggest a potential application of the sensor in measuring water content in oil. Based on this study, the future goal is to achieve the overall flexibility of the sensor, improve system integration, and expand its application to wearable devices.

**Author Contributions:** Conceptualization, L.X.; methodology, L.X.; investigation, L.X.; data curation, L.X.; writing—original draft preparation, L.X.; writing—review and editing, L.X.; Instruction in signal processing and MEMS design, Z.-m.Q. All authors have read and agreed to the published version of the manuscript.

**Funding:** This work was funded by the Beijing Municipal Natural Science Foundation, grant number 4224095.

**Institutional Review Board Statement:** Not applicable.

**Informed Consent Statement:** Not applicable.

**Data Availability Statement:** Data are contained within the article.

**Conflicts of Interest:** The authors declare no conflict of interest.

## References


1. Wu, G.; Xiong, L.; Dong, Z.; Liu, X.; Cai, C.; Qi, Z.M. Development of highly sensitive fiber-optic acoustic sensor and its preliminary application for sound source localization. *J. Appl. Phys.* **2021**, *129*, 164504. [CrossRef]
2. Lorenzo, S.; Solgaard, O. Acoustic Localization with an Optical Fiber Silicon Microphone System. *IEEE Sens. J.* **2022**, *22*, 9408–9416. [CrossRef]
3. Wang, J.; Ai, F.; Sun, Q.; Liu, T.; Li, H.; Yan, Z.; Liu, D. Diaphragm-based optical fiber sensor array for multipoint acoustic detection. *Opt. Express* **2018**, *26*, 25293–25304. [CrossRef] [PubMed]
4. Wang, Y.; Wan, S.; Zhang, S.; Yu, J. Speaker Recognition of Fiber-Optic External Fabry-Perot Interferometric Microphone Based on Deep Learning. *IEEE Sens. J.* **2022**, *22*, 12906–12912. [CrossRef]
5. Miles, R.N.; Degertekin, F.L.; Cui, W.; Su, Q.; Homentcovschi, D.; Banser, F. A biologically inspired silicon differential microphone with active Q control and optical sensing. In *Proceedings of Meetings on Acoustics*; AIP Publishing: Montreal, QC, Canada, 2013.
6. Bicen, B.; Jolly, S.; Jeelani, K.; Garcia, C.T.; Hall, N.A.; Degertekin, F.L.; Su, Q.; Cui, W.; Miles, R.N. Integrated Optical Displacement Detection and Electrostatic Actuation for Directional Optical Microphones with Micromachined Biomimetic Diaphragms. *IEEE Sens. J.* **2009**, *9*, 1933–1941. [CrossRef]
7. Bilaniuk, N. Optical microphone transduction techniques. *Appl. Acoust.* **1997**, *50*, 35–63. [CrossRef]
8. Dong, J.; Lu, R. Dual-loop Sagnac interferometer with a geometric phase shifter for quadrature phase bias locking. *Opt. Lett.* **2019**, *44*, 5422–5425. [CrossRef]
9. Dong, Z.; Hu, X.; Ren, D.; Xiong, L.; Liu, X.; Deng, X.; Cai, C.; Qi, Z.M. Judgment and Compensation of Deviation of the Optical Interferometric Sensor's Operating Point from the Interferometer Quadrature Point. *J. Light. Technol.* **2021**, *39*, 7008–7017. [CrossRef]
10. Ren, D.; Liu, X.; Zhang, M.; Gao, R.; Qi, Z.M. Low-Frequency Bi-Directional Microphone Based on a Combination of Bionic MEMS Diaphragm and Fiber Acousto-Optic Transducer. *IEEE Sens. J.* **2021**, *21*, 14655–14665. [CrossRef]
11. Ni, W.; Lu, P.; Fu, X.; Zhang, W.; Shum, P.P.; Sun, H.; Yang, C.; Liu, D.; Zhang, J. Ultrathin graphene diaphragm-based extrinsic Fabry-Perot interferometer for ultra-wideband fiber optic acoustic sensing. *Opt. Express* **2018**, *26*, 20758–20767. [CrossRef]
12. Donati, S. Developing self-mixing interferometry for instrumentation and measurements. *Laser Photonics Rev.* **2012**, *6*, 393–417. [CrossRef]
13. Xiong, L.; Zhang, M.; Dong, Z.; Hu, X.; Cai, C.; Qi, Z.M. Integrated Grating Interferometric Microphone. *Acta Opt. Sin.* **2023**, *43*, 0723001.
14. Kim, D.; Hall, N.A. Towards a sub 15-dBA optical micromachined microphone. *J. Acoust. Soc. Am.* **2014**, *135*, 2664–2673. [CrossRef] [PubMed]
15. Wu, S.; Wang, L.; Chen, X.; Zhou, B. Flexible Optical Fiber Fabry–Perot Interferometer Based Acoustic and Mechanical Vibration Sensor. *J. Light. Technol.* **2018**, *36*, 2216–2221. [CrossRef]
16. Li, C.; Zhen, H.; Xiaogang, S. Infrasound detection by laser diode self-mixing interferometry. *Chin. Opt. Lett.* **2013**, *11*, 021201.

17. Kuntzman, M.L.; Garcia, C.T.; Onaran, A.G.; Avenson, B.; Kirk, K.D.; Hall, N.A. Performance and Modeling of a Fully Packaged Micromachined Optical Microphone. *J. Microelectromech. Syst.* **2011**, *20*, 828–833. [CrossRef]
18. Okandan, M.; Hall, N.; Bicen, B.; Garcia, C.; Degertekin, F.L. Optical Microphone Structures Fabricated for Broad Bandwidth and Low Noise. In Proceedings of the 2007 IEEE SENSORS, Atlanta, GA, USA, 28–31 October 2007; pp. 1472–1475.
19. Li, Y.; Zuo, M.; Tao, J. An SOI-MEMS Acoustic Sensor Based on Optical Grating Interferometer. *IEEE Sens. J.* **2023**, *23*, 4757–4762. [CrossRef]
20. Zhang, M.; Wu, G.; Ren, D.; Gao, R.; Qi, Z.-M.; Liang, X. An Optical MEMS Acoustic Sensor Based on Grating Interferometer. *Sensors* **2019**, *19*, 1503. [CrossRef]
21. Milleri, N.D.; Onaran, G.; Wiesbauer, A.; Baschirotto, A. MEMS optical microphone based on light phase modulation. In Proceedings of the IEEE 52nd European Solid-State Device Research Conference (ESSDERC), Milan, Italy, 19–22 September 2022; pp. 289–292.
22. Hall, N.A.; Okandan, M.; Littrell, R.; Bicen, B.; Degertekin, F.L. Micromachined optical microphone structures with low thermal-mechanical noise levels. *J. Acoust. Soc. Am.* **2007**, *122*, 2031–2037. [CrossRef]
23. Hall, N.A.; Bicen, B.; Jeelani, M.K.; Lee, W.; Qureshi, S.; Degertekin, F.L.; Okandan, M. Micromachined microphones with diffraction-based optical displacement detection. *J. Acoust. Soc. Am.* **2005**, *118*, 3000–3009. [CrossRef]
24. Cao, Z.; Zhang, X. Experiments and theory of thermally-induced stress relaxation in amorphous dielectric films for MEMS and IC applications. *Sens. Actuators A Phys.* **2006**, *127*, 221–227. [CrossRef]
25. Choi, J.C.; Choi, Y.C.; Chang, S.J.; Kim, K.S.; Lee, T.W.; Lee, J.H. Experimental results of adjusting membrane stress and sacrificial layer thickness for targeting the pull-in voltage of capacitive MEMS microphone structure. In Proceedings of the 7th IEEE Electron Devices Technology & Manufacturing Conference (EDTM), Seoul, Republic of Korea, 7–10 March 2023; pp. 1–3.
26. Gao, G.; Mao, D.; Jiang, R.; Li, Z.; Liu, X.; Lei, B.; Bian, J.; Wu, S.; Fan, B. Investigation of Photoelastic Property and Stress Analysis for Optical Polyimide Membrane through Stress Birefringence Method. *Coatings* **2020**, *10*, 56. [CrossRef]
27. Beranek, L.L.; Mellow, T.J. (Eds.) Chapter 5—Microphones. In *Acoustics: Sound Fields and Transducers*; Academic Press: Oxford, UK, 2012; pp. 199–240.
28. Zhang, W.; Lu, P.; Qu, Z.; Fan, P.; Sima, C.; Liu, D.; Zhang, J. High Sensitivity and High Stability Dual Fabry-Perot Interferometric Fiber-Optic Acoustic Sensor Based on Sandwich-Structure Composite Diaphragm. *IEEE Photonics J.* **2021**, *13*, 1–13. [CrossRef]
29. Giovanni, M.D. *Flat and Corrugated Diaphragm Design Handbook*, 1st ed.; Routledge: New York, NY, USA, 1982; pp. 130–133.
30. Wiranata, A.; Ohsugi, Y.; Minaminosono, A.; Mao, Z.; Kurata, H.; Hosoya, N.; Maeda, S. A DIY Fabrication Approach of Stretchable Sensors Using Carbon Nano Tube Powder for Wearable Device. *Front. Robot. AI* **2021**, *8*, 773056. [CrossRef] [PubMed]
31. Kreja, I.; Sabik, A. Equivalent single-layer models in deformation analysis of laminated multilayered plates. *Acta Mech.* **2019**, *230*, 2827–2851. [CrossRef]
32. Gao, S.; Zhou, Z.; Huang, Z.; Feng, L. High-resolution micro-optical accelerometer with an electromagnetic driver: Design and analysis. *Appl. Opt.* **2021**, *60*, 7989–7994. [CrossRef] [PubMed]
33. Guilbaud-Massereau, V.; Celerier, A.; Machet, J. Study and improvement of the adhesion of chromium thin films deposited by magnetron sputtering. *Thin Solid Films* **1995**, *258*, 185–193. [CrossRef]
34. Li, X.; Zhu, H.; He, L.; Zhang, X.; Niu, F.; Feng, X. Discussion on the diaphragm tension measurement of condenser microphones. In Proceedings of the 27th International Congress on Sound and Vibration, Online, 11–16 July 2021.
35. Hu, X.; Yue, Y.; Cai, C.; Qi, Z.-M. Temperature-robust optical microphone with a compact grating interferometric module. *Appl. Opt.* **2023**, *62*, 6072–6080. [CrossRef]
36. Zhang, B.C.; Shi, Y.H.; Mao, J.; Huang, S.Y.; Shao, Z.B.; Zheng, C.J.; Jie, J.S.; Zhang, X. Single-Crystalline Silicon Frameworks: A New Platform for Transparent Flexible Optoelectronics. *Adv. Mater.* **2021**, *33*, 2008171. [CrossRef]
37. Hayber, S.E.; Tabaru, T.E.; Keser, S.; Saracoglu, O.G. A Simple, High Sensitive Fiber Optic Microphone Based on Cellulose Triacetate Diaphragm. *J. Light. Technol.* **2018**, *36*, 5650–5655. [CrossRef]
38. Wang, J.; Wu, H.; Jia, B. All-fiber-optic acoustic sensor array for real-time sound source localization. *Appl. Opt.* **2017**, *56*, 3347–3353. [CrossRef] [PubMed]
39. Mao, Z.; Zhao, J.; Xuan, W.; Wang, W.; Luo, J.; Xie, J. Distilling determination of water content in hydraulic oil with a ZnO/glass surface acoustic wave device. *Microsyst. Technol.* **2017**, *23*, 1841–1845. [CrossRef]
40. Kim, J.; Lee, S.; Bae, G.Y.; Moon, W. Evaluation of Polymer Materials for MEMS Microphones. *J. Microelectromech. Syst.* **2022**, *31*, 673–682. [CrossRef]
41. Chen, L.H.; Chan, C.C.; Yuan, W.; Goh, S.K.; Sun, J. High performance chitosan diaphragm-based fiber-optic acoustic sensor. *Sens. Actuators A Phys.* **2010**, *163*, 42–47. [CrossRef]

sensors

MDPI

Article

# Assessment of the Influence of the Geometrical Shape of the Damper on the Efficiency of an Ultrasonic Operation Piezoelectric Transducer

M. S. Vechera [1], S. I. Konovalov [2,*], R. S. Konovalov [2], B. I [3], V. M. Tsaplev [2], A. D. Soloveva [2] and J. Lee [4]

1 LLC "Constanta US", Ogorodny Lane 21, Saint Petersburg 198097, Russia; vms@constanta.ru
2 Department of Electroacoustics and Ultrasonic Engineering, Saint Petersburg Electrotechnical University "LETI", Prof. Popova 5, Saint Petersburg 197022, Russia; rskonovalov@etu.ru (R.S.K.); valery@convergences-fr.ru (V.M.T.); adsoloveva@stud.etu.ru (A.D.S.)
3 Department of Smart Manufacturing, Changwon National University, Changwon National University Road 20, Changwon 51140, Republic of Korea; ee.boris@changwon.ac.kr
4 School of Mechanical Engineering, Changwon National University, Changwon National University Road 20, Changwon 51140, Republic of Korea; jaesun@changwon.ac.kr
* Correspondence: sikonovalov.eut@gmail.com; Tel./Fax: +7-(812)-234-37-26

**Abstract:** The results of a study on the geometric shape of the damper on the efficiency of the ultrasonic piezoelectric transducer are presented. In particular, a damper in the form of a truncated cone is considered, the generatrix of which has an inclination angle $\alpha$ relative to the diameter of the piezoceramic plate. The shape of the damper in the form of a truncated cone is chosen based on the a priori assumption that this helps to increase the path of the wave in the damper material due to numerous reflections in it. A criterion for the efficiency of damper operation is proposed. The optimal (from the point of view of the damper efficiency) value of the angle $\alpha$ was determined theoretically and experimentally. The technology of its production is described. Satisfactory agreement between the results of theoretical and experimental studies was noted.

**Keywords:** nondestructive testing; piezoelectric transducer; damper

## 1. Introduction

The damper is one of the most critical structural elements of an ultrasonic piezoelectric transducer (PET). It is designed to increase the transmitter bandwidth and reduce the duration of transient processes. The mechanical damping of the active element of the probe is most often used to achieve this goal, and it is a universal means for suppressing the inertial properties of PETs [1–14]. Mechanical damping is currently used in the manufacture of single ultrasonic piezoelectric transducers with a large aperture (NDT, biomedical diagnostics). The damper introduces active losses into the oscillatory system. The presence of a reactive component disrupts the efficiency of its operation. A purely active input acoustic impedance of the damper is achieved only when the influence of ultrasonic waves reflected from its rear side is excluded. This implies a requirement for the damper material, which must be with a large ultrasonic attenuation coefficient. This is important because the damper has very limited dimensions. In addition, the damper material must have a high specific acoustic impedance, ideally the same as piezoceramics. The back side of the damper often has scatterers, from which the waves are reflected in different directions. As a result, the length of their path inside the material significantly increases. This leads to greater signal attenuation and does not allow reflected waves to hit the piezoelectric element directly after reflection from the rear end.

An analysis of the publications shows that dampers with powder fillers are currently the most widely used. Here, a fine powder of a material with a high specific acoustic impedance (e.g., tungsten) is introduced into the matrix (epoxy resin, sealant, polyurethane).

The papers [9,10] present specific materials used in such damper designs, the relationships between components, etc. The wide distribution of these types of dampers was probably facilitated by the well-established manufacturing technology, which achieved a high degree of identity of the properties of the transducers. In addition, the manufacturing technology of this type of damper does not require significant time or labor costs, which is an essential circumstance in the mass production of probes.

According to [10], the operation of a damper of this type becomes most effective when the size of the absorber particles is comparable with the wavelength, especially with composite fillers. Thus, in addition to powder filler with high specific acoustic impedance, to ensure the scattering of ultrasonic waves, a material with sharply different acoustic impedance value, e.g., crumb rubber, can also be introduced into the damper material. The end part of the damper can be manufactured, for example, of a spherical shape, with the center offset relative to the damper axis, which helps to increase scattering and eliminates the influence of reflected waves. In [10], profiled type dampers are also presented, having different shapes of the end surface, as well as dampers of complex shape, combining, for example, a horn part and a ball with a recess in the upper part. Typically, such dampers are made of metals with a specific acoustic impedance close to that of piezoceramics. From the point of view of the authors of this paper, the disadvantage of probes with such dampers is their ability to work only with plates of a specific diameter effectively. For plates with different diameters, it is necessary to select the shape of the cone generatrix again.

One can mention the damper [15] of the ultrasound transducer, the manufacturing of which consists of placing the piezoplate into a cast form, which is then filled with melted damping material with filler and then cooled until curing. The disadvantage of this method is the inability to obtain a damper with a characteristic impedance close to the acoustic impedance of piezoelement. This is because increasing the percentage of the filler to increase the impedance of the damper leads to the growth of the viscosity and heterogeneity of the entire mass. This complicates its filling and gluing to the piezoelement. The insufficient attenuation of ultrasound waves in such dampers leads to the necessity of increasing the size of the entire transducer. In addition, it causes some doubts by repeating the parameters of the PET equipped with a damper of this type. A similar drawback can be attributed to the damper described in [16]. The method of its manufacturing is that on the back of the fat–fat and prolonged in a modified resin, heated above the boiling temperature of the water, the piezoelement is poured into the filler to a height of at least half-waves at the resonant frequency of the transducer, poured with a binder and then cured.

The influence of this deficiency on the properties of the damper is significantly reduced in its design, presented in [17], which describes a two-layer damper, which is performed in two stages. First, a tiny sublayer of the damping composition is applied to the surface of the piezoelement (epoxy resin with fine tungsten powder). Further, after polymerization, the mentioned sublayer is poured with a layer of polyurethane with tungsten powder. After that, the transducer is centrifuged, which ultimately allows one to achieve a smooth decrease in the damper impedance as it moves from the surface of the piezoelement. This helps to reduce the influence of reflections from the back of the damper. However, the presence of the second layer entails the increase in its height.

In [18], a method of manufacturing a damper is described, which consists of the following. A layer of damping material is applied to the surface of a piezoelectric element fixed on a piezoelement. The specific impedance of the damping material is close to the impedance of the piezoelement. Next, the damping material is heated to the melting point. The layer of melt hardens on the surface of the piezoelement; a damper is formed without intermediate gluing layers. Then, the temperature is reduced, and the powder of the same damping material is introduced into the form and pressed. Due to the gradual decrease in the melt temperature, part of the powder melts, part is compressed, and a damper is formed with a heterogeneous macrostructure in height with a high ultrasound attenuation. The disadvantage of the described damper is the difficulty of obtaining the identity of the PET parameters.

Some other technologies of damper manufacturing [17,19] are known. These technologies increase the damping properties of the damper because a damping mass is placed on a vibrator or centrifuge and subjected to vibration processing that helps the compaction of the damping mass. As a result, the heavy particles of the filler are concentrated near the area of the damper that adjoins the piezoelement.

The analysis of the publications allows us to argue that, when developing dampers, the major efforts of the authors have been aimed at studying their specific technological features. We have studied the influence of the form of the damper on the effectiveness of its operation to a much lesser extent. For example, it is known that dampers often have the form of a truncated cone; however, the issues related to determining the optimal geometric parameters of these types of dampers remain unexplored. In this paper, we discuss some issues related to this problem.

## 2. The Setting of the Problem

The setting of the problem can be formulated as follows. Figure 1a shows a piezoelectric plate, the back surface of which is in contact with the damper. The following designations are used in the figure: *1*—piezoplate; *2*—a damper that has the form of a truncated cone. The resulting cone has an angle of inclination $\alpha$ regarding the line corresponding to the diameter of the plate. The shape of the damper in the form of a truncated cone is selected based on the a priori assumption that this helps to increase the path of the wave inside the damper due to its numerous pores. It is obvious that the effectiveness of the damper depends on the value of the angle $\alpha$. The goal is to determine the optimal value of $\alpha$, in which the signal falling on plate 1 will be minimal due to the reflections inside the damper.

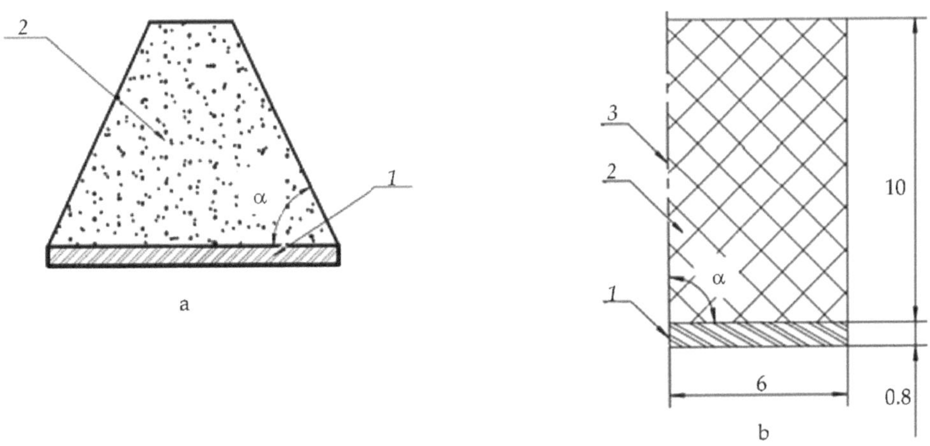


**Figure 1.** The setting of the problem: (**a**) sketch; (**b**) the geometry of the transducer; *1*—round piezoplate, *2*—damper, *3*—symmetry axis.

The solution to the problem was carried out in two stages (first, theoretically, by simulating a numerical experiment using the COMSOL Multiphysics 6.1 software, and second, subsequent comparison of the results at each of the stages).

For both the theoretical and experimental studies, PZT ceramics (Russian TsTS-19) was chosen as the material for the piezoplate. Its properties are described in [20]. The material of the damper is a mixture of the modified epoxy resin of KDA with the hardener of the ETAL-45M and the filler. As a filler, a fine powder of the PV-1 tungsten was used with an average particle size of 0.8. . .1.7 μm, made according to TU 14-22-143-2000 (1:1 by weight, and the mass fraction of the resin is indicated taking into account the hardener).

The parameters of the damper, which are necessary for the calculation, were determined experimentally. We studied these using a precision meter of the distribution rate and attenuation of the longitudinal and transverse waves of the production of SSRIE "Acoustics", which uses PET based on an active element from the lithium niobate, having resonant frequencies 1.25 and 2.5 MHz. The emitting–receiving system is axisymmetric, and the experimental setup consisted of a pulse generator and an oscilloscope. The attenuation was determined by measuring the ratio of pulse signal amplitudes on both sides of the plane-parallel damper specimen. The velocities of the longitudinal and transverse waves were determined through measuring the time of the pulse signal traveling through the specimen.

The result of this experiment for measuring the damper parameters is shown in Table 1.

**Table 1.** Damper parameters.

| Longitudinal Wave Velocity $c_l$, m/s | Transverse Wave Velocity $c_t$, m/s | Density $\rho$, kg/m$^3$ | Attenuation of Longitudinal Waves $\delta_l$, dB/mm | Attenuation of Transverse Waves $\delta_t$, dB/mm |
|---|---|---|---|---|
| $2083 \pm 36$ | $950 \pm 25$ | $2218 \pm 45$ | at frequency 1.25 MHz<br>$0.7 \pm 0.05$<br>at frequency 2.5 MHz<br>$0.75 \pm 0.03$ | at frequency 1.25 MHz<br>$1.0 \pm 0.06$<br>at frequency 2.5 MHz<br>$2.5 \pm 0.08$ |

While performing the theoretical and experimental research, the plate was excited by a signal in the form of a single-period meander with amplitude $V_{\text{in}} = 200$ V and duration $\tau = 1/f_0$, where $f_0$—the resonant frequency of piezoplate.

## 3. Numerical Modeling

Figure 1b shows the initial geometry adopted for the theoretical solution of the problem using the COMSOL Multiphysics 6.1. The figure shows the plate's geometric dimensions and the damper's height. The geometry of the problem under consideration is axisymmetric, which should be noted. As a result, the figure shows only the half of the model. It is worth noting that a truncated cone with an angle $\alpha = 90°$ transforms into a cylinder. This version of the damper is shown in the figure.

To simulate the operation of the PET, Solid Mechanics, Electrostatic, and the multiphysics Piezoelectric Effect Modules were used to describe the propagation of elastic waves. To describe an electric circuit connected to a piezoelement, an Electrical Circuit Module was used. The size of the mesh was selected from the Courant–Friedrichs–Lewy criterion [21], which establishes the dependence between the spatial $\Delta x$ and time steps $\Delta t$ and the velocities of the longitudinal $c_l$ or transverse $c_t$ waves:

$$\text{CFL} = \frac{\Delta t}{\Delta x} c_{l,t}. \tag{1}$$

In this model, the spatial step is equal to one-twelfth of the wavelength at the central frequency, the coefficient CFL $= 0.1$, and the transverse wave velocity is chosen as a wave velocity, since at the same frequency the length of the transverse wave is less than that of the longitudinal one. The mesh consisted of 7800 elements (changing depending on the angle) and 74,396 degrees of freedom, and the time step from (1) is 7 µs. The angle $\alpha$ during the calculation varied from 60° to 90° with step 2°.

To evaluate the effectiveness of the damper, the following criterion was used: exciting the plate with an electric signal of a given amplitude and duration causes an acoustic signal to emit into the damper. The same plate receives the reflected signal. The ratio of the amplitudes of these two signals for the various values of the angle $\alpha$ gives the desired dependence.

The result of the numerical simulation is presented in Figure 2. It shows the ratio of the amplitudes of electrical voltages at the surface of the piezoelement: the voltage from the reflected wave $V_{\text{out}}$ to the applied to the piezoplate $V_{\text{in}}$, depending on the angle $\alpha$ of inclination of the side wall of the damper (forming a truncated cone). The abscissa axis is for the inclination in the generatrix in degrees, and the ordinate axis is for the ratio $A = V_{out}/V_{in}$. The data in the figure indicate that the best result at $\alpha$ = 64° ± 66° at the frequency 2.5 MHz can be achieved. Indeed, at this value of the angle $\alpha$, the parameter $A$ reaches the value 0.0001. For comparison, one can cite data from the same Figure 2a: if $\alpha = 60°$ and $\alpha = 70°$, then the values of $A$ are 0.33 and 0.14, respectively. At the frequency 1.25 MHz (Figure 2b), the optimal value will be $\alpha = 66°$, and $A = 0.36$.

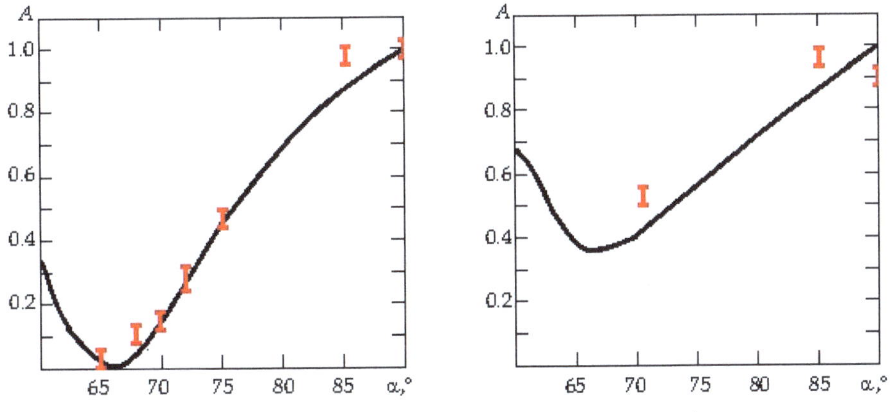


**Figure 2.** (**a**) Change in the coefficient $A$ as the function of the angle of inclination $\alpha$ for the excitation frequency 2.5 MHz; (**b**) change in the coefficient $A$ as the function of the angle of inclination $\alpha$ for the excitation frequency 1.25 MHz.

The analysis of the data given in Figure 2 allows us to argue that with $\alpha$ = 64° ± 66°, one can obtain the best result to achieve the minimum amplitudes of the reflected signals.

For a more detailed study of the processes in the damper, we investigated the output electrical voltage of the piezoelement, depending on the time. Figure 3 presents the change in the electrical voltage $V$ at the output of the piezoplate over time at various values of the angle of inclination of the forming truncated cone $\alpha$.

If the dynamics of changes in the amplitudes of the reflected signals at the output of the piezoplate, as a function of the angle $\alpha$ at a frequency of 2.5 MHz (Figure 3f–j), is considered, the existence of a certain value of the angle $\alpha$, at which the damper works most efficiently can be noted. Within the interval $\alpha$ = 60° ± 66° the signals *1* and *2* practically disappear when $\alpha$ = 64° ± 66°. The level of multiplication of the reflected signals also reduced practically to zero. The further growth of $\alpha$ within the range 66° ± 90° leads to an increase in the signal amplitude due to the multiple overlapping of pulses inside the damper. Similar processes are observed at 1.25 MHz (Figure 3a–e). The signal amplitudes are higher than at 2.5 MHz, although the noise level is higher, which may indicate a weaker attenuation of the multiplied reflected and transformed waves in the damper.

The increase in the signal amplitude caused by multiple reflections inside the damper at $\alpha$ < 64° ± 66° can be explained, because the back reflecting wall of the damper begins to "function as a reflective plane". Indeed, the analysis of the signal fronts inside the damper showed that, at $\alpha$ < 64° ± 66°, the signal is first reflected from the back of the damper and does not change due to multiple reflections from the generatrix.

As an example, Figure 4 shows changes in the wavefront in time for the frequency 1.25 MHz for $\alpha$ = 64° and $\alpha$ = 66°. One can note that near the back of the damper, the signal receives multiple reflections.

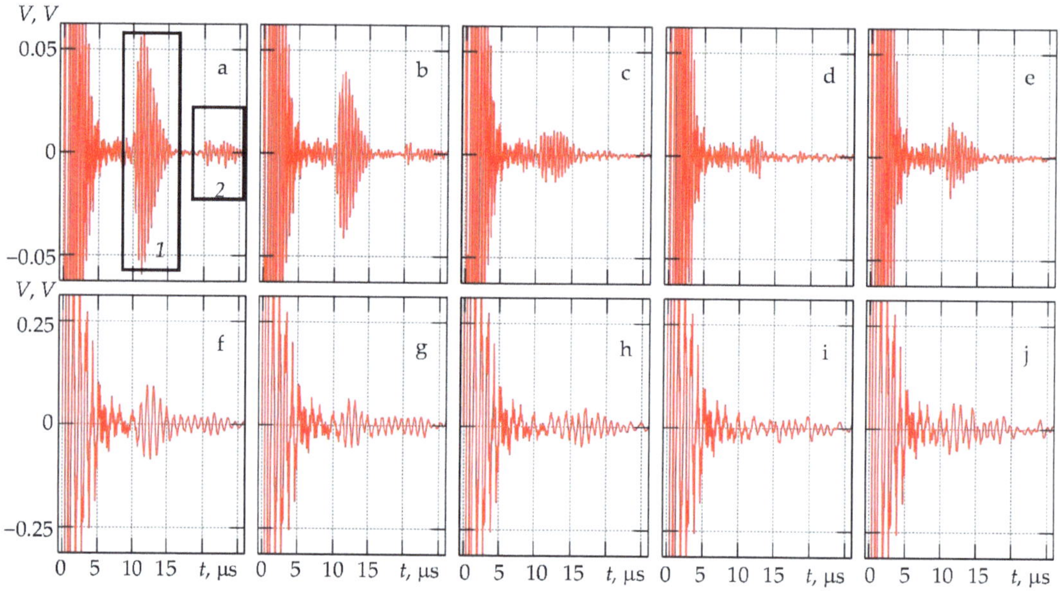


**Figure 3.** Piezoelement output voltage at 1.25 MHz at the angles of inclination α: (**a**) 60°, (**b**) 62°, (**c**) 64°, (**d**) 66°, and (**e**) 68° and at 2.5 MHz at the angles of inclination of α: (**f**) 60°, (**g**) 62°, (**h**) 64°, (**i**) 66°, and (**j**) 68°. *1*—signal reflected from the back side of the damper, *2*—signal caused by re-reflections inside the damper.

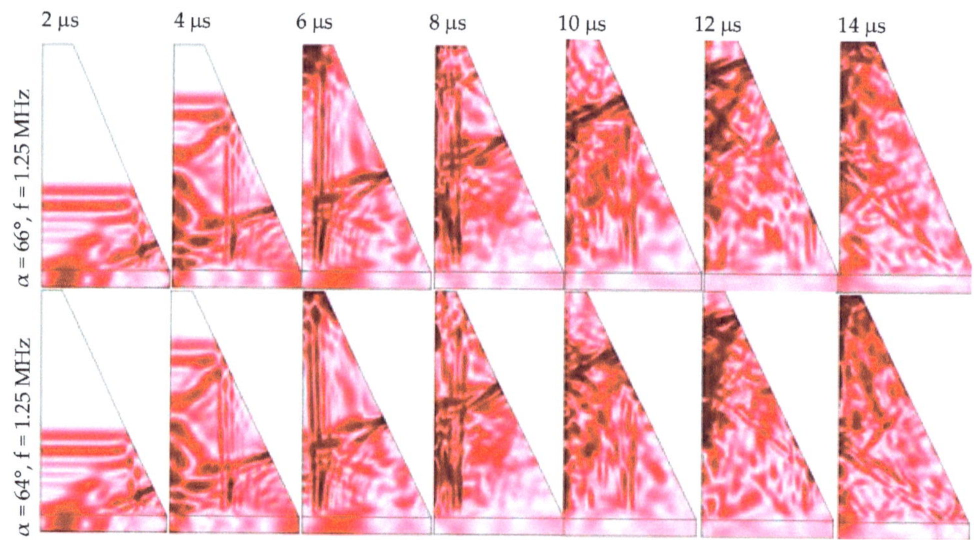


**Figure 4.** Changes in the signal front over time at 1.25 MHz for α = 64° and α = 66°.

## 4. Experimental Research

For experimental verification of the calculated values, the piezoceramics TsTS-19 with 2.5 and 1.25 MHz resonance frequencies and 12 mm diameter are used. To minimize the errors caused by the spread of the parameters of the piezoelectric elements, all the piezoelements were selected in groups in terms of the resonance frequency and capacity.

The spread of the resonance frequencies during the experiment did not exceed ± 10 kHz of the average value for the group, and the capacity did not exceed 30% of the average.

Before use, the piezoplates were selected to obtain the same resonance frequency. To achieve this, a signal generator AFG1022 (TEKTRONIX, Beaverton, OR, USA) and an oscilloscope MDO3012 (TEKTRONIX, Beaverton, OR, USA) were used. Figure 5 shows the experimental setup. The generator applied the continuous sinusoidal signal of 10 V to the electrodes of the piezoplate clamped in the equipment. The frequency step of the exciting signal was 1 kHz. The change in the signal amplitude was monitored using an oscilloscope. The minimum amplitude of the signal on the oscilloscope corresponded to the resonance frequency. The frequency deviation of each plate was no more than 10 kHz.

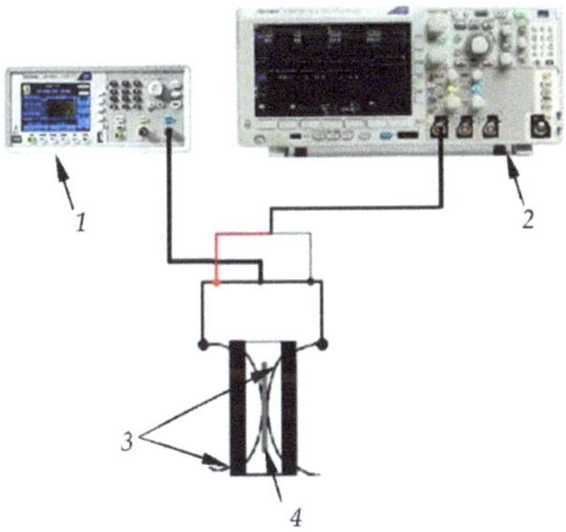


**Figure 5.** The experimental setup to determine the resonant frequencies of piezoplates: *1*—signal generator; *2*—oscilloscope; *3*—clamping electrodes; *4*—piezoplate.

We manufactured the specimens using 3D printing technology from water-soluble PVA plastic. The equipment consisted of an injection mold in the form of a truncated cone with a hole in top for filling the damping mass. The base of the form was a circle with a radius equal to the radius of piezoplate. Such manufacturing technology is rather cheap, easy, and quick with regard to preparing forms, with a lack of technological maps, the possibility of automatic printing, and no requirements for the consistency of control. In addition, using a 3D printer, one can create a shape of any size and configuration. To fix the piezoceramics and forms, double-sided foamed tape was used.

The damping composition was prepared according to the previously described technology (a mixture of the modified epoxy resin of KDA with the hardener of the ETAL-45M and the filler), and after degassing, it was poured into molds. The filling was carried out using a syringe without a needle. After 48 h, the forms with a polymerized damper were removed from the adhesive tape and placed into water to dissolve the mold. It took about 3–4 h to soften and partially dissolve the mold. The rest of the mold, which did not dissolve during this time, was removed mechanically.

According to the described technology, 30 samples were made using piezoplates with the frequency of 2.5 MHz and 9 samples with 1.25 MHz. Figure 6a shows photos of the piezoplates with injection molds, and Figure 6b shows the piezoplates with dampers.

Specimens having the slope of the cone generatrix 90°, 85°, 75°, 72°, 70°, 68°, and 66° were made using the 2.5 MHz plates—three specimens for each value of the angle of inclination, which was made to prevent random errors in manufacturing.

Specimens with 1.25 MHz piezoplates with an inclination of cone generatrix 90°, 85°, and 70° were also made in triplicate.

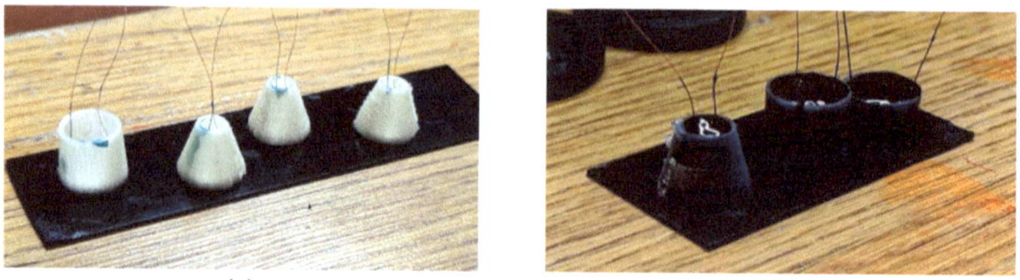

(**a**) (**b**)

**Figure 6.** Photos: (**a**) piezoplates with injection molds for various angles α; (**b**) piezoplates with dampers.

All these prepared experimental PETs were studied in the same way, as under the theoretical investigations. They were excited by a signal in the form of a single-period meander at the frequency of the piezoplate resonance. The signal amplitude was 200 V, generated by the ultrasonic flaw detector UCD-50. Signals from the back free wall of the damper came from the digital output of the flaw detector and were recorded using UDOscill v2.3.0.4 software. Figure 2 shows how changes in the ratio of the maximum amplitude of the signal were reflected from the back of the damper to the amplitude of the probing signal, depending on the angle of inclination (see red marks). The red marks indicate the range of values A for a particular angle of inclination α obtained experimentally. This allows one to compare the theoretical and experimental data. Excellent agreement between the simulated and experimental data is observed at the resonant frequency of 2.5 MHz. The points corresponding to the experimental data differ from the calculated not more than about 8% (see Figure 2a). When 1.25 MHz piezoplates were used, the mismatch rose to 15%.

## 5. Conclusions

The paper describes the experiment and the simulation of the influence of the damper's geometric form on the effectiveness of its work. The following conclusions can be formulated:

1. The criterion for assessing the effectiveness of the damper in the form of the ratio of amplitudes of signals reflected from the free end of the damper and the probing impulse was proposed. This criterion is convenient for the operational control of PET properties during their production.
2. The FEM model of the PET with a conical damper was developed and experimentally confirmed.
3. The technology for the damper manufacturing with water-soluble mold was developed and tested.

The effectiveness of a conical damper with the tungsten powder filler mixed with the resin of KDA in a ratio of 1:1 turned out to be higher in comparison with the damper of a cylindrical shape. The damper, made from exactly such a mixture, can be effectively used to manufacture transducers with a frequency not less than 2.5 MHz. For lower frequencies, selecting another damping composition with other fillers or its other ratio is necessary. The maximum efficiency of the conical damper is achieved at the inclination of the cone generatrix is within α = 64°. . .66°.

To further advances in this area, it would be beneficial to focus on the following topics:

- Evaluation of the effectiveness of dampers with more complex shapes (for example, having a recess on the back side);
- Estimation of the amplitude and duration of the output acoustic signal depending on the shape of the damper.

**Author Contributions:** Conceptualization, S.I.K.; Methodology, S.I.K. and V.M.T.; Software, B.I., A.D.S. and J.L.; Validation, M.S.V., R.S.K., B.I. and A.D.S.; Formal analysis, S.I.K., R.S.K., B.I. and V.M.T.; Investigation, M.S.V., R.S.K., B.I., V.M.T., A.D.S. and J.L.; Data curation, M.S.V., R.S.K., B.I., A.D.S. and J.L.; Writing—original draft, M.S.V., S.I.K. and V.M.T.; Writing—review & editing, S.I.K. and R.S.K.; Visualization, M.S.V., B.I., V.M.T. and J.L.; Supervision, S.I.K.; Project administration, S.I.K. All authors have read and agreed to the published version of the manuscript.

**Funding:** The work was carried out with the financial support of the Russian Science Foundation (grant No. 22-22-20014) and the St. Petersburg Science Foundation (Agreement No. 16/2022 of 14 April 2022). This work was supported by a National Research Foundation of Korea (NRF) grant funded by the Korea government (MOE) (NRF-2022R1I1A3069291).

**Institutional Review Board Statement:** Not applicable.

**Informed Consent Statement:** Not applicable.

**Data Availability Statement:** Data are contained within the article.

**Acknowledgments:** The authors would like to thank the members of the electroacoustics and ultrasonic engineering of the Saint Petersburg Electrotechnical University "LETI" for useful discussions.

**Conflicts of Interest:** The authors declare no conflict of interest.

## References

1. Sachse, W.; Hsu, N.N. Ultrasonic Transducers for Materials Testing and Their Characterization. In *Physical Acoustics;* Mason, W.P., Thurston, R.N., Eds.; Academic Press: New York, NY, USA, 1979; Chapter 4, Volume 14, pp. 277–406. Available online: https://www.sciencedirect.com/science/article/pii/B9780124779143500098 (accessed on 1 December 2023).
2. Radmanovich, M.D.; Mancic, D.D. *Designing and Modeling of the Power Ultrasonic Transducers;* University of Nis, Faculty of Electronics: Nis, Serbia, 2004; 198p, ISBN 86-80135-87-9. Available online: http://alejandria.ccm.itesm.mx/biblioteca/digital/liele/dtpput.pdf (accessed on 1 January 2020).
3. Sherman, C.H.; Butler, J.L. *Transducers and Arrays for Underwater Sound;* Springer Science + Business Media, LLC.: New York, NY, USA, 2007; Chapter 2–4, 610p, ISBN 978-0-387-33139-3. Available online: https://www.springer.com/gp/book/9781441921987 (accessed on 1 December 2023).
4. Papadakis, E.P.; Oakley, C.G.; Selfridge, A.R.; Maxfield, B. Fabrication and characterization of transducers. In *Physical Acoustics;* Thurston, R.N., Pierce, A.D., Papadakis, E.P., Eds.; Academic Press: New York, NY, USA, 1999; Volume 24, pp. 43–134. Available online: https://www.sciencedirect.com/science/article/pii/S0893388X99800244 (accessed on 1 December 2023).
5. Prokic, M. *Piezoelectric Transducers Modeling and Characterization;* MPI: Dättwil, Switzerland, 2004; 266p. Available online: http://alejandria.ccm.itesm.mx/biblioteca/digital/liele/pcmc.pdf (accessed on 1 January 2020).
6. Kikuchi, Y. (Ed.) *Ultrasonic Transducers;* Corona Pub. Co.: Tokyo, Japan, 1969; 407p.
7. Uchino, K. (Ed.) *Advanced Piezoelectric Materials. Science and Technology,* 2nd ed.; Woodhead Publishing: Sawston, UK, 2017; 830p, ISBN 978-0-85709-065-2. Available online: https://www.sciencedirect.com/science/article/pii/B9780081021354099872 (accessed on 1 December 2023).
8. Nakamura, K. (Ed.) *Ultrasonic Transducers. Materials and Design for Sensors, Actuators and Medical Applications;* Woodhead Publishing: Sawston, UK, 2012; 722p, ISBN 978-1-84569-989-5. Available online: https://www.sciencedirect.com/science/book/9781845699895 (accessed on 1 December 2023).
9. Ermolov, I.N. (Ed.) *Ul'trazvukovye Preobrazovateli dlya Nerazrushayushchego Kontrolya (Ultrasonic Piezoelectric Transducers for Nondestructive Testing);* Mashinostroenie: Moscow, Russia, 1986; 280p. (In Russian)
10. Domarkas, V.I.; Kazhis, R.-I.Y. *Kontrol'no-Izmeritel'nye P'ezoelektricheskie Preobrazovateli (Verification and Measurement Piezoelectric Transducers);* Mintis: Vilnus, Lithuania, 1974; 256p. (In Russian)
11. Gitis, M.B. Transducers for pulsed ultrasonic flaw detection—Fundamental theoretical aspects. *Sov. J. Nondestruct. Test.* **1981**, *17*, 131–145. Available online: https://www.researchgate.net/publication/253702177_Transducers_for_pulsed_ultrasonic_flaw_detection_-_Fundamental_theoretical_aspects_Review (accessed on 1 December 2023).
12. Korolev, M.V.; Karpel'son, A.E. *Shirokopolosnye Ul'trazvukovye P'ezopreobrazovateli (Wideband Ultrasonic Piezoelectric Transducers);* Mashinostroenie: Moscow, Russia, 1982; 157p. (In Russian)
13. Konovalov, S.I.; Kuz'menko, A.G. *Osobennosti Impul'snykh Rezhimov Raboty Elektroakusticheskikh P'ezoelektricheskikh Preobrazovatelei (Peculiarities of Pulsed Operating Modes of Electroacoustic Piezoelectric Transducers);* Politekhnika: St. Petersburg, Russia, 2014; 294p, ISBN 978-5-7325-1045-4. (In Russian). Available online: http://elibrary.ru/item.asp?id=25146021 (accessed on 1 December 2023).
14. Rathod, V.T. A Review of Acoustic Impedance Matching Techniques for Piezoelectric Sensors and Transducers. *Sensors* **2020**, *20*, 4051. [CrossRef] [PubMed]
15. Trzaskos, C.R. Ultrasonic Transducer and Process to Obtain High Acoustic Attenuation in the Backing. U.S. Patent US4382201A, 27 April 1981. Available online: https://patents.google.com/patent/US4382201A/en (accessed on 1 December 2023).

16. Maryin, N.S. Sposob Izgotovleniya Dempfera dlya Ul'trazvukovogo Preobrazovatelya (The Method of Manufacturing a Damper for an Ultrasound Transducer). Patent RU94025902, 10 May 1996. (In Russian). Available online: https://i.moscow/patents/ru94025902a1_19960510 (accessed on 1 December 2023).
17. Kashirin, V.A.; Konovalov, S.I.; Stepanov, B.G. Ul'trazvukovoy Preobrazovatel' (Ultrasonic Transducer). Patent RU2150109, 27 May 2000. (In Russian). Available online: https://i.moscow/patents/ru2150109c1_20000527 (accessed on 1 December 2023).
18. Kurgacheva, I.G.; Kulemin, E.A.; Shkarlet, O.M.; Kuklev, A.V.; Sukhorukov, D.V. Sposob Izgotovleniya Dempfera Ul'trazvukovogo Preobrazovatelya (The Method of Manufacturing the Damper of the Ultrasound Transducer). Patent USSR 888029, 7 March 1987. (In Russian). Available online: https://findpatent.ru/patent/129/1295328.html (accessed on 1 December 2023).
19. Alferov, Y.E.; Dronov, A.N.; Sagetelan, G.R. P'yezoelektricheskiy Preobrazovatel' Ul'trazvukovogo Diagnosticheskogo Zonda (Piezoelectric Transducer of an Ultrasound Diagnostic Probe). Patent RU2253191, 27 May 2005. (In Russian). Available online: https://www.freepatent.ru/patents/2253191 (accessed on 1 December 2023).
20. Pugachev, S.I. (Ed.) *Piezokeramicheskie Preobrazovateli (Piezoceramic Transducers: Handbook of Methods of Measuring and Calculating the Parameters)*; Sudostroenie: Leningrad, Russia, 1984; 256p. (In Russian)
21. Courant, R.; Friedrichs, K.; Lewy, H. Über die partiellen Differenzengleichungen der mathematischen Physik. *Math. Ann.* **1928**, *100*, 32–74. Available online: https://link.springer.com/content/pdf/10.1007/BF01448839.pdf (accessed on 1 December 2023). [CrossRef]

sensors

MDPI

Article

# Application of the Lamb Wave Mode of Acoustic Emission for Monitoring Impact Damage in Plate Structures

**Baochun Xu *, Jiacai Huang * and Yachun Jie**

Nanjing Institute of Technology, Nanjing 211167, China; y00450220301@njit.edu.cn
* Correspondence: baochun_xu@njit.edu.cn (B.X.); huangjiacai@126.com (J.H.)

**Abstract:** The impact acoustic emission (AE) of plate structures is a transient stress wave generated by local materials under impact force that contains the state information of the impacted area. If the impact causes damage, the AE from material damage will be superimposed on the impact AE. Therefore, this paper details the direct extraction of damage-induced AEs from impact AEs for the health monitoring of plate structures. The damage-induced AE was analysed based on various aspects, including the cut-off range and propagation speed characteristics of the Lamb wave mode, the correlation between the force direction and the Lamb wave mode, and the impact damage process. According to these features, the damage-induced AE wave packets were extracted and verified via impact tests on epoxy glass fibreboards. The results demonstrated the feasibility of the proposed method for determining whether an impact causes damage via the direct extraction of the damage-induced AE from the impact AE.

**Keywords:** health monitoring of plate structures; impact; damage; acoustic emission; Lamb wave

## 1. Introduction

Plate structures are widely used for sealing high-speed transportation systems such as aircraft, satellites, and high-speed rails. They might become damaged during service due to working loads and the impact of foreign objects, and even minor damage can lead to catastrophic events under harsh working conditions. Thus, health-monitoring systems for plate structures have been developed to forestall such events. Among them, acoustic emission (AE) technology is widely used because AEs are very sensitive in detecting damage-coupled material fracturing. Further, the non-stationary nature of AEs is associated with damage types, and AEs carry rich damage information [1–3]. Therefore, AE-based monitoring is extensively applied in composite performance testing and structural health monitoring. Romhány et al. [4] reviewed relevant papers from 1991 to 2017 on the use of AEs for characterising polymer composites; they found that acoustic source localisation and failure assessment of structural materials based on AEs were considered an ideal method for monitoring the health of composite structures rather than a mere testing approach for such materials. Fotouhi et al. [5] pointed out that the online monitoring of crack propagation based on AEs is an ideal approach for acquiring accurate fracture mechanical parameters of materials. Saeedifar and Zaroucha [6] reviewed the relevant literature on the damage characterisation of laminated composite materials (spanning from 1975 to 2020); they concluded that using AE technology in damage characterisation is advantageous due to its sensitivity to changes in the health states of structural materials. The above-mentioned studies show that AEs carry information on material performance and structural health. Hence, health information can be better obtained through structural health monitoring based on AEs.

In the cited works, methods such as eliminating or lowering the interference of load changes on damage-induced AEs with approximately static or low-frequency alternating loads, or reducing the interference between different damage mechanisms by setting specific experimental conditions to only produce or focus on certain damage, were adopted. Using these approaches, reliable damage-induced AE information can be obtained to study the mapping relationship between damage mechanisms and characteristics. These are all suitable experimental designs. Some damage processes, however, are beyond control in real environments, such as damage inflicted on composite plate structures under the impact of foreign objects during service. The previous studies on the impact damage monitoring of composite plate structures mostly focused on positioning technologies and impact flaw detection. Impact stress waves (i.e., AEs) are the optimal choice for determining an impact's location. In plate structures, an acoustic source is identified by a monitoring system based on the propagation speed and time of arrival (TOA) of the AE [7–10]. The health of the impacted part can be diagnosed using other methods since AEs are applied in impact positioning in most cases. Petrucci and Dhakal [11,12] applied a quasi-static-load bending force to a polymer composite plate structure behind the impact plate structure, causing stress concentration at the damage location that might have led to an increase in damage and damage-induced AEs. The type of impact damage in plate structures can be determined by analysing their characteristics. In addition, the proposed methods are suitable for studying the material properties of plate structures and the corresponding damage development under loads after an impact. In studies on the post-impact damage of composite plate structures, researchers such as Ying and Dziendzikowski [13,14] diagnosed the impact damage in plate structures by generating $A_0$-mode Lamb waves in plates with actuators and collecting the scattered waves caused by the damage. Researchers such as Nardi and Frieden [15–17] showed that the degree and type of damage in plate structures can be identified according to the change in their vibration frequency characteristics before and after the impact damage is in inflicted. The post-impact monitoring method can undoubtedly avoid the interference of impact stress waves on defect detection, allowing one to effectively inspect invisible damage in composite plates; however, this method must be performed under specific conditions, such as with a continuous and stable tensile force, under a bending force, or in an offline status, with these option preventing online monitoring. The best strategy for the online health monitoring of plate structures is to evaluate impact damage by using impact AEs. Okafor et al. [18] conducted relevant research in 2001, reporting that the increase in AE energy alongside kinetic energy without damage was significantly higher than these increments accompanying the damage that occurred through connecting the signal energy generated by the impact with the kinetic energy of the impact. Bruno et al. [19] calculated damage parameter (DPs) using the wave packet characteristic parameter of the first arrival sensor for evaluating the delamination area resulting from a high-speed impact in research concerning the evaluation of the high-speed impact damage of composite plate structures. Saeedifar et al. [20] found that the signal characteristics of the high-frequency band (the filtering frequency band was 100–900 kHz) of low-speed-impact AEs are basically consistent with the AE characteristics of quasi-static loads, proving that they can indicate the impact damage type in the case of low-speed impacts in composite plate structures. These studies demonstrate that impact AEs carry damage information. In studies by Okafor, Bruno et al., and Saeedifar et al. [18–20], damage evaluation was completely dependent on the impact AE characteristics; in the first two cases, the research was conducted as per the energy characteristics of the AEs generated via kinetic energy, while such an evaluation focused on the characteristics of the impact-generated AE in the last case. Moreover, the impact plate structures of external objects are uncontrollable in engineering applications. The characteristic change in AEs depends on factors such as impact location, impact energy, and propagation distance; thus, factors unrelated to damage might affect damage monitoring. This paper suggests that eliminating the interference of AE characteristic changes caused by non-damage factors in damage monitoring is worth studying.

The influence of non-damage factors can be minimised if damage-induced AEs can be directly extracted for the health monitoring of plate structures. Neither the systematic analysis of the difference between damage and external impact AEs nor a sound basis and technical route for extracting damage-induced AEs [21,22] are available in the literature, although some studies have pointed out that the $S_0$ mode of damage activation can be extracted for locating damage via empirical mode decomposition. The separate extraction of damage-induced AEs in accordance with the above-mentioned works on the characteristics of high- and low-speed impacts is difficult. There are various factors to consider. An impact is a dynamic process beyond control, which might lead to diversified damage [11–17]. When damage-induced AEs have low energy and the AE energy generated by the impact behaviour is dominant, the abnormality of damage related to health cannot be derived from the amplitude when these two energies are superimposed [7,17–19]. Regarding the generation mechanisms of AEs, there is no difference between the impact of external objects and damage-induced AEs since they are both stress waves with the same frequency domain characteristics, without discrepancies in the form of propagation in the plate [4,6]. In this case, judging whether a structure is damaged based on changes in characteristics such as frequency, amplitude, and waveform does not yield convincing evidence [18–20]. The above-mentioned AE-based impact-damage-monitoring methods of plate structures are based on the characteristics of damage-induced AEs under a static load, but they cannot provide a basis for health diagnosis through the theoretical analysis of the characteristics of plate structures with the help of signal information extraction technologies.

In this paper, the feasibility of extracting damage information from impact AEs is demonstrated through the theoretical analysis of the AE characteristics in a plate, and this also forms the basis for selecting an appropriate signal-information-processing technology. Establishing a theory that can describe the characteristics of plate stress waves (AE) with universal significance and considers the influence of plate structure parameters on these waves is unrealistic. Nonetheless, the theoretical analysis of the stress wave characteristics in a plate for a certain plate structure is feasible. Therefore, this manuscript discusses the relationship between the force direction and the characteristics of wave velocity for the Lamb wave mode as a form of stress wave propagation in thin plate structures, which are used commonly in engineering, to interpret the AE waveform characteristics in the plate according to the instantaneous energy release of AEs. An appropriate signal-processing method is also proposed to extract the damage-induced AEs from the impact AEs as per the above characteristics for the health monitoring of plate structures.

## 2. Analysis of the Relationship between the Lamb Wave Mode and Force Direction of AEs in Plate Structures

An AE is a stress wave that propagates as a Lamb wave in thin plate structures. The relationship between the stress wave mode and force direction in a thin plate was analysed according to the Lamb wave dispersion curve; the conclusions were derived from the mechanical equations of plate structures and then verified via a lead-breaking test.

### 2.1. Analysis of the Lamb Wave Mode Characteristics in Plates

Lamb waves are stress waves propagating in a structure with two free parallel planes, and their wave characteristics are expressed by the Rayleigh–Lamb equation, which determines whether the Lamb wave mode is multi-mode or dispersion. There are at least two Lamb wave modes, with their propagation speeds related to the frequency at any activation frequency. A Lamb wave is divided into a symmetric (S) wave and an anti-symmetric (A) wave based on the phase relationship of the mass points on the surface of an object. Compared with isotropic material plates, composite material plates present complex wave characteristics that depend on the corresponding fibre material, fibre laying direction, and resin material. However, the main characteristics of Lamb waves are basically consistent. The relationship between the group velocity and frequency–thickness product ($fh$, in MHz·mm) of Lamb waves in an aluminium plate structure [21] is shown in Figure 1 to clarify their characteristics in the plate; the multi-mode and dispersion characteristics for these Lamb waves carry rich information with a high application value for plate structure health monitoring. A well-characterised Lamb wave can be selected for plate structure health monitoring; this is conducive to determining its mode and wave speed and better serves a given application.

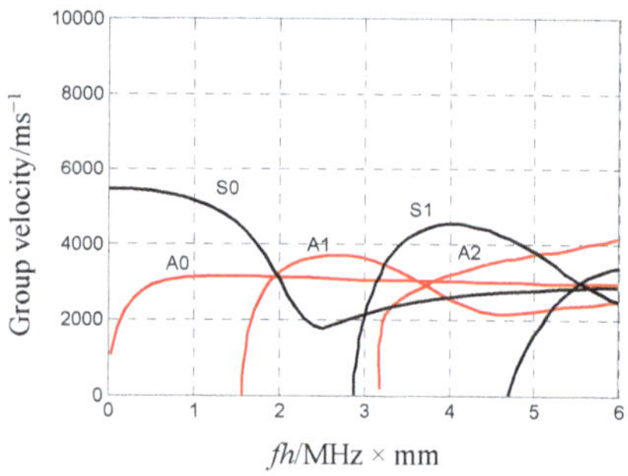


**Figure 1.** Relationship between group velocity and frequency–thickness product of Lamb waves in aluminium plates.

Furthermore, if the frequency remains unchanged, the plate will become thinner as $fh$ decreases. In this case, the labels $S_0$ and $A_0$ in Figure 1 indicate that the Lamb wave mode has been cut short. The propagation speed of $S_0$ and $A_0$ varies significantly in this mode's cut-off range. The speed change of the $S_0$ mode is not highly correlated with the change in $fh$ when $fh < 1$, and the slight variation in the whole range can be approximated as a constant value. In contrast, the speed change of the $A_0$ mode is highly correlated with the change in $fh$ and increases along with $fh$, showing a significant frequency dispersion.

Moreover, the characteristics of the Lamb wave mode in this mode's cut-off range can also be found in plate structures consisting of other materials, such as composites. The $S_0$ and $A_0$ modes of a Lamb wave can be observed in a plate structure when the $fh$ of the Lamb wave is in the modes' cut-off ranges. Under this condition, the $A_0$ mode waveform should have a high frequency in the front and a low frequency behind it, indicating dispersion if the excitation source is a broadband signal. In comparison, the propagation speed of the $S_0$ mode tends to be constant, and its waveform should maintain the initial propagation of the generation source. The $S_0$ mode propagation speed is almost independent of $fh$ in this mode's cut-off range; therefore, some studies recommend its use for locating the acoustic source [4,22,23].

### 2.2. Analysis of the Relationship between Force Direction and Wave Velocity Characteristics

An AE in a plate is an elastic wave whose vibration corresponds to classical mechanical theory. Hence, its characteristics can be analysed based on mechanical equations. The characteristics of waves generated by forces perpendicular and parallel to a plate are defined by the balance equation of force and the motion equation of particles [24,25]. Assuming that the plate has a thickness of $h$, the force being applied to its structure unit is $F_1$ in the vertical direction and $F_2$ in the horizontal direction (Figure 2).

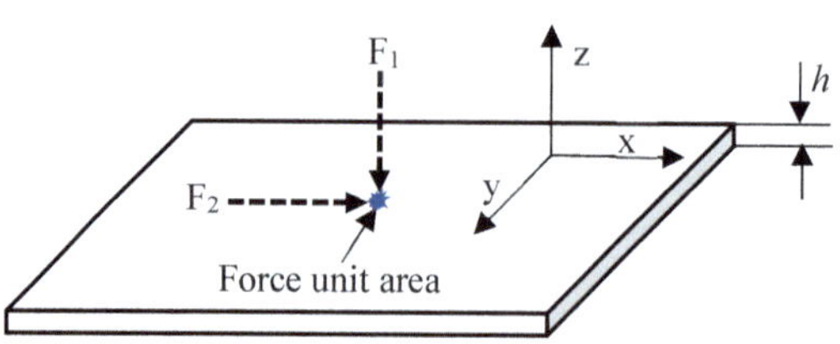


**Figure 2.** Force-activating waves in a plate structure.

If the force direction is perpendicular to the plate's surface, the balance equation of the plate can be simplified as

$$D\Delta^2\zeta - P = 0 \tag{1}$$

where $D$ is the bending strength of the plate, $P$ is the force acting on the unit area of the plate surface, $\zeta$ is the displacement per unit area under the force, and $\Delta$ is the Laplacian Operator, which is $\Delta = d^2/dx^2$ under one-dimensional conditions. According to the definition of bending strength, the following equations can be obtained from the relationship between force and displacement:

$$D = Eh^3/12\left(1-\sigma^2\right) \tag{2}$$

and

$$P = -\rho h \partial^2\zeta/\partial t^2 \tag{3}$$

Here, $E$ is Young's modulus, $\sigma$ is the Poisson's ratio, $\rho$ is the plate density, and $\rho h$ is the mass per unit area. By substituting Equations (2) and (3) into Equation (1), the free vibration of the plate can be expressed as

$$\rho h \partial^2\zeta/\partial t^2 + Eh^3/12\left(1-\sigma^2\right)\Delta^2\zeta = 0 \tag{4}$$

The solution to Equation (4) is discussed within the harmonic range. Hence, $\zeta$ can be written as $\zeta_0 exp[j(kx-\omega t)]$, where $\zeta_0$ is constant. By discussing a one-dimensional problem, $\Delta^2\zeta$ and $\partial^2\zeta/\partial t^2$ in Equation (4) can be written as

$$\Delta^2\zeta = \partial^4\zeta/\partial t^4 = k^4\zeta_0 exp[j(kx-\omega t)] \tag{5}$$

and

$$\partial^2\zeta/\partial t^2 = -\omega^2\zeta_0 exp[j(kx-\omega t)] \tag{6}$$

By substituting Equations (5) and (6) into Equation (4), the last can be simplified as

$$\rho(-\omega^2) + Eh^2k^4/12\left(1-\sigma^2\right) = 0 \tag{7}$$

$$\omega = hk^2\left[E/12\rho\left(1-\sigma^2\right)\right]^{1/2} \tag{8}$$

or

$$k = \left[12\rho\omega^2\left(1-\sigma^2\right)/Eh^2\right]^{1/4} \tag{9}$$

The wave phase velocity ($c = \omega/k$) can be derived from Equation (9) as follows:

$$c = \left[Eh^2/12\rho\left(1-\sigma^2\right)\right]^{1/4}\sqrt{\omega} = \left[E/12\rho\left(1-\sigma^2\right)\right]^{1/4}\sqrt{\omega h} \tag{10}$$

The wave group velocity ($c_g = \partial\omega/\partial k$) can be obtained from Equations (8) and (9) as follows:

$$\begin{aligned} c_g = 2hk\left(E/12\rho\left(1-\sigma^2\right)\right)^{1/2} &= 2h\left[12\rho\omega^2\left(1-\sigma^2\right)/Eh^2\right]^{1/4}\left[E/12\rho\left(1-\sigma^2\right)\right]^{1/2} \\ &= \left[4E/3\rho\left(1-\sigma^2\right)\right]^{1/4}\sqrt{\omega h} \end{aligned} \tag{11}$$

The generated wave propagates along the axis direction if the force direction is parallel to the plate surface. However, the mass–point displacement propagates in both the y-axis and x-axis directions. Thus, the motion equations of the two directions can be presented as

$$(\rho/E)\partial^2u_x/\partial t^2 = \left[1/\left(1-\sigma^2\right)\right]\partial^2u_x/\partial x^2 + [1/2(1+\sigma)]\partial^2u_x/\partial y^2 + [1/2(1-\sigma)]\partial^2u_y/\partial x\partial y \tag{12}$$

and

$$(\rho/E)\partial^2 u_y/\partial t^2 = \left[1/\left(1-\sigma^2\right)\right]\partial^2 u_y/\partial y^2 + [1/2(1+\sigma)]\partial^2 u_y/\partial x^2 + [1/2(1-\sigma)]\partial^2 u_x/\partial x\partial y \tag{13}$$

If the wave propagating along the x-axis is considered alone while neglecting coupling, the following can be obtained:

$$\partial^2 u_x/\partial t^2 = \left[E/\rho\left(1-\sigma^2\right)\right]\partial^2 u_x/\partial x^2 \tag{14}$$

and

$$\partial^2 u_y/\partial t^2 = [E/2\rho(1+\sigma)]\partial^2 u_y/\partial x^2 \tag{15}$$

Then, the wave velocity of particle displacement along the x-axis is expressed as

$$c_{ext} = \left(E/\rho\left(1-\sigma^2\right)\right)^{1/2} \tag{16}$$

while that along the y-axis is given by

$$c_t = (E/2\rho(1+\sigma))^{1/2} \tag{17}$$

The propagation speed of the wave generated by the force in the plate is related to the force direction, and according to the results obtained from the above equations, there are, theoretically, three waves with different velocity characteristics. The propagation speed of the wave generated by the force perpendicular to the plate direction is shown in Equation (11), and the corresponding wave velocity is associated with the frequency. The propagation speed of the wave generated by the force parallel to the plate is shown in Equations (16) and (17), and the wave velocity is independent of the frequency. Note worthily, the characteristics of the plate structure material discussed above are isotropic or approximately isotropic. If the characteristics of the plate structure material are anisotropic, the expression of the wave speed will differ, but the relationship between the propagation speed and frequency of the wave excited by the force in different directions is still valid [26,27].

### 2.3. Analysis of the Relationship between Force Direction and AE Mode

An AE is a transient elastic wave generated by the dynamic change of local areas after a structure is subjected to external conditions, such as the impact of external objects and alternating loads. Therefore, AEs carry health status information about a structure pertaining to the acoustic source. However, an AE is a transient elastic wave with an uncontrollable behaviour and a wide frequency band, and its spreading within a plate is in the form of multi-mode or dispersion Lamb waves. In other words, the information carried by an AE about the health status of a plate structure cannot be interpreted easily due to its complex waveform. Hence, this study's authors considered that basic support from theoretical analysis is essential for interpreting such information in a specific and reliable manner. However, the adoption of a universally applicable theory is not realistic since the characteristics of AE propagation in a plate are related to its structural parameters, for which there are complex types. The Lamb wave has a mode cut-off range with only two basic modes, $S_0$ and $A_0$, according to the Lamb wave mode characteristics analysed above. If the Lamb waveform is in this range, the corresponding AE information is simple and not influenced by other modes. Thin plate structures have a certain application value [7,13]. Thus, in this study, a plate structure with a thickness not greater than 2 mm was considered. Since the AE signals in plate structures are broadband, i.e., mainly below 500 kHz [6], the product of frequency and plate thickness in this study is less than 1 MHz·mm, which is within the mode cut-off range. That is, an AE waveform and its carried information can be interpreted based on the propagation characteristics of the $S_0$ and $A_0$ modes.

According to the definition of an AE, an acoustic source in a plate structure originates from the action of a force, which might result from external events, such as the impact of external objects, or internal events, such as plate fracturing. The different characteristics of wave velocity generated by vertical and parallel forces are proven by Equations (11), (16) and (17). The AEs generated in the plate structure under the action of a force should also conform to this characteristic, and it propagates as a Lamb wave in the plate. Further characteristic analysis shows that the relationship between the force direction and Lamb wave mode can be established for the plate structure. Equation (17) refers to a shear horizontal (SH) wave according to particle vibration form and wave velocity, but an SH wave cannot be obtained from the conversion of the piezoelectric effect because its energy is low in the signal collected by a piezoelectric device [28,29]. Moreover, such an SH wave neither overlaps with the Lamb wave mode nor participates in its conversion [25]. Therefore, the SH wave is ignored in an AE, conforming to the methods adopted in previous works [4,26,30,31]. Hence, the corresponding relationship that must be discussed is only that between the stress wave (see Equations (11) and (16)) and the Lamb wave mode. The $fh$ in a thin plate structure is within the mode cut-off range of the Lamb wave, as explained above. Equations (11) and (16) regarding force activation in the thin plate should correspond to the $S_0$ and $A_0$ modes since, based on the interpretation of the Lamb wave mode, these are the only AE modes in the thin plate. In Equation (11), the velocity of the bending wave is related to the frequency. Equation (16) refers to an extended wave that is a symmetric mode of the Lamb wave, which is independent of frequency according to the particle displacement in Equation (11). Figure 3 displays the $S_0$ (when $fh < 1$ MHz·mm) and $A_0$ modes of a 2 mm thick aluminium plate and epoxy–polyester fibreglass in the mode cut-off range. The wave velocity of the $A_0$ mode monotonically increases along with the frequency, exhibiting frequency dispersion; the wave velocity of the symmetric wave mode, in contrast, is almost constant. Based on this, Equation (11) agrees with the $A_0$ mode characteristics of Lamb waves, while Equation (16) is consistent with the characteristics of their $S_0$ mode.

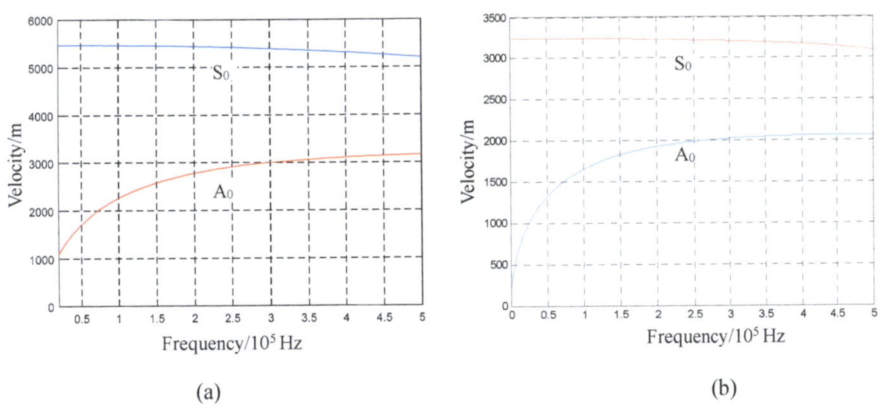


**Figure 3.** Zero-order-mode group velocity curve of the Lamb wave in 2 mm thick plate structures when the frequency is below 500 kHz (**a**) in an aluminium plate and (**b**) in epoxy–polyester fibreglass.

Therefore, the AEs generated by the force that is perpendicular to the plate surface and those generated by the force parallel to the plate correspond to the $A_0$ and $S_0$ modes, respectively.

### 2.4. Mode Verification of AE Force Excitation in the Plate Structure

The force direction in the plate structure determining the Lamb wave mode of an AE can be explained theoretically based on the analysis provided above. This was verified by studying a 2 mm thick aluminium plate and a composite plate. The force-generated AE in the plate structure was stimulated through the common lead-breaking test as follows [4,31]. A piece of lead was broken by pressing its core on the end face and surface of the plate to simulate AE activation via forces parallel and perpendicular to the plate surface, respectively. The sensor adopted was a piezoelectric ceramic sensor (PZT) with a diameter and thickness of 8 mm and 0.5 mm, respectively. It was pasted on the plate structure surface and had a signal acquisition frequency of $10^7$ Hz. Various PZTs were placed on the propagation path, in line with the lead breaking position, to observe the change in the AE waveform of each lead piece breaking in the propagation process. The lead-breaking position on the upper surface of the plate structure was close to the end face to ensure that this distance was equal to the distance from the lead-breaking positions to the sensor surface.

Figure 4a,b illustrate the lead-breaking test conducted on an aluminium plate, which was 48 cm in length and width, including the positioning of the PZTs on the aluminium plate; PZT1 and PZT2 were 20 cm apart and 14 cm away from the plate edge. Figure 4c,d display the AE signals generated via breaking lead, with $10^4$ acquisition points. The signal acquired from lead breaking at the end face of the plate structure consisted of sub-graphs I and II, which corresponded to the AE signals collected from PZT1 and PZT2, respectively. In sub-graph I, the AE impact wave reaching PZT1 is indicated by Block 1; it remained in the impact state while reaching PZT2 after a 20 cm propagation, as shown by Block 2 in sub-graph II. Considering wave reflection and attenuation, the wave packet that reached the sensor first was the least likely to be contaminated, with the frequency component being the closest to the initial state. Hence, the time–frequency information of the first 5000 points in sub-graph I of Figure 4c was analysed using a wavelet transform, as shown in Figure 4e. The main frequency components were distributed in the range below 500 kHz when an AE was generated via lead breaking at the end face (Block 1 in Figure 4c). The impact wave in Box 1 was composed of abundant frequency components, which kept the wave in a state similar to that detected by PZT1 after propagation within a certain distance and the arrival of PZT2, although there are abundant frequency components of the impact wave in Block 1. This shows that the propagation speed of waves with different frequencies is consistent in an AE, and that is why the initial impact waveform can be maintained in the propagation. If the maximum points of the wave packet indicated by Block 1 and Block 2 in Figure 4c are taken as the reference points, the time difference between the two impact wave packets is approximately $3.69 \times 10^{-5}$ s. Based on this, the wave velocity can be calculated as 5420 m/s, and the distance between the two PZTs is 20 cm. This is close to the $S_0$ mode wave velocity described in Figure 3a. The propagation characteristics and speed of wave propagation indicated by Blocks 1 and 2 in Figure 4c indicate that the $S_0$ mode is the AE generated via lead breaking at the end face of the aluminium plate.

Figure 4d shows the AE signal obtained when the lead was broken on the upper surface of the plate structure near the end face. The AE generated via lead breaking, which is a transient event, is transient and a part of the broadband spectrum, which can be verified via the lead breaking experiment at the end face (Figure 4c,e). The waveform generated through lead breaking on the upper surface of the aluminium plate first emits a high frequency and then a low frequency after reaching PZT1 after propagating for some distance, according to sub-graph I in Figure 4d, which is more evident after reaching PZT2 via 20 cm propagation, as shown in sub-graph II. The waveform characteristics in Figure 4d show that the propagation speed of the AE generated through lead breaking on the upper surface is characterised by frequency dispersion and frequency dependency. The first 5000 points of the PZT1 signal were used for a wavelet time–frequency analysis. As shown in Figure 4f, the AE frequency ranges from 0 to 100 kHz, and the time–frequency of the signal is distributed in the shape of an arc, with the high-frequency wave reaching PZT1 before the low-frequency one. This experiment proves that the AE signals collected by PZT1 and PZT2 are consistent with the characteristics of the $A_0$ mode shown in Figure 3a. Based on the time–frequency analysis in Figure 4f, the main frequency band of the signal is lower than 100 kHz, with the thickness of the aluminium plate being 2 mm, and the AE generated via lead breaking is in the mode cut-off range. Apparently, in this experiment, the $A_0$ mode corresponded to the AE generated via lead breaking on the upper surface of the aluminium plate.

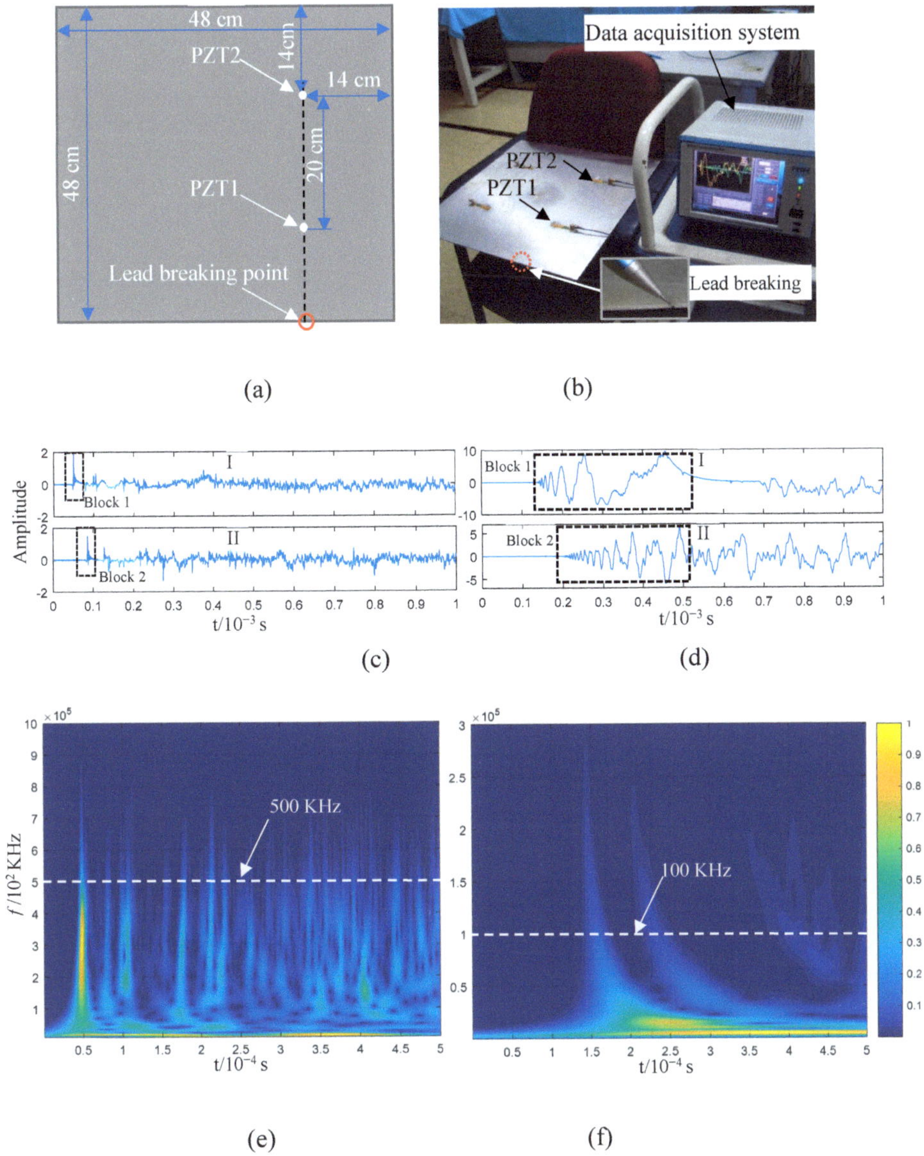


**Figure 4.** Lead-breaking test conducted on an aluminium plate structure: (**a**) experimental layout and (**b**) setting; (**c**,**d**) signals collected from the piezoelectric ceramic sensors PZT1 and PZT2 at the end face (**c**) and upper surface (**d**); (**e**,**f**) wavelet time–frequency analysis of the first 5000 points of the PZT1 signal, with lead breaking at the end face (**e**) and upper surface (**f**).

Figure 5 illustrates the lead-breaking test performed on the 2 mm thick epoxy fibreglass plate, whose side length was ~60 cm. Three PZTs were placed on the plate's centre at intervals of 4.7 cm along a straight line with respect to the lead-breaking point; PZT1 was placed ~5.0 cm away from the plate end face. The waveform indicated by Block 1 in Figure 5c is distributed in a frequency band below 500 kHz; it remains almost unchanged after the wave packets marked in Blocks 1 to 3 pass through PZT1, PZT2, and PZT3. This indicates that the propagation speed of the varying frequency components of the waveform was consistent. If the maximum points of the wave packets in Figure 5c are taken as the reference points, the propagation time from PZT1 to PZT3 is ~$2.72 \times 10^{-5}$ s, with a distance of 9.4 cm. Based on this, the wave velocity can be calculated as 3460 m/s. The characteristics and propagation speed of the wave packets in Figure 5c are perfectly consistent with the $S_0$ mode in Figure 3b. According to these results, the AE generated via lead breaking at the end face of the epoxy fibreglass plate corresponded to the $S_0$ mode.

The AE waveform generated via lead breaking at the upper surface and collected by PZT1 was not fully expanded due to the short distance (Figure 5d). It continued to spread for some distance, eventually reaching PZT2 and PZT3. Then, it fully expanded, and the high-frequency component was clearly faster than the low-frequency one. The primary frequency of the signal was distributed in the frequency range of 0–100 kHz (Figure 5f). The waveform variation in the propagation process shown in sub-graphs I–III of Figure 5d indicates that the AE was characterised by dispersion. Considering that the $fh$ in the plate is in the cut-off range of the Lamb wave mode, it can be concluded that the $A_0$ mode generated an AE at the lead-breaking surface of the epoxy fibreglass plate.

The propagation characteristics of the Lamb wave velocity of the $S_0$ and $A_0$ modes can perfectly explain the AE waveform generated via lead breaking based on the above tests on two plate structures. They also confirm that the $S_0$ mode wave velocity of the Lamb wave in the plate structure was approximately constant under the effect of a low frequency–thickness product, while the $A_0$ mode wave velocity exhibited significant dispersion. The conclusion in the previous subsection is supported by these test results. That is, in thin plate structures, the AE generated by the force parallel to the plate surface is in the $S_0$ mode, while that generated by the force perpendicular to it is in the $A_0$ mode. This conclusion provides strong support for extracting damage-induced AEs from plate structure impacts.

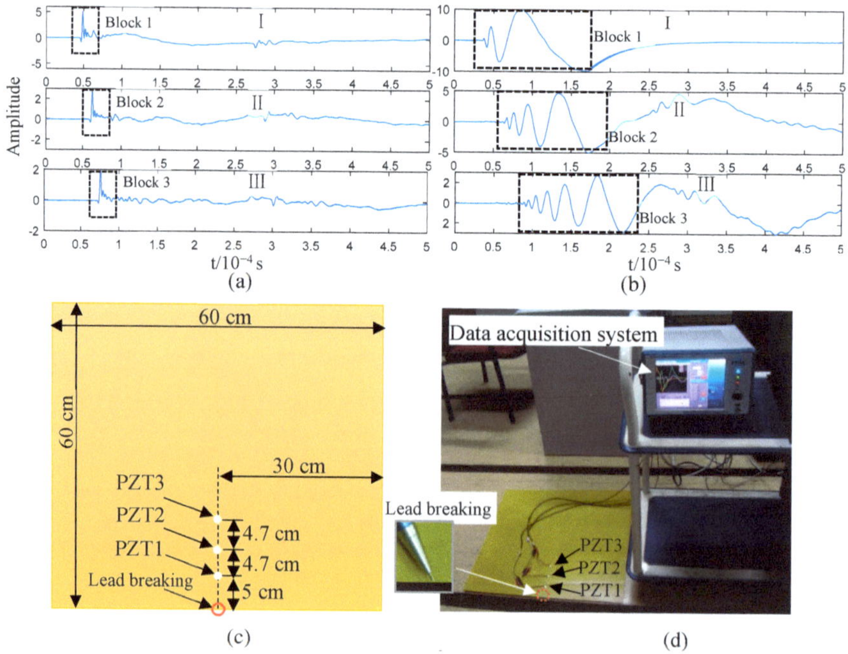


**Figure 5.** *Cont.*

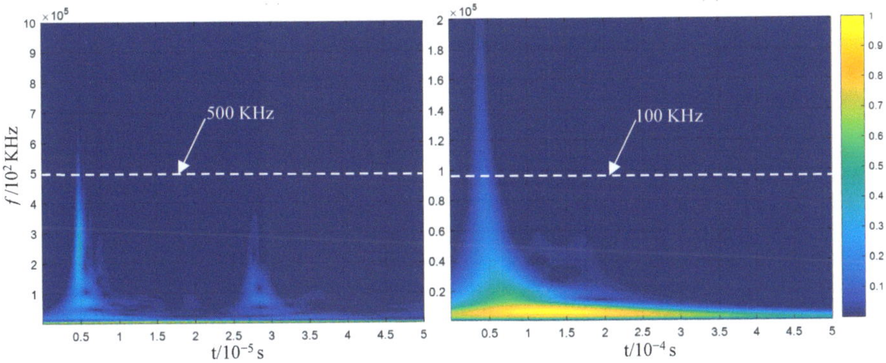


(e) (f)

**Figure 5.** Lead-breaking test conducted on epoxy fibreglass plate structure: (**a**) experimental layout and (**b**) setting; (**c**,**d**) signal collected from the piezoelectric ceramic sensors PZT1, PZT2, and PZT3 at the end face (**c**) and on the upper surface (**d**); (**e**,**f**) wavelet time–frequency analysis of the first 5000 points of the PZT1 signal with lead breaking at the end face (**e**) and upper surface (**f**).

## 3. Analysis of AE Mode Information in Plate Structures

The force acting on the plate structure is the AE source, and it can be divided into an external force (OP) and an internal force (IP). The OP refers to the impact of external matter, such as ice, stones, and birds, on the plate structure, while IP indicates material fracturing, such as matrix fractures, fibre fractures, and adhesive failures, in the plate. The OP-generated AE is dominated by the $A_0$ mode, while the IP-generated AE is dominated by the $S_0$ mode [26,30,31]. The results of this study are consistent with these findings. However, the reasons for drawing this conclusion should be discussed in detail to illustrate the rationality of the technical route adopted for research purposes. Regarding the IP, with random fracture direction and size, the generated AE should not be dominated by a certain mode. Some damage types are dominated by a certain mode, but multiple damage types coexist under the material deformation, extrusion, and fracturing resulting from the impact. In general, the damage caused by an impact has the same magnitude in the parallel and perpendicular directions simultaneously, generating the $S_0$ and $A_0$ mode waves. Nonetheless, various AE frequency components of the $S_0$ mode maintain the same waveform in the propagation process, with concentrated energy and abundant high-frequency components due to the instantaneous nature of AEs and the mode approaching the constant propagation speed under a low frequency–thickness product. Researchers can visually observe the symbolic waveform to facilitate the study of its characteristics, but no symbolic waveform can be determined for tracking research since there is frequency dispersion in the propagation of the $A_0$ mode. Specifically, frequency components are distributed on the time-axis from high to low, with the energy distributed among various frequency components and different waveforms at varied positions on the same propagation path. In this way, researchers might ignore the symbolic waveform since their attention is caught by the $S_0$ mode with an unchanged waveform during propagation. Therefore, the authors conclude that the AE generated by an IP is dominated by the $S_0$ mode. The fact that OP-generated AEs are dominated by the $A_0$ mode might be related to the plate's design. The plate structure's surface, in general, is flat and smooth. When it is impacted by external objects, the component force of the impact parallel to the plate surface is relatively small or even negligible due to the surface's flatness and smoothness. In an AE, the wave energy (amplitude) of the $A_0$ mode is dominant since the plate structure is subjected to a force perpendicular to its surface, and the wave energy of the $S_0$ mode is smaller. In that case, only the $A_0$-mode waveform with high amplitude can be observed, leading to a failure to recognise the $S_0$-mode waveform. This is different from the AE generated by impact damage. Based on the above analysis, the author of this paper considers it reasonable to conclude that the OP- and IP-generated AEs are dominated by the $A_0$ and $S_0$ modes, respectively, after excluding the specific force actions that can induce certain types of damage under laboratory conditions.

According to the above analysis, the obtained AE signal is the superposition of the OP- and IP-generated AEs if the plate damage is caused by the impact of external objects. The amplitude of the IP-generated AE (damage) is far less than that of the OP-generated one, considering that external matter can only damage a plate structure with high kinetic energy. Moreover, there is essentially no difference between the two acoustic sources. Diagnosing whether damage is caused directly by the waveform variation of the impact—as well as determining, without sufficient analysis, that an impact AE is the cause, even if it is judged from a certain characteristic quantity of the impact AE—is difficult. This AE generated by damage might be recognised if the impact AE is interpreted based on various factors, such as the characteristics of the mode's wave velocity, the mode's frequency band, the relationship between force direction and mode, and the damage process induced by the impact in the mode cut-off range. When an external object impacts the plate structure, its kinetic energy is absorbed by the latter and converted into potential energy, leading to the deformation of the plate structure. This energy conversion process is realised by a force, more specifically, an external force, as shown in Figure 6. The generated internal force will cause fracturing if the deformation exceeds the bearing capacity of the plate structure, as shown in sub-graph II of Figure 6. There is a time difference between the appearance of the external and internal forces; undoubtedly, there is also a time difference for the corresponding AE signals. Note that the $S_0$ mode wave cannot be superimposed with the $A_0$ mode wave when the two modes are generated simultaneously since the wave velocity of the former is greater than that of the latter. That is, the OP-generated AE is dominated by the $A_0$ mode. Even if there is a low-energy $S_0$-mode wave, this cannot be superimposed on the $A_0$ mode wave. As the IP occurs later than the OP, the IP-generated AE also arises later than the OP-generated AE; however, the $S_0$ mode velocity is greater than the $A_0$ one and far greater than the low-frequency component of the $A_0$ mode. In this case, the AE of the $S_0$ mode generated alongside damage might be superimposed with that generated by OP. The AE of the $S_0$ mode extracted from that of the $A_0$ mode generated by an impact can only be generated by damage. Since damage-induced AEs continue to occur in the impact process, which is beyond control, the corresponding $S_0$-mode wave should also be extensively distributed in the OP-generated AE waveform.

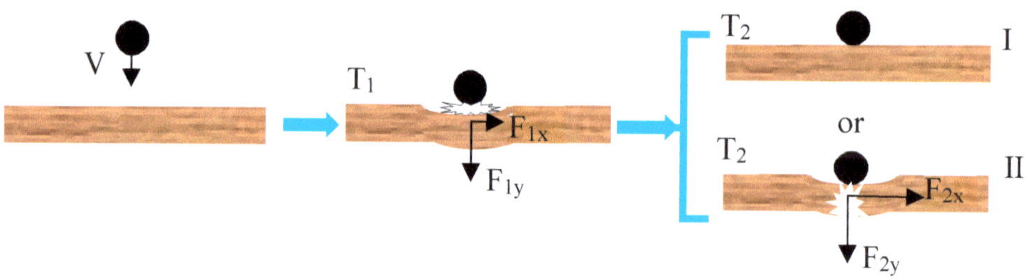


**Figure 6.** Process of a plate structure experiencing an impact.

The following conclusions can be drawn from the above considerations. First, the waveforms of the $S_0$ and $A_0$ modes generated simultaneously by OP are separated since the velocity of the former exceeds that of the latter. Therefore, the $S_0$-mode wave extracted from the AE of the $A_0$ mode generated by the OP can only be the damage-induced AE. Moreover, the $S_0$-mode wave has concentrated energy and several high-frequency components, whereas the $A_0$-mode wave features dispersion, with the energy distributed in the low-frequency range; the different frequency band characteristics of the two mode waves are conducive to their separation. Moreover, the $A_0$ mode of the OP-generated AE has an absolute position in terms of energy, although its AE is also induced by damage. In this case, the evidence regarding whether the $A_0$-mode wave is superimposed in the impact AE as the indicator of damage is not convincing. Hence, this paper proposes that the superimposition of the $S_0$ mode in the impact AE wave should be considered as the indicator of damage inflicted on the plate structure. Appropriate signal-processing technology must also be selected to extract indicators of damage. With its instantaneous nature, the impact AE wave is characterised by several high-frequency components. Since the $S_0$ mode has weak frequency dispersion under a low-thickness–frequency product, the impact waveform in the AE propagation mode is almost unchanged with concentrated high-frequency energy. Since the $A_0$ mode has a strong frequency dispersion, its AE energy is distributed in the low-frequency range, with little high-frequency energy. A high-frequency filter was selected here to facilitate the illustration of the $S_0$ mode components in the AEs. Furthermore, the time sequence information of the high-frequency wave packet extracted from the AEs was used for recognising the Lamb wave pattern and confirming the acoustic source, and its original phase information was retained. Therefore, the $S_0$-mode signal components of the AEs were extracted in this study using zero-phase high-pass filtering technology.

## 4. Application of Impact AE in Plate Structure Health Monitoring

Damage information can be extracted from an impact AE wave packet through multi-angle interpretation under the effect of the low-frequency–thickness product, according to the above-stated conclusions. These conclusions were successively verified by processing experimental data on plate structure damage based on the proposed technical route. Three tests were conducted: the first test aimed to verify the AE mode generated by plate structure extrusion, while the second and third ones were plate structure impact tests. The data from the last two tests were then interpreted with respect to various aspects based on the proposed technical route; this enabled the assessment of the feasibility of extracting $S_0$-mode information from AEs as an indicator of damage.

### 4.1. Analysis of the AE of Plate Impact Damage

The AEs generated by impact damage inflicted on the plate structure were observed since it they are related to the feasibility of the proposed technical route. The damage-induced AE results from the fracturing of the plate structure, according to the impact damage process described above. Hence, the analysis of the AE generated by impact damage is equivalent to the direct analysis of the AE generated by material fracture. The deformation and extrusion of the plate structure, with local absorption of impact kinetic energy, were simulated in a test under the application of a local force; to eliminate the interference caused by OP, the applied force was close to the static state. The AEs that propagated in the plate resulted from its rupture, separation, and fracture, and these characteristics are similar to the damage-induced AEs caused by impacts. A 2 mm thick epoxy fibreglass plate was tested by using PZTs as sensors (Figure 7); PZT1 was placed 10 cm away from PZT2, and the force area was located along the extension line of the two PZTs.

Local damage occurred after the plate structure's edge, which was roughly 8 cm away from PZT1, was stressed (Figure 7b). The AE signals collected by PZT1 and PZT2 are shown in Figure 7c,e; the waveform indicated by Block 1 is a shock waveform that remained almost unchanged during its propagation from PZT1 to PZT2. Then, the first 5000 points of these signals were taken for the wavelet time–frequency analysis, and the results are shown in Figure 7d,f. The frequency components of the impact waveform are distributed in the frequency range below 500 kHz. If the maximum points of the two impact wave packets are taken as the reference points, their corresponding propagation time is 25.4 μs. With a propagation distance of 10 cm, the calculated propagation speed of the wave packet is ~3900 m/s. The shock wave packet indicated by Block 1 in Figure 7c,e is a part of the $S_0$ mode from the perspectives of wave propagation change, $fh$, and wave velocity. The measured wave velocity (3900 m/s) in this mode is rather large (similar to the wave velocity in the mode cut-off range in Figure 3b) because the AE's acoustic source was not strictly in a straight line with respect to the two sensors. Figure 7b shows that the damaged area deviating from the straight line of the two sensors resulted from the force area deviation due to uneven gripping. Moreover, the damage was inflicted over an area; thus, even if it passed through the straight line between the two sensors, the location of the acoustic source on this line could not be strictly guaranteed. Therefore, the time difference from the measured wave packet to the two sensors is less than the theoretical time difference, resulting in a higher calculated velocity. The anisotropy of the composite materials in the epoxy plate is another reason for this; although the propagation velocity of the Lamb wave in the plate can be approximately isotropic, it differs between different propagation directions.

The waves indicated by Block 2 in Figure 7c,e represent propagation dispersion, and the corresponding propagation speed decreased with the decline in frequency; this behaviour is consistent with the feature of the $A_0$ mode under the influence of a low-frequency–thickness product (shown in Figure 3b). The propagation velocity of the wave in Block 2 was then estimated for further verification. The local maximum of the time–frequency curve was set as the reference point by considering the points with a frequency of 11 KHz, which are the two points marked in Figure 7d,f. The time difference between the two points is ~1.485 μs, with a propagation distance of 10 cm. Further, the wave velocity is ~670 m/s. The speed approaches the wave velocity of 680 m/s, corresponding to 11 KHz (Figure 3b). The dispersion characteristics and wave velocity prove that the mode in Box 2 is $A_0$.

According to the above test, there exist $S_0$ and $A_0$ modes in the AEs when the 2 mm thick epoxy fibreglass plate is partially fractured by the extrusion material. Further, both modes exhibit obvious waveforms. The results show that the damage caused by the local extrusion of the plate structure has the comprehensive characteristics of multiple damage types, and the resulting force exists in two directions simultaneously; no force is absolutely dominant in a given direction. Therefore, the conclusion concerning the simultaneous existence of $S_0$ and $A_0$ modes in AEs generated through the impact damage of the thin plate structure is consistent with practical situations.

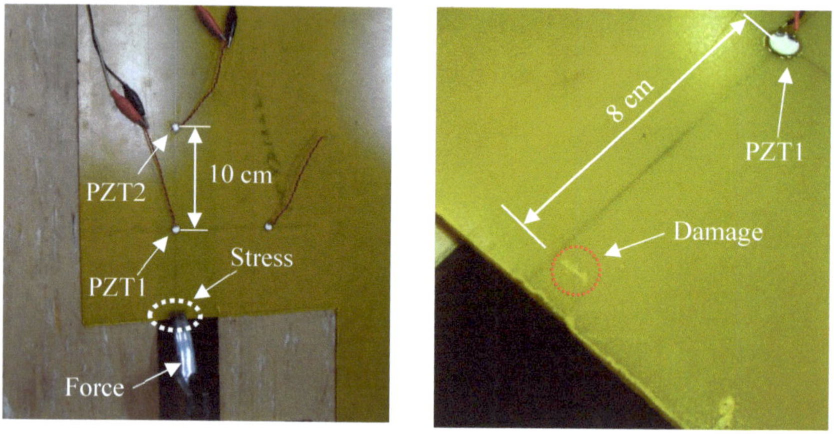


(a) (b)

**Figure 7.** *Cont.*

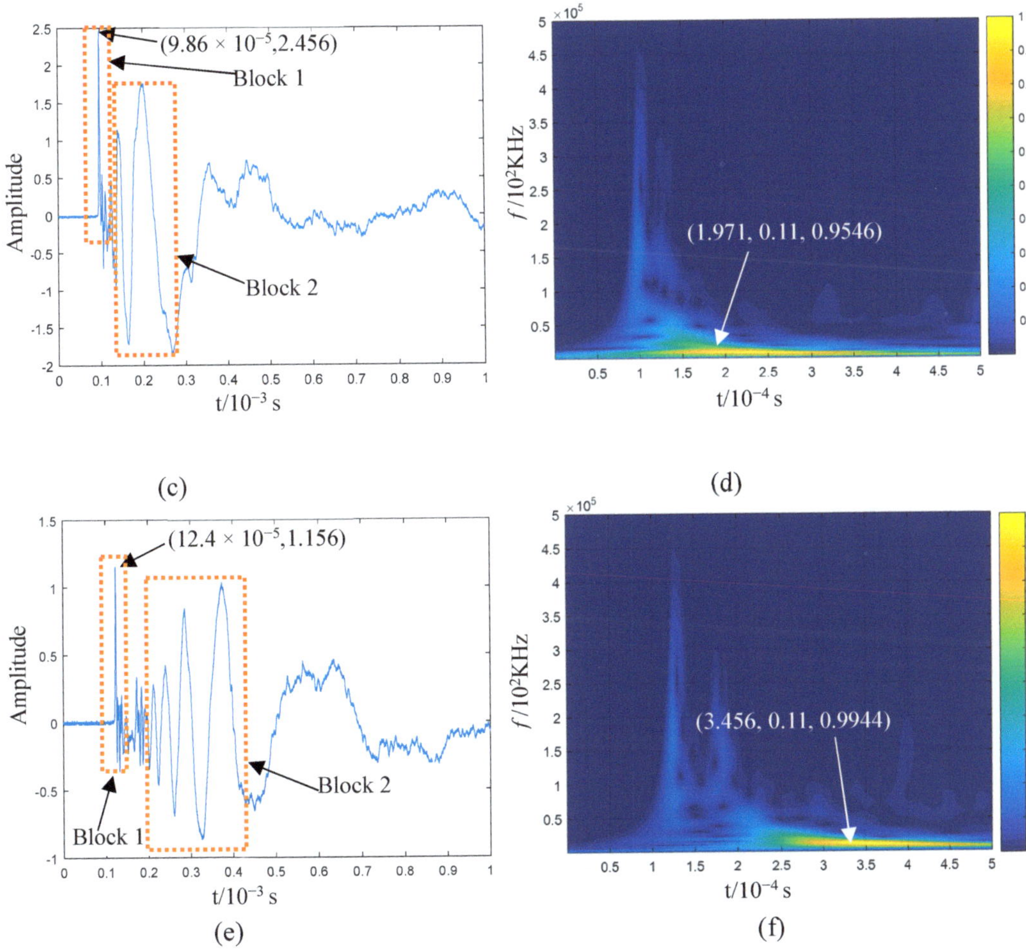


**Figure 7.** Acoustic emission (AE) signals generated via the local damage of the epoxy fibreglass plate and its wavelet time–frequency analysis: (**a**) local force of epoxy plate and piezoelectric ceramic layout; (**b**) damage area on the epoxy plate; (**c**) AE signal obtained by PZT1 and (**d**) corresponding wavelet time–frequency analysis of the first 5000 points; (**e**) AE signal obtained by PZT2 and (**f**) corresponding wavelet time–frequency analysis of the first 5000 points.

### *4.2. Extraction of Damage Information from Impact AE in Thin Plate Structures*

The above-mentioned results show that the impact damage of a thin plate structure is caused by a combination of AE modes. Two tests were successively conducted, one without damage in the impact and the other with impact damage. Then, by processing the test data according to the proposed technical route, the $S_0$ mode information was extracted from the impact AE to diagnose whether there was damage. The impact non-damage test layout is displayed in Figure 5b. The tested epoxy fibreglass plate had a thickness of 2 mm; the impact point was 20 cm away from PZT3 on the straight line between the three PZTs, which were spaced 4.7 cm apart. The sampling rate was $10^7$, with $10^4$ sampling points. An impact hammer was used in the first test (Figure 8). Due to the light weight of the hammer, no damage was visually observed at the impact point.

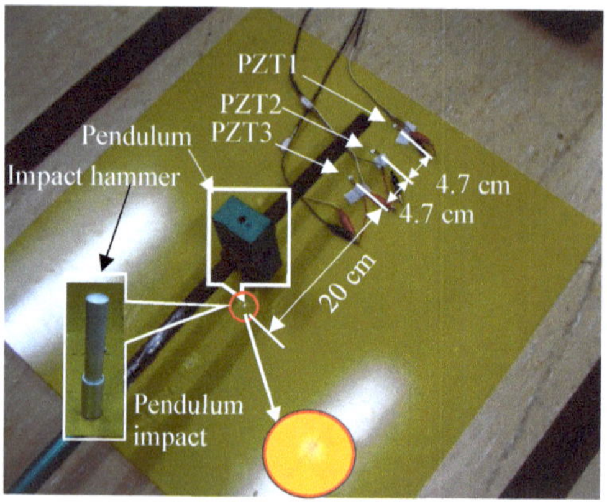


**Figure 8.** Layout of the impact damage test.

The resulting AE is shown in Figure 9a. No impact damage was observed on the plate structure through visual inspection, although the AE amplitude exceeded the voltage input range of the detection system (as shown in the insets of Figure 9a). The wavelet time–frequency analysis (Figure 9b,d) showed the dispersion characteristics of the wave in the signals collected by all the PZTs. However, the $S_0$-mode waveform could not be identified in the time domain and time–frequency plots. The high-frequency components of the AE's $S_0$ mode were abundant according to the above analysis. Those in Figure 9a could be extracted to suppress the interference of low-frequency waves. On this basis, the mode information of the impact AEs could be analysed. The lower frequency limit of high-pass zero-phase filtering was set as 100 kHz. The filtering results are shown in Figure 10a, with the sub-graphs I–III presenting the filtering results of the three sensor signals.

If attenuation in the propagation process is considered, the AE was the component of the closest original signal obtained by PZT3. Therefore, the sub-graph I in Figure 10a was employed for a time–frequency analysis (Figure 10b). The wave packets in the $F_1$ and $F_2$ block diagrams in sub-graphs I and II are high-frequency components introduced via overload; they are not required for the analysis. The wave packets $C_2$ and $C_3$ represent the waveform when the wave packet $C_1$ passes through PZT2 and PZT1. The wave packet velocity can be calculated if the extreme point (indicated by an arrow in the figure) is taken as the reference point (with the X-axis corresponding to $0.95 \times 10^{-3}$ s, $1.236 \times 10^{-3}$ s, and $1.532 \times 10^{-3}$ s), as shown in Table 1. The frequency bands of the three wave packets ($C_1$, $C_2$, and $C_3$) are within 100 and 150 kHz, according to the frequency-domain coordinates corresponding to the $C_1$ wave packet in Figure 10b. Based on Figure 3b, the wave velocity range of the $A_0$ mode in this frequency band ranges from 1657 to 1830 m/s. The propagation velocities of $C_1$, $C_2$, and $C_3$ calculated in Table 1 are consistent with this range, and the distribution of the three wave packets on the time axis is also consistent with dispersion curve 1 in Figure 9b–d. Therefore, the wave packets including $C_1$, $C_2$, and $C_3$ probably correspond to the $A_0$ mode. In the test experiment illustrated in Figure 8, the type of impact was OP, and the AE generated by the impact was dominated by the $A_0$ mode and occurred for the first time. Hence, the first-mode wave packet in the high-pass filter signal (wave packets $C_1$, $C_2$, and $C_3$ in Figure 10a) was generated by the impact.

**Table 1.** Calculation of the propagation speed of wave packets $B_n$ and $C_n$ (n = 1, 2, 3).

| Wave Packet | Lower Label ($10^{-7}$ s) | | | Speed between Two Sensors (m/s) | | Mode |
|---|---|---|---|---|---|---|
| | 1 | 2 | 3 | PZT3 and PZT2 | PZT2 and PZT1 | |
| B | 369 | 507 | 639 | 3406 | 3561 | $S_0$ |
| C | 956 | 1236 | 1532 | 1680 | 1590 | $A_0$ |

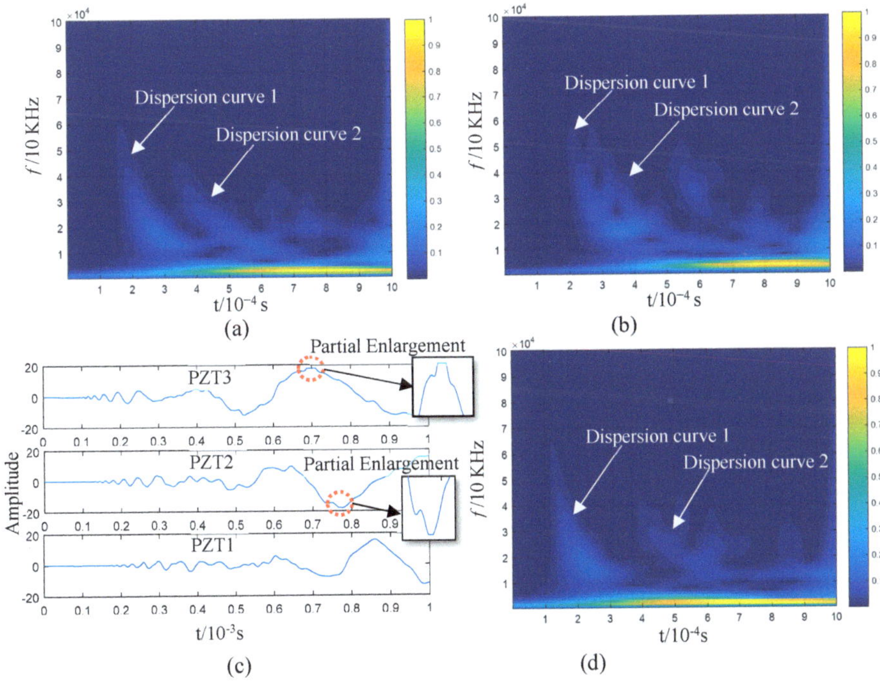


**Figure 9.** Acoustic emission (AE) and time–frequency analysis of an epoxy fibreglass plate after impact: (**a**) impact AE signal obtained using three piezoelectric ceramics; (**b**–**d**) wavelet time–frequency analysis of PZT3 (**b**), PZT2 (**c**), and PZT1 (**d**) signals.

The reason why the OP-generated AE is dominated by the $A_0$ mode was discussed in the previous section, and we do not intend to ignore the $S_0$ mode generated simultaneously because its energy is small. In the test illustrated in Figure 9, the component force parallel to the plate surface cannot be zero since a small amount of the $S_0$ mode wave packet exists in the impact AE. These two modes occur simultaneously when the plate is impacted according to the previous analysis. Since the $S_0$ mode propagation speed is greater than that of the $A_0$ mode, the two mode waves start to separate when they are generated, and the $S_0$ mode wave packet is always ahead of the $A_0$ mode wave packet on the time axis. A trace of the $S_0$ mode wave packet can be found by carefully analysing the waveforms before $C_1$, $C_2$, and $C_3$. The waveforms marked in Blocks $B_1$–B3 and their enlarged views are displayed in Figure 10a, showing their similarity. If the midpoint of these waves (the point indicated by an arrow in the insets) is taken as a reference point, the corresponding propagation velocity of the wave packet is ~3400 m/s (Table 1). The speed of the $S_0$ mode in Figure 3b is 3250 m/s. That is, the actual value matches the theoretical ones if the influences of noise, waveform attenuation, and propagation path on velocity are considered. Therefore, it has been confirmed that the above three waveforms correspond to the $S_0$ mode according to the propagation velocity of the $B_n$ (n = 1,2,3) wave packets. However, this is not enough proof to demonstrate that the $B_1$–$B_3$ wave packets resulted from the impact; it must be proven that the $B_1$–$B_3$ and $C_1$–$C_3$ wave packets originated from the impact. The $A_0$-mode wave packets generated simultaneously were calculated according to the reference point time of the $S_0$-mode wave packets $B_1$–$B_3$ when the propagation distance and the velocity of the two modes were known. If the coordinates of these reference points fall within the $C_1$–$C_3$ wave packet period, the $B_1$–$B_3$ and $C_1$–$C_3$ wave packets will likely be generated from the impact. The actual propagation velocity between PZT3 and PZT1 (as shown in Table 1) can be obtained as the velocity of the $S_0$ and $A_0$ modes. The propagation distance is the distance from the impact position to the sensor (Figure 8). The time difference corresponding to the wave packet reference points of the impact-generated $A_0$ mode can be determined by calculating the TOA differences of the two mode wave packet reference points by referring to the point marked in the $B_1$–$B_3$ wave packets in Figure 10a. The results are summarised in Table 2. The time coordination of the $A_0$-mode reference point in Table 2 was compared with the time domain of the corresponding wave packets $C_1$–$C_3$ (Figure 11). The reference time of the $S_0$-mode wave packets $B_1$–$B_3$ and the corresponding time point of the calculated $A_0$-mode wave packets are indicated in Figure 11; the time area of the $A_0$-mode wave packets calculated using the $S_0$-mode ones is highly coincident with the $C_1$–$C_3$ wave packets. Therefore, the $S_0$-mode wave packets $B_1$–$B_3$ and the $A_0$-mode wave packets $C_1$–$C_3$ in Figure 10a likely originated simultaneously from the impact. The $S_0$- and $A_0$-mode wave packets in the AEs generated by the impact of external objects, which were separated at the moment of generation due to their different propagation velocities, were verified via the analysis of the experimental results. The $S_0$-mode wave packets will not be superimposed with the $A_0$ mode ones in the AE generated via the impact of external objects since their propagation velocity is greater than that of the $A_0$-mode wave packets.

**Table 2.** Reference point time of $S_0$ mode calculated based on its reference point.

| Sensors | Distance between Impact Position and Sensor (m) | Figure 3b Wave Velocity (m/s) | | Reference Point of $S_0$ Mode ($10^{-7}$ s) | Calculation Results | |
|---|---|---|---|---|---|---|
| | | $S_0$ | $A_0$ | | Time Difference of Arrival Sensor in Two Modes ($10^{-7}$ s) | Reference Point of $A_0$ Mode ($10^{-7}$ s) |
| PZT3 | 0.200 | 3482 | 1632 | 369 | 651 | 1020 |
| PZT2 | 0.247 | | | 507 | 804 | 1311 |
| PZT1 | 0.294 | | | 639 | 957 | 1596 |

Besides the typical Class II wave packets $B_1$–$B_3$ and $C_1$–$C_3$, Figure 10a also shows $D_1$–$D_3$ and $E_1$–$E_3$ wave packets with obvious amplitudes as well as other low-amplitude clutter (marked by dotted lines in sub-graphs II and III). $D_1$–$D_3$ are considered to be the reflected waves of $C_1$–$C_3$ from the time sequence distribution and velocity. Compared with $D_2$ and $D_3$, $D_1$ has an abnormal waveform change, which was likely caused by clutter interference. The propagation direction of $E_1$–$E_3$ should be from PZT3 to PZT1 based on the time series layout. The propagation velocity of $E_n$ measured using the reference point is between the $S_0$ and $A_0$ modes. A disturbance could not be proven in this paper since the wave packet is late in its time sequence, with abnormal amplitude changes, and no further discussion of it will be provided. No obvious $S_0$-mode wave packets were observed when considering the impact AE wave packets; in particular, no $S_0$-mode wave packets were discovered after the $C_1$–$C_3$ ones. The $A_0$-mode wave packets in the impact AE result from the impact of external objects. The analysis results in Figure 10a indicate that this impact did not damage the plate structure according to the mode characteristics of damage and external impact AE described in the previous section; this is consistent with the impact test conditions.

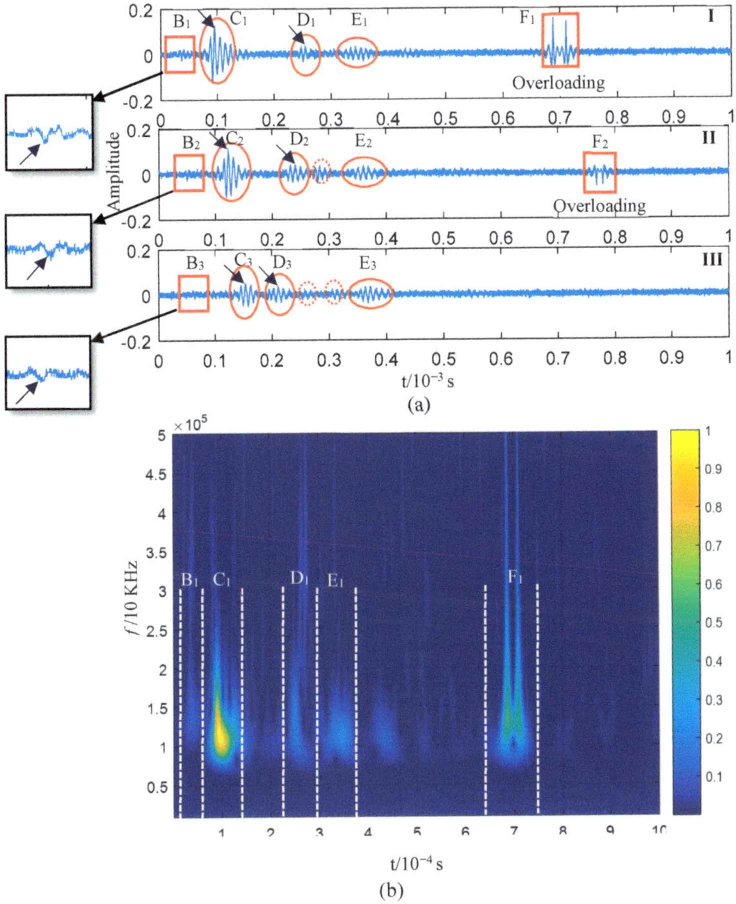


**Figure 10.** (**a**) High-pass zero-phase filtering of acoustic emission signals obtained using three piezoelectric ceramics, and (**b**) wavelet time–frequency analysis of sub-graph I shown in (**a**).

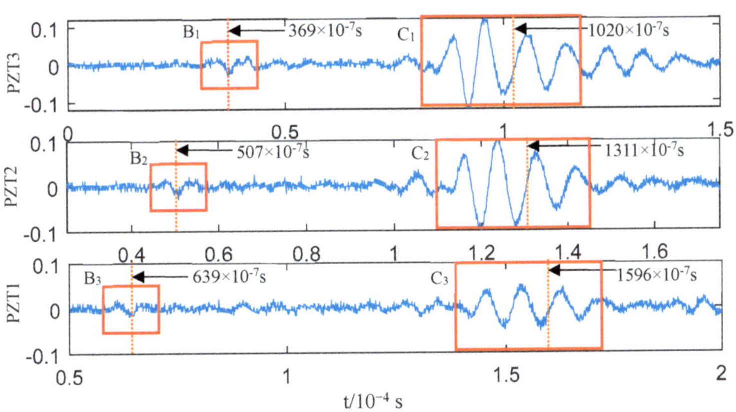


**Figure 11.** $A_0$-mode time point calculated according to the $S_0$-mode time point: the figure shows partially enlarged views of the first two wave packets ($B_1$–$B_3$, $C_1$–$C_3$) in sub-graphs I–III of Figure 10a.

When the impact tool was replaced by a pendulum mass for analysing the impact at the same position, damage could be observed on the epoxy fibreglass plate, as shown in Figure 8. The corresponding signals collected by the three PZTs are illustrated in Figure 12; these results are similar to those in Figure 9a in terms of the waveform. The AE generated by the impact of external matter was dominant, and the signal was overloaded with the waveform shown in dispersion. The difference is that the damage caused by the pendulum impact was visible. The high-frequency components of the three sensor signals (Figure 12a) were extracted using a zero-phase high-pass filter with a lower cut-off frequency of 100 kHz to further analyse the health information carried by the AEs generated by the pendulum impact; the results are shown in Figure 12b.

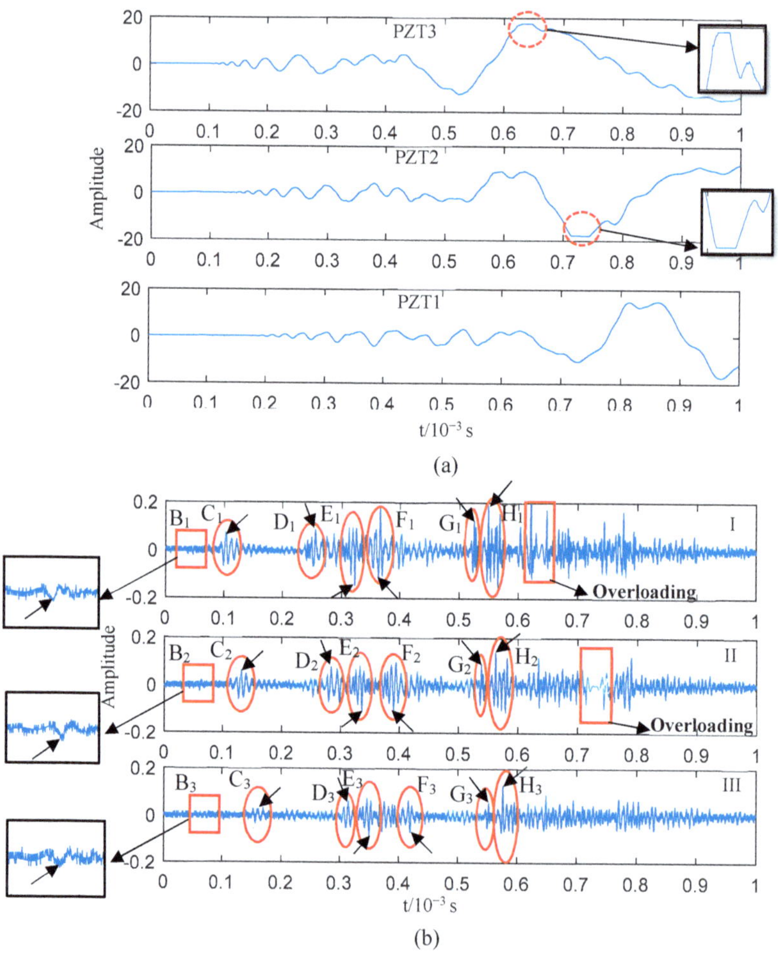


**Figure 12.** (**a**) Acoustic emission signals generated by pendulum impact and detected using the three piezoelectric ceramic devices. (**b**) The corresponding high-pass zero-phase filtering results.

Compared with Figure 10a, the high-frequency signal obtained via filtering shown in Figure 12b has more wave packets. The sequence of wave packets in the three PZT signals is denoted as $B_n$–$H_n$. To confirm the modes of these wave packets, their propagation speeds were calculated. The typical points of each series of wave packets were selected (and marked with arrows in Figure 12b) as reference points for calculating the wave velocity, as shown in Table 3. According to the results, the $B_n$ wave packets were in the $S_0$ mode, while the $C_n$ ones were in the $A_0$ mode. The acoustic source of the $B_n$ and $C_n$ wave packets was the pendulum impact rather than the damage. This was explained in detail in the last impact test data analysis. In general, the high-frequency wave packets of these two modes could reach the sensor since the AEs were generated under the impact of external objects on the plate surface the first time. The wave velocity calculated for the $D_n$ wave packets based on the reference points is consistent with the $A_0$ mode; their propagation direction is along the line drawn from PZT3 to PZT1 according to the time corresponding to the $D_1$, $D_2$, and $D_3$ wave packets, excluding the reflection of $B_n$ wave packets from this propagation direction. The amplitude of the $D_n$ wave packets is not lower than that of the $B_n$ ones. In this case, the $D_n$ wave packets reflecting as $B_n$ ones can also be excluded. Therefore, their source is different from that of the $B_n$ wave packets, which can only be IP. It can be speculated, then, that the acoustic source of the $D_n$ wave packets is delamination since there is no obvious $S_0$ mode before them.

**Table 3.** Wave packet velocity calculation, pattern recognition, and determination of acoustic sources in pendulum impact.

| Wave Packet | Reference Point Coordinates of the Wave Packet ($10^{-7}$ s) | | | Wave Velocity between PZT3 and PZT2 (m/s) | Wave Velocity between PZT2 and PZT1 (m/s) | Mode | Sound Source |
|---|---|---|---|---|---|---|---|
| | Subscript 1 | Subscript 2 | Subscript 3 | | | | |
| B | 477 | 611 | 741 | 3507 | 3615 | $S_0$ | Impact |
| C | 1026 | 1306 | 1605 | 1679 | 1572 | $A_0$ | Impact |
| D | 2558 | 2842 | 3144 | 1655 | 1556 | $A_0$ | Damage |
| E | 3221 | 3355 | 3490 | 3508 | 3481 | $S_0$ | Damage |
| F | 3634 | 3902 | 4194 | 1754 | 1610 | $A_0$ | Damage |
| G | 1610 | 5385 | 5522 | 3561 | 3431 | $S_0$ | Damage |
| H | 5492 | 5623 | 5762 | 3588 | 3381 | $S_0$ | Damage |

Table 3 shows that the wave velocity of the $E_n$ wave packets was approaching 3500 m/s, which is close to that of the $S_0$ mode; this indicates that the $E_n$ wave packets were in the $S_0$ mode. The $E_n$ wave packets were generated by the force applied parallel to the plane according to the relationship between the force direction and Lamb wave mode described before. The pendulum impacted the plate in a nearly perpendicular direction under the test conditions. The amplitude varies significantly between the $E_n$ wave packets with the impacted $S_0$-mode wave ($B_n$ wave packets). This means that the $E_n$ component from the plate direction of the pendulum impact can be nearly excluded; hence, the $E_n$ wave packets come from impact damage, either matrix or fibre breakage. For the remaining $F_n$, $G_n$, and $H_n$ wave packets, according to the wave velocity in Table 3, it can be confirmed that the $F_n$ ones belong to the $A_0$ mode and the others belong to the $S_0$ mode. The $A_0$-mode $F_n$ wave packets previously had $S_0$-mode wave packets, which could only occur due to some kind of plate damage rather than generation via pendulum impact from the perspective of time sequence and amplitude. The $S_0$-mode $G_n$ and $H_n$ wave packets could merely have occurred due to damage rather than the pendulum impact. The typical wave packet ($B_n$–$H_n$) mode in the sequence diagram in Figure 12 has been confirmed. In addition, other wave packets may stem from damage. However, the acoustic source might be interfered with by the reflected waves at the edge since it is not situated along a straight line with respect to the three sensors; thus, it could not be recognised in the three sensor signals. Hence, it is not advantageous to further discuss the above modes.

The high-frequency components of the AEs reveal the acoustic source activity in the plate more clearly based on a comparison of the data before and after the impact AE filtering, as shown in Figures 9, 10 and 12. The impact AEs are dominated by the $A_0$ mode according to the impact wave test data shown in Figure 10a. The number of wave packets of the two modes is limited, with an orderly distribution at high frequencies (>100 kHz). The plate structure damage caused by the impact is a continuous process with multiple events and various damage types as per the distribution of the high-frequency wave packets in Figure 12b. Regarding the time sequence, damage-induced AEs occur significantly later than impact AEs. The high-frequency components, the wave packet mode ($S_0$ or $A_0$), and the distribution of the AE wave packet time sequence can be obtained by analysing the data from two tests for the multi-angle interpretation and accurate identification of damage-induced wave packets. The feasibility of the monitoring method for diagnosing the damage of thin plate structures has been proven since damage-induced AEs can be directly extracted from impact AEs.

## 5. Conclusions

AEs are essential for monitoring the health of plate structures. This paper proposed the direct extraction of damage-induced AEs from AE impacts in the damage diagnosis of thin plate structures because the monitoring method based on AE characteristics is susceptible to non-damage factors. Since there is no difference between impact and damage-induced AEs in nature, this paper recognised the two AEs by interpreting multiple factors, such as the mode ($S_0$ or $A_0$), amplitude, and time sequence distribution of the wave packets with the help of high-frequency-band zero-phase filtering based on the relationship between Lamb mode and force direction. This research consisted of the following four aspects.

1. The AE Lamb wave modes in thin plate structures were studied. The AE frequency in the plate was distributed below several hundred kHz, and the AE Lamb waves in the thin plate were within the mode cut-off range, with only two lowest-order modes ($S_0$ and $A_0$). The propagation velocity of the $S_0$ mode was almost unchanged within the mode cut-off range, while that of the $A_0$ mode showed significant dispersion.

2. The propagation velocities and modes of the stress waves generated by forces perpendicular and parallel to the plate surface were analysed using the mechanical equations of plate structures. The analysis showed that the stress wave generated by the force perpendicular to the plate surface corresponded to the $S_0$ mode, while that generated by the parallel force corresponded to the $A_0$ mode. The results of lead-breaking tests conducted on aluminium and epoxy fibreglass plates support this conclusion.

3. The difference between impact- and damage-induced AE in plate structures was analysed. First, the two AEs are separated in the time sequence. The plate structure undergoes deformation after absorbing kinetic energy until finally fracturing. Impact AEs arose before damage-induced AEs in the collected data. Second, the two AE modes have different characteristics. The smooth surface of the plate structure bears the impact of external objects in its vertical direction. The process of impact-induced damage is uncontrollable and characterised by multiple damage types, with forces acting in the parallel and perpendicular directions of the plate surface. As a result, impact AEs are dominated by the $A_0$ mode, whereas both the $S_0$ and $A_0$ modes are significant in damage-induced AEs.

4. This paper proposed considering the $S_0$ mode in impact AEs as an indication of whether the impact causes damage to a thin plate structure. This is because the $S_0$ mode in OP-generated AEs only occurs before the $A_0$ mode. Damage-induced AEs take place later than impact AEs. Thus, the generated $S_0$- and $A_0$-mode waves can only be superimposed with the $A_0$ mode low-frequency waves of the impact AEs. Therefore, if the $S_0$ mode wave packets extracted in the AE occur later than the OP-generated $A_0$-mode ones, the AE only occurs when the plate structure is damaged, and it is not an isolated event. The above conclusions were verified by analysing the impact test data of an epoxy fibreglass plate.

This study was performed in the cut-off range of the Lamb wave mode, which is applicable for thin (<2 mm) plate structures. If the frequency–thickness product exceeds the mode cut-off range, further discussions on the impact damage diagnosis method are recommended. The AE Lamb wave mode confirmed by the wave velocity was leveraged. The epoxy fibreglass plate used in the test is not significantly anisotropic; that is, it can be considered approximately isotropic for data analysis. If the fibre-laying angle of the composite plate is highly associated with the wave velocity, a detailed Lamb wave dispersion curve is required for discussing the AE mode.

**Author Contributions:** Methodology, B.X.; Software, Y.J.; Validation, B.X.; Investigation, Y.J.; Writing—original draft, B.X.; Writing—review & editing, B.X.; Project administration, J.H.; Funding acquisition, J.H. All authors have read and agreed to the published version of the manuscript.

**Funding:** This work is supported by the Natural Science Research Foundation of Jiangsu Provincial Universities (Grant No. 20KJA510007) and the Research Foundation of Nanjing Institute of Technology (Grant No. CKJB202208).

**Institutional Review Board Statement:** Not applicable.

**Informed Consent Statement:** Not applicable.

**Data Availability Statement:** Not applicable.

**Conflicts of Interest:** The authors declare no conflict of interest.

## References

1. Hamam, Z.; Godin, N.; Fusco, C.; Doitrand, A.; Monnier, T. Acoustic Emission Signal Due to Fiber Break and Fiber Matrix Debonding in Model Composite: A Computational Study. *Appl. Sci.* **2021**, *11*, 8406. [CrossRef]
2. Yousefi, J.; Najfabadi, M.A.; Toudeshky, H.H.; Akhlaghi, M. Damage evaluation of laminated composite material using a new acoustic emission Lamb-based and finite element techniques. *Appl. Compos. Mater.* **2018**, *25*, 1021–1040. [CrossRef]
3. Xu, X.; Zhou, Y.; Chen, W.; Gao, Y.; Fu, Q.; Liu, X.; Feng, C. Macro-Microscopic Deterioration Behavior of Gypsum Rock after Wetting and Its Constitutive Model Based on Acoustic Emission. *Minerals* **2022**, *12*, 1168. [CrossRef]
4. Romhány, G.; Czigány, T.; Karger-Kocsis, J. Failure Assessment and Evaluation of Damage Development and Crack Growth in Polymer Composites Via Localization of Acoustic Emission Events: A Review. *Polym. Rev.* **2017**, *57*, 397–439. [CrossRef]
5. Fotouhi, M.; Najafabadi, M.A. Acoustic emission-based study to characterize the initiation of delamination in composite materials. *J. Thermoplast. Compos. Mater.* **2016**, *29*, 519–537. [CrossRef]
6. Saeedifar, M.; Zarouchas, D. Damage characterization of laminated composites using acoustic emission: A review. *Compos. Part B Eng.* **2020**, *195*, 108039. [CrossRef]
7. Gorgin, R.; Wang, Z.; Wu, Z.; Yang, Y. Probability based impact localization in plate structures using an error index. *Mech. Syst. Signal Process.* **2021**, *157*, 107724. [CrossRef]
8. Tobias, A. Acoustic-emission source location in two dimensions by an array of three sensors. *Non-Destr. Test.* **1976**, *9*, 9–12. [CrossRef]
9. Kundu, T. Acoustic source localization. *Ultrasonics* **2014**, *54*, 25–38. [CrossRef]
10. Zhao, G.; Hu, H.; Li, S.; Liu, L.; Li, K. Localization of impact on composite plates based on integrated wavelet transform and hybrid minimization algorithm. *Compos. Struct.* **2017**, *176*, 234–243. [CrossRef]
11. Petrucci, R.; Santulli, C.; Puglia, D.; Nisini, E.; Sarasini, F.; Tirillò, J.; Kenny, J.M. Impact and post-impact damage characterisation of hybrid composite laminates based on basalt fibres in combination with flax, hemp and glass fibres manufactured by vacuum infusion. *Compos. Part B Eng.* **2015**, *69*, 507–515. [CrossRef]
12. Dhakal, H.N.; Sarasini, F.; Santulli, C.; Tirillò, J.; Zhang, Z.; Arumugam, V. Effect of basalt fibre hybridisation on post-impact mechanical behaviour of hemp fibre reinforced composites. *Compos. Part A Appl. Sci. Manuf.* **2015**, *75*, 54–67. [CrossRef]
13. Tie, Y.; Zhang, Q.; Hou, Y.; Li, C. Impact damage assessment in orthotropic CFRP laminates using nonlinear Lamb wave: Experimental and numerical investigations. *Compos. Struct.* **2020**, *236*, 111869. [CrossRef]
14. Dziendzikowski, M.; Kurnyta, A.; Dragan, K.; Klysz, S.; Leski, A. In situ Barely Visible Impact Damage detection and localization for composite structures using surface mounted and embedded PZT transducers: A comparative study. *Mech. Syst. Signal Process.* **2016**, *78*, 91–106. [CrossRef]
15. Nardi, D.; Lampani, L.; Pasquali, M.; Gaudenzi, P. Detection of low-velocity impact-induced delaminations in composite laminates using Auto-Regressive models. *Compos. Struct.* **2016**, *151*, 108–113. [CrossRef]
16. Frieden, J.; Cugnoni, J.; Botsis, J.; Gmür, T. Low energy impact damage monitoring of composites using dynamic strain signals from FBG sensors—Part II: Damage identification. *Compos. Struct.* **2012**, *94*, 593–600. [CrossRef]
17. Pérez, M.A.; Pernas-Sánchez, J.; Artero-Guerrero, J.A.; Serra-López, R. High-velocity ice impact damage quantification in composite laminates using a frequency domain-based correlation approach. *Mech. Syst. Signal Process.* **2021**, *147*, 107124. [CrossRef]
18. Okafor, A.C.; Otieno, A.W.; Dutta, A.; Rao, V.S. Detection and characterization of high-velocity impact damage in advanced composite plates using multi-sensing techniques. *Compos. Struct.* **2001**, *54*, 289–297. [CrossRef]
19. Zamorano, B.; Elvin, N. Passive high-speed impact damage assessment in composite panels using embedded piezoelectric sensors. *J. Intell. Mater. Syst. Struct.* **2016**, *27*, 104–116. [CrossRef]

20. Saeedifar, M.; Najafabadi, M.A.; Zarouchas, D.; Toudeshky, H.H.; Jalalvand, M. Barely visible impact damage assessment in laminated composites using acoustic emission. *Compos. Part B Eng.* **2018**, *152*, 180–192. [CrossRef]
21. Liu, Z. Ultrasonic Lamb Waves in Nondestructive Testing. *NDT* **1999**, *21*, 409–413.
22. Xu, B.; Wu, J.; Wang, M. Study of modal acoustic emission to monitor the impact damage in a composite plate. *J. Vibroeng.* **2017**, *19*, 3335–3348. [CrossRef]
23. Xu, B.; Yuan, S.; Wang, M.; Qiu, L. Dtetermining impact induced damage by lamb wave mode extracted by EMD method. *Measurement* **2015**, *65*, 120–128.
24. Ziola, S.M.; Gorman, M.R. Source location in thin plates using cross-correlation. *J. Acoust. Soc. Am.* **1991**, *90*, 2551–2556. [CrossRef]
25. Geng, R.; Sheng, G.; Liu, S. A Study on Modal Acoustic Emission Theory. *NDT* **2002**, *24*, 302–306.
26. Surgeon, M.; Wevers, M. One sensor linear location of acoustic emission events using plate wave theories. *Mater. Sci. Eng. A* **1999**, *265*, 254–261. [CrossRef]
27. Gorman, M.R.; Ziola, S.M. Plate waves produced by transverse matrix cracking. *Ultrasonics* **1991**, *29*, 245–251. [CrossRef]
28. Ribichini, R.; Cegla, F.; Nagy, P.B.; Cawley, P. Study and comparison of different EMAT configurations for SH wave inspection. *IEEE Trans. Ultrason. Ferroelectr. Freq. Control.* **2011**, *58*, 2571–2581. [CrossRef]
29. Nakamura, N.; Ogi, H.; Hirao, M.; Nakahata, K. Mode conversion behavior of SH guided wave in a tapered plate. *NDT E Int.* **2012**, *45*, 156–161.
30. Jeong, H.; Jang, Y.-S. Wavelet analysis of plate wave propagation in composite laminates. *Compos. Struct.* **2000**, *49*, 443–450. [CrossRef]
31. Dahmene, F.; Yaacoubi, S.; El Mountassir, M.; Bendaoud, N.; Langlois, C.; Bardoux, O. On the modal acoustic emission testing of composite structure. *Compos. Struct.* **2016**, *140*, 446–452. [CrossRef]

sensors MDPI

*Systematic Review*

# Systematic Evaluation of Ultrasonic In-Line Inspection Techniques for Oil and Gas Pipeline Defects Based on Bibliometric Analysis

**Jie Huang [1,2], Pengchao Chen [2], Rui Li [2], Kuan Fu [2], Yanan Wang [2], Jinyao Duan [2] and Zhenlin Li [1,*]**

[1] College of Mechanical and Storage and Transportation Engineering, China University of Petroleum (Beijing), Beijing 102249, China; 2022315301@student.cup.edu.cn
[2] General Research Institute, China Oil & Gas Pipeline Network Corporation, Langfang 065000, China; chenpc@petrochina.com.cn (P.C.); kjlirui@petrochina.com.cn (R.L.); fukuan@pipechina.com.cn (K.F.); wangyn07@pipechina.com.cn (Y.W.); duanjy01@pipechina.com.cn (J.D.)
* Correspondence: zhenlinli@263.net

**Abstract:** The global reliance on oil and gas pipelines for energy transportation is increasing. As the pioneering review in the field of ultrasonic defect detection for oil and gas pipelines based on bibliometric methods, this study employs visual analysis to identify the most influential countries, academic institutions, and journals in this domain. Through cluster analysis, it determines the primary trends, research hotspots, and future directions in this critical field. Starting from the current global industrial ultrasonic in-line inspection (ILI) detection level, this paper provides a flowchart for selecting detection methods and a table for defect comparison, detailing the comparative performance limits of different detection devices. It offers a comprehensive perspective on the latest ultrasonic pipeline detection technology from laboratory experiments to industrial practice.

**Keywords:** oil and gas; pipeline; in-line inspection; ultrasonic testing; bibliometrics; defects

## 1. Introduction

Oil and natural gas account for 57.5% of the global primary energy consumption [1]. Pipelines, integral to the transportation of these resources, are distinguished by their capacity for transporting large volumes over long distances with minimal energy loss [2–5]. They are crucial in connecting the upstream and downstream sectors of the oil and gas industry and are essential in long-distance transportation. By the end of 2022, the total operational mileage of global oil and gas pipelines was approximately 2 million km, with an additional 26,708 km under construction, and it is projected to reach 2.2 million km by 2025 [6].

Data from the Pipeline and Hazardous Materials Safety Administration (PHMSA) of the United States Department of Transportation reveal that between 2003 and 2022, there were 12,785 significant pipeline incidents in the United States. These incidents resulted in 274 deaths and 1120 injuries. The average cost of each incident was around 541 million USD, leading to a total loss of 10.8 billion USD [7]. This underscores the growing need for enhanced inspection of oil and gas pipelines to prevent severe accidents, such as leaks and explosions, which can lead to environmental pollution and human casualties.

Pipeline in-line inspection (ILI) commonly employs technologies such as magnetic flux leakage (MFL), eddy current testing (EC), and ultrasonic testing (UT) [8–11]. As a prevalent and reliable method, magneto-electric composite internal inspection combines MFL and EC and dominates about 90% of the inspection market. Despite its widespread use, this method faces challenges like structural complexity (with some sections weighing up to 4 tons), limited effectiveness in detecting planar defects (particularly at girth welds), and sensitivity to inspection speed and lift-off distance. In contrast, ultrasonic testing, a wave-based method, is notable for its strong directionality and penetration power, making it more effective for identifying pipeline wall defects [12]. However, existing literature reviews generally discuss UT as one of the non-destructive testing techniques for oil and gas pipelines. For instance, Feng et al. integrate practical inspection data to elucidate the application of both conventional ultrasonics and electromagnetic ultrasonics in the detection of circumferential welds in oil and gas pipelines from the perspectives of mechanism, quantitative methods, and inspection reliability [11,13,14]. Alternatively, focusing on specific aspects of ultrasonics, researchers like Zang et al. delve into issues such as dispersion, multimode propagation, and attenuation in ultrasonic-guided wave technology. Methods to address these challenges from a theoretical standpoint are proposed, with a comprehensive overview provided of guided wave excitation devices [15]. Andika, on the other hand, outlines the application of machine learning signal processing methods in handling the high-volume, high-velocity, and diverse data generated using ultrasonic in-line inspection (ILI). This includes preprocessing, learning algorithms, outputs, and evaluation metrics [16,17]. There is a lack of systematic discussion that spans from a macro perspective (i.e., integrating practical industrial inspection scenarios with a vertical analysis of existing inspection technologies for oil and gas pipelines) to a meso perspective (i.e., a horizontal comparative analysis of ultrasonic testing technologies and the internal connections between detection methods and technologies) and down to a micro perspective (i.e., ultrasonic data analysis and signal processing). Bibliometric methods, based on clustering algorithms, offer an effective means to avoid subjective biases in literature selection [1,18–20]. Research indicates that there is currently no comprehensive review of ultrasonic internal inspection for oil and gas pipelines based on bibliometric methods.

The remainder of this paper is structured as follows. Section 2 introduces the data sources and research methods used in this study, detailing the primary process of literature analysis. Section 3, utilizing visualization analysis techniques, provides an overview of renowned research institutions and journals in the field of ultrasonic inspection of oil and gas pipelines worldwide. In the Section 5, a systematic review of the current research hotspots in the field of ultrasonic testing is conducted based on the results of literature clustering analysis. Starting from the current global industrial level of ultrasonic in-line inspection (ILI) detection, a flowchart for selecting detection methods and a table comparing detectable defects are provided. Furthermore, a detailed comparison of the performance limits of different detection devices is presented. The latest technologies in ultrasonic pipeline inspection are discussed from a holistic perspective, spanning from laboratory experimentation to industrial implementation. Finally, Section 6 summarizes and suggests recommendations for future research directions.

## 2. Research Methodology and Data Analyses

To accurately, objectively, and comprehensively reveal the research achievements in ultrasonic defect detection for oil and gas pipelines, this study employs data from the Web of Science (WOS) Core Collection, specifically the Science Citation Index Expanded and the Social Sciences Citation Index. The initial information retrieval was conducted using the search strategy designed in the Web of Science Advanced Search: (((TS = (ultraso* AND (test* OR inspect* OR detect*))) AND TS = (pipe* OR tube)) AND TS = (defect* OR flaw* OR CRACK OR SCC OR discontinue)) AND TS = (medic* OR patient* OR child* OR parent* OR women), resulting in the collection of 782 papers. The use of an asterisk "*" after keywords indicates multiple similar words with the same prefix (e.g., "pipe*" represents "pipe", "pipeline", etc.), expanding the search scope to encompass as many articles related to the research topic as possible.

Due to potential algorithmic limitations of the WOS, the dataset may include literature irrelevant to the central theme. To address that, a rigorous screening process was applied, involving the review of titles and abstracts, and extending to full articles, when necessary, to filter out duplicates and non-relevant papers. Additionally, a snowball search method was employed to expand the dataset, which was then subjected to a second round of screening. Our research indicates that ultrasonic defect detection in oil and gas pipelines emerged as a significant area of study since the 1990s. Consequently, this paper focuses on papers, conference proceedings, and reviews published from 1992 onwards, totaling 350 documents for bibliometric analysis. The process of dataset creation and the structure of this paper are depicted in Figure 1.

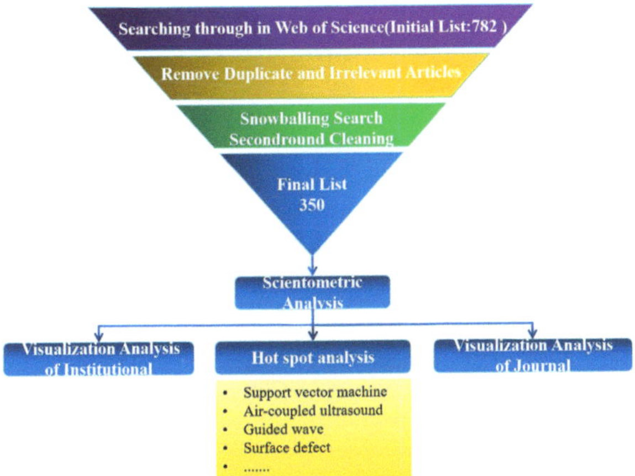


**Figure 1.** Dataset Creation Process and Research Content.

After finalizing the data collection, the dataset underwent a bibliometric analysis using scientific mapping tools, specifically VOSviewer 1.6.15 and CiteSpace 6.1.R2. The analysis outcomes facilitated a detailed review of the critical research areas in ultrasonic testing for oil and gas pipelines.

## 3. Visualization Analysis

### 3.1. Visualization Analysis of Institutions and Countries

In the macroscopic research system, scientific institutions represent the smallest unit and the primary executors of research activities. To a certain extent, the number of such institutions indicates the countries leading in the field. Table 1 displays a ranking of the top ten countries globally in the ultrasonic field, sorted based on the quantity of their research institutions.

**Table 1.** The Most Influential Countries in the Global Ultrasonic Field (Institutions).

| Country | Institutions | Burst | Centrality |
|---|---|---|---|
| China | 61 | 3.26 | 0.36 |
| England | 48 | 0 | 0.73 |
| Germany | 33 | 2.1 | 0.27 |
| USA | 32 | 1.78 | 0.9 |
| Italy | 27 | 0 | 0.63 |
| Republic of Korea | 23 | 0 | 0.34 |
| Japan | 17 | 2.08 | 0.16 |
| Brazil | 14 | 0 | 0.12 |
| India | 13 | 1.87 | 0.19 |
| France | 12 | 2.44 | 0.22 |

China, the UK, and Germany occupy the top three positions on the list. To ensure objectivity in ranking, additional indicators such as 'burst' and 'centrality' were also considered. "Burst", a feature in CiteSpace, detects sudden changes in information and, when combined with the number of authors at the macro level of a country, effectively reveals the regional development level in the field. This sudden change indicates the following:

- During this period, institutions from the country experienced a sharp increase in citation frequency.
- The institutions in the country have effectively addressed crucial issues in the field.

"Centrality", on the other hand, represents the role of a node as a bridge within a network. In social networks, centrality is used to identify "boundary spanners". Nodes with higher centrality values play a more significant bridging role between other nodes. These institutions are highlighted with purple circles (nodes with centrality $\geq 0.1$). Nodes with high intermediation centrality often connect different clustering paths and are referred to in CiteSpace as "Turning Points".

In Table 2, countries in the field of ultrasonics are re-ranked from high to low based on indicators such as "Institutions", "Burst", and "Centrality". By integrating data from Table 3 and Figures 2 and 3, it is observed that over time, the research focus in ultrasonic testing for oil and gas pipelines has gradually shifted from Europe and America to Eurasia. The primary reason is that Europe and America, being early adopters of pipelines for oil and gas transport, have already completed the exploration phase of pipeline integrity testing. They have established reliable and comprehensive detection systems and have formally entered the phase of applying ultrasonics. In contrast, the Eurasian region is still in the nascent stages. Particularly in China, with the gradual implementation of the "High-Quality Development 2.0" and "Smart Pipeline Network" concepts, there has been a surge in demand for ultrasonic testing in pipeline inspection, spurring significant research in the field in recent years. However, from the perspective of "centrality", the research in ultrasonic testing in the Eurasian region still lags behind the established positions of Europe and America.

**Table 2.** Global Influential Countries in the Ultrasonic Field Based on Different Criteria.

| Standard | Institutions | Burst | Centrality |
|---|---|---|---|
| ↓ | China | China | USA |
| | England | France | England |
| | Germany | Germany | Italy |
| | USA | Japan | China |
| | Italy | India | Republic of Korea |
| | Republic of Korea | USA | Germany |
| | Japan | England | France |
| | Brazil | Italy | India |
| | India | Republic of Korea | Japan |
| | France | Brazil | Brazil |

**Table 3.** The Most Influential Institutions in the Global Ultrasonic Field.

| Institutions | Total Link Strength | Documents | Citations | Country |
|---|---|---|---|---|
| University of Warwick | 12 | 7 | 334 | UK |
| Federal Institute for Materials Research and Testing | 14 | 4 | 207 | Germany |
| Seoul National University | 4 | 3 | 185 | Republic of Korea |
| Newcastle University | 16 | 6 | 138 | UK |
| University of Palermo | 34 | 6 | 127 | Italy |
| Kaunas University of Technology | 6 | 9 | 119 | Lithuania |
| Brunel University London | 30 | 8 | 118 | UK |
| University of Electronic Science and Technology of China | 9 | 6 | 108 | China |
| Cranfield University | 5 | 3 | 92 | UK |
| City University of Hong Kong | 16 | 6 | 90 | China |

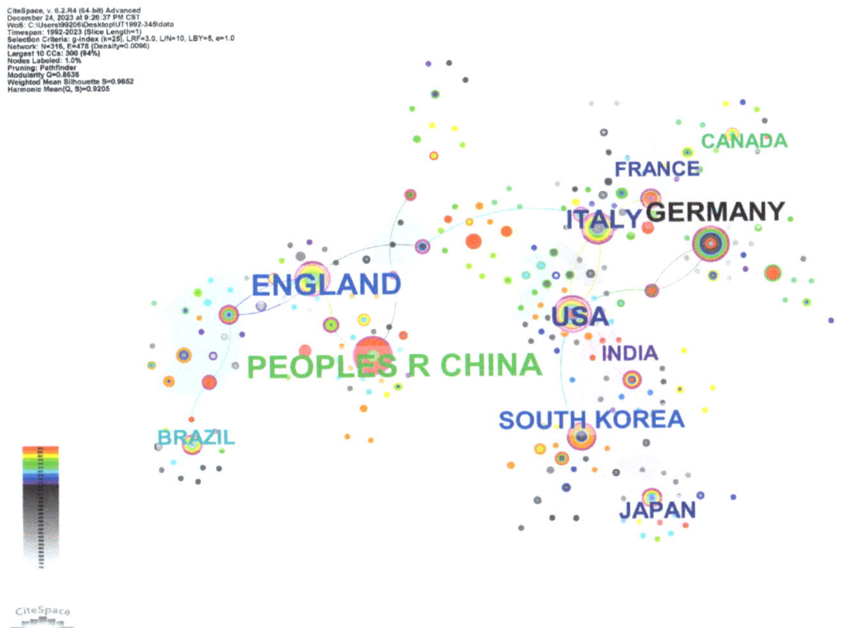


**Figure 2.** Different Country Cooperation Networks.

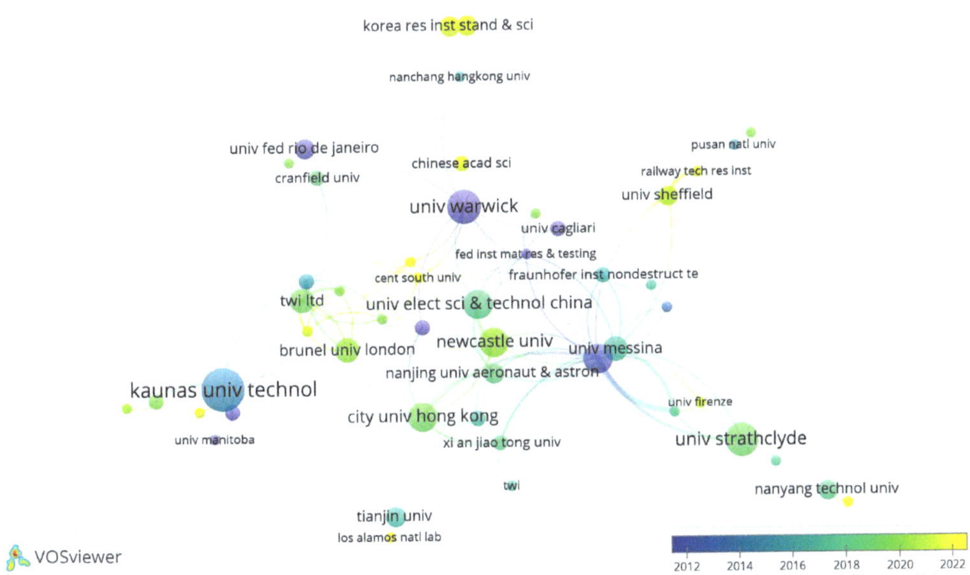


**Figure 3.** Different Institution Cooperation Networks.

## 3.2. Visualization Analysis of Journals

As illustrated, publications were visualized and analyzed using VOSviewer. In contrast to traditional metrics based solely on the number of published documents, this analysis employs "total link strength" to represent a journal's influence. This approach provides a more comprehensive perspective for assessing journal impact, as it considers not only the volume of publications but also their inter-citation relationships. The thickness of the lines indicates the strength of the connections, while from a temporal perspective, it can signify the influence of one journal on another.

Figure 4 reveals that ultrasonic testing has gradually shifted from theoretical research to practical applications over time. This transition reflects the natural progression of scientific research, moving from theoretical exploration to real-world applications. Identifying specific application scenarios for ultrasonics is becoming a research trend, suggesting that ultrasonic technology may find broader applications in the future. This is an exciting trend, as it implies that our research findings are more likely to be transformed into practical technologies and applications, bringing tangible benefits to society.

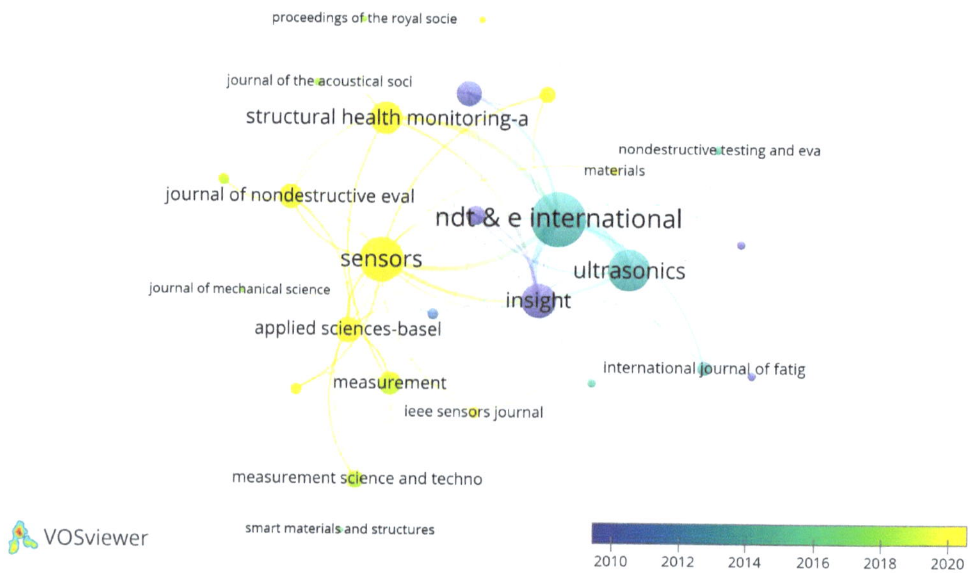


**Figure 4.** Different Journal Cooperation Networks.

According to the data, the top ten influential journals are displayed in Table 4, with *NDT & E International*, *Sensors* and *Ultrasonics* occupying the top three positions. These journals, recognized as leading publications in the field of ultrasonics, have made significant theoretical contributions to the field of ultrasonics and influencing subsequent research.

**Table 4.** The Most Influential Journals in the Global Ultrasonic Field.

| Journals | Total Link Strength | Documents | Citations |
|---|---|---|---|
| *NDT & E International* | 220 | 49 | 2055 |
| *Sensors* | 199 | 64 | 618 |
| *Ultrasonics* | 123 | 75 | 715 |
| *Insight* | 106 | 56 | 709 |
| *Measurement* | 94 | 37 | 680 |
| *Structural Health Monitoring* | 82 | 32 | 440 |
| *Applied Sciences-Basel* | 74 | 26 | 478 |
| *Journal of Nondestructive Evaluation* | 74 | 22 | 347 |
| *Optik* | 72 | 21 | 301 |
| *Applied Physics A-Materials Science & Processing* | 58 | 16 | 365 |

## 4. Types of Defects and Traditional ILI Methods in Oil and Gas Pipelines

### 4.1. Common Types of Defects in Oil and Gas Pipelines

As global energy demand grows, the significance of oil and gas pipeline systems becomes increasingly pronounced [21]. The health of pipelines directly impacts the reliability and safety of the energy supply. Ultrasonic testing technology has been closely associated with concepts such as "condition assessment", "defect detection", and "pipeline safety" since the last century.

Defects, as the core of pipeline condition assessment, are crucial for ensuring the safety and reliable operation of oil and gas pipeline systems. Pipeline defects can be broadly categorized into three types based on their origins: manufacturing defects arising during prefabrication, welding defects and geometric defects occurring during pipeline construction, and corrosion defects formed during usage. However, the current pipeline construction quality is high with rare manufacturing defects.

Based on the shape of defects, oil and gas pipeline defects can be classified into planar defects and volumetric defects (details in Table 5).

**Table 5.** Common Defect Types and Causes.

| Type | Kind | | Cause | Location |
|---|---|---|---|---|
| Planar defect | Crack | Fatigue Crack | Fatigue cracks are cracks formed in pipes under the action of alternating or cyclical loads, such as tension, compression, or bending. | Weld & Body |
| | | Stress Corrosion Cracking (SCC) | Stress corrosion cracks are formed in pipelines under the simultaneous influence of stress and environmental corrosion. | |
| | | Hydrogen-Induced Cracking (HIC) | Metals absorb hydrogen atoms. When subjected to stress, hydrogen causes a change in the material's atomic structure, making it more susceptible to fracture. | |
| | Infusion | | During welding, incomplete fusion occurs when the welding material fails to fully melt and fuse with the base material. This can result in the presence of cracks or unjoined areas in the weld, thereby reducing the strength and reliability of the weld. | Weld |
| | Underpenetration | | Incomplete penetration occurs when welding material fails to fully penetrate through to the base material during welding. This results in the presence of gaps in the weld, which may create weak points, especially under pressure loads. | Weld |
| | Undercut | | The arc melts the edge of the base material at the weld seam, leaving a groove. Undercut weakens the load-bearing cross-section of the joint, making it susceptible to stress concentration. | Weld |
| Volumetric defect | Porosity | | During the welding process, gas does not escape in time and forms voids inside or on the surface of the weld metal. | Weld & Body |
| | Inclusion | | Inclusions or foreign substances present in the interior of the weld metal or fusion line. | Weld & Body |
| | Corrosion | | Pipeline damage caused by chemical environmental corrosion or stray currents affecting the material. | Weld & Body |
| | Dent | | | Body |

(1) Volumetric defects encompass porosity, inclusions and corrosion, posing relatively low risks, with the primary failure mode being plastic failure.
(2) Planar defects include lack of penetration, lack of fusion, cracks, and undercutting. The defects at the root weld are the main cause of pipeline cracking and are also a focal point of research in non-destructive testing of circumferential weld defects.
(3) Structural discontinuity defects include irregular weld seam profiles, misalignment, and poor formation, leading to stress concentration and reduced material performance, further exacerbating defect propagation [22–24].

Corrosion in pipelines typically manifests as localized corrosion, primarily caused by pitting in buried pipelines or coating defects or dissolution. The corrosion induced by coating detachment and extensive corrosion can be precisely detected and quantified. However, pitting, as a particular case of corrosion, remains challenging in oil and gas pipeline inspection, as it can easily evolve into pinholes and lead to crack initiation, posing a higher risk of failure for high-pressure natural gas pipelines.

Pitting is a chemical process involving four stages: passive film breakdown, pit initiation, metastable pitting, and pit propagation and growth. Due to factors such as temperature, acidity, environmental conditions, and biological factors, corrosion often exhibits randomness, making it difficult to estimate corrosion rates. To address the challenge of unpredictable corrosion rates, scholars have developed corrosion rate models based on comprehensive coating information, the effectiveness of cathodic protection, specific corrosion rates of certain soils, and other pipeline characteristics. These models predict corrosion rates and depths using probability and statistical methods, including regression models and Markov process models [25].

(1) Regression models can easily estimate future corrosion defect depths in pipelines without requiring extensive mathematical knowledge for application. However, this model relies on a large amount of data and cannot account for the randomness of pitting phenomena.
(2) Markov process models consider the physical and chemical characteristics of the environment and accurately reflect the randomness of localized corrosion defect growth. However, their use requires specialized knowledge and programming skills, which may limit their widespread application.

Moreover, continuous electromagnetic in-line inspection is considered an ideal method to determine corrosion rates. However, it is challenging to determine corrosion rates using this information due to the complex pipeline environment and variations in sensor calibration methods before each inspection. Locating the same corrosion defect between two consecutive inspections may be a daunting task. It is worth mentioning that while accurate corrosion rate prediction through in-line inspection is challenging, it still provides essential information on pipeline body and wall defects required for building more accurate probabilistic and statistical models. At the same time, accurate corrosion rate models can optimize in-line inspection tasks, greatly reducing costs for pipeline operators.

### 4.2. Conventional ILI Methods for Oil and Gas Pipelines

The predominant in-line inspection methodologies in the oil and gas pipeline industry currently encompass MFL, EC and UT.

#### 4.2.1. Electromagnetic Testing

(a) Magnetic flux leakage testing

The principle of MFL is depicted in Figure 5. Upon applying an external magnetic field that induces magnetic saturation in the pipeline, defects within the pipe wall that disrupt the magnetic flux lead to distortion and external leakage of this flux. This results in an MFL field being formed on the pipe's surface. Triaxial Hall sensors are employed to detect this field, enabling the acquisition of detailed information regarding the defects' distribution and dimensions [10,26,27]. MFL technology stands as a highly regarded and reliable method for pipeline inspection within the sector [28], commanding approximately 90% of the total market equipment share. Contrasting with ultrasonic testing, which is restricted to liquid pipelines, MFL is capable of detecting volumetric defects in both the interior and exterior of pipelines within a specified wall thickness limit (up to 30 mm). Furthermore, it requires less stringent cleanliness conditions for pipelines than ultrasonic testing.

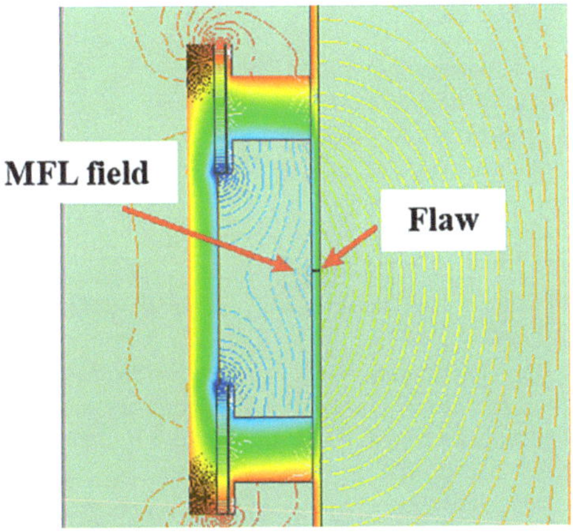


**Figure 5.** Principle of Magnetic Flux Leakage Testing.

**Limitations:** As depicted in Table 6, MFL exhibits a reduced efficacy in detecting planar defects, particularly at girth welds. It is notably less effective, or even incapable, of identifying narrow or closed cracks, such as those caused by stress corrosion or hydrogen-induced cracking [29–31].

**Table 6.** Types of Internal Inspectors and Detection Capabilities.

| Type | Kind | | MFL | UTWM [a] | UTCD [a] | EMAT | Geometric Inspection | EC |
|---|---|---|---|---|---|---|---|---|
| Planar defect | Crack | Fatigue Crack | × | × | √+ | √+ | × | √− |
| | | Stress Corrosion Cracking (SCC) | × | × | √+ | √+ | × | √− |
| | | Hydrogen-Induced Cracking (HIC) | × | × | √+ | × | × | √− |
| | Infusion | Straight Weld | × | × | √+ | √+ | × | √ |
| | | Spiral Weld | √+ | × | × | × | × | √ |
| | | Girth Weld | √+ | × | √+ | × | × | × |
| | Underpenetration | | √ | √ | √ | √ | × | √ |
| | Undercut | | × | | √ | √ | × | √ |
| Volumetric defect | Porosity | | √ | × | √+ | × | × | × |
| | Inclusion | | √− | √+ | √− | √− | × | × |
| | Corrosion | Internal | √+ | √+ | × | × | √− | √− |
| | | External | √+ | √+ | × | × | × | × |
| Geometry | Dent | | √ | √ | × | × | √+ | × |
| | Distortion | | √ | √ | × | × | √+ | × |

[a]: Only applied in liquid environment; ×: Undetectable; √: Detectable; √−: Detectable with Limitations; √+: Detectable and Quantifiable.

(b) Eddy current testing

Eddy current testing operates by generating an eddy current magnetic field using an excitation coil, which interacts with the original magnetic field, thereby altering the complex impedance of the testing coil. This impedance alteration is utilized to detect and identify defects on the pipeline's inner wall and is extensively applied in pipeline crack detection.

**Limitations:** EC's effectiveness is limited by the skin and lift-off effects, restricting its use to the surface or near-surface inspection of conductive materials. Moreover, extracting diagnostic signals poses significant challenges and makes it difficult to accurately determine the type and size of defects [32–35]. Consequently, eddy current testing is often used in conjunction with MFL, facilitating more accurate localization of defects.

(c) Advancements in electromagnetic inspection

(1) Advancements in Electromagnetic Inspection Equipment

In the realm of equipment, ROSEN and Baker Hughes have independently developed triaxial ultra-high-definition magnetic flux leakage internal detectors. These smart PIGs, integrated with the modules of geometric deformation and IMU, possess the capability to navigate through a diameter of 1.5 times their own and can identify open cracks exceeding 0.5 mm and pinholes of at least 5 mm, achieving resolutions of 1.6 mm circumferentially and 1 mm axially. The China Oil and Gas Pipeline Network Corporation has made significant strides in pipeline internal inspection technology, creating a triaxial ultra-high-definition electromagnetic composite internal detector specifically for 1219 mm diameter pipelines. This advanced PIG is equipped with MagEC ultra-high-definition electromagnetic composite probes, sophisticated mechanical geometric deformation sensors, and differential electromagnetic eddy current sensors. It has demonstrated a detection rate of up to 90% for pinhole defects larger than 3 mm in diameter at girth welds and 80% for open cracks that are wider than 0.5 mm, longer than 24 mm, and extend to a depth of 50% of the wall thickness.

(2) Advancements in Electromagnetic Inspection Research

From a research perspective, the focus on electromagnetic composite detection has predominantly revolved around interpreting detection data. Cheng et al. employed the YOLOv5 and Vision Transformer (ViT) algorithms for pipeline defect inspection and classification, exploring the influence of different model architectures on defect identification performance. Comparative analysis indicates that the composite algorithm surpasses the standalone YOLOv5 algorithm in terms of accuracy in classifying pipeline defects, while maintaining its capability for high-precision defect detection [36]. Y. Shen and colleagues leveraged magnetic flux leakage (MFL) signals and convolutional neural network (CNN) technology to forecast the dimensions and locations of corrosion defects in steel pipes. Their research underscores the promising application of CNN models in enhancing pipeline integrity management [29]. Bin Liu introduced a stress factor in developing a numerical analytical model for the internal detection of complex MFL defects in pipelines. His research delved into the variations in magnetic signals induced by different sizes of defect end face segments, analyzed the patterns of MFL signal distribution under various geometric dimensions and stress conditions at the defect end faces, and evaluated their impact on the characteristics of magnetic signals [31]. Additionally, Jianhua Pan and colleagues implemented an advanced CLIQUE algorithm for marking defect regions in segmented pipelines. They then utilized the SSA_BP neural network for extracting and classifying three-dimensional MFL feature signals from these marked regions. The findings from their study reveal that this approach enhances the efficiency of defect marking and provides a more detailed analysis of the marked areas [37].

4.2.2. Ultrasonic Testing

Ultrasonic testing is a technique that employs ultrasonic waves to detect macroscopic defects, characterize changes in material structure and mechanical properties, and measure the geometric features of a workpiece [38]. This method boasts advantages such as high detection speed, precision, directivity, sound wave energy, and strong penetration capability [39]. Ultrasonics, as a means of detecting metal defects, was first proposed by Sokolov of the Soviet Union in 1929, evolving subsequently into transmission testing and pulse-echo testing methods [40]. As shown in the Table 6, ultrasonic in-line testing primarily addresses issues related to pipeline cracks and metal loss. Crack detection involves the recognition and quantification of crack location and size by generating shear waves in the pipe wall using ultrasonic probes arranged at a 45° angle [41].

Currently, companies worldwide with the capability to develop complete sets of ultrasonic equipment, such as ROSEN, TDW, and China National Petroleum Corporation (CNPC), possess electromagnetic ultrasonic crack detection capabilities, with roughly equivalent detection capacities. Under speeds not exceeding 2.5 m/s, they can detect cracks as small as 50 mm in length and 2 mm in depth, with an error margin of ±1 mm in depth and ±20 mm in length. Compared to Baker Hughes and NDT Global's UltraScan CDP/DUO and EVO Eclipse UCx, traditional ultrasonic methods using coupling agents have significant advantages in terms of detection depth and speed.

(a) Ultrasonic Phased Array

NDT Global's PROTON, developed on the principle of ultrasonic phased array technology, employs a PAUT system to design phase delays for multiple probes. This design allows for the editing of the direction and focus of the ultrasonic beam, enabling imaging and detection at various depths and angles. It is capable of detecting not only metal corrosion and crack defects but also stress corrosion cracks, fatigue cracks, and welding crack defects [42].

(b) Time-of-Flight Diffraction (TOFD)

Another commonly used method for crack detection is the Time of Flight Diffraction (TOFD) method, first proposed by Silk in 1977 [43]. This method relies on the diffraction of ultrasonic waves by "corners" and "ends" within the internal defects, differing fundamentally from the traditional pulse-echo method, which depends on direct reflection signals [44–46]. As depicted in Figure 6, TOFD typically employs a dual probe arrangement with one transmitter and one receiver on the same side. This method significantly reduces the time required to scan a weld, as it does not necessitate raster scanning at each position.

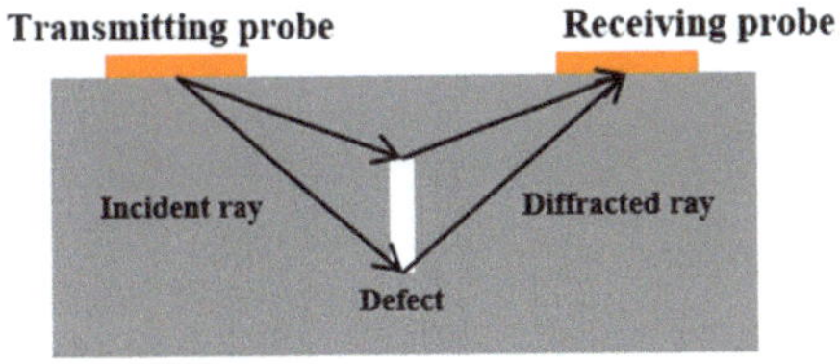


**Figure 6.** Principle of TOFD.

(c) Ultrasonic Pulse-Echo Technique

In addition to the abovementioned applications, another form of ultrasonic testing in pipeline inspection involves emitting ultrasonic pulse waves perpendicular to the pipe wall. This method calculates the pipeline's wall thickness by measuring the pulses' echo times reflected off the inner and outer pipeline surfaces. This approach, introduced to the market in the early to mid-1980s, is instrumental in identifying and quantifying volumetric defects [47,48]. NDT Global's EVO 1.0 UMp+ ultrasonic internal detector (NDT Global, Houston, TX, USA), currently deployed in petroleum pipeline inspection services, exemplifies this technology. It can quantify pipeline wall thickness and crack detection with an accuracy of ±0.4 mm when the data reliability is at 90% [49].

**Limitations:** These ultrasonic testing methods still rely on a coupling agent to address the significant impedance mismatch at the propagation interface. Oil, serving as an effective coupling agent, has facilitated the widespread application of this technology in liquid pipelines. In contrast, traditional ultrasonic testing faces challenges in natural gas pipelines, limiting its application. It is often used only as a supplementary method, performing external inspections of girth welds or spiral welds before commissioning or during the excavation verification stage, rather than being applicable for internal pipeline inspections.

### *4.3. In-Line Inspection (ILI) Tool Selection and Calibration Methods*

#### 4.3.1. In-Line Inspection (ILI) Tool Selection Methods

To address the inspection of different types of defects and the varying applicability of various inspection techniques, as shown in Figure 7, a pipeline risk assessment to determine suitable internal inspection techniques is conducted initially. Industrial experimental safety certification of internal inspection equipment, including but not limited to vibration, pressure resistance, strength, and detection accuracy tests, is performed to verify whether the selected detector's performance meets the qualification requirements. Simultaneously, issues such as dents caused by improper construction and pipeline deformations caused by soil displacement significantly restrict conventional internal inspection tasks. Consequently, pigging and diameter measurement tasks are conducted two or three times before internal inspection to obtain constraint point information and carry out necessary modifications. Next, the selection of technology is conducted by assessing its suitability and capability, considering factors such as defect types, compatibility with other techniques, ease of operation, etc. Finally, an evaluation is conducted to determine the technology to be adopted, followed by data analysis. Field excavation verification is used to assess the detection quality of in-line inspection (ILI) tools, specifically their accuracy.

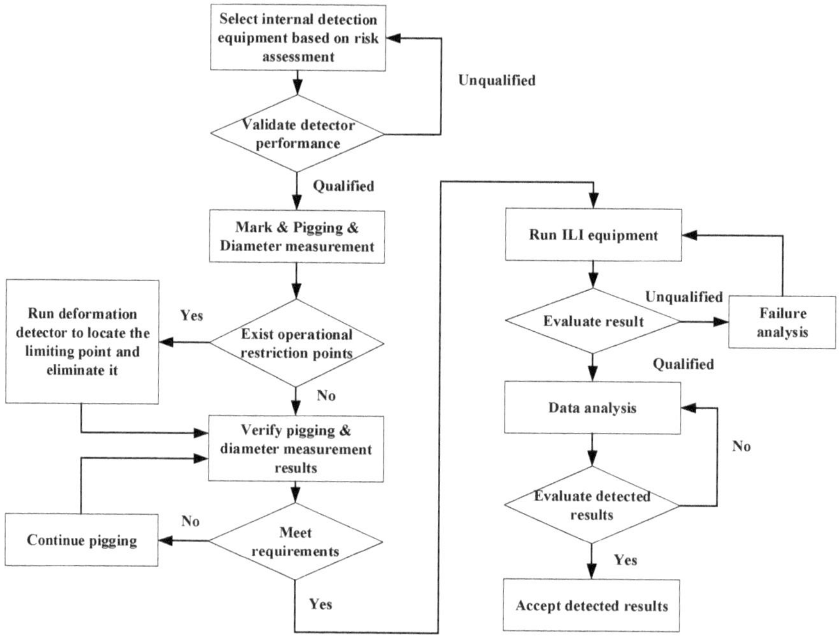


**Figure 7.** In-Line Inspection Process.

#### 4.3.2. In-Line Inspection (ILI) Tool Calibration Methods

As is well known, both in-line inspection (ILI) tools and field measuring instruments are subject to measurement errors, and calibration helps estimate the true magnitude of these errors. The performance of ILI tools can be assessed by statistically comparing ILI data with field excavation data to ensure the reliability of the data. It is crucial to conduct reliability and risk assessments based on the probability distribution of pipeline defects or the failure probability of individual defects. However, the calibration of ILI tools still faces challenges in the following aspects [50]:

(1) Measurement Errors: Determining the measurement errors of ILI tools and field instruments, including systematic errors (such as constant bias and non-constant bias) and random errors.
(2) Calibration Experiment: Conducting calibration experiments to compare ILI readings with field measurement depths and estimate the true size of corrosion metal loss and its associated errors.
(3) Model Verification: Checking for the presence of non-linear regression, variance heterogeneity, outliers, and non-normality of measurement errors in the model.
(4) Performance Evaluation: Evaluating the performance of ILI tools when significant measurement errors exist in field instruments and determining the number of unsuccessful field verifications.

To address these issues, scholars have proposed error assessment methods based on probability and statistics. Among them, the traditional errors-in-variables (EIV) method is used to handle cases where both measuring instruments have errors, but it requires knowledge of the variance ratio of errors or one of the variances. The Grubbs and Jaech estimators are used to estimate the variance of measurement errors, but they may produce negative variances or unreasonable variance values in practical applications. Compared to the former two, a variant method combining V-Wald and V-Jaech provides a more accurate estimation of the true defect depth and the variance of measurement errors. It distinguishes between the sampling distribution caused by non-constant bias and similar measurement tool errors. Even in cases where measurement errors are similar, it can provide reliable calibration results, which are difficult to achieve in traditional models.

## 5. Analysis of Research Hotspots

Keywords reflect the interrelationships among various themes explored in the literature, briefly summarizing the research content. Analyzing keywords is instrumental in gaining insights into the research hotspots of a field. To mitigate potential errors from using a single software, both VOSviewer 1.6.15 and CiteSpace 6.1.R2 were employed to generate the keyword co-occurrence maps shown in Figure 8a,b. These maps exhibit good consistency between the two software tools. In these maps, the size of each node or the intensity of its color represents the frequency of the keyword's occurrence. As the frequency increases, the node's circle enlarges, or the area's color intensifies.

As demonstrated in Figure 9, cluster analysis conducted using CiteSpace 6.1.R2 identified four current research hotspots in the field of ultrasonic defect detection for long-distance oil and gas pipelines: support vector machine, air-coupled ultrasound, guided wave testing, and surface defects inspection.

Figure 10 reveals that since 1992, ultrasonic testing has emerged as a means of non-destructive evaluation. During 2005–2015, ultrasonic technology saw widespread application in the field of defect inspection in oil and gas pipelines. However, its use in natural gas pipelines was mostly limited to external defect detection at welds during construction. In response to the limitations of magnetic flux leakage detection, particularly in identifying planar defects at girth welds, the existing methods and capabilities were found inadequate for the practical needs of the oil and gas engineering sector. This led to increased research focus on ultrasonic testing in pipelines, rapidly advancing the technology and theory of ultrasonic defect detection.

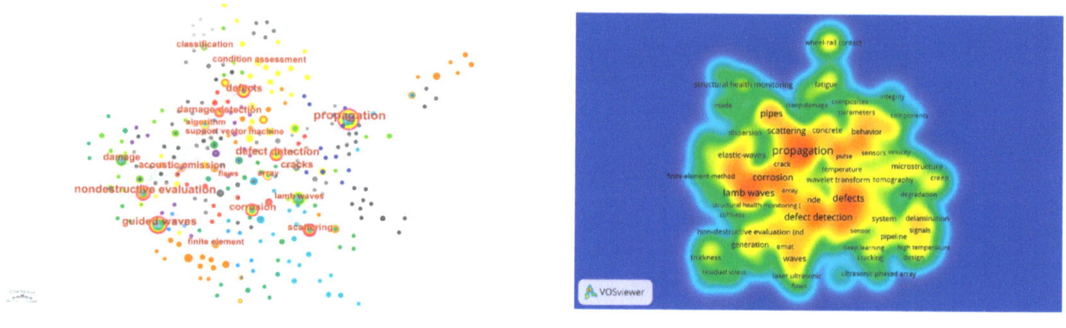


(**a**) (**b**)

**Figure 8.** Keyword Visualization Analysis. (**a**) Citespace; (**b**) VOSviewer.

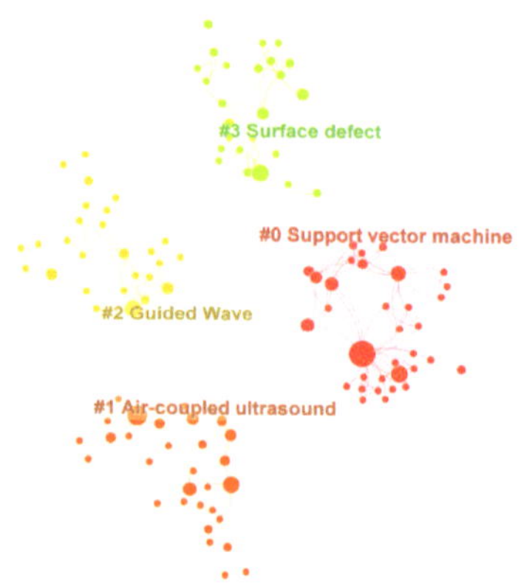


**Figure 9.** Keyword Clustering Analysis.

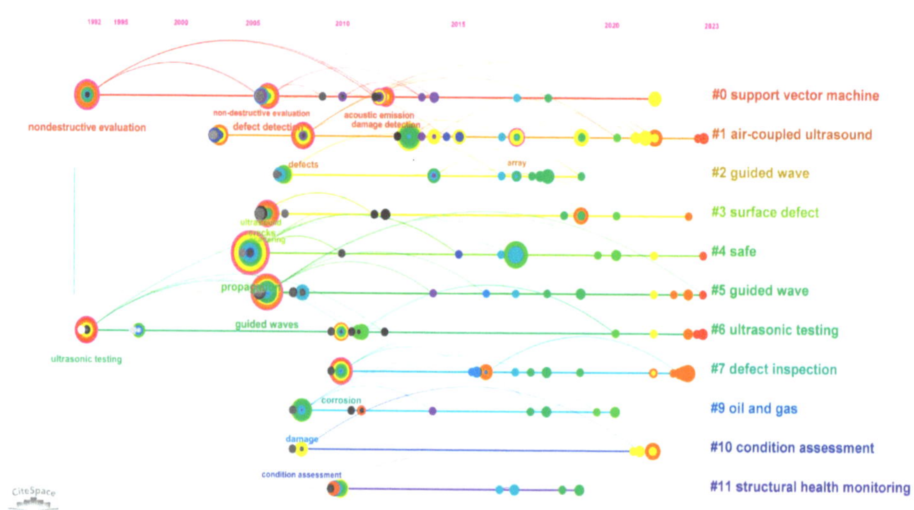


**Figure 10.** Keyword Temporal Map.

From 2015 to the present, there has been an explosive development of new acoustic testing techniques. Current ultrasonic testing in oil and gas pipelines focuses on overcoming challenges such as the inapplicability of coupling agents in natural gas pipelines (where planar defects are more problematic than in oil pipelines) and the application of artificial intelligence and data mining for the analysis and interpretation of existing ultrasonic data. This explains why concepts like support vector machine, air-coupled ultrasound, and guided waves have become hot topics in the field.

### 5.1. Surface Defect

#### 5.1.1. Conventional Surface Inspection Methods

Surface defect detection has always been a core focus of pipeline inspection. Conventional surface inspection methods include penetrant testing (PT), magnetic particle testing (MPT), and eddy current testing (as shown in Table 7). Compared to ultrasonic in-line inspection, PT and MPT offer higher sensitivity but require the removal of pipeline coatings. They also demand high cleanliness of the test surface and are not suitable for in-line inspection [3]. As mentioned in the previous section, EC has matured in internal inspection applications. However, due to the skin and lift-off effects, this technology is quite sensitive to speed and lift-off [51–53].

**Table 7.** Comparison of Nondestructive Testing Methods for Surface Defects.

| | Penetration Testing | Magnetic Particle Testing | Eddy Current Testing |
|---|---|---|---|
| Principle | Capillary phenomenon | Magnetic force | Electromagnetic induction |
| Range | Any non-porous material | Ferromagnetic materials | Conductive materials |
| Position | surface opening defects | surface or near-surface defects | Surface |
| Sensitivity | High | High | Low |
| Speed | Slow | Fast | Fast, can be automated |
| Effect of Defect Orientation on Detection Probability | Unaffected by defect orientation | Affected by defect orientation, easily detects defects perpendicular to the direction of magnetic lines | Affected by defect orientation, easily detects defects perpendicular to the direction of eddy currents |
| Effect of Surface Roughness on Detection Probability | Rougher the surface, lower the probability of detection | Affected, but less than penetration testing | Greatly affected |

#### 5.1.2. Ultrasonic Surface Inspection Techniques

UT can effectively address the abovementioned issues. However, during the propagation of ultrasonic waves, interference can create a series of uneven sound pressure zones near the wave source, known as the near-field zone. The presence of this area can lead to inaccurate quantitative assessment of surface defects. Additionally, tiny cracks on the inner surface of the pipeline wall can easily mask larger defects below, affecting the detection of larger defects. To address this issue, scholars from around the world have conducted extensive research. Considering the near-field zone's impact on detection, J.M. Ha et al. proposed an ultrasonic detection method based on autoencoders to inspect defects within the dead zone of the probe. This method can identify subtle deviations caused by defects. To validate the model's effectiveness, aluminum blocks with near-surface defects were subjected to B-scan testing. The results showed that the proposed method outperforms traditional gate-based detection methods in identifying the size and location of near-surface defects [54]. Jun He and colleagues presented a quantitative approach for detecting surface anomalies using Rayleigh waves generated by lasers. Experimental results demonstrate that this method outperforms standing wave energy or reflected wave energy techniques, particularly in imaging vertical and inclined defects [55]. Similar challenges are also encountered in ultrasonic phased array testing. Tian and YK established a mathematical model combining background subtraction with the square difference algorithm based on linear acoustic theory to extract the echo features of near-surface defects. Simulation and experimental results show that this model can effectively extract the echo features of near-surface defects, achieving positioning and quantitative accuracy values of 0.2 mm and 0.3 mm, respectively, for near-surface defects [56].

### 5.2. *Guided Wave Inspection Technique*

Guided wave ultrasonic testing (GWUT) is a non-destructive testing method that utilizes the propagation characteristics of ultrasonic waves in solids to detect internal structures, defects, or variations in materials [57]. Standard testing methods include "pulse-echo" and "pitch–catch" modes. Unlike bulk ultrasonic waves propagating in the direction of wall thickness, guided waves propagate axially along the pipeline, as illustrated in Figure 11. With only a single or a few probes, it is possible to effectively conduct long-distance screening of cross-sectional or axial damage in critical areas of the pipeline [58,59]. Guided wave testing, emerging as an external inspection method for non-piggable pipeline segments, began gaining prominence in the 1970s [60,61]. This technique allows for screening pipeline defects over tens of meters by merely removing a portion of the coating layer [62].

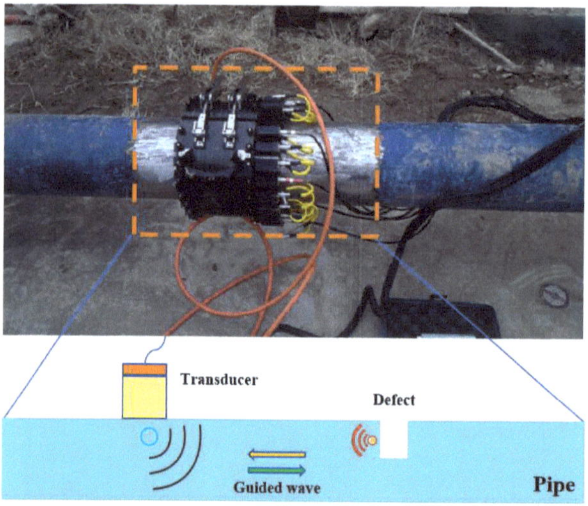


**Figure 11.** Principle of Guided Waves.

#### 5.2.1. Potential-Based Pipeline External Inspection Methods

Furthermore, there is another category of pipeline defect inspection methods based on changes in electrical potential. As Table 8 illustrates, these methods generally require manual patrolling, heavily relying on the operator's subjective judgment. They are highly susceptible to environmental disturbances and incapable of ascertaining the state of coating delamination. Guided wave testing can detect coating delamination through changes in the dispersion characteristics of torsional waves caused by local coating [63]. However, its application in buried pipelines is severely constrained by the need for localized excavation and coating removal, limiting the widespread adoption of this technology [64].

**Table 8.** Pipeline External Inspection Methods Without Excavation.

| Method | Coating Delamination | Size | Location | Severity | External Interference | Cathodic Protection Evaluation | Operator | Pipeline Inspection | Current Source |
|---|---|---|---|---|---|---|---|---|---|
| Pipeline-to-Soil Potential Survey (P/S) | × | × | × | × | ✓ | ✓ | ✓ | ✓ | GP/G |
| Close-Interval Potential Survey (CIPS) | × | × | × | ✓ | ✓ | ✓ | ✓ | ✓ | GP/G |
| Direct Current Voltage Gradient (DCVG) | × | ✓ | ✓ | ✓ | × | × | ✓ | ✓ | D |
| Pearson | × | ✓ | ✓ | × | ✓ | ✓ | ✓ | ✓ | A |
| Pipe Current Attenuation Method | × | × | ✓ | × | ✓ | ✓ | ✓ | ✓ | A |
| Frequency Domain Reflectometry (FDR) | × | × | × | × | ✓ | ✓ | ✓ | ✓ | A |

Cathodic Protection System—GP. Direct Current—D. Alternating Current—A. Ground—G.

#### 5.2.2. Non-Contact Ultrasonic-Guided Wave Testing

The advent of novel non-contact ultrasonic testing technologies has expanded the possibilities of using guided waves for internal pipeline inspection. Non-contact ultrasonics can be categorized into electromagnetic ultrasonics, laser ultrasonics, and air-coupled ultrasonics. Their principles, advantages, and limitations are shown in Table 9.

**Table 9.** Comparison of Non-Contact Ultrasound-Guided Wave Techniques.

| Technology Type | Principle of Detection | Advantages | Limitations |
|---|---|---|---|
| Laser Ultrasonic-Guided Wave Testing | Utilizes laser pulses to induce stress pulses in the test piece through thermoelastic or ablation effects, generating ultrasonic wave signals. | Non-contact detection suitable for high-temperature, high-pressure, toxic environments; high sensitivity allows for inspection on complex structures. | Efficiency of laser conversion to ultrasonic signals may be low; Sigh-power lasers may damage the surface of the specimen; sensitivity of detection may be suboptimal. |
| Electromagnetic Ultrasonic-Guided Wave Testing | Employs a probe to emit ultrasonic-guided waves, using the time difference of reflections from the inner and outer walls of the pipeline to determine wall thickness and corrosion. | Capable of long-distance detection, convenient operation, and minimally affected by external factors, such as temperature, pressure, and internal flow media. | Direct measurement of wall thickness is not possible; sensitive to defects in wall depth and circumferential width, only axial length of defects can be measured within a certain range. |
| Air Coupled Ultrasonic-Guided Wave Testing | Uses air as the coupling medium, transmitting and receiving ultrasonic waves through air-coupled transducers to detect material defects. | Non-contact detection without the need for a coupling agent, suitable for high-temperature or inaccessible environments; capable of inspecting very thin workpieces. | Due to the attenuation of ultrasound by air and impedance differences at the air–solid interface, there is significant reflection and low conversion efficiency of ultrasonic waves, resulting in a potentially low SNR. |

(a) Laser Ultrasonic-Guided Waves Testing

Owing to high maintenance costs and factors like surface roughness, speed, and environmental vibrations, LUGWT is predominantly used in precision industries, such as aerospace for defect detection, and remains largely experimental in pipeline applications [65]. For instance, JH et al. [55] utilized laser-generated guided waves to assess pipeline corrosion and successfully evaluated the location and size of defects in two-dimensional scan images.

(b) Electromagnetic Ultrasonic-Guided Waves Testing

In industrial in-line inspections, the applications of non-contact ultrasonic-guided waves are presently limited to EmatScan CD and ROCDeMAT-c developed by ROSEN and P II, as well as NDT Global's ARTscan intelligent PIG as shown in Table 10. Electromagnetic ultrasonic guided waves operate through a process where alternating electric currents induce high-frequency eddy currents on the surface of a pipeline. When subjected to an external magnetic field, these eddy currents stimulate the generation of guided waves [66,67].

**Table 10.** Types of Internal Inspectors and Inspection Capabilities.

| Corporation | Product | Crack (Length × Depth) | Operation Speed | Wall Thickness (mm) | Depth Sizing | Length Sizing | Medium Type | Orientation to Pipe Axis | Min. Bend Radius |
|---|---|---|---|---|---|---|---|---|---|
| P II | UltraScan CD EDGE | 15 × 1 | ≤2.5 m/s | 5~10 | ±0.7 | ±7.5 | liquid | 0° | 1.5 D |
| | UltraScan CDP/DUO | 25 × 1 | ≤5 m/s | 5~13 | ±0.7 | ±7.5 | liquid | 0° | 1.5 D |
| | EmatScan CD | 50 × 2 | ≤2.5 m/s | 7~13 | ±1.1 | ±10 | G/L | 0° | 1.5 D |
| ROSEN | ROCDeMAT-c | 40 × 2 | ≤2.5 m/s | 0~20 mm | ±0.15 t | ±20 | G/L | ±18° | 1.5 D |
| TDW | SpirALL | 30 × 2 | ≤2.5 m/s | 0~13 mm | ±1 | ±10 | G/L | ±10° | 1.5 D |
| CNPC | / | 50 × 2 | ≤2.5 m/s | 7~13 | ±1.1 | ±10 | liquid | 0° | 1.5 D |
| NDT Global | PROTON | 20 × 1 | ≤1.4 m/s | 7~13 | ±1 | ±10 | liquid | ±10° | 3 D |
| | EVO Eclipse UCx | 20 × 1 | ≤4 m/s | 0~13 mm | ±1 | ±10 | liquid | ±10° | 1.5 D |

To deepen understanding of these technologies, scholars worldwide have conducted extensive research [68–70]. Nurmalia proposed an EMAT pipeline inspection technology based on high-order torsional-guided waves T(0,2), finding that phase measurement as a quantitative detection method holds considerable potential [71]. Liu introduced a novel flexible EMAT transducer for generating L(0,2)-guided waves in pipelines, demonstrating its effectiveness in corrosion detection in this mode [72]. Masahiko also developed an EMAT technique to detect corrosion defects on steel pipes' external surface, basing axial defects' location and depth assessment on the amplitude and phase shift responses of the round-trip signals in SH0 and SH1 modes. The results indicated that the SH1 mode is more sensitive to defects than the SH0 mode [73].

**Limitations:** In summary, non-contact ultrasonic-guided wave testing primarily utilizes Lamb and surface waves. The circumferential propagation characteristic of Lamb waves makes this technology particularly effective in detecting stress corrosion cracks in natural gas pipelines and is sensitive to crack defects at straight weld seams. However, it faces challenges in detecting minute cracks at girth welds [74,75]. Due to the multimodal nature, dispersion, and long-distance attenuation properties of ultrasonic-guided waves, the testing typically involves using broadband low-frequency modulated pulses as the excitation signal. This approach, while effective, also leads to limitations in terms of defect resolution and the accuracy of defect localization in ultrasonic-guided wave testing.

### 5.3. Air-Coupled Ultrasound

Air-coupled ultrasonics represents a novel non-contact ultrasonic testing technology, utilizing air as the coupling medium. Air-coupled ultrasonics boasts distinct features such as non-contact operation, large stand-off distance, and low power consumption. This technology is suitable for ferromagnetic materials and applies to natural gas pipelines and medium- to low-pressure pipelines where traditional ultrasonics and magnetic flux leakage methods are ineffective. Thus, it offers a broader range of applicability [76–78].

Air-coupled ultrasonics, compared to contact or immersion ultrasonics, experience a significant reduction in sensitivity, approximately 80 dB lower [79,80]. This energy attenuation primarily arises from three factors: ultrasonic wave attenuation in air, substantial reflection at the air–solid interface, and the efficiency of ultrasonic transducer conversion. The inherent acoustic attenuation in air and surface reflection remains an unavoidable natural phenomenon in ultrasonics, leading to low transducer efficiency and prolonged pulse reverberation due to the significant impedance mismatch.

To enhance signal quality, current research in the industry is focused on two main areas:

- Optimization of air-coupled ultrasonic probe structures and materials.

Air-coupled probes can be categorized into piezoelectric and capacitive types. Capacitive transducers operate by applying an excitation voltage between a metallized film and a conductive substrate, causing the film to vibrate under electrostatic action, thereby generating ultrasonics at specific frequencies. D.W. Schindel and others, through comparative studies, found that capacitive ultrasonic probes offer a broad frequency response and good damping, effectively addressing the issue of high central frequency present during ultrasonic wave excitation in piezoelectric ultrasonics [81]. However, due to their high cost and strong environmental dependency, capacitive probes remain largely experimental and have not yet seen widespread industrial application [82]. In 1995, Hutchins et al. achieved the fabrication of regular air layers on the conductive substrate of electrostatic transducers through etching, demonstrating these probes' excellent bandwidth performance [83].

Advances in piezoelectric ultrasonic transducers have also been achieved, especially in their structural design. To improve resonance effects, one to three connected composite sensors have been developed, reducing the impedance of sensor materials and enhancing efficiency and coupling performance [84]. Selecting superior matching layer materials to allow more energy transmission through the air into the test object is a key factor in realizing these improvements. Alvarez-Arenas and others at the Spanish CSIC Institute of Acoustics, after researching various material characteristics, identified nylon as ideal materials (with mixed cellulose ester and polyvinylidene fluoride being suitable for frequencies above 2 MHz). This research largely resolved the issue of selecting matching materials and was the first to study the variation of material attenuation coefficients with frequency [65].

Piezoelectric probes, owing to their higher power output, have long dominated the commercial ultrasonic probe market. Systems developed by companies like Probe Corporation in Japan, and Ultran, PAC, and SONOTEC in Germany, which are based on piezoelectric transducers, are examples. Research in areas such as focused air-coupled transducers and defect detection in composite materials and lithium-ion batteries has been conducted by institutions like Beihang University, Nanjing University, and the Chinese Academy of Sciences' Institute of Acoustics. However, there remains a lack of development in equipment specifically for internal inspection of oil and gas pipelines.

- Signal encoding

Hutchins et al. [85] utilized pulse compression technology to encode excitation signals, using capacitive sensors to generate broadband chirp signals in air for measuring solid samples. Their results demonstrated that this signal processing technique significantly enhances the signal-to-noise ratio (SNR) and detection precision of air-coupled testing, validating the feasibility of pulse compression technology in improving air-coupled ultrasonic performance. The team also conducted a comparative study of existing signal encoding techniques, concluding that the optimal choice of modulated signals depends on the available bandwidth and type of measurement [86]. Garcia and colleagues proposed using Golay sequences to encode Lamb waves excited by air-coupled ultrasonics, finding that the SNR of ultrasonic signals under Golay encoding improved by 21 dB compared to non-encoded excitation [87]. Additionally, Tang et al. introduced phase-encoded excitation and pulse compression techniques, effectively raising the SNR of received signals by over 10 dB [88]. Li and colleagues suggested using P4 polyphase sequences to encode excitation in air-coupled piezoelectric transducer-based non-destructive testing systems. Their mixed signal processing method increased the SNR by 12.11 dB and improved the time-domain resolution by approximately 35% [89].

Despite these advancements in probe design and signal encoding, the effectiveness of air-coupled ultrasonics in metal detection remains limited, particularly with high-speed operation and vibration impacts of internal detectors. Air-coupled ultrasonics was first proven in 1973 for generating Lamb waves in metal plates. Since then, Lamb waves have been extensively used in various materials, including fiber-reinforced polymer composites [90,91]. In the industrial sector, NDT GLOBAL is currently the only company applying air-coupled ultrasonic testing in the internal inspection of metal pipelines. Their developed ARTscan system (typically 400 kHz–1.2 MHz, with pressure above 7 MPa) merges low-frequency broadband signals with resonant and guided wave technologies. This not only allows for highly precise measurements of metal loss in pipelines ($\pm$0.4 mm) and the detection and classification of thick-walled defects but also facilitates comprehensive geometric measurements. This advancement represents a significant step in applying air-coupled ultrasonics in metal pipeline inspection. Compared to immersion ultrasonic, which requires coupling agents and high cleanliness, ART has a 50% deformation pass-through capability (compared to 10% for magnetic flux leakage) and 1.5D high pass-through ability, allowing for pipeline inspection without pigging. It means that when inspecting these mainline oil and gas pipelines, oil and gas suppliers can avoid reducing the internal pressure and flow velocity of the pipelines, thereby avoiding a significant decrease in the transportation volume of oil and gas during pipeline internal inspection operations. Taking Central Asia–China and China–Myanmar natural gas pipelines as examples, the direct losses caused by the reduction in throughput during a single pipeline internal inspection operation can amount to hundreds of millions of dollars.

### *5.4. Support Vector Machine*

As the technology of ultrasonic internal inspection for pipelines continues to advance, the influx of a large amount of data imposes higher demands on signal processing methods. Signal processing of ultrasonic internal inspection signals can help reduce signal complexity and extract useful information [92,93]. Some common ultrasonic signal processing methods and data process flow are shown in Figure 12.

The support vector machine (SVM) is a supervised learning model used for classification and regression analysis. It has shown excellent performance in data mining and pattern recognition tasks in inspecting oil and gas pipelines. SVM can accurately handle linear problems; for complex nonlinear issues, data can be mapped to a high-dimensional space through nonlinear transformation functions [94–96]. To reduce computational costs, kernel functions can be used instead of nonlinear transformation functions [97]. Like SVM, other machine learning (ML) methods, such as decision trees, random forests, and naive Bayes have been used. These ML methods can extract useful features from acoustic signals and perform accurate classification and defect detection based on these features [98–100]. However, the class of traditional machine learning algorithms represented by support vector machines may not perform as well as deep learning when dealing with complex or large data models. They require manual feature engineering and may also rely on domain-specific expert knowledge. They are generally less suitable for handling high-dimensional data and nonlinear relationships. They are more suitable for small samples and nonlinear tasks but less suitable for large-scale datasets and multi-classification tasks [101–103].

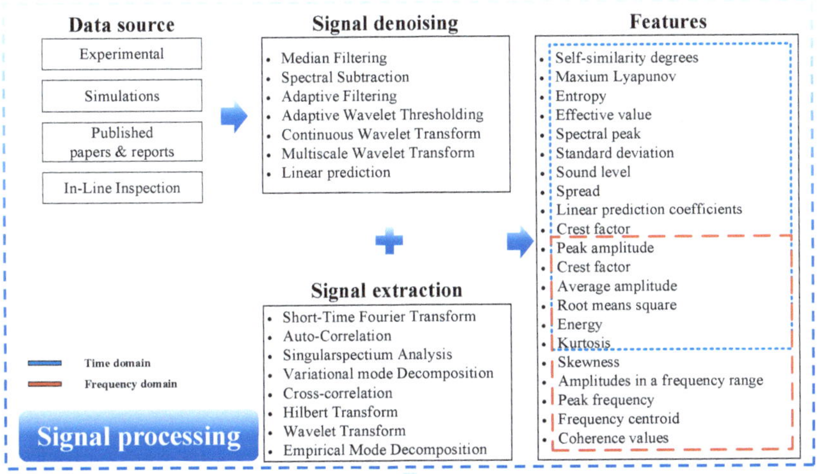


**Model training**

| Method | Description | Applications |
|---|---|---|
| **Traditonal ML** | | |
| Support Vector Macnines | Uses hyperplanes to separate data nto classes | Anomaly detection<br>Classifying defect types |
| Decision Trees | Hierarchical structure for classification | Identifying defect categories<br>Feature selection |
| Random Forest | Ensemble of decision trees for robustness | Improving accuracy<br>Handling noisy data |
| Naive Bayes | Probabilistic model based on Bayes theorem | Probability based on anomaly detection |
| **DeepLearning** | | |
| Convolutional NNs (CNN) | Specialized for image data with convolution | Defect localization<br>Object detection |
| Recurrent NNS (RNN) | Suitable for sequence data and time series | Anomaly detection<br>Pipeline health monitoring |
| LongShort-Term Memory(LSTM) | Type of RNN designed to mitigate vanishingand exploding gradien issues | Capturing temporal dependencies<br>Sequences of defect occurrence |
| Autoencoders | Learn efficient data representatons | Dimensionality reduction<br>Anomaly detection |
| Generative Adversaria Networks(GANs) | Generate data similar to training samples using a generator discriminator setup | Data augmentation<br>Synthetic defect generation |
| Transformer | Process sequential data using attention | Capturing dependencies in sequences<br>Anomaly detection |

**Figure 12.** Ultrasonic Signal Analysis Workflow.

Deep learning methods have gained significant attention with the increasing volume of data and the growing complexity of ultrasonic signals. Common deep learning networks include convolutional neural networks (CNN) [104], graph neural networks (GNNs) [98], recurrent neural networks (RNN) [105], long short-term memory (LSTM) [106], and AutoEncoder [107]. It is evident that data-driven techniques dominated by machine learning (ML) methods have demonstrated significant advantages in ultrasonic in-line inspection compared to physical models [108]. Their primary applications fall into the following three categories:

- Defect Classification:

Using deep learning algorithms to classify defects in the pipe body and weld area has been a prominent research topic in the ultrasonic inspection of oil and gas pipelines. Traditional defect classification methods struggle to differentiate between different types of defects accurately. Deep learning methods like CNN and LSTM often excel in handling complex signals, extracting temporal features, and improving detection accuracy. For example, Bettayeb et al. used wavelet transformation to extract feature vectors related to defects on the pipe body and weld, such as cracks, porosity, or inclusions. These feature vectors contain two-dimensional information about defects, and an artificial neural network (ANN) trained with a backpropagation algorithm was used to classify these feature vectors. The results showed that combining wavelet transformation and ANN could significantly suppress noise levels and improve defect classification accuracy [109]. Building on this, Sambath et al. analyzed 1084 samples using similar principles and achieved a reasonable classification rate of 94% [110]. Guo et al. proposed an image-based deep learning defect classification method. They used a gated recurrent unit fully convolutional network (GRU-FCN) to extract temporal features from A-scan ultrasound signals. The training, validation, and test datasets comprised 3600 ultrasound waveforms collected in experiments. The results were compared with LSTM, GRU, and ResNet, revealing that GRU-FCN achieved higher accuracy [111].

- Defect Characterization:

Defect characterization has been a critical basis for transitioning from non-destructive testing (NDT) to non-destructive evaluation (NDE). The accuracy of quantifying defects directly influences the effectiveness of non-destructive evaluation. C. Guo et al. proposed a novel residual vision transformer (Res-ViT) architecture based on deep residual networks (ResNet) and visual transformers (ViT). They conducted experiments on elliptical defects with inclination angles of 60° at different noise levels. Compared to the principal component analysis and nearest neighbor method, the root mean square error (RMSE) of the defect size was reduced by 61% [112]. Miorelli et al. introduced a CNN model for automatically locating and sizing defects from guided ultrasonic wave data. The deep learning model was trained using both simulated and experimental data, and the experiments demonstrated the model's adaptability to real-world environments, achieving an accuracy of up to 90% [113]. Bai et al. compared the classical Bayesian inversion method proposed by Miorelli and a CNN regression model. In this study, the classical Bayesian method exhibited higher accuracy and lower uncertainty in defect characterization, but it introduced more discreteness due to the model's inherent uncertainty [114].

- Data Preprocessing:

In practical experiments or industrial applications, data often come with noise due to environmental factors or equipment characteristics, which can affect subsequent signal analysis. Data preprocessing aims to improve the quality of ultrasonic detection data. Data preprocessing techniques include but are not limited to image denoising, feature recognition and extraction, and data compression. Noise reduction primarily aims to improve the signal-to-noise ratio (SNR). Yang et al. designed a lightweight denoising network called the global interactive attention lightweight denoising network (GIALDN) for analyzing vibration signals and locating internal defects in CFRP laminates. In GIALDN, a threshold-based denoising method was used to eliminate noise-related features and enhance feature discriminability. The results showed that GIALDN achieved a location accuracy of 98.68%, which was more than 15% higher than VGGnet11 and FaultNet, and outperformed LSTM, RNN, Rsenet18, SEresnet18, and Densenet121 [115]. For ultrasonic defect detection, which typically involves a larger volume of data compared to conventional eddy current testing, compressing input data into latent features can replace the indiscriminate retention of raw data. Kesharaju proposed a feature selection method based on the genetic algorithm (GA) and fully convolutional neural network (FCNN). This method used a subset of preselected features as input to the FCNN model. The results showed that the performance improved by 94% compared to the model based on principal component analysis (PCA) [116].

**Limitations:** Currently, most machine learning methods are primarily in the theoretical validation stage, with only a few models being applied in industrial practices, and their performance is not ideal. The main challenge lies in the difficulty of obtaining a sufficient amount of high-quality data due to the confidentiality and sensitivity of oil and gas pipeline data. Therefore, most research based on machine learning methods is conducted using simulated data or laboratory data for defect or anomaly detection analysis, making it challenging to adapt to real-world pipeline inspection data.

## 6. Potential Challenges and Opportunities

Considering the current testing technology, it is evident that we are still in the early stages of ultrasonic non-destructive evaluation. Therefore, in the foreseeable future, internal ultrasonic inspection of oil and gas pipelines will continue to face the following challenges and opportunities:

(1) Exploration of novel sensor designs and materials for enhanced sensitivity and resolution in UT inspections.
(2) Research into multi-modal sensor arrays combining UT with other NDT techniques (e.g., electromagnetic acoustic transducers or distributed fiber optic sensors) for complementary defect characterization.
(3) Miniaturization of sensors for improved accessibility to challenging pipeline geometries and locations.
(4) Addressing the impact of velocity and vibration on the accuracy and precision of detection from both theoretical and sensor optimization perspectives to achieve ultrasonic inspection at the velocity of the conveying medium.
(5) Development of novel encoding algorithms to mitigate artifacts and enhance the sensitivity of ultrasound imaging.
(6) Air-coupled testing holds promising applications in the inspection of oil and gas pipelines because of its ability to detect cracks and metal loss defects in thick-walled pipelines that magnetic flux leakage testing may not detect.
(7) Integration of machine learning algorithms for optimized encoding parameter selection, real-time adaptive imaging and corrosion quantification [117].
(8) Exploration of novel materials with tailored acoustic properties for matching layers to optimize acoustic impedance matching and minimize signal loss at transducer interfaces.
(9) The enhancement of generalization and transfer learning capabilities of ultrasonic data analysis models established based on simulation and laboratory in industrial application environments.

## 7. Conclusions

In this study, we conducted a bibliometric analysis based on 350 ultrasonic testing-related publications from the Web of Science (WOS) database since 1992. Utilizing data visualization techniques, we identified the most influential countries, institutions, and publications in the field globally. Our analysis of the data level elucidates the potential developmental reasons behind these trends. We observed that Western developed countries, represented primarily by the UK and the USA, continue to maintain a central position in the realm of ultrasonic testing. Due to industrial development needs, China has shown rapid progress in ultrasonics in recent years, and the focal point of ultrasonic testing is progressively shifting from theoretical aspects to practical scene applications.

To mitigate the impact of personal biases on this research, we relied on cluster analysis and timeline methods to pinpoint the current research hotspots in ultrasonic testing for oil and gas pipelines. This approach enabled us to delineate the research tasks in various stages of oil and gas pipeline ultrasonic testing. Building on the hotspot analysis, we also presented the ongoing opportunities and challenges in this field.

Like ultrasonic testing technology itself, which possesses unique advantages and limitations, our study is not without its constraints. For instance, due to the limitations of the database, not all relevant literature in the field may be covered. However, the data sourced from WOS ensures a comprehensive collection and analysis of core research findings in the ultrasonic testing domain.

**Author Contributions:** For this research articles, J.H. is in charge of methodology, writing and conceptualization. P.C. is in charge of review, supervision, and funding acquisition and formal analysis. R.L. is in charge of data curation, supervision and field inspection. K.F., J.D., Y.W. and Z.L.is in charge of data curation, supervision and verification. All authors have read and agreed to the published version of the manuscript.

**Funding:** This research was funded by National Key R&D Program of China, grant number 2022YFB3207600; National Key R&D Program of China, grant number 2022YFC3070100; the Science and Technology Research Project of PipeChina, grant number WZXGL202104; the Science and Technology Research Project of PipeChina, grant number AQWH202203.

**Conflicts of Interest:** Author P.C., R.L., K.F., J.D., Y.W. was employed by General Research Institute, China Oil & Gas Pipeline Network Corporation. The remaining authors declare that the research was conducted in the absence of any commercial or financial relationships that could be construed as a potential conflict of interest.

## References

1. Chen, C.; Li, C.; Reniers, G.; Yang, F. Safety and security of oil and gas pipeline transportation: A systematic analysis of research trends and future needs using WoS. *J. Clean. Prod.* **2021**, *279*, 123583. [CrossRef]
2. Bin Ali, K.; Abdalla, A.N.; Rifai, D.; Faraj, M.A. Review on system development in eddy current testing and technique for defect classification and characterization. *IET Circuits Devices Syst.* **2017**, *11*, 338–351. [CrossRef]
3. Nestleroth, J.B. Pipeline in-line inspection challenges to NDT. *Insight—Non-Destr. Test. Cond. Monit.* **2006**, *48*, 524. [CrossRef]
4. Le, D.V.-K.; Chen, Z.; Rajkumar, R. Multi-sensors in-line inspection robot for pipe flaws detection. *IET Sci. Meas. Technol.* **2020**, *14*, 71–82. [CrossRef]
5. Khan, A.; Qurashi, A.; Badeghaish, W.; Noui-Mehidi, M.N.; Aziz, M.A. Frontiers and Challenges in Electrochemical Corrosion Monitoring; Surface and Downhole Applications. *Sensors* **2020**, *20*, 6583. [CrossRef] [PubMed]
6. Reports › EGIG. Available online: https://www.egig.eu/reports (accessed on 22 January 2024).
7. PHMSA. Data and Statistics Overview. Available online: https://www.phmsa.dot.gov/data-and-statistics/pipeline/data-and-statistics-overview (accessed on 23 January 2024).
8. Xie, F.; Wu, X.; Zhang, H.; Zhang, X. Research on pulsed eddy current automatic detection technology for weld defects of pressure vessel cylinder. *Measurement* **2021**, *176*, 109081. [CrossRef]
9. Quarini, J.; Shire, S. A Review of Fluid-Driven Pipeline Pigs and their Applications. *Proc. Inst. Mech. Eng. Part E J. Process Mech. Eng.* **2007**, *221*, 1–10. [CrossRef]
10. Yu, G.; Liu, J.; Zhang, H.; Liu, C. An Iterative Stacking Method for Pipeline Defect Inversion with Complex MFL Signals. *IEEE Trans. Instrum. Meas.* **2020**, *69*, 3780–3788. [CrossRef]
11. Feng, Q.; Li, R.; Nie, B.; Liu, S.; Zhao, L.; Zhang, H. Literature Review: Theory and Application of In-Line Inspection Technologies for Oil and Gas Pipeline Girth Weld Defection. *Sensors* **2016**, *17*, 50. [CrossRef]
12. Felice, M.V.; Fan, Z. Sizing of flaws using ultrasonic bulk wave testing: A review. *Ultrasonics* **2018**, *88*, 26–42. [CrossRef]
13. Parlak, B.; Yavasoglu, H. A Comprehensive Analysis of In-Line Inspection Tools and Technologies for Steel Oil and Gas Pipelines. *Sustainability* **2023**, *15*, 2783. [CrossRef]
14. Ma, Q.; Tian, G.; Zeng, Y.; Li, R.; Song, H.; Wang, Z.; Gao, B.; Zeng, K. Pipeline In-Line Inspection Method, Instrumentation and Data Management. *Sensors* **2021**, *21*, 3862. [CrossRef] [PubMed]
15. Zang, X.; Xu, Z.-D.; Lu, H.; Zhu, C.; Zhang, Z. Ultrasonic guided wave techniques and applications in pipeline defect detection: A review. *Int. J. Press. Vessels Pip.* **2023**, *206*, 105033. [CrossRef]
16. Cantero-Chinchilla, S.; Wilcox, P.D.; Croxford, A.J. Deep learning in automated ultrasonic NDE—Developments, axioms and opportunities. *NDT E Int.* **2022**, *131*, 102703. [CrossRef]
17. Yang, Z.; Yang, H.; Tian, T.; Deng, D.; Hu, M.; Ma, J.; Gao, D.; Zhang, J.; Ma, S.; Yang, L.; et al. A review on guided-ultrasonic-wave-based structural health monitoring: From fundamental theory to machine learning techniques. *Ultrasonics* **2023**, *133*, 107014. [CrossRef]
18. Fan, H.; Tariq, S.; Zayed, T. Acoustic leak detection approaches for water pipelines. *Autom. Constr.* **2022**, *138*, 104226. [CrossRef]
19. Yuan, J.; Mao, W.; Hu, C.; Zheng, J.; Zheng, D.; Yang, Y. Leak detection and localization techniques in oil and gas pipeline: A bibliometric and systematic review. *Eng. Fail. Anal.* **2023**, *146*, 107060. [CrossRef]

20. Xue, Z.; Xu, Y.; Hu, M.; Li, S. Systematic review: Ultrasonic technology for detecting rail defects. *Constr. Build. Mater.* **2023**, *368*, 130409. [CrossRef]
21. Huang, S.; Peng, L.; Sun, H.; Li, S. Deep Learning for Magnetic Flux Leakage Detection and Evaluation of Oil & Gas Pipelines: A Review. *Energies* **2023**, *16*, 1372. [CrossRef]
22. Xie, S.; Duan, Z.; Li, J.; Tong, Z.; Tian, M.; Chen, Z. A novel magnetic force transmission eddy current array probe and its application for nondestructive testing of defects in pipeline structures. *Sens. Actuators Phys.* **2020**, *309*, 112030. [CrossRef]
23. She, S.; Chen, Y.; He, Y.; Zhou, Z.; Zou, X. Optimal design of remote field eddy current testing probe for ferromagnetic pipeline inspection. *Measurement* **2021**, *168*, 108306. [CrossRef]
24. Lei, H.; Huang, Z.; Liang, W.; Mao, Y.; Que, P.W. Ultrasonic pig for submarine oil pipeline corrosion inspection. *Russ. J. Nondestruct. Test.* **2009**, *45*, 285–291. [CrossRef]
25. Velázquez, J.C.; Hernández-Sánchez, E.; Terán, G.; Capula-Colindres, S.; Diaz-Cruz, M.; Cervantes-Tobón, A. Probabilistic and Statistical Techniques to Study the Impact of Localized Corrosion Defects in Oil and Gas Pipelines: A Review. *Metals* **2022**, *12*, 576. [CrossRef]
26. Dai, L.S.; Feng, Q.S.; Sutherland, J.; Wang, T.; Sha, S.Y.; Wang, F.X.; Wang, D.P. Application of MFL on Girth-Weld Defect Detection of Oil and Gas Pipelines. *J. Pipeline Syst. Eng. Pract.* **2020**, *11*, 04020047. [CrossRef]
27. Peng, X.; Anyaoha, U.; Liu, Z.; Tsukada, K. Analysis of Magnetic-Flux Leakage (MFL) Data for Pipeline Corrosion Assessment. *IEEE Trans. Magn.* **2020**, *56*, 6200315. [CrossRef]
28. Ege, Y.; Coramik, M. A new measurement system using magnetic flux leakage method in pipeline inspection. *Measurement* **2018**, *123*, 163–174. [CrossRef]
29. Shen, Y.; Zhou, W. Development of a convolutional neural network model to predict the size and location of corrosion defects on pipelines based on magnetic flux leakage signals. *Int. J. Press. Vessels Pip.* **2024**, *207*, 105123. [CrossRef]
30. Wu, D.; Liu, Z.; Wang, X.; Su, L. Composite magnetic flux leakage detection method for pipelines using alternating magnetic field excitation. *NDT E Int.* **2017**, *91*, 148–155. [CrossRef]
31. Liu, B.; Liang, Y.; He, L.; Lian, Z.; Geng, H.; Yang, L. Quantitative study on the propagation characteristics of MFL signals of outer surface defects in long-distance oil and gas pipelines. *NDT E Int.* **2023**, *137*, 102861. [CrossRef]
32. Wang, Y.; Fan, M.; Cao, B.; Ye, B.; Wen, D. Measurement of coating thickness using lift-off point of intersection features from pulsed eddy current signals. *NDT E Int.* **2020**, *116*, 102333. [CrossRef]
33. Piao, G.; Guo, J.; Hu, T.; Deng, Y.; Leung, H. A novel pulsed eddy current method for high-speed pipeline inline inspection. *Sens. Actuators Phys.* **2019**, *295*, 244–258. [CrossRef]
34. Pourazizi, R.; Mohtadi-Bonab, M.A.; Szpunar, J.A. Investigation of different failure modes in oil and natural gas pipeline steels. *Eng. Fail. Anal.* **2020**, *109*, 104400. [CrossRef]
35. Xin, J.; Chen, J.; Li, C.; Lu, R.; Li, X.; Wang, C.; Zhu, H.; He, R. Deformation characterization of oil and gas pipeline by ACM technique based on SSA-BP neural network model. *Measurement* **2022**, *189*, 110654. [CrossRef]
36. Chen, P.; Li, R.; Fu, K.; Zhong, Z.; Xie, J.; Wang, J.; Zhu, J. A cascaded deep learning approach for detecting pipeline defects via pretrained YOLOv5 and ViT models based on MFL data. *Mech. Syst. Signal Process.* **2024**, *206*, 110919. [CrossRef]
37. Pan, J.; Gao, L. A novel method for defects marking and classifying in MFL inspection of pipeline. *Int. J. Press. Vessels Pip.* **2023**, *202*, 104892. [CrossRef]
38. Uhlig, S.; Alkhasli, I.; Schubert, F.; Tschöpe, C.; Wolff, M. A review of synthetic and augmented training data for machine learning in ultrasonic non-destructive evaluation. *Ultrasonics* **2023**, *134*, 107041. [CrossRef]
39. Cinquin, M.; Castaings, M.; Hosten, B.; Brassier, P.; Pérès, P. Monitoring of the moisture content in carbon-epoxy plates using Lamb waves. *NDT E Int.* **2005**, *38*, 37–44. [CrossRef]
40. Yu, Y.; Safari, A.; Niu, X.; Drinkwater, B.; Horoshenkov, K.V. Acoustic and ultrasonic techniques for defect detection and condition monitoring in water and sewerage pipes: A review. *Appl. Acoust.* **2021**, *183*, 108282. [CrossRef]
41. Chen, P.; Li, R.; Fu, K.; Zhao, X. Research and Method for In-line Inspection Technology of Girth Weld in Long-Distance Oil and Gas Pipeline. *J. Phys. Conf. Ser.* **2021**, *1986*, 012052. [CrossRef]
42. Drinkwater, B.W.; Wilcox, P.D. Ultrasonic arrays for non-destructive evaluation: A review. *NDT E Int.* **2006**, *39*, 525–541. [CrossRef]
43. Gang, T.; Sheng, Z.Y.; Tian, W.L. Time resolution improvement of ultrasonic TOFD testing by pulse compression technique. *Insight* **2012**, *54*, 193–197. [CrossRef]
44. Cunningham, L.J.; Mulholland, A.J.; Tant, K.M.M.; Gachagan, A.; Harvey, G.; Bird, C. The detection of flaws in austenitic welds using the decomposition of the time-reversal operator. *Proc. R. Soc.-Math. Phys. Eng. Sci.* **2016**, *472*, 20150500. [CrossRef]
45. Zhang, J.; Yu, P.; Gang, T. Measurement of the ultrasonic scattering matrices of near-surface defects using ultrasonic arrays. *Nondestruct. Test. Eval.* **2016**, *31*, 303–318. [CrossRef]
46. Shakibi, B.; Honarvar, F.; Moles, M.D.C.; Caldwell, J.; Sinclair, A.N. Resolution enhancement of ultrasonic defect signals for crack sizing. *NDT E Int.* **2012**, *52*, 37–50. [CrossRef]
47. Beller, M.; Uzelac, N.I.; Barbian, A. Combined Quantitative In-Line Inspection of Pipelines for Metal Loss and Cracks. In Proceedings of the 2006 International Pipeline Conference, Calgary, AB, Canada, 25–29 September 2006; Volume 2: Integrity Management; Poster Session; Student Paper Competition. pp. 857–864.

48. Willems, H.; Kopp, G.; Meinzer, T. Recent advancements in ultrasonic in-line inspection. In Proceedings of the 19th Pipeline Technology Conference, Berlin, Germany, 23–25 May 2016.
49. EVO Platform. Available online: https://www.ndt-global.com/platforms/evo-platform/ (accessed on 22 January 2024).
50. Caleyo, F.; Alfonso, L.; Espina-Hernández, J.H.; Hallen, J.M. Criteria for performance assessment and calibration of in-line inspections of oil and gas pipelines. *Meas. Sci. Technol.* **2007**, *18*, 1787–1799. [CrossRef]
51. Chang, J.; Chu, Z.; Gao, X.; Soldatov, A.I.; Dong, S. A magnetoelectric-ultrasonic multimodal system for synchronous NDE of surface and internal defects in metal. *Mech. Syst. Signal Process.* **2023**, *183*, 109667. [CrossRef]
52. Yuan, X.; Zhao, J.; Li, W.; Yin, X.; Li, X.; Zhao, J.; Fan, M. Novel phase reversal feature for inspection of cracks using multi-frequency alternating current field measurement technique. *Mech. Syst. Signal Process.* **2023**, *186*, 109857. [CrossRef]
53. Duan, Z.; Xie, S.; Huang, L.; Zhao, R.; Tian, M.; Liu, T.; Pei, C.; Chen, Z. Quantitative sizing of compound location defects based on PECT-EMAT hybrid testing methods. *Mech. Syst. Signal Process.* **2022**, *178*, 109267. [CrossRef]
54. Ha, J.M.; Seung, H.M.; Choi, W. Autoencoder-based detection of near-surface defects in ultrasonic testing. *Ultrasonics* **2022**, *119*, 106637. [CrossRef]
55. He, J.; Liu, X.; Cheng, Q.; Yang, S.; Li, M. Quantitative detection of surface defect using laser-generated Rayleigh wave with broadband local wavenumber estimation. *Ultrasonics* **2023**, *132*, 106983. [CrossRef]
56. Tian, Y.; Xu, G.; Wei, L.; Zhou, G.; Lin, Y.; Fan, Q.; Gu, X. Phased array ultrasonic S-scan testing of near-detection-surface defects based on a background subtraction algorithm. *Mater. Res. Express* **2022**, *9*, 036507. [CrossRef]
57. Mirmahdi, E.; Khamedi, R.; Afshari, D.; Khosravi, M. Investigating the effects of defects and the effect of geometric anisotropy in stainless steel pipes: Phased array ultrasonic test, SH-wave. *J. Pipeline Sci. Eng.* **2023**, *3*, 100140. [CrossRef]
58. Olisa, S.C.; Khan, M.A.; Starr, A. Review of Current Guided Wave Ultrasonic Testing (GWUT) Limitations and Future Directions. *Sensors* **2021**, *21*, 811. [CrossRef]
59. El Mountassir, M.; Yaacoubi, S.; Dahmene, F. Reducing false alarms in guided waves structural health monitoring of pipelines: Review synthesis and debate. *Int. J. Press. Vessel. Pip.* **2020**, *188*, 104210. [CrossRef]
60. Vanaei, H.R.; Eslami, A.; Egbewande, A. A review on pipeline corrosion, in-line inspection (ILI), and corrosion growth rate models. *Int. J. Press. Vessel. Pip.* **2017**, *149*, 43–54. [CrossRef]
61. Kim, Y.Y.; Kwon, Y.E. Review of magnetostrictive patch transducers and applications in ultrasonic nondestructive testing of waveguides. *Ultrasonics* **2015**, *62*, 3–19. [CrossRef]
62. Heinlein, S.; Cawley, P.; Vogt, T.K. Reflection of torsional T(0,1) guided waves from defects in pipe bends. *NDT E Int.* **2018**, *93*, 57–63. [CrossRef]
63. Wilcox, P.D.; Lowe, M.; Cawley, P. The effect of dispersion on long-range inspection using ultrasonic guided waves. *NDT E Int.* **2001**, *34*, 1–9. [CrossRef]
64. Demma, A.; Cawley, P.; Lowe, M.; Roosenbrand, A.G. The reflection of the fundamental torsional mode from cracks and notches in pipes. *J. Acoust. Soc. Am.* **2003**, *114*, 611–625. [CrossRef] [PubMed]
65. Álvarez-Arenas, T. Acoustic impedance matching of piezoelectric transducers to the air. *IEEE Trans. Ultrason. Ferroelectr. Freq. Control* **2004**, *51*, 624–633. [CrossRef]
66. Ribichini, R.; Cegla, F.; Nagy, P.B.; Cawley, P. Study and Comparison of Different EMAT Configurations for SH Wave Inspection. *IEEE Trans. Ultrason. Ferroelectr. Freq. Control* **2011**, *58*, 2571–2581. [CrossRef] [PubMed]
67. Kim, H.W.; Kwon, Y.E.; Cho, S.H.; Kim, Y.Y. Shear-Horizontal Wave-Based Pipe Damage Inspection by Arrays of Segmented Magnetostrictive Patches. *IEEE Trans. Ultrason. Ferroelectr. Freq. Control* **2011**, *58*, 2689–2698. [CrossRef] [PubMed]
68. Wu, Y.; Wu, Y.; Zhang, T. A Type of EMAT for Simultaneous Detection of Two Components of Rayleigh Wave. *IEEE Sens. J.* **2021**, *21*, 22562–22568. [CrossRef]
69. Zhang, X.; Cheng, J.; Qiu, G.; Tu, J.; Song, X.; Cai, C. Shear horizontal circumferential wave EMAT design for pipeline inspection based on FEM. *Int. J. Appl. Electromagn. Mech.* **2020**, *64*, 913–919. [CrossRef]
70. Zhang, J.; Liu, M.; Jia, X.; Gao, R. Numerical Study and Optimal Design of the Butterfly Coil EMAT for Signal Amplitude Enhancement. *Sensors* **2022**, *22*, 4985. [CrossRef] [PubMed]
71. Nurmalia; Nakamura, N.; Ogi, H.; Hirao, M. EMAT pipe inspection technique using higher mode torsional guided wave T(0,2). *NDT E Int.* **2017**, *87*, 78–84. [CrossRef]
72. Liu, T.; Pei, C.; Cai, R.; Li, Y.; Chen, Z. A flexible and noncontact guided-wave transducer based on coils-only EMAT for pipe inspection. *Sens. Actuators Phys.* **2020**, *314*, 112213. [CrossRef]
73. Hirao, M.; Ogi, H. An SH-wave EMAT technique for gas pipeline inspection. *NDT E Int.* **1999**, *32*, 127–132. [CrossRef]
74. Fakih, M.A.; Chiachío, M.; Chiachío, J.; Mustapha, S. A Bayesian approach for damage assessment in welded structures using Lamb-wave surrogate models and minimal sensing. *NDT E Int.* **2022**, *128*, 102626. [CrossRef]
75. Zhang, D.; Cui, C.; Zhang, X.; Jiang, Z.; Zhang, X.; Bao, Y. Monitoring fatigue cracks in rib-to-deck joints of orthotropic steel deck using ultrasonic Lamb waves. *Thin-Walled Struct.* **2023**, *189*, 110922. [CrossRef]
76. Chimenti, D.E. Review of air-coupled ultrasonic materials characterization. *Ultrasonics* **2014**, *54*, 1804–1816. [CrossRef]
77. Wang, G.; Cheng, Q.; Zhao, W.; Liao, Q.; Zhang, H. Review on the transport capacity management of oil and gas pipeline network: Challenges and opportunities of future pipeline transport. *Energy Strategy Rev.* **2022**, *43*, 100933. [CrossRef]
78. Álvarez-Arenas, T. Simultaneous determination of the ultrasound velocity and the thickness of solid plates from the analysis of thickness resonances using air-coupled ultrasound. *Ultrasonics* **2010**, *50*, 104–109. [CrossRef]

79. Zhang, X.; Kinnick, R.R.; Greenleaf, J.F. Air-coupled ultrasound stimulated optical vibrometry for resonance analysis of rubber tubes. *Ultrasonics* **2009**, *49*, 26–30. [CrossRef]
80. Bente, K.; Rus, J.; Mooshofer, H.; Gaal, M.; Grosse, C.U. Broadband Air-Coupled Ultrasound Emitter and Receiver Enable Simultaneous Measurement of Thickness and Speed of Sound in Solids. *Sensors* **2023**, *23*, 1379. [CrossRef]
81. Schindel, D.W.; Hutchins, D.A.; Grandia, W.A. Capacitive and piezoelectric air-coupled transducers for resonant ultrasonic inspection. *Ultrasonics* **1996**, *34*, 621–627. [CrossRef]
82. Livings, R.; Dayal, V.; Barnard, D. Damage Detection in a Multi-Layered, Multi-Material Composite Using Air-Coupled Ultrasonic Resonance Imaging. *J. Nondestruct. Eval.* **2016**, *35*, 55. [CrossRef]
83. Hutchins, D.; Wright, W.; Schindel, D. Ultrasonic measurements in polymeric materials using air-coupled capacitance transducers. *J. Acoust. Soc. Am.* **1994**, *96*, 1634–1642. [CrossRef]
84. Kelly, S.P.; Farlow, R.; Hayward, G. Applications of through-air ultrasound for rapid NDE scanning in the aerospace industry. *IEEE Trans. Ultrason. Ferroelectr. Freq. Control* **1996**, *43*, 581–591. [CrossRef]
85. Gan, T.; Hutchins, D.; Billson, D.; Schindel, D. The use of broadband acoustic transducers and pulse-compression techniques for air-coupled ultrasonic imaging. *Ultrasonics* **2001**, *39*, 181–194. [CrossRef] [PubMed]
86. Hutchins, D.; Burrascano, P.; Davis, L.; Laureti, S.; Ricci, M. Coded waveforms for optimised air-coupled ultrasonic nondestructive evaluation. *Ultrasonics* **2014**, *54*, 1745–1759. [CrossRef]
87. Garcia-Rodriguez, M.; Yañez, Y.; Garcia-Hernandez, M.J.; Salazar, J.; Turo, A.; Chavez, J.A. Application of Golay codes to improve the dynamic range in ultrasonic Lamb waves air-coupled systems. *NDT E Int.* **2010**, *43*, 677–686. [CrossRef]
88. Tang, J.; Zhu, W.; Qiu, X.; Song, A.; Xiang, Y.; Xuan, F.-Z. Non-contact phase coded excitation of ultrasonic Lamb wave for blind hole inspection. *Ultrasonics* **2022**, *119*, 106606. [CrossRef]
89. Li, H.; Zhou, Z. Application of P4 Polyphase codes pulse compression method to air-coupled ultrasonic testing systems. *Ultrasonics* **2017**, *78*, 57–69. [CrossRef]
90. Yamaguchi, Y.; Sato, Y. Simultaneous nondestructive estimation of thickness and longitudinal wave velocity of adhesive layers in adhesive joints through air-coupled ultrasonic testing. *NDT E Int.* **2023**, *138*, 102905. [CrossRef]
91. Luo, K.; Liu, Y.; Liang, W.; Chen, L.; Yang, Z. Rapid damage reconstruction imaging of composite plates using non-contact air-coupled Lamb waves. *NDT E Int.* **2024**, *143*, 103047. [CrossRef]
92. Sparse ultrasonic guided wave imaging with compressive sensing and deep learning. *Mech. Syst. Signal Process.* **2022**, *178*, 109346. [CrossRef]
93. Dipietrangelo, F.; Nicassio, F.; Scarselli, G. Structural Health Monitoring for impact localisation via machine learning. *Mech. Syst. Signal Process.* **2023**, *183*, 109621. [CrossRef]
94. Layouni, M.; Hamdi, M.S.; Tahar, S. Detection and sizing of metal-loss defects in oil and gas pipelines using pattern-adapted wavelets and machine learning. *Appl. Soft Comput.* **2017**, *52*, 247–261. [CrossRef]
95. Cruz, F.C.; Simas Filho, E.F.; Albuquerque, M.C.S.; Silva, I.C.; Farias, C.T.T.; Gouvêa, L.L. Efficient feature selection for neural network based detection of flaws in steel welded joints using ultrasound testing. *Ultrasonics* **2017**, *73*, 1–8. [CrossRef] [PubMed]
96. Rachman, A.; Zhang, T.; Ratnayake, R.M.C. Applications of machine learning in pipeline integrity management: A state-of-the-art review. *Int. J. Press. Vessels Pip.* **2021**, *193*, 104471. [CrossRef]
97. Sun, H.; Ramuhalli, P.; Jacob, R.E. Machine learning for ultrasonic nondestructive examination of welding defects: A systematic review. *Ultrasonics* **2023**, *127*, 106854. [CrossRef] [PubMed]
98. Latête, T.; Gauthier, B.; Belanger, P. Towards using convolutional neural network to locate, identify and size defects in phased array ultrasonic testing. *Ultrasonics* **2021**, *115*, 106436. [CrossRef] [PubMed]
99. Akram, N.; Isa, D.; Rajkumar, R.; Lee, L. Active incremental Support Vector Machine for oil and gas pipeline defects prediction system using long range ultrasonic transducers. *Ultrasonics* **2014**, *54*, 1534–1544. [CrossRef] [PubMed]
100. de Moura, E.P.; Siqueira, M.H.S.; da Silva, R.R.; Rebello, J.M.A.; Calôba, L.P. Welding defect pattern recognition in TOFD signals Part 1. Linear classifiers. *Insight—Non-Destr. Test. Cond. Monit.* **2005**, *47*, 777–782. [CrossRef]
101. Rautela, M.; Senthilnath, J.; Moll, J.; Gopalakrishnan, S. Combined two-level damage identification strategy using ultrasonic guided waves and physical knowledge assisted machine learning. *Ultrasonics* **2021**, *115*, 106451. [CrossRef]
102. Abdar, M.; Pourpanah, F.; Hussain, S.; Rezazadegan, D.; Liu, L.; Ghavamzadeh, M.; Fieguth, P.; Cao, X.; Khosravi, A.; Acharya, U.R.; et al. A review of uncertainty quantification in deep learning: Techniques, applications and challenges. *Inf. Fusion* **2021**, *76*, 243–297. [CrossRef]
103. Medak, D.; Posilović, L.; Subašić, M.; Budimir, M.; Lončarić, S. Automated Defect Detection from Ultrasonic Images Using Deep Learning. *IEEE Trans. Ultrason. Ferroelectr. Freq. Control* **2021**, *68*, 3126–3134. [CrossRef]
104. Kim, K.; Kim, K.S.; Park, H.-J. Multi-Branch Deep Fusion Network-Based Automatic Detection of Weld Defects Using Non-Destructive Ultrasonic Test. *IEEE Access* **2023**, *11*, 114489–114496. [CrossRef]
105. Cheng, X.; Ma, G.; Wu, Z.; Zu, H.; Hu, X. Automatic defect depth estimation for ultrasonic testing in carbon fiber reinforced composites using deep learning. *NDT E Int.* **2023**, *135*, 102804. [CrossRef]
106. Huang, L.; Hong, X.; Yang, Z.; Liu, Y.; Zhang, B. CNN-LSTM network-based damage detection approach for copper pipeline using laser ultrasonic scanning. *Ultrasonics* **2022**, *121*, 106685. [CrossRef] [PubMed]
107. Zhang, F.; Wang, L.; Ye, W.; Li, Y.; Yang, F. Ultrasonic lamination defects detection of carbon fiber composite plates based on multilevel LSTM. *Compos. Struct.* **2024**, *327*, 117714. [CrossRef]

108. Shao, W.; Sun, H.; Wang, Y.; Qing, X. A multi-level damage classification technique of aircraft plate structures using Lamb wave-based deep transfer learning network. *Smart Mater. Struct.* **2022**, *31*, 075019. [CrossRef]
109. Bettayeb, F.; Rachedi, T.; Benbartaoui, H. An improved automated ultrasonic NDE system by wavelet and neuron networks. *Ultrasonics* **2004**, *42*, 853–858. [CrossRef]
110. Sambath, S.; Nagaraj, P.; Selvakumar, N. Automatic Defect Classification in Ultrasonic NDT Using Artificial Intelligence. *J. Nondestruct. Eval.* **2011**, *30*, 20–28. [CrossRef]
111. Guo, Y.; Xiao, Z.; Geng, L.; Wu, J.; Zhang, F.; Liu, Y.; Wang, W. Fully Convolutional Neural Network with GRU for 3D Braided Composite Material Flaw Detection. *IEEE Access* **2019**, *7*, 151180–151188. [CrossRef]
112. Guo, C.; Li, M.; Xu, J.; Bai, L. Ultrasonic characterization of small defects based on Res-ViT and unsupervised domain adaptation. *Ultrasonics* **2024**, *137*, 107194. [CrossRef] [PubMed]
113. Miorelli, R.; Fisher, C.; Kulakovskyi, A.; Chapuis, B.; Mesnil, O.; D'Almeida, O. Defect sizing in guided wave imaging structural health monitoring using convolutional neural networks. *NDT E Int.* **2021**, *122*, 102480. [CrossRef]
114. Bai, L.; Le Bourdais, F.; Miorelli, R.; Calmon, P.; Velichko, A.; Drinkwater, B.W. Ultrasonic Defect Characterization Using the Scattering Matrix: A Performance Comparison Study of Bayesian Inversion and Machine Learning Schemas. *IEEE Trans. Ultrason. Ferroelectr. Freq. Control* **2021**, *68*, 3143–3155. [CrossRef]
115. Yang, B.; Zhang, Y.; Wang, S.; Xu, W.; Xiao, M.; He, Y.; Mo, F. A global interactive attention-based lightweight denoising network for locating internal defects of CFRP laminates. *Eng. Appl. Artif. Intell.* **2022**, *116*, 105436. [CrossRef]
116. Kesharaju, M.; Nagarajah, R. Feature selection for neural network based defect classification of ceramic components using high frequency ultrasound. *Ultrasonics* **2015**, *62*, 271–277. [CrossRef] [PubMed]
117. Liu, Y.; Tan, X.; Bao, Y. Machine learning-assisted intelligent interpretation of distributed fiber optic sensor data for automated monitoring of pipeline corrosion. *Measurement* **2024**, *226*, 114190. [CrossRef]

MDPI AG
Grosspeteranlage 5
4052 Basel
Switzerland
Tel.: +41 61 683 77 34

*Sensors* Editorial Office
E-mail: sensors@mdpi.com
www.mdpi.com/journal/sensors

www.ingramcontent.com/pod-product-compliance
Lightning Source LLC
Chambersburg PA
CBHW041425210926
51850CB00013B/3354
* 9 7 8 3 7 2 5 8 4 4 8 7 6 *